Artificial cells, cell engineering and therapy

Related titles:

Biomaterials, artificial organs and tissue engineering
(ISBN 978-1-85573-737-2)
Biomaterials are materials and devices that are used to repair, replace or augment the living tissue and organs of the human body. The purpose of this wide-ranging introductory textbook is to provide an understanding of the needs, uses and limitations of materials used in the human body, and to explain the biomechanical principles and biological factors involved in achieving the long-term stability of replacement parts in the body. The book examines the industrial, governmental and ethical factors involved in the use of artificial materials in humans and discusses the principles and applications of engineering of tissue to replace body parts. The approach necessarily incorporates a wide range of reference material because of the complex multidisciplinary nature of the fields of biomedical materials, biomechanics, artificial organs and tissue engineering. There is an accompanying CD-ROM providing supplementary information and illustrations.

Surfaces and interfaces for biomaterials
(ISBN 978-1-85573-930-7)
This book presents the current level of understanding on the nature of a biomaterial surface, the adaptive response of the biomatrix to that surface, techniques used to modify biocompatibility, and state-of-the-art characterisation techniques to follow the interfacial events at that surface.

Sterilisation of tissues using ionising radiations
(ISBN 978-1-85573-838-6)
This volume contains the proceedings of the Cellucon Trust/International Atomic Energy Agency meeting held at the North East Wales Institute in September 2003 on the use of ionising radiation for tissue sterilisation. Existing methods and processing for sterilising tissues are proving, in many instances, inadequate. Infections have been transmitted from the graft to the recipient, and attention has now been drawn to the need for a reliable end-sterilisation method which does not damage the functionality of the final tissue. This volume identifies the best method of using radiation technology to assist in the production of safe tissue allografts. Tissue bankers, users of tissues for transplantation and regulators who oversee safety will all find this book an invaluable source of reference.

Details of these and other Woodhead Publishing materials books and journals, as well as materials books from Maney Publishing, can be obtained by:

- visiting our web site at www.woodheadpublishing.com
- contacting Customer Services (e-mail: sales@woodhead-publishing.com; fax: +44 (0) 1223 893694; tel.: +44 (0) 1223 891358 ext. 130; address: Woodhead Publishing Limited, Abington Hall, Abington, Cambridge CB21 6AH, England)

Artificial cells, cell engineering and therapy

Edited by
S. Prakash

CRC Press
Boca Raton Boston New York Washington, DC

WOODHEAD PUBLISHING LIMITED
Cambridge England

Published by Woodhead Publishing Limited, Abington Hall, Abington, Cambridge CB21 6AH, England
www.woodheadpublishing.com

Published in North America by CRC Press LLC, 6000 Broken Sound Parkway, NW, Suite 300, Boca Raton, FL 33487, USA

First published 2007, Woodhead Publishing Limited and CRC Press LLC

British Library Cataloguing in Publication Data
A catalogue record for this book is available from the British Library.

Library of Congress Cataloging in Publication Data
A catalog record for this book is available from the Library of Congress.

Woodhead Publishing Limited ISBN 978-1-84569-036-6 (book)
Woodhead Publishing Limited ISBN 978-1-84569-307-7 (e-book)
CRC Press ISBN 978-0-8493-9111-8
CRC Press order number: WP9111

The publishers' policy is to use permanent paper from mills that operate a sustainable forestry policy, and which has been manufactured from pulp which is processed using acid-free and elementary chlorine-free practices. Furthermore, the publishers ensure that the text paper and cover board used have met acceptable environmental accreditation standards.

Project managed by Macfarlane Production Services, Dunstable, Bedfordshire, England (macfarl@aol.com)
Typeset by Godiva Publishing Services Limited, Coventry, West Midlands, England
Printed by TJ International Limited, Padstow, Cornwall, England

Contents

Contributor contact details

(* = main contact)

Editor
Dr Satya Prakash
Department of Biomedical Engineering
Faculty of Medicine
Lyman Duff Building
3775 University Street
McGill University
Montreal QC
H3A 2B4
Canada

E-mail: satya.prakash@mcgill.ca

Chapter 1
Satya Prakash*, Jasmine R. Bhathena and Hongmei Chen
Department of Biomedical Engineering
Faculty of Medicine
Lyman Duff Building
3775 University Street
McGill University
Montreal QC
H3A 2B4
Canada

E-mail: satya.prakash@mcgill.ca

Chapter 2
Kevin Poon, Victor Castellino and Yu-Ling Cheng*
Department of Chemical Engineering and Applied Chemistry
University of Toronto
200 College Street
Toronto
Ontario
M5S 3E5
Canada

E-mail: ylc@chem-eng.utoronto.ca
ylc@ecf.utoronto.ca

Chapter 3
Gillian Barratt
Centre d'Etudes Pharmaceutiques
UMR CNRS 8612
Université Paris-Sud 11
5 Rue JB Clément
92296 Chatenay-Malabry Cedex
France

E-mail: gillian.barratt@u-psud.fr

Chapter 4
Weiping Yu* and Daishui Su
Celsion Corporation
10220 Old Columbia Road
Columbia
MD 21046

USA

E-mail: wyu@celsion.com

Chapter 5

Carly Griffin, Deyzi Gueorguieva, Amanda McLachlan-Burgess, Mallika Somayajulu-Nitu and Siyaram Pandey*
Department of Chemistry & Biochemistry
227-1 Essex Hall
University of Windsor
401 Sunset Avenue
Windsor, Ontario
N9B 3P4
Canada

E-mail: spandey@uwindsor.ca

Chapter 6

Rony Atoui, Ray C. J. Chiu and Dominique Shum-Tim*
Division of Cardiothoracic Surgery
The Montreal General Hospital
McGill University Health Center
1650 Cedar Avenue
Suite C9-169
Montreal, Quebec
H3G 1A4
Canada

E-mail: dshumtim@yahoo.ca

Chapter 7

K. A. Corsi and J. Huard*
Stem Cell Research Center
Children's Hospital of Pittsburgh
4100 Rangos Research Center
3460 Fifth Avenue
Pittsburgh
PA 15213-2582
USA

E-mail: jhuard+@pitt.edu

Chapter 8

Satya Prakash*, Jasmine Bhathena and Aleksandra Malgorzata Urbanska
Department of Biomedical Engineering
Faculty of Medicine
Lyman Duff Building
3775 University Street
McGill University
Montreal, Quebec
H3A 2B4
Canada

E-mail: satya.prakash@mcgill.ca

Chapter 9

Vivek Dixit
Head, Liver Biosupport & Tissue Engineering
Department of Medicine
Division of Digestive Diseases
675 Circle Drive South
MRL Room 1240
Los Angeles
CA 90024-7019
USA

E-mail: dixit@ucla.edu

Chapter 10

Dr Murray Potter
Department of Pathology and Molecular Medicine
McMaster University
1200 Main Street
W. Hamilton
Ontario
L8S 4J9
Canada

E-mail: mpotter@hhsc.ca

Chapter 11
J. W. Kawiak*, L. H. Granicka, A. Weryński and J. M. Wójcicki
Institute of Biocybernetics and Biomedical Engineering
Polish Academy of Sciences
4 Trojdena Str
02-109 Warsaw
Poland

E-mail: Jan.Wojcicki@ibib.waw.pl
jkawiak@cmkp.edu.pl

Chapter 12
Thomas Ming Swi Chang
Artificial Cells and Organs Research Centre
Department of Physiology, Medicine and Biomedical Engineering
Faculty of Medicine
McGill University
3655 Drummond Street
Montreal
Quebec
H3G 1H6
Canada

E-mail: artcell.med@mcgill.ca

Chapter 13
Jun Luo, Ray C-J Chiu and Dominique Shum-Tim,*
The Montreal General Hospital
McGill University Health Center
1650 Cedar Avenue
Suite C9-169
Montreal
Quebec
H3G 1A4
Canada

E-mail: dshumtim@yahoo.ca

Chapter 14
Race L. Kao*, Charles E. Ganote, D. Glenn Pennington and I. William Browder
Professor and Chair of Excellence
Department of Surgery
James H. Quillen College of Medicine
East Tennessee State University
P. O. Box 70575
Johnson City
TN 37604
USA

E-mail: kao@etsu.edu

Chapter 15
Satya Prakash*, Trisna Lim and Wei Ouyang
Department of Biomedical Engineering
Faculty of Medicine
Lyman Duff Building
3775 University Street
McGill University
Montreal QC
H3A 2B4
Canada

E-mail: satya.prakash@mcgill.ca

Chapter 16
Shimon Efrat
Department of Human Genetics and Molecular Medicine
Sackler School of Medicine
Tel Aviv University
Ramat Aviv
Tel Aviv
69978
Israel

E-mail: sefrat@post.tau.ac.il

Chapter 17
Xing-Guo Mei
Studio 6th
Beijing Institute of Pharmacology and Toxicology
27 Tai Ping Road
Haidian District
Beijing 100850
PR China

E-mail: xg_mei@yahoo.com

Chapter 18
Binglan Yu and Thomas Ming Swi Chang*
Artificial Cells and Organs Research Centre and Departments of Physiology
Medicine and Biomedical Engineering
Faculty of Medicine
McGill University
Montreal
Quebec
H3G 1Y6
Canada

E-mail: artcell.med@mcgill.ca

Chapter 19
L. Eduardo Cruz* and Silvia P. Azevedo
Cryopraxis Criobiologia Ltd
Pólo de Biotecnologia do Rio de Janeiro
Avenida 24, s/nº
Cidade Universitária
Ilha do Fundão
CEP 21.941-590 – Rio de Janeiro
Brazil

E-mail: dirplan@cryopraxis.com.br
ec@silvestrelabs.com.br

Chapter 20
D. Amre*, Ramila D. Amre and Satya Prakash*
Department of Biomedical Engineering
Faculty of Medicine
Lyman Duff Building
3775 University Street
McGill University
Montreal QC
H3A 2B4
Canada

E-mail: satya.prakash@mcgill.ca

Chapter 21
William C. Zamboni
University of Pittsburgh Cancer Institute
Hillman Cancer Center
Research Pavilion, Room G.27c
5117 Centre Ave
Pittsburgh
PA 15213
USA

E-mail: zamboniwc@msx.upmc.edu

Preface

Artificial cell, cell engineering and therapy are emerging technologies that promise to have a significant impact on the practice of medicine in the coming years. They offer potential for the treatment of conditions such as cancer, bone marrow regenerations, myocardial regenerations, kidney failure, diabetes, liver failure, myocardial gene therapy, inflammatory bowel diseases, blood substitutes, and cancer. Based on novel concepts the field has generated interest and attention among researchers, the medical community, patient associations, and the media. There is a plethora of ongoing research. Information on these developments and their importance with regards to clinical applications is, however, dispersed and complex. This book strives to amalgamate the current state-of-the-art knowledge-base in an accessible format in order to give the reader a comprehensive bird's-eye view of the domain.

This book does not attempt to target a specific audience. It is aimed at researchers, physicians, scientists, students, as well as other healthcare professionals who are interested in recent developments on different aspects of the field. It is also anticipated that the book will aid healthcare authorities in the planning, development and execution of long-term healthcare strategies.

The aim of the book is to cover and provide comprehensive knowledge about all aspects of artificial cell, cell engineering and therapy. The book consists of 21 state-of-the-art chapters, grouped into four parts.

Part 1: The artificial cell. This section commences with an introductory chapter that takes the reader through the basic elements concerning the history, concepts, design methods, current status and future prospective of artificial cells. The subsequent chapters provide details on artificial cell hydrophilic polymers in targeted drug delivery, liposomes and nano capsules. The finer aspects underlying the preparation of targeted formulations, potential utility in various clinical conditions, current status of the knowledge, etc. are covered.

Part II: Cell engineering. This part of the book commences with a cutting-edge chapter describing the mechanisms leading to apoptosis and how the phenomenon could be exploited for clinical applications. This is followed by a chapter describing the use of bone marrow stromal cells in myocardial regeneration and cell signaling. Stem cells and musculoskeletal tissue engineering are covered in subsequent chapters.

Part III: Artificial cells for cell therapy. This part begin with a chapter on artificial cells for oral delivery of live bacterial cells for therapy, followed by a chapter on artificial cell liver transplant detailing current status of knowledge in bacterial cell-based therapies and use of artificial cells in liver transplants. Chapters on artificial cell detailing novel approach to gene therapy approach and capillary artificial devices for therapy follow. These chapters outline potentials of artificial cell in cell therapy procedures for various clinical conditions, current status of the knowledge in the field and future prospective.

Part IV: The clinical relevance of artificial cells and cell engineering. The last section of the book attempts to capture and integrate the information presented in earlier sections and provide a perspective to the reader on the various applications of artificial cells in medicine. A number of different areas are covered: the role of artificial cells in medicine with emphasis on artificial cells as blood substitutes, use of bone marrow cells in myocardial regeneration therapy, myocardial regeneration tissue engineering and therapy, potential use of artificial cells in renal failure and other kidney diseases, cell therapy using engineered live cells for treatment of diabetes, drugs delivery system for active brain targeting, enzyme therapy, artificial cell for inflammatory bowel disease and carrier-mediated artificial-cell targeted drug delivery for cancer therapy.

It is now evident that artificial cells, cell engineering and therapy constitute a domain at the cutting edge of science and provide fascinating research that is posed to improve human health. This domain will provide innovative and novel therapeutic options. Certainly, extensive basic and clinical research is still required to facilitate the translation of experimental findings to clinical practice. To achieve these objectives will require the establishment of new platforms of communication between research and clinical scientists, and public and private support systems. It is expected that this book will contribute in some measure to bringing together physicians, molecular biologists, biochemists, chemists, pharmaceutical scientists, policy makers, industrial partners and other interested parties on a common platform, encouraging the better exploitation of a field that shows tremendous promise.

This book will be essential reading to many. Readers with more focused research objectives will find some chapters consistent with their interests. In general, through state-of-the-art contributions from renowned researchers in this rapidly evolving discipline, this book attempts to take its readers through many of the frontiers of current research in the domain. The book has many thoughtful and carefully presented chapters that provide important information on the subject unavailable in this form elsewhere.

Dr Satya Prakash
McGill University
Montreal, Canada

Acknowledgements

The preparation of this book would not have been possible without the outstanding work of the contributing authors. During the production of this book, I received great help from my colleagues. I would like to express my thanks to my colleagues, especially, to Dr D. Shum-Tim, Dr R. C. Chiu, Dr D. Amre, Dr V. Dixit, Dr W.P. Yu, Dr W. C. Zamboni, Dr S.R. Pandey, Dr J. M. Wójcicki, Dr H. L. Galiana and Dr T. M. S. Chang for their encouragement and support.

I am also grateful to Mr Bob Sitton for his editorial support of the entire book. In addition, I would like to thank Ms Debbie Tranter, commissioning editor, and Ms Laura Bunney, the project editor, for her care and patience. Special thanks go to the project coordinator Ms Melanie Cotterell for her excellent work and support.

I would like to acknowledge the great help that I received during the production of this book from my former and present graduate students. All past and current students are acknowledged. I would, however, specifically like to thank Mr Mitchell L. Jones, Ms Jasmine R. Bhathena, Mr Christopher J. Martoni, Ms Alexandra M. Urbanska, Ms Maryam Mirzaei, Ms Trisna Lim, Ms Olabisi Lawuyi, Ms Terrence Metz, Ms Fatemeh Afkhami, Ms Hongmei Chen, and Dr Wei Ouyang, who provided many data and plots for this book. Some of these students are listed as co-authors in various chapters of the book.

I would also like to acknowledge financial support from Canadian Institute of Health Research (CIHR) for my biomedical technology, artificial cell, cell engineering and therapy research program. Research investigator awards from FRSQ and CIHR are also acknowledged. A discovery grant from Natural Science and Engineering Research Council (NSERC), Canada and research grants form FQRNT, DFC, CFI, FRSQ, Micropharma and other public and private research agencies are acknowledged.

My wife, Dr Ghada Maria Saddi, and my sons, Nilay and Avi, have tolerated my indulgence in this book. My wife, my sons and my parents have always given me the necessary emotional support throughout the preparation of this book.

Finally, I would like to thank the excellent reviewers for their many suggestions which greatly improved the quality of the book.

Satya Prakash

Part I

The artificial cell

1

Introduction to artificial cells: concept, history, design, current status and future

S PRAKASH, J R BHATHENA and H CHEN,
McGill University, Canada

1.1 Introduction: concept and history

The emergence of cellular life is one of the major transitions in evolution. The existence of a cell boundary allows metabolism and genetic information to be part of a well-defined component. Theoretical models and experimental data support the idea that simple protocells should be obtainable from simple systems of coupled reactions dealing with these three topics. The building of an artificial cell would be a fundamental breakthrough in our understanding of life, its origins and evolution, not to mention a wide array of potential medical and technological applications.

Artificial cells – ultrathin polymeric or biological membranes of cellular dimensions – were first prepared in the laboratory of T.M.S. Chang at McGill University in Canada in the 1960s.[1] Artificial cell microencapsulation is used to encapsulate biologically active materials in specialized ultrathin semipermeable polymer membranes.[1,2] The polymer membrane protects encapsulated materials from harsh external environments while at the same time allowing for the metabolism of selected solutes capable of passing into and out of the microcapsule. In this manner, the enclosed material (live bacteria, DNA, proteins, drugs, etc.) can be retained inside and separated from the external milieu, making microencapsulation particularly useful for biomedical and clinical applications.[3–5] Since the 1980s, microencapsulation research has made great strides in developing approaches for the controlled release of therapeutic agents, targeted delivery of drugs, bacterial cells, mammalian cells, DNA and other nucleic acids, proteins, etc. (Table 1.1) to the host. An encapsulation membrane serves as an immunobarrier, allows the bi-directional exchange of small molecules including nutrients, wastes, selected substrates and products, and prevents the passage of large substances such as cells, immunocytes and antibodies (Fig. 1.1).[57,58]

It is also possible to prepare artificial cells in the molecular, nano- or even macro-dimensions. Cell encapsulation promises immuno-isolation, which has initiated a flurry of research into bioartificial organs and tissue engineering,

Table 1.1 Concept and relevance of artificial cells

Year	Innovation	Reference
1957	First polymeric artificial cells developed.	6
1964	The first scientific publication describing the principle of artificial cells, including methods of preparation, *in vitro* and *in vivo* studies and potential areas of biotechnological and medical applications. Polymeric artificial cells containing enzymes and haemoglobin developed. Intermolecularly crosslinked protein produced and conjugation of haemoglobin to polymer achieved.	1
1965–1966	Extrusion drop technique for encapsulating intact cells for immuno-isolation developed. Multi-compartment artificial cells developed.	7–9
1966	Silastic artificial cells and microspheres containing protein produced. Artificial cells containing magnetic materials with other materials produced. The first report on the use of biodegradable membrane microcapsules and microparticles as delivery systems for drugs and biotechnology products. This forms the basis of many of the present approaches using biodegradable microcapsules, microparticles, nanocapsules, nanoparticles and others.	10, 11
1966–1969	Artificial cells with ultra-thin membranes, and which contain adsorbents for use in hemoperfusion developed.	11–13
1968	Implanted enzyme containing artificial cells used for enzyme therapy in acatalesmic mice. The first scientific report of the implantation of polymeric artificial cells containing an enzyme for replacement of the deficient gene in a mouse model of an inborn error of metabolism.	14
1970–1975	First clinical use of artificial cells in patients (for hemoperfusion).	11, 15, 16
1971	Implanted enzyme-containing artificial cells used for lymphosarcoma suppression. Glutaraldehyde-crosslinked haemoglobin used to form soluble polyhemoglobin.	17, 18
1972	Crosslinked protein–lipid membrane artificial cells with transport carrier produced.	8, 19
1976	Biodegradable polylactide microcapsules and microparticles containing proteins and hormones produced.	20
1977–1978	Artificial cells containing multi-enzyme systems with cofactor recycling developed.	20–22
1980	Alginate–polylysine–alginate encapsulated islets implanted into diabetic rats. Artificial cell membrane with Na^+K^+-ATPase produced. The first report on the laboratory demonstration of encapsulation of islets and the ability of this to maintain a normal blood glucose level when implanted.	5, 19

Table 1.1 Continued

Year	Innovation	Reference
1986	Artificial cell membrane that excludes small hydrophilic molecules but which is permeable to large lipophilic molecules produced. Novel finding of extensive enterorecirculation of amino acids, which could allow oral therapy with artificial cells containing enzymes to selectively remove specific unwanted systemic amino acids.	23–29
1989	First clinical use of artificial cells containing enzymes in a patient with Lesch–Nyhan disease	30, 31
1994	Biodegradable polymeric nano-artificial red blood cells developed. First clinical report of the use of encapsulated islets in a diabetic patient.	32–35
1996	Hollow polymeric fiber for encapsulation of genetically engineered cells developed. Encapsulated bacteria lower high plasma urea levels to normal in uremic rats with induced kidney failure.	36, 37
1999	Artificial cell microcapsules containing genetically engineered *Escherichia coli* DH5 cells to lower plasma potassium, phosphate, magnesium, sodium, chloride, uric acid, cholesterol, and creatinine developed. Microcapsules as bio-organs for somatic gene therapy.	4, 38–41
2000–2003	Artificial cells co-encapsulating hepatocytes and adult stem cells developed.	42–44
2001	Monitoring of the mechanical stability of various types of microcapsules, predicting the performance of microcapsules *in vivo*, and quality control of microcapsules during scale-up productions developed.	45
2002	Treatment of hemophilia B in mice with non-autologous somatic gene therapeutics. Novel approach to tumor suppression with microencapsulated recombinant cells.	46, 47
2003	Antiangiogenic cancer therapy with microencapsulated cells.	48
2004	First report of microencapsulated genetically engineered lactobacilli for lowering cholesterol. Combined immunotherapy and antiangiogenic therapy of cancer with microencapsulated cells.	49, 50
2005	First studies of artificial cell microcapsules as an alternative to liver cell transplants for the treatment of liver failure. Novel targeted drug delivery using polymeric microcapsule for treatment of Crohn's and inflammatory bowel disease (IBD). New effective characterization of microcapsules using genipin. Encapsulation of recombinant cells with a novel magnetized alginate for magnetic resonance imaging.	38, 39, 51–56

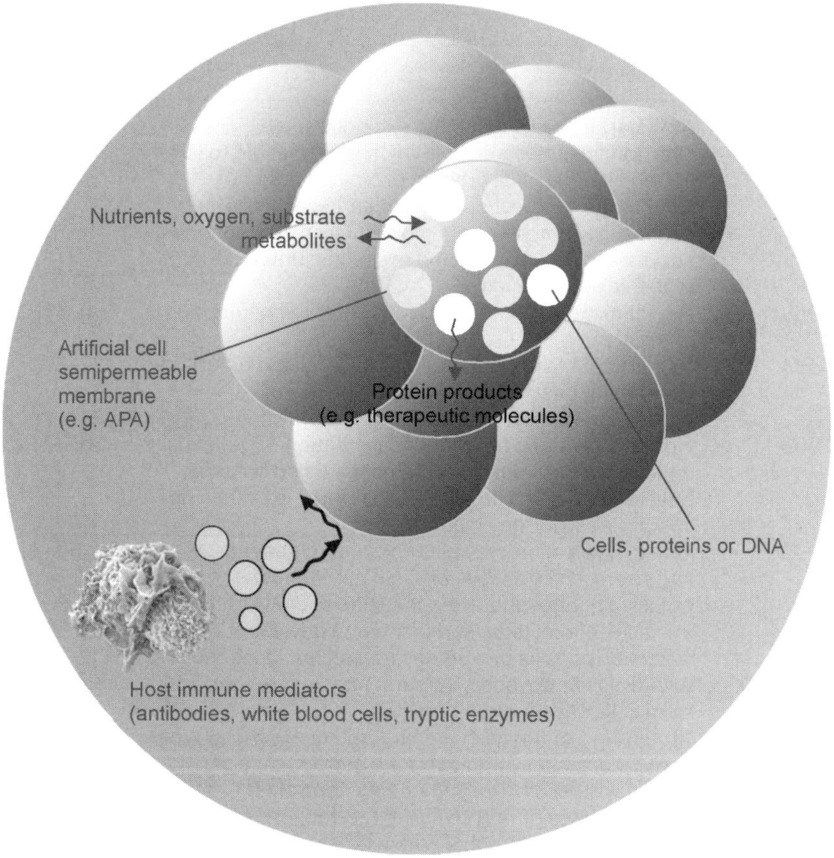

1.1 The basic concept of artificial cells.

while the prospect of encapsulation increasing long-term *in vivo* cell survivability has opened new avenues for both targeted and recurrent therapeutic drug delivery systems. Recent advances in molecular biology, cell biology, biotechnology, nanotechnology and other areas have resulted in rapid developments in this area for basic research and for gene therapy, enzyme therapy, cell therapy, blood substitutes, liver support systems and other areas. Significant advances have made the translation of this concept to clinical use – for example, in the treatment of type 1 diabetes using encapsulated islets – increasingly practicable. The potential therapeutic applications of encapsulated cells are enormous,[7,59–67] preferentially include replacement organ functions, correction of hormone/enzyme deficiencies, treatment of cancers, central nervous system (CNS) diseases and other disorders (Table 1.2).

Accompanying these myriad avenues of research has been the development of a multitude of polymers from which to choose when encapsulating a given cell

Table 1.2 Examples of promising therapeutic applications by artificial cell microencapsulation technology

Disorder	Enclosed materials and functions	Route of administration	Reference
Diabetes	Pancreatic islets for insulin secretion	Implantation	35, 68–74
Renal failure	*E. coli* DH5 transfected with the urease gene; metabolic induced *Lactobacillus delbrueckii*, for urea removal	Oral delivery	37, 75, 76
Liver failure	Hepatocytes for liver function support	Implantation	54, 77–79
Hemophilia	Recombinant C2C12 myoblasts secreting factor IX	Implantation	46, 80
Hyperlipidemia and atherosclerosis	Recombinant CHO-E3 cells secreting wild-type apoE3 protein	Injection	81
Dwarfism	Non-autologous C2C12 cells secreting human growth hormone	Implantation	82
MPS VII	Recombinant 2A-50 fibroblasts secreting β-glucuronidase	Injection	83
ADA deficiency	Human fibroblasts expressing ADA	Implantation	84
Cancers	Varied recombinant cells secreting endostatin, cytokines, antibodies, etc. for tumor suppression	Implantation	50, 85–92
CNS diseases (Parkinson's, Huntington's, ALS, chronic pain, etc.)	Recombinant cells secreting CNTF, GDNF, dopamine, CSF-1, and β-endophin; chromaffin cells	Implantation	36, 93–102
IBD and Crohn's disease	*Lactobacillus bifidobacterium* to alleviate IBD diseases; targeted delivery of encapsulated thalidomide	Oral delivery	103–105
Hypercholesteraemia	*Pseudomonas pictorum, Lactobacillus plantarum 80* for cholesterol lowering	*In vitro* study	49, 106

Abbreviations: ADA, adenosine deaminase; ALS, amyotrophic lateral sclerosis; CNS, central nervous system; CNTF, ciliary neurotrophic factor; CSF-1, colony stimulating factor-1; GDNF, glial cell line-derived neurotrophic factor; IBD, inflammatory bowel disease; MPS VII, mucopolysaccharidosis type VII (neurodegenerative diseases/enzyme deficiency).

species. In order to successfully encapsulate a cell while maintaining its function, a polymer with appropriate mechanical strength, pore size, capsule uniformity, degradation characteristics, biocompatibility and low immunogenicity must be chosen. Conflicting reports of success illustrate the complexity of finding an appropriate polymer: variables include not only the type of cell to be encapsulated, but also the environment and duration to which that cell will be exposed.

The long-term vision is to commercialize biotechnology applications for successively more sophisticated versions of an artificial bionanomachine: the artificial cell. Research is now being carried out with a view to achieve this vision by commercializing biomedical applications of the artificial cell platform, starting with one of the simplest of human cells, the red blood cell.

1.2 Designing an artificial cell

Designing microcapsules for use in transplantation or for oral administration requires a subtle balance which optimizes capsule robustness while maintaining cell viability and biocompatibility. The use of different membranes allows for variations in permeability, mass transfer, mechanical stability, biocompatibility and buffering capability that can be exploited to fit a desired application. The mass transport properties of a membrane are critical since the influx rate of molecules, essential for cell survival and proliferation, and the outflow rate of metabolic waste ultimately determine the viability of encapsulated cells. For example, while a certain minimum pore size and maximum wall thickness are required for adequate diffusion of nutrients and effusion of wastes, an over-permeable capsule will facilitate the infiltration of deleterious immunoglobulins and/or complement components,[107] leading to the eventual degradation of the capsule. Conversely, an overly thick capsule would starve the cells of both oxygen and nutrients leading to a necrotic core, with important immunogenic consequences. A fine balance is thus required between capsule thickness and permeability.

Frequently, the membrane permeability is defined by the molecular weight cut-off (MWCO), the maximum molecular weight of a molecule that may freely pass through the pores of the capsule membrane (Fig. 1.2). The MWCO of orally delivered microcapsules must allow for passage of substrates from the gastrointestinal (GI) tract as well as unwanted metabolites from the plasma and then either facilitate the subsequent removal of the altered molecule or provide for its storage.[110] Though very promising, cell encapsulation is not yet applied in routine clinical practice. For successful exploitation, a microcapsule must meet a set of stringent demands. Finding a suitable material and method to construct a microcapsule is challenging. First, the material should be biocompatible, non-toxic and benign to the immune system. Second, the process for cell encapsulation should be mild enough to preserve adequate cell viability and retain a high initial seeding density of cells. Third, microcapsules should be able to bear shear stress, culture media or other environmental constraints. Finally, capsule parameters including size, mechanical strength, permeability, surface topography, etc., and other host-related factors have to be adjusted to obtain optimum clinical features. Table 1.3 summarizes the main considerations on selecting suitable material and approach for cell encapsulation.

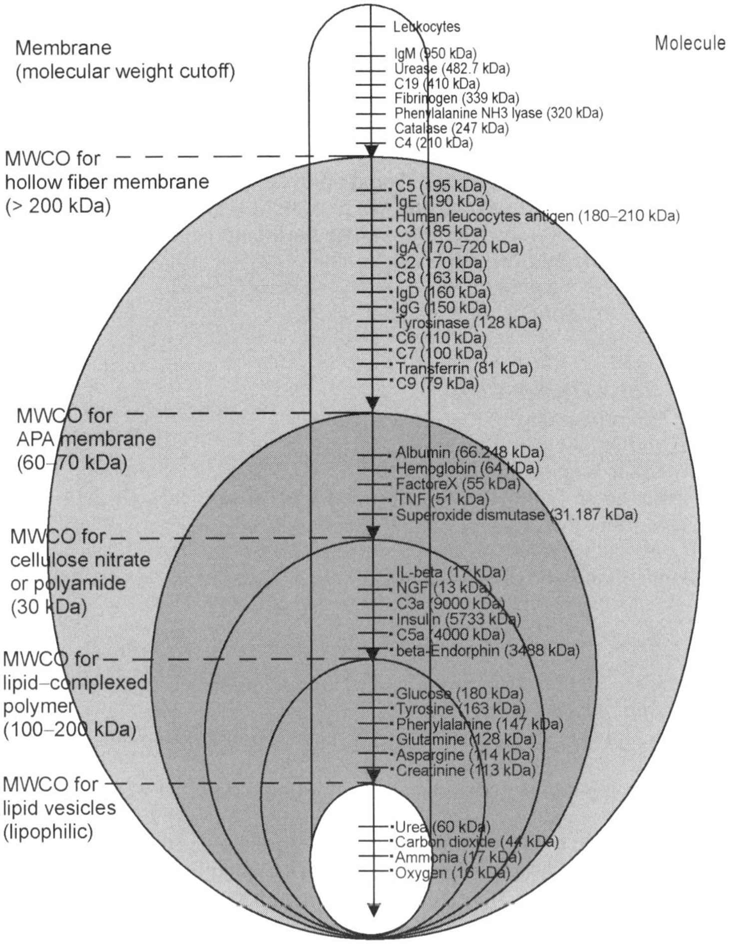

1.2 Molecular cut-off of different types of microcapsule membranes. The molecular weights of various cells, enzymes, antibodies, complement components, proteins, peptides and metabolites are listed on the right. Abbreviations: C2-9 and C19, various components of the complement cascade; Ig, immunoglobulin; IL-1, interlukin 1; NGF, nerve growth factor.[60,108,109]

1.3 Microcapsule design membrane materials

Recognizing the complexities inherent to the process, a variety of materials has been investigated for use in microencapsulation (Fig. 1.3). The major polymers currently under investigation for the encapsulation of live cells include alginate, hydroxyethyl methacrylate-methyl methacrylate (HEMA-MMA) and other hydrogels, polyacrylonitrile-polyvinyl chloride (PAN-PVC), hollow fiber membranes (HFMs), siliceous encapsulates, cellulose membranes and molecular variations on each of the above.

Table 1.3 Selection considerations of biomaterials and methods used for cell encapsulation

Material characteristics/chemistry
- Availability of clinical grade materials with reproducible characteristics
- Potential impurities and leachable residues, e.g. endotoxin, solvent, additive, initiator, cross-linker, pore-forming agent, precipitating agent
- Biocompatibility, non-cytotoxicity to both host and encapsulated cells
- Non-thrombogenity if contacted with blood, non-tumorigenity
- Must not trigger host immune response
- No interference with cell functions *in vitro* and *in vivo*
- Sterilization options for materials
- Cost

Formulation and processing
- Ease of processing
- Reproducibility of critical desired features
- Complete encapsulation
- Lack of harsh chemicals, temperatures or pH needed for synthesis and crosslinking
- Maintenance of cell survival and function during processing and storage (for pre-formed device)
- Sterilization during processing and after manufacture
- Ease of sealing

Artificial cell properties
- Geometry, surface morphology and charge, dimensions scaled between species and implant sites
- Adequate mechanical integrity to withstand handling, application and retrieval (if needed), minimize defects
- Provide suitable extracellular microenvironment for cell growth and proliferation
- Highly selective permeability (high diffusion in the low MW nutrient range and low diffusion in the high MW immunoglobulin range)
- Adequate stability of critical properties (e.g. membrane transport and resistance to biodegradation) in the host environment for the implant lifetime
- Appropriate biodegradability if desired
- Alternation of the host physiology of the biological fluids/tissue environment (material itself and potential degraded substance)

1.3 (opposite) Photomicrographs of (A) HepG2 mammalian cells encapsulated in alginate–chitosan–polyethylene glycol (PEG) – poly-L-lysine–alginate (ACPPA) microcapsules in minimum essential medium (MEM) 5 days after coating, empty ACPPA microcapsules in physiological solution. Capsule size: $450 \pm 30\,\mu$m. (Magnification: 6.5×).[53] (B) Alginate-poly-L-lysine–alginate (APA) membrane system for encapsulation and oral delivery of empty microcapsules and microcapsules containing *E. coli* DH5 cells.[37,57] (C) APA microcapsules for oral delivery of thalidomide and photomicrograph of alginate–chitosan microcapsule (right) under 250× magnification.[55] (D) Encapsulated HepG2 cells in alginate–poly-L-lysine–PEG–alginate (APPA) microcapsules 1 day after encapsulating; magnification: 10×, size: 450 μm and alginate–chitosan–PEG (ACP) microcapsules 1 week after encapsulating; magnification: 10×, size: 450 μm.[51] (E) CLSM (confocal laser scanning microscopy) images of genipin crosslinked microcapsules under normal and fluorescent light. Bars represent 200 μm.[111]

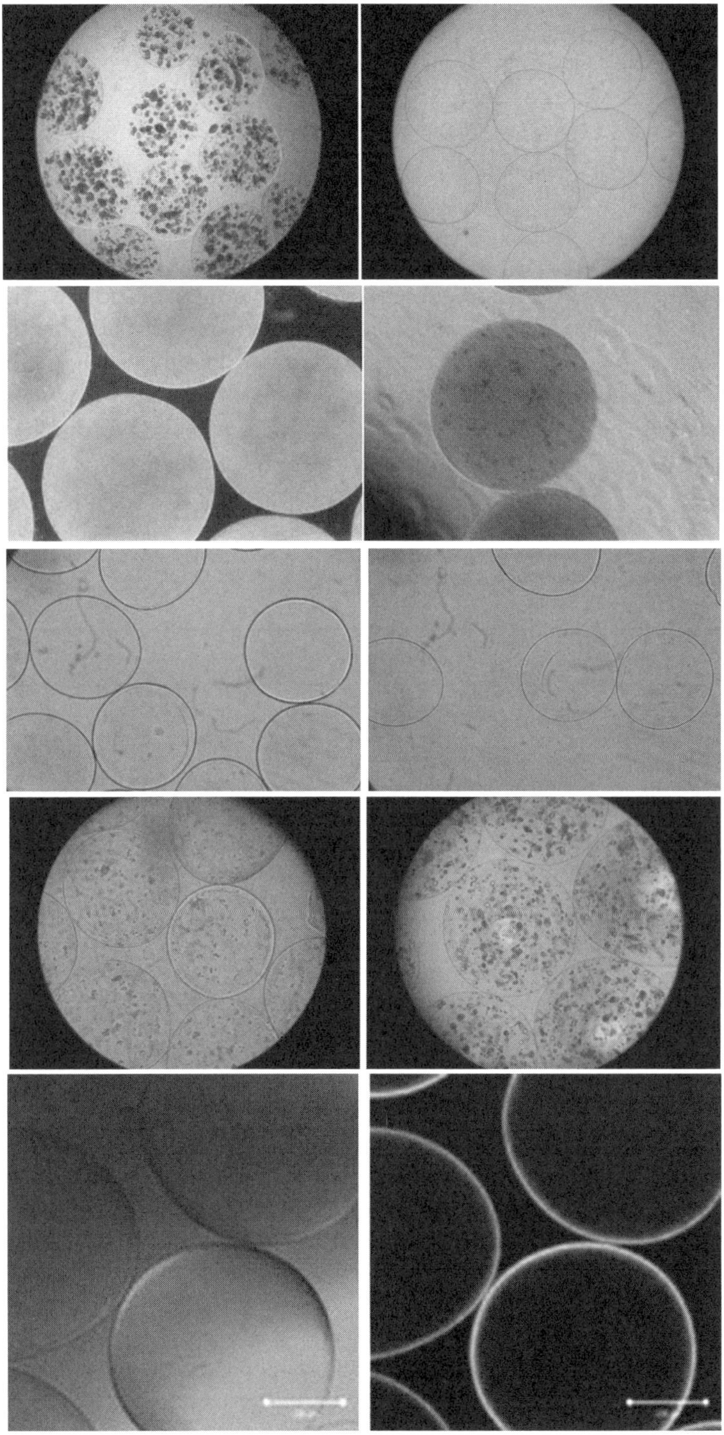

(a)

(b)

(c)

(d)

(e)

The chief polymers used for the encapsulation can broadly be classified into two groups: thermoplastic polymers and hydrogel polymers. Thermoplastic polymers are water insoluble and include HEMA-MMA and polyethylene glycol (PEG). While these capsules are stable, the diffusion properties of water-soluble nutrients appear to be restricted, which could potentially compromise long-term cell viability *in vivo*.[112] In contrast, hydrogels have been favored for cellular microencapsulation due to their appealing qualities. Hydrogels require mild encapsulation conditions which are important to retain cell viability;[112] they exhibit malleable and pliable features that reduce mechanical or frictional irritation to surrounding tissue and their hydrophilic nature minimizes protein adsorption and cell adhesion by eliminating interfacial tension with surrounding fluids and tissues.[110] In addition, they allow for significant permeability of low molecular mass nutrients and metabolites necessary for cellular survival.[110] A wide variety of biomaterials have been investigated for artificial cell formulation. Table 1.4 lists

Table 1.4 Characteristics of some polyelectrolytes/biomaterials used for cell encapsulation

Polymer	Type	Source
Alginate	Polyanion	Natural
Carboxymethylcelluose	Polyanion	Natural
Carrageenan	Polyanion	Natural
Cellulose sulfate	Polyanion, permanent charged	Natural
Heparin	Polyanion, permanent charged	Natural
Pectinate	Polyanion	Natural
Xanthan	Polyanion	Natural
Polyacrylic acid	Polyanion	Synthetic
Poly(styrene sulfonate)	Strong polyanions, permanent charged	Synthetic
Chitosan	Weak polycation	Natural
Poly(L-lysine)	Weak polycation	Natural
Hydroxyethyl cellulose	Weak polycation	Synthetic
Poly(allylamine) hydrochloride	Weak polycation	Synthetic
Poly(diallyldimethyl ammonium) chloride	Strong polycation, permanent charged	Synthetic
Poly(ethyleneimine)	Polycation	Synthetic
Poly(methylene co-guanidine) (PMCG) hydrochloride	Polycation, oligomeric	Synthetic
Poly(vinylamine) hydrochloride	Weak polycation	Synthetic
Protamine sulfate	Polycation	Synthetic
Agarose	Thermally gelled neutral polymer	Natural
Polyacrylonitrile (PAN)-polyvinyl chloride (PVC)	Thermoplastic polymer	Synthetic
Polyacrylates (PMMA-MAA-HEMA)	Thermoplastic polymer	Synthetic
Polyvinyl alcohol (PVA)	Polymeric hydrogel	Synthetic
Polyethylene glycol (PEG)	Thermoplastic polymer	Synthetic

some of the polyelectrolyte membrane materials that have been successfully used for cell encapsulation.

Alginate is a natural polymer derived from kelp. It is the most widely used polymer for cell encapsulation, possessing unique characteristics that render it especially suitable in combination with other polymers.[113,114] As a result of its exceptional cell compatibility and optimal gelling properties, alginate is ideal for cell immobilization. The molecular structure of alginate consists of β-D-mannuronic acid and α-L-guluronic acid residues. Variations in the ratios of the two molecules as well as the kinetics of the synthesis reaction facilitate the control of polymer properties such as pore size and degradation characteristics.[114]

Table 1.5 Artificial cell encapsulation for delivery

Microcapsule membrane polymer type	Microcapsule membrane features		Reference
	Strengths	Limitations	
APA (alginate–poly(L-lysine)–alginate)] + Variants (PEG, Ba^{+2}, Ca^{+2})	Acceptable food additive Cell and tissue compatible Mild reaction conditions Low cost Ease of control over parameters Starch useful as prebiotic probiotic Short/medium-term mechanical stability Flexible permselectivity Established synthesis protocols Low immunogenicity when PEGylated.	Susceptible to acid Reduced mechanical stability during lactic fermentation Insufficient immunoprotection Susceptible to long-term Ca^{+2} loss, consequent mechanical instability Structurally rigid Must be PEGylated to prevent fibrotic overgrowth	68, 116–123
A-PMCG-A (alginate–poly(methylene-co-guanidine)–alginate)	Better mechanical stability than Ca^{+2}/A, Ba^{+2}/A, BAPA, PMCG Cheaper than poly-L-lysine (PLL) Capsule size/permeability independently adjustable	Immunogenicity Long-term mechanical stability yet to be determined	124, 125
KC/LBG (kappa-carrageenan/locust bean gum gel)	Strong, rigid gel Acid resistant Thermoreversible Good results in cryo-preservation studies	Less biocompatible than alginate Potassium ions damaging to cells and potentially host electrolyte composition Insufficient immuno-protection	126, 127

Table 1.5 Continued

Microcapsule membrane polymer type	Microcapsule membrane features: Strengths	Microcapsule membrane features: Limitations	Reference
CAP (cellulose acetate phthalate)	Resistant to gastric acid conditions Established enteric coating material for controlled release Readily dissolves in mildly acidic to neutral environment of small intestine	Harsh reaction conditions (HCl) limits probiotic viability during membrane formation Membrane is non-porous Limited access to substrate during storage	128, 129
CS/A/PMCG (cellulose sulfate/sodium alginate/ poly(methylene-co-guanidine))	Short-term applications negate long-term mechanical stability and biocompatibility concerns	Encapsulated cells sensitive to alginate purity	124, 130–132
Gellan gum/ xanthan gum	Acid resistant Stabilized by calcium ions Easy to mix bacterial suspension with gum prior to gelation Economical processing Retention of cellular viability in pasteurized yogurt No shrinkage in lactic and acetic acid solutions	Gellan gum requires high setting temperature Acid survival dependent on strain Some reports indicate poor viability in storage Insufficient immuno-protection	133
Agarose	Prolonged stability in storage Cell and tissue compatible Mild reaction conditions Narrow size distribution of beads Low cost/readily available	Limited mechanical stability Insufficient immuno-protection due to cellular protrusion and absence of permselective layer	134
Starch	Natural adhesion (prebiotic) Exhibits good spray-drying properties allowing for easy scale-up and economical processing Small-sized microparticles with excellent cell coverage	Limited protection against acid stress, protease activity and pancreatin Poor survival in foods Nature of adhesion dependent on strain (presence of cell surface protein)	135

Table 1.5 Continued

Microcapsule membrane polymer type	Microcapsule membrane features: Strengths	Microcapsule membrane features: Limitations	Reference
HEMA-MMA (hydroxymethyl-acrylate-methyl methacrylate)	Insolubility in aqueous solutions confers greater mechanical stability	Non-adherent membrane properties require co-encapsulation with matrix to facilitate anchorage-dependent cell adhesion/growth	
Multi-layered HEMA-MMA-MAA	Exceptional design flexibility Independent adjustment of mechanical stability, permselectivity Promising compatibility with blood-contact applications	Single-layered capsules possess insufficient mechanical stability Immunogenicity yet to be determined Synthesis protocol more complex than other designs	
PAN-PVC (poly(acrylo-nitrile-vinyl-chloride))	Established mechanical stability, permselectivity Good biocompatibility	Molecular-weight cut-offs currently in question Long-term immuno-genicity not yet established	
AN-69 (acrylonitrile/sodium meth-allylsulfonate)	Good mechanical stability, permselectivity Amitogenic Large-scale encapsulation (~50 million cells/minute) now possible	Immunogenicity not well established	
PEG/PD_5/PDMS (poly(ethylene glycol)/poly (pentamethyl-cyclopenta-siloxane)/poly(dimethyl-siloxane))	Good mechanical stability PDMS confers excellent oxygen permeability	Long-term fibrogenicity not ideal for cell encapsulation	
PDMAAm (poly (*N,N*-dimethyl acrylamide))	Improved mechanical stability when crosslinked with telechelic stars	Oxygen permeability inferior to copolymers with PDMS	
Siliceous encapsulates	Simple synthesis mechanism confers high design flexibility	Questionable toxicity, immunogenicity	

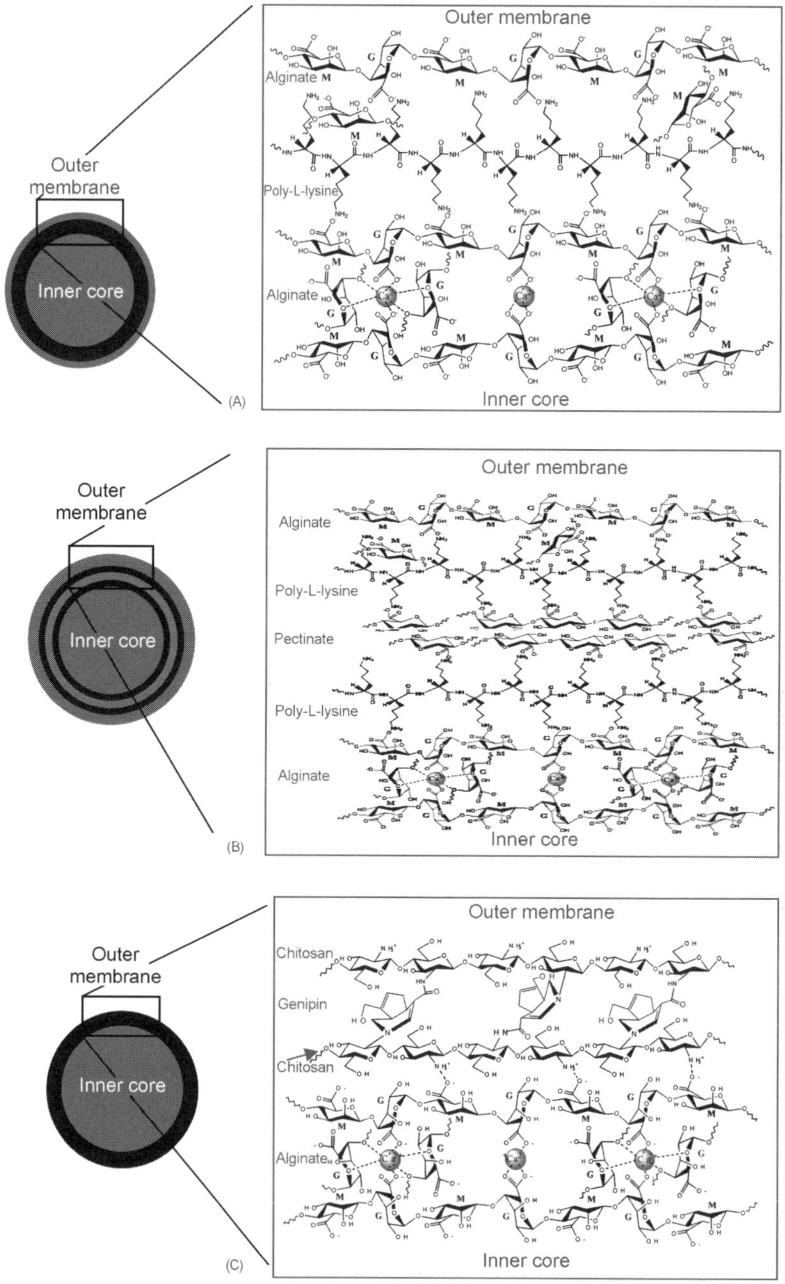

1.4 Microcapsule membrane design: (A) Alginate–poly-L-lysine–alginate (APA) membrane concept and layer-by-layer molecular structure;[108] (B) artificial cell alginate–poly-L-lysine–pectinate–poly-L-lysine–alginate (APPPA) microcapsule membrane structure;[136] and (C) schematic molecular structure of the genipin crosslinked alginate–chitosan (GCAC) microcapsule.[111,137,138]

Divalent cations such as Ca^{+2}, Ba^{+2}, and Sr^{+2} have typically been cross-linked with alginate to induce gel formation via a sol-gel transformation process.[114,115]

A host of other formulations have been proposed for cell entrapment; alternative encapsulation membrane systems have been designed to attempt to improve on certain physicochemical limitations in the classical alginate system (Table 1.5). Each membrane, however, is not without disadvantages as improved mechanical durability often compromises biocompatibility and vice versa. Ideal carriers are non-toxic, readily available and inexpensive. Cell loading, cell viability in the support material and protection from environmental stresses must also be assessed. The molecular design strategy and the schematic diagram for selected microcapsules are shown in Fig. 1.4.

1.4 The diversity of artificial cell preparation methods

To implement the technology of artificial cell encapsulation, the choice of matrix/membrane materials and formulation methods is critical as it dictates the resulting performance of artificial cells. Numerous microencapsulation techniques, fundamentally different in the nature of entrapment mechanism, have been developed for cell encapsulation. A typical encapsulation process starts with a scheme to generate a controlled-size droplet from a liquid cell suspension, followed by a rapid solidification or gelation to stabilize the droplet and further interfacial process, if needed, to obtain a solid microcapsule membrane surrounding the droplet. Gelation or solidification can occur through a change in temperature for thermoreversible gels, or by formation of an insoluble complex via chemical or ionic crosslinking. The commonly used techniques encompass:

- ionic gelation (e.g. calcium-alginate beads),
- complex coacervation (e.g. alginate–poly-L-lysine–alginate (APA) microcapsules),
- interfacial precipitation (e.g. HEMA-MMA system),
- phase separation (e.g. gelatin and agar capsules),
- solvent extraction or evaporation method (e.g. spray-drying for probiotics encapsulation).

Encapsulation by interfacial polymerization and by alginate gelation represent two main axes of research in biomolecule immobilization. Other methods described in the literature are presented below.

1.4.1 Microencapsulation by polyelectrolyte complexation

The simplest method to form a physical membrane barrier around living cells utilizes the interaction of oppositely charged polymers.[110] Since charged polymers exhibit solubility in water, they constitute an advantageous system that is compatible with the cellular environment. While natural and synthetic polymers can be used for this purpose, the structural integrity and biocompatibility of the resulting microcapsule depend strongly on the choice of polymer. Natural polymers are more cell compatible, react under milder conditions and allow for the encapsulation of fragile cells.[110,139] In contrast, synthetic polymers, which are less compatible, can be prepared in large quantities with high purity, which results in greater homogeneity with respect to capsule composition and permeability.[110] The alginate–PLL (poly-L-lysine) system uses the natural polymer alginate. APA microcapsules are produced by the immobilization of individual cells or tissue in an alginate droplet that is then hardened by gelation in a Ca^{2+}-rich solution.[139] After gelation in calcium chloride, the beads are washed in a PLL solution to form a membrane that is permselective and immunoprotective. Lastly, the capsules are washed and suspended in a solution of alginate to bind all positively charged PLL residues still present at the capsule surface.[2] The APA system employing polyelectrolyte complexation has proven advantageous as its aqueous-based, relatively mild encapsulation conditions do not compromise cell viability. Furthermore, modifications have been developed that have improved mechanical stability and biocompatibility. Nevertheless, other techniques have been attempted, mainly to address physicochemical limitations found in APA capsules.[123]

1.4.2 Artificial cell microencapsulation by agarose

Agarose is a readily available neutral polymer that exhibits temperature-sensitive water solubility that can be utilized for cell entrapment.[140] Experimentally, an agarose/cell suspension is transformed into liquid microbeads and subsequently hardened by a reduction in temperature. In order to eliminate cellular protrusion, an additional agarose layer can be added. With regards to applications, agarose beads appear to be better suited for allo-transplantation rather than xeno-transplantation as the membrane is not sufficiently immunoprotective against certain xeno-reactive antibodies.[110]

1.4.3 Artificial cell microencapsulation based on interfacial phase inversion

This process utilizes water-insoluble, thermoplastic polymers rather than hydrogels. Experimentally, microencapsulation is performed by extruding a suspension of cells to form a liquid core, immersing the cell droplet in a polymer solution to form a liquid shell and finally extracting the polymer solvent in order to precipitate a solid shell.[110] The potential advantages of encapsulating with thermoplastic polymers include easy scale-up, better capsule durability and more control over desired properties.[110] However, as mentioned above, the diffusion properties of water-soluble nutrients appear to be limited, which could potentially compromise long-term cell viability *in vivo*.

1.4.4 Artificial cell microencapsulation by conformal coating techniques

Conformal coating eliminates unused space in a microcapsule core by forming a membrane barrier directly on the cell mass. Many different methods have been employed; however, the technique generally involves the centrifugation of the cell mass through a solution of the coating polymer in a biocompatible solvent.[141] This technique theoretically improves mass transport and increases cell packing. Furthermore, it is directly applicable to entrapping cell clusters rather than single cells: entrapping single cells would result in a high ratio of membrane material to cell mass.[141] Therefore, it would appear that this process is ideal for the encapsulation of pancreatic islets, a cluster of cells that relies heavily on mass transport. However, there is little evidence of reported *in vivo* success in the literature.

In interfacial polymerization, the diamines may be replaced by materials of diverse structures (e.g. proteins, polysaccharides and chitosan). Operating conditions may then be modified to pH ~7 without use of toxic solvents. Transacylation avoids the use of toxic crosslinking agents. The encapsulation of cells by a mild interfacial chemical process then becomes possible. Alternatives to alginate gel have been proposed. Most of them are based on thermal gelation. The beads are produced through dispersion of a sol–gel in vegetable oil and gelified by lowering the temperature. A commonly used material for entrapment is kappa-carrageenan, more stable than alginate beads in biological media. But the high temperatures (45–50 ºC) used in the process are unfavorable especially for immobilization of plant and animal cells. Consequently more than 80% of cell encapsulation processes is still carried out using alginate as the matrix for immobilization. Hydrogel beads provide the immobilization of the cells but not a protection. Alginate beads may be coated by simple suspension of the alginate beads in PLL solution. A 10–30 μm membrane is formed around the beads. The conditions for the formation of this membrane are very mild. Chitosan has been proposed to replace PLL but most research is still based on PLL, especially for encapsulation of animal cells and artificial organs. More recently, immobilization inside a membrane was obtained by dropwise addition of the alginate solution into chitosan solution (interfacial coacervation). A membrane, very similar to the previous case, will be formed at the interface. The use of strongly charged polymers instead of weakly charged ones allows the formation of improved membranes.

1.5 Production of artificial cells: microencapsulation technologies

A challenge involved in the production of uniform capsules is ensuring excellent repeatability and reproducibility both within and between batches. The technique used to produce capsules (for example alginate–PLL–alginate) is based on the entrapment of individual cells or tissue in an alginate droplet that is then transformed into a rigid bead by gelation in a Ca^{2+}-rich solution. Stepwise, the cells to be encapsulated are first suspended in a solution of alginate that has been purified. The viscous alginate–cell suspension is then passed through a needle using a syringe pump. Before exiting the needle, compressed air is administered to shear the alginate–cell solution into relatively homogeneous droplets. The alginate–cell droplets are then extruded through the needle tip and allowed to fall in a gently stirred ice-cold calcium chloride solution, where they are transformed into rigid beads as described above. After gelation in calcium chloride, the beads are suspended in a PLL solution to form a membrane that is permselective and immunoprotective. By varying PLL concentration, alginate concentration or contact time, the porosity of the membrane can be manipulated. Lastly, the capsules are washed and suspended in a solution of alginate to bind all positively charged PLL residues still present at the capsule surface. This step is necessary to improve capsular durability and biocompatibility. Also, if a liquefied core is desired, the final APA capsules can be washed in a sodium citrate solution, which chelates calcium and solubilizes the intracapsular alginate gel. Figure 1.5 displays an apparatus utilizing air flow to form droplets and a calcium chloride solution for gelation into microbeads.

In order to improve the mechanical stability and biocompatibility of the APA system, a number of modifications have been employed. First, high-viscosity alginate is recommended as it results in a decreased formation of tails or strains during droplet formation and in turn a more uniform structure to the beads. Second, alginate with an intermediate concentration of guluronic acid molecules is employed rather than low guluronic acid content (low-G) alginate which has been seen to swell and break after implantation. Third, covalent links are created between adjacent layers of APA microcapsule membranes using a hetero-bifunctional photoactivatable crosslinker N-5-azido-2-nitrobenzoyloxy-succinimide (ANB-NOS), and the resulting microcapsules are extremely resistant to chemical and mechanical stresses.[143] Fourth, an interesting strategy is to incorporate an inorganic component into microcapsule membrane via sol–gel approach and form alginate–silica complexes to provide additional mechanical strength of the APA capsules.[144,145] Alginate purification schemes have also been developed to substantially reduce host response. In order to address the potential for encapsulated cells protruding through the membrane, a two-step method of cell encapsulation was developed. Specifically, this method involves producing alginate microbeads containing entrapped cells and then

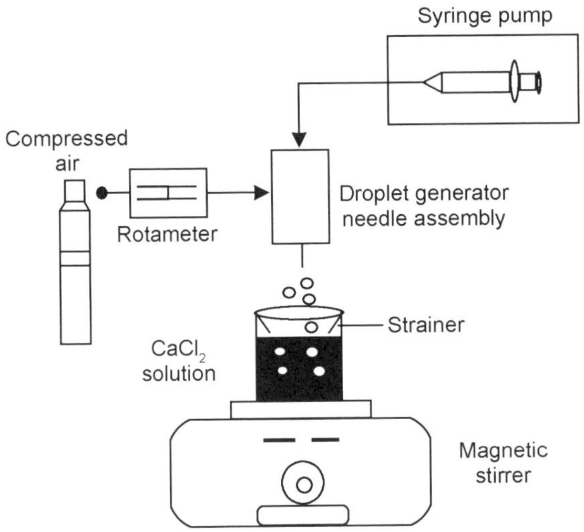

(a) Microencapsulation apparatus

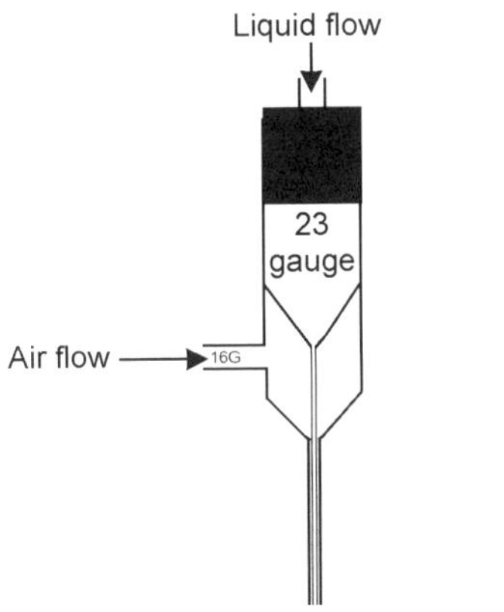

(b) Droplet needle assembly

1.5 Microencapsulation apparatus (A) and microencapsulation droplet assembly (B).[2,50,142]

subsequently entrapping the individual beads within a larger sphere of alginate. Subsequently, the larger sphere is then immersed in PLL and alginate as described above. Lastly, more technologically advanced encapsulators are now commercially available that utilize various technologies to produce capsules of smaller diameter with narrow size distribution. The adoption of automated machines for microencapsulation results in improved reproducibility in terms of shape, size and morphology. Droplets for bead formation can be formed using an assortment of techniques such as emulsification or coacervation, rotating disk atomization, airjet, electrostatic dripping, mechanical cutting and the vibrating nozzle method.

Large-scale microcapsule production has been initiated by a Swiss-based company Inotech with the Encapsulator®, the first commercially available instrument for the controlled polymer microencapsulation of drugs, animal cells, plant cells, microorganisms and enzymes which combines the faculties of defined bead diameter, sterile working conditions and high productivity (Fig. 1.6). The Encapsulator's technique is based on a harmonically vibrating nozzle. The laminar liquid jet is broken into droplets by vibration. The machine is capable of producing microcapsules of uniform spherical shape, with a selected diameter between 400 and 2000 μm, under sterile conditions, without harm to cells or microbes, with a high throughput volume and low dead volume and with high cell viability. It is most suitable for cell encapsulation as it produces monodisperse

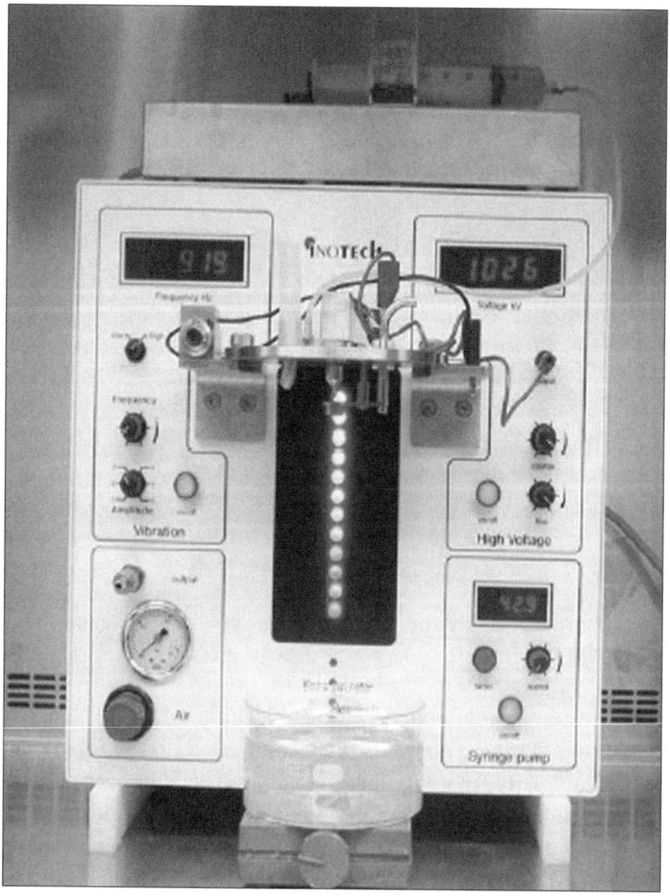

1.6 Inotech's Encapsulator® is a bench-top unit that has a broad range of applications in biotechnology and medicine.[60]

beads with a small diameter that prevents diffusion limitations; the narrow size distribution prevents cell necrosis. The JetCutter system from geniaLab® is an electrostatic droplet generator with a holder for multiple needles to allow rapid automated production of microcapsules. Bead generation is achieved by cutting a solid jet of fluid coming out of a nozzle by means of rotating cutting wires into cylindrical segments which then form beads, owing to surface tension, on their way to a hardening device. Since the velocity of the fluid and the speed of the cutting tool are kept constantly the produced beads show a very narrow distribution of particle size. The diameter of the particles that have been produced by the JetCutter ranges from about 120 μm to greater than 3 mm.

1.6 Artificial cell microcapsule membrane characterization

Whether employing the alginate-PLL-alginate membrane or alternative designs, the capsule properties play an integral role in the performance of encapsulated cells. The basis for analyzing the effectiveness of a capsule can thus be broken down into four important properties: permeability and mass transport, mechanical stability, immunoprotection and biocompatibility. An optimal balance must be achieved that permits the bidirectional diffusion of nutrients, oxygen, metabolites, metabolic waste and therapeutic protein products while simultaneously exluding immune cells and tryptic enzymes. Generally, membrane permeability can be explained by a thermodynamic parameter, known as the equilibrium partition coefficient and a kinetic parameter, known as the diffusion coefficient. Both of these parameters are strongly influenced by the type and size of solute, the interactions between solute and membrane and the membrane thickness. The mechanical stability and durability of microcapsules is important in order to maintain capsular integrity in long-term applications. As the driving force behind the idea of microencapsulating cells, sufficient immunoprotection and biocompatibility are necessary for capsules to be successful in clinical applications. The host immune system response to microencapsulated cells can be divided into two categories: response to encapsulated cells, which can be triggered by insufficient immunoprotection, and response to the capsular membrane, which is instigated by inadequate biocompatibility.

A European working group tested alginate beads, as well as microcapsules based on alginate, cellulose sulfate and polymethylene-co-guanidine in regards to water activity, bead or capsule size, mechanical resistance and transport behavior.[146] The water activity and mechanical resistance were observed to increase with bead and capsule size. Transport properties (ingress) were assessed using a variety of low molar mass and macromolecular probes. For the membrane-containing microcapsules, larger membrane thickness, observed for the larger capsules, retarded ingress. The authors recommend that permeability be assessed using either a large range of probes or a broad molar mass standard, with measurements at one or two molar masses insufficient to simulate the behavior in application. Furthermore, the various methods of assessing transport properties agree, in ranking, for the beads and capsules characterized, with gels having smaller radii being less permeable. For microcapsules, the permeation across the membrane dominates the ingress, and thicker membranes have lower permeability.[146]

For many applications microcapsules are subjected to mechanical stresses exerted by their environments that may induce deformation and potential breakup, enhance maintaining integrity of the microcapsules containing live cells is critical for effective *in vivo* immuno-isolation and success of therapeutic applications. The mechanical stability is perhaps more important for oral administration of artificial cells because of the exposure to harsh and mechanically disruptive environments present in the GI tract.[122] Even though mechanical property has been recognized as a limiting factor,[147] its quantitative assessment is demanding owing to the small size and fragility of the microcapsules.[148,149] A commonly used approach involves measuring the failure of capsules under uniaxial loading.[110] Mechanical compression tests are used to estimate elasticity and rupture of capsules, with the sensitivity of the force transducer required to be tuned from μN to tens of N to fit the anticipated sample strength.[146,150] Gauthier *et al.*[151] evaluated the mechanical strength of PVA beads (600–900 μm in diameter) by single-bead unconfined compression using Mach-1 Mechanical Tester (manufactured by BioSyntech Inc. Canada).[152] In recent years a number of new approaches, such as osmotic pressure test,[45] detection of fluorescent markers,[149] scanning acoustic microscopy,[153] combination of atomic force microscopy (AFM) and confocal laser scanning microscopy (CLSM),[154] have been developed and quantitative values for the capsule strength reported.

AFM, a technique originally used to characterize the surfaces of insulating crystals, also provides the possibility of examining membrane surfaces.[155,156] It has been shown that AFM, based on measuring the inter-atomic force interaction between a probe and the sample, is a powerful tool for investigation of nonconductive samples under a wide variety of operating conditions.[157–159] In particular, several reports have demonstrated the interesting application of AFM in imaging soft (hydro) gel samples in liquids.[157,160,161] In this manner, three-dimensional information about the membrane topology can be obtained and the surface roughness quantified as a function of treatment. Consequently, AFM measurements can be used to define the optimum conditions for minimizing surface inhomogeneities introduced by the microencapsulation process. The advantages of the AFM technique include the possibility of carrying out measurements in a buffer solution without extensive sample preparation, as well as obtaining quantitative measurements of surface roughness. Samples almost maintain their original shapes during the imaging process. The surface morphology and the roughness parameters obtained clearly verify the proper way to substantially improve the smoothness of the membrane surface.

In CLSM the light from out-of-focus structures is faded out, and, provided the material is sufficiently translucent, a non-destructive view through the capsule wall is shown. By using different fluorescence labels, the unambiguous identification of several compounds is possible.[111] By collecting several coplanar cross-sections, a three-dimensional reconstruction of the inspected objects can be performed. Hence, computational image analysis allows the visualization and quantification of structures not only on the surface, but also inside the material. CLSM is a useful tool for the characterization of microcapsules. Compared with ordinary light microscopy and the technically even more demanding electron microscopy, CLSM provides additional information such as the three-dimensional localization and quantification of the encapsulated phase, as well as the polymer composition of the wall material.[162] The encapsulated phase can be identified and exactly localized in all three spatial dimensions. A visualization of the polymer distribution throughout the capsule wall is possible and, in particular, allows the detection of an inhomogeneous distribution of some capsule wall polymers.[111] By applying adequate procedures of image analysis and fluorescence labeling, CLSM may allow determination of the encapsulation rate of microcapsules without the need for any destruction, extraction and chemical assays. The results obtained by this optical method for microcapsules correlate very well with those from chemical analysis.[162]

The advanced nuclear magnetic resonance (NMR) imaging technique together with CLSM allows routine evaluation of capsule performance. The use of NMR technology is an impressive breakthrough in the visualization of details of the alginate crosslinks.[155] It was not possible hitherto to visualize low concentrations of polysaccharides by using proton density NMR-imaging because of the surrounding free water. But the T_1 weighted sequence selected in the study for imaging enhances the contrast of the microcapsule image by suppressing the proton signals of the aqueous surroundings.

The use of NMR to study carbohydrate polymers such as agarose[163] and alginate[164–167] has been shown feasible under various experimental conditions. Previous work reported magnetic resonance imaging (MRI) studies on large-size alginate capsules with diameters >2 mm.[164–168] Others have reported the imaging of 800–900 μm alginate beads with a more powerful MRI research tool using a 17.6 T vertical magnet.[169] Shen *et al.* have now shown a novel approach to improve the sensitivity for detection under normal clinical conditions by incorporating nanoscale, magnetic particles to enhance the MRI signal intensity of these microcapsules.[170] Their work shows a method for monitoring 300–400 μm microcapsules suitable for gene therapy applications under a clinically relevant setting. Furthermore, the incorporation of ferrofluid, permitting clear imaging of these microcapsules under *in vitro* and *in vivo* conditions, allows quantitative tracking of the amounts of implanted microcapsules. Reconstruction of serial two-dimensional images into three-dimensional graphics (www.kitware.com) extends the range of linearity so that this approach is feasible for three-dimensional quantitative non-invasive monitoring.

For many reasons it is of interest to analyze and to visualize the deposition/distribution of the polymers involved in the microencapsulation. Commonly used visualization techniques for microparticles and microcapsules include conventional light microscopy (LM) and scanning electron microscopy (SEM).[171–174] LM is always impeded by the scattered or emitted light from structures outside the optical focal plane. SEM, in contrast, usually requires a relatively complex sample pretreatment (e.g. gold sputtering) and does not allow visualization of the inner structures of objects. Mechanical sections of embedded particles allow inspection of the particle wall structure, but it is not possible to visualize the encapsulated phase which is often lost during the microscopy sample pre-treatment. Although the membrane thickness and coating density substantially affect the properties and performance of the microcapsules, these parameters are difficult to quantify. Discrepant results were reported; for example, Ma *et al.*[175] and Ross and Chang[176] found that APA microcapsules had a wall thickness of 11–13 μm, whereas others reported a membrane thickness of 40–120 μm[177–179] using prior labeling methods. Other destructive methods such as cross-sectioning or mass measurement are time and labor consuming.[175,180] Previous attempts to visualize and assess microcapsule membranes have, therefore, had many shortcomings.

To overcome these limitations, Chen *et al.* recently described a new CLSM method using fluorogenic genipin treatment without prior labeling of polymers.[111,137] This method enables the visualization and quantification of polyamine microcapsule membrane rapidly, easily and effectively (Fig. 1.7). Furthermore, using this approach only a little sample treatment is required; it is thus possible to quickly determine the membrane thickness and the density of polyamine membrane during the process of development and optimization, on a routine basis so as to facilitate the understanding and improvement of microcapsule performance. This novel approach can be used for the characterization of a variety of other microcapsule formulations and biomaterials.

1.7 Artificial cells: current status and future prospective

The process of engineering vehicles for delivery of therapeutics has taken various forms, falling broadly into two categories: viral and non-viral. The former, widely employing the use of various viruses and retroviruses to carry plasmid DNA to host cells, faces a number of problems, including toxicity, immune/inflammatory responses, gene control and targeting issues, as well as the ever-present danger of the virus recovering its ability to cause disease. Perhaps the worst fear is that the gene may mis-target and trigger uncontrolled division, resulting in cancer. In an attempt to avoid such difficulties, researchers

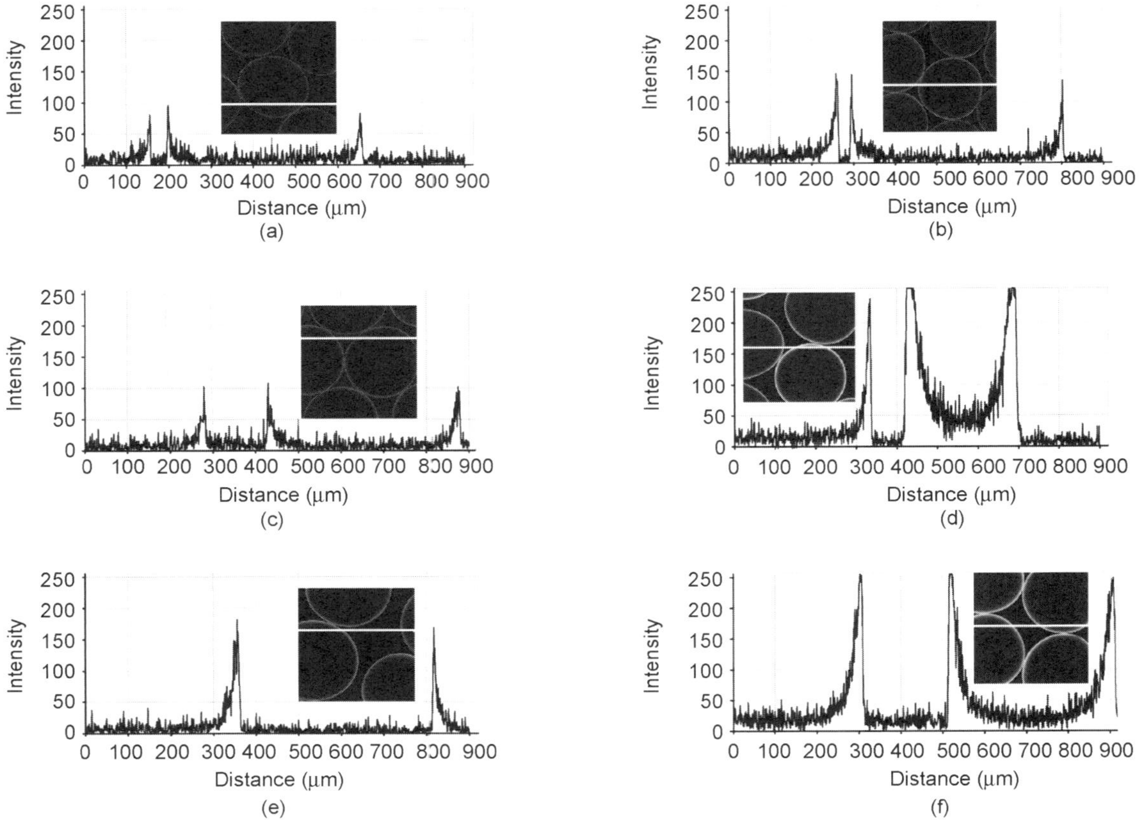

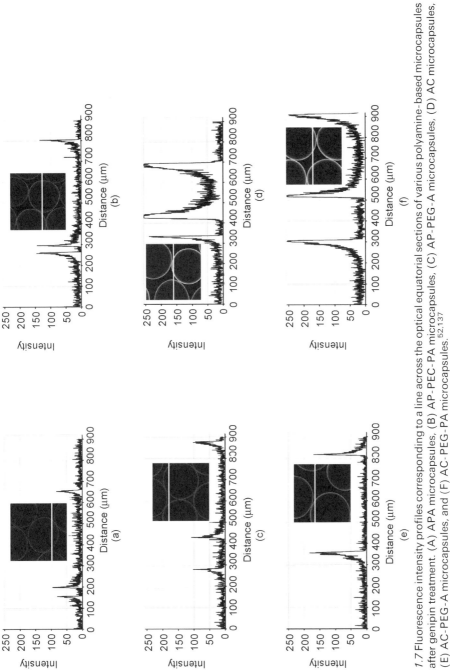

1.7 Fluorescence intensity profiles corresponding to a line across the optical equatorial sections of various polyamine-based microcapsules after genipin treatment. (A) APA microcapsules, (B) AP-PEC-PA microcapsules, (C) AP-PEG-A microcapsules, (D) AC microcapsules, (E) AC-PEG-A microcapsules, and (F) AC-PEG-PA microcapsules.[52,137]

are working to design their own carriers from various synthetic and biomaterials. The goal is to create an administrable artificial cell carrying a recombinant gene with the size, strength and permeability to travel to a directed site, avoid an immune response and efficiently deliver its therapeutic cargo. The methods employed, thus far, range from biomaterial-encapsulated cells to durable synthetic polymersomes. Medical applications demonstrate the potential and flexibility of microencapsulated cell technology in treating pathological conditions.[66,181] Nevertheless, proof of principle in treating certain diseases has not yet translated into widespread clinical applicability.[109,181] Therefore, capsule design must be ameliorated with specific attention focused on improving membrane properties.

A substantial challenge related to the biomaterials used in cell encapsulation has been the lack of clinical-grade polymers. Another challenge involves the production of uniform capsules with excellent repeatability and reproducibility both within and between batches. The discovery of suitable immune-compatible polycations represents another principal area of study. Still other challenges involve the assessment of the exact dosage and MWCO value, as well as the overall biocompatibility of the system.[136] A number of issues need to be carefully evaluated when selecting suitable cell types for immobilization. Indeed, encapsulation requires an appropriate source of functional cells. The principal controversy surrounds the potential risk of inadvertent transfer to humans of animal viruses present in the xenotransplant, and many forums have concluded that research should proceed with allotransplantation over xenotransplantation.[51,53,77] The choice of transplantation site is another important consideration. Here, it is necessary to weigh issues such as the safety and possibility of retransplantation (peritoneal cavity, subcutaneous transplantation) against proximity to the circulation (intrahepatic transplantation or membranes supporting vascularization). Still another pertinent issue is that permanent graft survival of encapsulated cells has never been reported. It has been proposed that encapsulation of more than one type of materials in combination may attain improved therapeutic benefits by joint effects.[42,66,182] The use of membrane stabilizing agents, such as hydrocortisone, glucose and glycerol, was recently reported to improve encapsulated cell viability over extended periods of time.[183] Encapsulation of islets isolated from rats with a compound membrane composed of phosphorylcholine-containing polymers and cellulose acetate led to improved biocompatibility in both *in vitro* and *in vivo* studies[184] probably because of low protein adsorption and reduced thrombus formation of the membrane used.

With advances in the science of encapsulated cell therapies, regulatory authorities have been gradually adjusting their policies to accommodate these new therapeutic approaches. In the United States, for example, all islet transplant studies (and presumably all future encapsulation-type clinical studies) will be regulated by the US Food and Drug Administration (FDA) under an investigational new-device submission. For now, Europe will probably rely on

FDA guidelines, because specific regulations in this field are presently lacking. Recently, the US Pharmacopeia and National Formulary included a new therapeutic category for cell-based products, which constitutes a significant step toward accepting this technology and encouraging clinical trials. A major ethical concern surrounding the use of microencapsulated cells is to ensure that patients are treated with a technology that demonstrates a clearly proven biosafety based on standardized protocols and procedures. In this regard, it is important to avoid poorly conducted studies that put individuals at unnecessary risk and unfairly raise hopes and expectations.

Technological and biological limitations, as well as ethical, political and regulatory obstacles, must be overcome if the promise of cell encapsulation technology is to be realized.

Cell encapsulation represents an alternative non-viral approach for the long-term delivery of therapeutic products. A wide spectrum of cells and tissues can be immobilized, enhancing the potential applicability of this strategy in the treatment of multiple diseases.[181] Therefore, diverse solutions are clearly required to address the inherent limitations with this technology. Nevertheless, if optimized, microencapsulated cell therapy could prove to be a clinically appropriate and flexible therapy for the future treatment of disease. Some of the important considerations for consistent clinical success of cell encapsulation include a source of functional cells; a biocompatible, as well as mechanically and chemically stable, membrane of a suitable permeability cut-off value that provides immune protection, functional performance, biosafety, and long-term survival.

1.8 Conclusions and future trends

At the start of the twenty-first century, research is poised at the interphase of material chemistry, polymer science and nanomaterial science. Artificial cells now range from macro-dimensions, to micro-dimensions, to nano-dimensions, and to molecular dimensions. The entry of hybrid conjugates that combine synthetic or natural polymers with proteins, DNA, live cells or drugs into clinical development has established artificial cell therapeutics as a credible option for pharmaceutical and medicine development. Although polymer chemistry can produce an almost infinite number of structures, the key to their successful medical application has been optimization of structure in light of a defined biological rationale. Various types of artificial cells can be designed with special features required for live cell or drug delivery. Cell encapsulation represents an alternative non-viral approach for the long-term delivery of therapeutic products. A wide spectrum of cells and tissues can be immobilized, enhancing the potential applicability of this strategy in the treatment of multiple diseases such as diabetes, kidney failure, liver malfunctions, etc.[181] The future of artificial cells, and the potential of achieving a membrane system that is much closer to ideal, are firmly rooted in improved processing techniques and enhanced knowledge of polymer chemistry. In choosing novel polymers, it is essential that they be standardized, specifically in terms of composition, purity grade, reaction conditions and source reproducibility.[111] Furthermore, the manufacturing process should be such that scaling-up can be achieved while not compromising quality control. This step is essential for allowing cell microencapsulation to be more applicable to clinical situations. Growing expertise in this multi-disciplinary field is helping to design even more sophisticated technologies with bioresponsive elements. For example, nanotechnology[185] using a layer-by-layer self-assembly technique may fulfill the stringent requirements for cell encapsulation in a more satisfactory manner by the formation of nano-capsules or nano-film coating with tailored properties.[186] Microfabrication techniques[187] may also surmount challenges associated with conventional biologics delivery for therapy. Asymmetrical, drug-loaded microfabricated particles with specific ligands linked to the surface are proposed for improving oral bioavailability of peptide drugs. Other promising applications include micromachined silicon membranes to create implantable capsules containing pancreatic islets for immuno-isolation in possible treatment for diabetes. These interdisciplinary approaches will accelerate the clinical realization of the many exciting potential applications, bringing us towards a new era in which polymeric artificial cells are combined with advances in biotechnology and molecular biology to deliver new therapeutics. There is real hope that, once optimized, these artificial cell constructs will provide the enhanced treatments that are so urgently sought for life-threatening and incapacitating diseases.

1.9 Acknowledgements

The authors would like to acknowledge the support of research grants from the Canadian Institute of Health Research (CIHR) and the Natural Sciences and Engineering Research Council (NSERC) of Canada. Authors acknowledge the contributions of A. Kulamarva and D. Kim to the preparation of the manuscript.

1.10 References

1. Chang TMS. Semipermeable microcapsules. *Science* 1964; 146(364):524–525.
2. Prakash S, Chang TMS. Preparation and *in-vitro* analysis of microencapsulated genetically-engineered *Escherichia coli* Dh5 cells for urea and ammonia removal. *Biotechnology and Bioengineering* 1995; 46(6):621–626.
3. Sefton MV, May MH, Lahooti S, Babensee JE. Making microencapsulation work: conformal coating, immobilization gels and *in vivo* performance. *Journal of Controlled Release* 2000; 65(1–2):173–186.
4. Chang PL. Encapsulation for somatic gene therapy. *Annals of the NY Academy of Science* 1999; 875:146–158.

5. Lim F, Sun AM. Microencapsulated islets as bioartificial endocrine pancreas. *Science* 1980; 210(4472):908–910.
6. Chang TMS. Hemoglobin Corpuscles. Report of a research project for Honours Physiology. 1957. (Medical Library, McGill University, 1957.)
7. Chang TMS. Semipermeable Aqueous Microcapsules. Ph.D. Thesis. McGill Univ. Montreal. 1965.
8. Chang TMS. *Artificial Cells*. 1972. Thomas, Springfield.
9. Chang TM, MacIntosh FC, Mason SG. Semipermeable aqueous microcapsules. I. Preparation and properties. *Canadian Journal of Physiological Pharmacology* 1966; 44(1):115–128.
10. Chang TMS. Biodegradable semipermeable microcapsules containing enzymes, hormones, vaccines, and other biologicals. *Journal of Bioengineering* 1976; 1:25–32.
11. Chang TM. Semipermeable aqueous microcapsules ('artificial cells'): with emphasis on experiments in an extracorporeal shunt system. *Transactions of the American Society of Artificial Internal Organs* 1966; 12:13–19.
12. Chang TMS, Malave N. The development and first clinical use of semipermeable microcapsules (artificial cells) as a compact artificial kidney. *Transactions of the American Society of Artificial Internal Organs* 1970; 16:141–148.
13. Chang TMS. Removal of endogenous and exogenous toxins by a microencapsulated absorbent. *Canadian Journal of Physiological Pharmacology* 1969; 47:1043–1045.
14. Chang TMS, Poznansky MJ. Semipermeable microcapsules containing catalase for enzyme replacement in acatalsaemic mice. *Nature* 1968; 218:242–245.
15. Chang TMS. Haemoperfusions over microencapsulated adsorbent in a patient with hepatic coma. *Lancet* 1972; 23:1731–1732.
16. Chang TMS. Microencapsulated adsorbent hemoperfusion for uremia, intoxication and hepatic failure. *Kidney International* 1975; 7:S387–S392.
17. Chang TMS. Stabilization of enzyme by microencapsulation with a concentrated protein solution or by crosslinking with glutaraldehyde. *Biochemical Biophysical Research Communications* 1971; 44:1531–1533.
18. Chang TMS. The *in vivo* effects of semipermeable microcapsules containing L-asparaginase on 6C3HED lymphosarcoma. *Nature* 1971; 229:117 118.
19. Rosental AM, Chang TMS. The incorporation of lipid and Na+-K+-ATPase into the membranes of semipermeable microcapsules. *Journal of Membrane Science* 1980; 6:329–338.
20. Campbell J, Chang TMS. Microencapsulated multi-enzyme systems as vehicles for the cyclic regeneration of free and immobilized coenzymes. *Enzymes Engineer* 1978; 3:371–377.
21. Grunwald J, Chang TMS. Nylon polyethyleneimine microcapsules for immobilizing multienzymes with soluble dextran-NAD+ for the continuous recycling of the microencapsulated dextran-NAD+. *Biochemical and Biophysical Research Communications* 1978; 81:565–570.
22. Cousidneau J, Chang TMS. Formation of amino acid from urea and ammonia by sequential enzyme reaction using a microencapsulated multi-enzyme system. *Biochemical and Biophysical Research Communications* 1977; 79:24–31.
23. Bourget L, Chang TMS. Microencapsulated phenylalanine ammonia-lyase in artificial cells for phenylalanine conversion. *Artificial Organs* 1984; 8(1):109.
24. Bourget L, Chang TMS. Artificial cell microencapsulated phenylalanine ammonia-lyase. *Applied Biochemistry and Biotechnology* 1984; 10:57–59.
25. Bourget L, Chang TMS. Phenylalanine ammonia-lyase immobilized in

semipermeable microcapsules for enzyme replacement in phenylketonuria. *FEBS Letters* 1985; 180(1):5–8.
26. Bourget L, Chang TMS. Phenylalanine ammonia-lyase immobilized in microcapsules for the depletion of phenylalanine in plasma in phenylketonuric rat model. *Biochimica et Biophysica Acta* 1986; 883(3):432–438.
27. Bourget L, Chang TMS. Effects of oral-administration of artificial cells immobilized phenylalanine ammonia-lyase on intestinal amino-acids of phenylketonuric rats. *Biomaterials Artificial Cells and Artificial Organs* 1989; 17(2):161–181.
28. Chang TMS, Bourget L, Lister C. A new rheory of enterorecirculation of amino-acids and its use for depleting unwanted amino-acids using oral enzyme artificial cells, as in removing phenylalanine in phenylketonuria. *Artificial Cells Blood Substitutes and Immobilization Biotechnology* 1995; 23(1):1–21.
29. Sipehia R, Bannard R, Chang TMS. Adsorption of large lipophilic molecules with exclusion of small hydrophilic molecules by microencapsulated activated charcoal formed by coating with polyethylene membrane. *Journal of Membrane Science* 1986; 29:277–286.
30. Palmour RM, Goodyer P, Reade T, Chang TMS. Microencapsulated xanthine oxidase as experimental therapy in LeschNyhan disease. *Lancet* 1989; 2:687–688.
31. Chang TMS. Preparation and characterization of xanthine oxidase immobilized by microencapsulation in artificial cells for the removal of hypoxanthine. *Biomaterials Artificial Cells and Artificial Organs* 1989; 17:611–616.
32. Chang TMS, Powanda D, Yu WP. Analysis of polyethylene-glycol-polylactide nano-dimension artificial red blood cells in maintaining systemic hemoglobin levels and prevention of methemoglobin formation. *Artificial Cells Blood Substitutes and Immobilization Biotechnology* 2003; 31:231–248.
33. Chang TMS, Yu WP. *Blood Substitutes: Principles, Methods, Products and Clinical Trials*, Vol. 2 ed. Chang TMS. 216–231. 2006. Karger, Basel.
34. Yu WP, Chang TMS. Submicron biodegradable polymer membrane hemoglobin nanocapsules as potential blood substitutes: a preliminary report. *Artificial Cells Blood Substitutes and Immobilization Biotechnology* 1994; 22:889–894.
35. Soonshiong P, Heintz RE, Merideth N, Yao QX, Yao ZW, Zheng TL *et al.* Insulin independence in a Type-1 diabetic patient after encapsulated islet transplantation. *Lancet* 1994; 343(8903):950–951.
36. Aebischer P, Schluep M, Deglon N, Joseph JM, Hirt L, Heyd B *et al.* Intrathecal delivery of CNTF using encapsulated genetically modified xenogeneic cells in amyotrophic lateral sclerosis patients. *Nature Medicine* 1996; 2(6):696–699.
37. Prakash S, Chang TMS. Microencapsulated genetically engineered live *E-coli* DH5 cells administered orally to maintain normal plasma urea level in uremic rats. *Nature Medicine* 1996; 2(8):883–887.
38. Prakash S, Chang TM. Artificial cells microencapsulated genetically engineered *E. coli* DH5 cells for the lowering of plasma creatinine *in-vitro* and *in-vivo*. *Artificial Cells Blood Substitutes and Immobilization Biotechnology* 2000; 28(5):397–408.
39. Prakash S, Chang TM. Artificial cell microcapsules containing genetically engineered *E. coli* DH5 cells for *in-vitro* lowering of plasma potassium, phosphate, magnesium, sodium, chloride, uric acid, cholesterol, and creatinine: a preliminary report. *Artificial Cells Blood Substitutes and Immobilization Biotechnology* 1999; 27(5–6):475–481.
40. Chang PL. Microcapsules as bio-organs for somatic gene therapy. *Annals of the New York Academy of Science* 1997; 831:461–473.

41. Chang PL, Shen N, Westcott AJ. Delivery of recombinant gene-products with microencapsulated cells *in-vivo*. *Human Gene Therapy* 1993; 4(4):433–440.
42. Liu Z, Chang TMS. Coencapsulation of stem cells and hepatocytes: *in vitro* conversion of ammonia and *in vivo* studies on the lowering of bilirubin in Gunn rats after transplantation. *International Journal of Artificial Organs* 2003; 26:491–497.
43. Liu Z, Chang TMS. Transplantation of coencapsulated hepatocytes and marrow stem cells into rats. *Artificial Cells Blood Substitutes and Immobilization Biotechnology* 2002; 30:99–112.
44. Liu Z, Chang TMS. Effects of bone marrow cells on hepatocytes: when co-cultured or co-encapsulated together. *Artificial Cells Blood Substitutes and Immobilization Biotechnology* 2000; 28:365–374.
45. Van Raamsdonk JM, Chang PL. Osmotic pressure test: a simple, quantitative method to assess the mechanical stability of alginate microcapsules. *Journal of Biomedical Materials Research* 2001; 54(2):264-271.
46. Van Raamsdonk JM, Ross CJ, Potter MA, Kurachi S, Kurachi K, Stafford DW *et al.* Treatment of hemophilia B in mice with nonautologous somatic gene therapeutics. *Journal of Laboratory and Clinical Medicine* 2002; 139(1):35–42.
47. Cirone P, Bourgeois JM, Austin RC, Chang PL. A novel approach to tumor suppression with microencapsulated recombinant cells. *Human Gene Therapy* 2002; 13(10):1157–1166.
48. Cirone P, Bourgeois JM, Chang PL. Antiangiogenic cancer therapy with microencapsulated cells. *Human Gene Therapy* 2003; 14(11):1065–1077.
49. Jones ML, Chen H, Ouyang W, Metz T, Prakash S. Microencapsulated genetically engineered *Lactobacillus plantarum* 80 (pCBH1) for bile acid deconjugation and its implication in lowering cholesterol. *Journal of Biomedicine Biotechnology* 2004; 2004(1):61–69.
50. Cirone P, Bourgeois JM, Shen F, Chang PL. Combined immunotherapy and antiangiogenic therapy of cancer with microencapsulated cells. *Human Gene Therapy* 2004; 15(10):945–959.
51. Haque T, Chen H, Ouyang W, Martoni C, Lawuyi B, Urbanska AM *et al.* Superior cell delivery features of poly(ethylene glycol) incorporated alginate, chitosan, and poly-L-lysine microcapsules. *Molecular Pharmacology* 2005; 2(1):29-36.
52. Chen HM, Wei OY, Bisi LY, Martoni C, Prakash S. Reaction of chitosan with genipin and its fluorogenic attributes for potential microcapsule membrane characterization. *Journal of Biomedical Materials Research Part A* 2005; 75A(4):917–927.
53. Haque T, Chen H, Ouyang W, Martoni C, Lawuyi B, Urbanska A *et al.* Investigation of a new microcapsule membrane combining alginate, chitosan, polyethylene glycol and poly-L-lysine for cell transplantation applications. *International Journal of Artificial Organs* 2005; 28(6):631–637.
54. Haque T, Chen H, Ouyang W, Martoni C, Lawuyi B, Urbanska AM *et al. In vitro* study of alginate–chitosan microcapsules: an alternative to liver cell transplants for the treatment of liver failure. *Biotechnology Letters* 2005; 27(5):317–322.
55. Metz T, Jones ML, Chen HM, Halim T, Mirzaei M, Haque T *et al.* A new method for targeted drug delivery using polymeric microcapsules: implications for treatment of Crohn's disease. *Cell Biochemistry and Biophysics* 2005; 43(1):77–85.
56. Shen F, Li AA, Gong YK, Somers S, Potter MA, Winnik FM et al. Encapsulation of recombinant cells with a novel magnetized alginate for magnetic resonance imaging. *Human Gene Therapy* 2005; 16(8):971–984.

57. Chang TMS, Prakash S. Therapeutic uses of microencapsulated genetically engineered cells. *Molecular Medicine Today* 1998; 4(5):221–227.
58. Orive G, Hernandez RM, Gascon AR, Calafiore R, Chang TMS, de Vos P *et al.* History, challenges and perspectives of cell microencapsulation. *Trends in Biotechnology* 2004; 22(2):87–92.
59. Senior K. Encapsulated cell technology provides new treatment options. *Drug Discovery Today* 2001; 6(1):6–7.
60. Prakash S, Bhathena J. Microencapsulation. In: Akay, M, editor, *Wiley Encylopaedia of Biomedical Engineering*. John Wiley & Sons, Inc., New Jersey, 2006.
61. Orive G, Hernandez RM, Gascon AR, Igartua M, Pedraz JL. Encapsulated cell technology: from research to market. *Trends in Biotechnology* 2002; 20(9):382–387.
62. Orive G. Cell microencapsulation technology for biomedical purposes: novel insights and challenges. *Trends in Pharmacological Sciences* 2003; 24(5):207–210.
63. Orive G, Hernandez RM, Gascon AR, Calafiore R, Chang TM, Vos PD *et al.* Cell encapsulation: promise and progress. *Nature Medicine* 2003; 9(1):104–107.
64. Chang TMS. Artificial cells for replacement of metabolic organ functions. *Artificial Cells, Blood Substitutes, and Immobilization Biotechnology* 2003; 31(2):151–161.
65. Chang TMS. Artificial cells for cell and organ replacements. *Artificial Organs* 2004; 28(3):265–270.
66. Chang TMS. Therapeutic applications of polymeric artificial cells. *Nature Reviews Drug Discovery* 2005; 4(3):221–235.
67. Chang PL. Encapsulation for somatic gene therapy. *Bioartificial Organs Ii: Technology, Medicine, and Materials* 1999; 875:146–158.
68. de Vos P, van Hoogmoed CG, van Zanten J, Netter S, Strubbe JH, Busscher HJ. Long-term biocompatibility, chemistry, and function of microencapsulated pancreatic islets. *Biomaterials* 2003; 24(2):305–312.
69. Zhou M, Chen D, Yao Q, Xia Z, Wang C, Zhu H. Microencapsulation of rat islets prolongs xenograft survival in diabetic mice. *Chinese Medical Journal (English)* 1998; 111(5):394–397.
70. Sun YL, Ma XJ, Zhou DB, Vacek I, Sun AM. Normailization of diabetes in spontaneously diabetic cynomologous monkeys by xenografts of microencapsulated porcine islets without immunosuppression. *Journal of Clinical Investigation* 1996; 98:1417–1422.
71. de Vos P, Hamel AF, Tatarkiewicz K. Considerations for successful transplantation of encapsulated pancreatic islets. *Diabetologia* 2002; 45(2):159–173.
72. de Groot M, Schuurs TA, van Schilfgaarde R. Causes of limited survival of microencapsulated pancreatic islet grafts. *Journal of Surgical Research* 2004; 121(1):141–150.
73. de Vos P, Marchetti P. Encapsulation of pancreatic islets for transplantation in diabetes: the untouchable islets. *Trends in Molecular Medicine* 2002; 8(8):363–366.
74. Soonshiong P, Feldman E, Nelson R, Komtebedde J, Smidsrod O, Skjakbraek G *et al.* Successful reversal of spontaneous diabetes in dogs by intraperitoneal microencapsulated islets. *Transplantation* 1992; 54(5):769–774.
75. Chow KM, Liu ZC, Prakash S, Chang TMS. Free and microencapsulated Lactobacillus and effects of metabolic induction on urea removal. *Artificial Cells Blood Substitutes and Immobilization Biotechnology* 2003; 31(4):425–434.
76. Ranganathan N, Dickstein J. Composition for alleviating symptoms of uremia in patients. PA/USA patent 6 706 263. 2004 Mar 2004.

77. Chang TMS. The role of artificial cells in cell and organ transplantation in regenerative medicine. *Panminerva Medica* 2005; 47(1):1–9.
78. Wang LS, Sun JH, Li L, Harbour C, Mears D, Koutalistras N *et al.* Factors affecting hepatocyte viability and CYPIA1 activity during encapsulation. *Artificial Cells Blood Substitutes and Immobilization Biotechnology* 2000; 28(3):215–227.
79. Umehara Y, Hakamada K, Seino K, Aoki K, Toyoki Y, Sasaki M. Improved survival and ammonia metabolism by intraperitoneal transplantation of microencapsulated hepatocytes in totally hepatectomized rats. *Surgery* 2001; 130(3):513–520.
80. Hortelano G, Chang PL. Gene therapy for hemophilia. *Artificial Cells Blood Substitutes and Immobilization Biotechnology* 2000; 28(1):1–24.
81. Tagalakis AD. Apolipoprotein E delivery by peritoneal implantation of encapsulated recombinant cells improves the hyperlipidaemic profile in apoE-deficient mice. *Biochimica et Biophysica Acta – Molecular and Cell Biology of Lipids* 2005; 1686(3):190–199.
82. AlHendy A, Hortelano G, Tannenbaum GS, Chang PL. Correction of the growth defect in dwarf mice with nonautologous microencapsulated myoblasts – an alternate approach to somatic gene-therapy. *Human Gene Therapy* 1995; 6(2):165–175.
83. Ross CJD, Bastedo L, Maier SA, Sands MS, Chang PL. Treatment of a lysosomal storage disease, mucopolysaccharidosis VII, with microencapsulated recombinant cells. *Human Gene Therapy* 2000; 11(15):2117–2127.
84. Chang PL, Van Raamsdonk JM, Hortelano G, Barsoum SC, MacDonald NC, Stockley TL. The *in vivo* delivery of heterologous proteins by microencapsulated recombinant cells. *Trends in Biotechnology* 1999; 17(2):78–83.
85. Bergers G, Hanahan D. Cell factories for fighting cancer. *Nature Biotechnology* 2001; 19(1):20–21.
86. Hann E, Reinartz S, Clare SE, Passow S, Kissel T, Wagner U. Development of a delivery system for the continuous endogenous release of an anti-idiotypic antibody against ovarian carcinoma. *Hybridoma* 2005; 24(3):133–140.
87. Hao SG, Su LP, Gu XL, Moyana T, Xiang J. A novel approach to tumor suppression using microencapsulated engineered J558/TNF-alpha cells. *Experimental Oncology* 2005; 27(1):56–60.
88. Lohr M, Hoffmeyer A, Kroger JC, Freund M, Hain J, Holle A *et al.* Microencapsulated cell-mediated treatment of inoperable pancreatic carcinoma. *Lancet* 2001; 357(9268):1591–1592.
89. Read TA, Farhadi M, Bjerkvig R, Olsen BR, Rokstad AM, Huszthy PC *et al.* Intravital microscopy reveals novel antivascular and antitumor effects of endostatin delivered locally by alginate-encapsulated cells. *Cancer Research* 2001; 61(18):6830–6837.
90. Read TA, Sorensen DR, Mahesparan R, Enger PO, Timpl R, Olsen BR *et al.* Local endostatin treatment of gliomas administered by microencapsulated producer cells. *Nature Biotechnology* 2001; 19(1):29–34.
91. Thorsen F, Read TA, Lund-Johansen M, Tysnes BB, Bjerkvig R. Alginate-encapsulated producer cells: a potential new approach for the treatment of malignant brain tumors. *Cell Transplantation* 2000; 9(6):773–783.
92. Visted T. Progress and challenges for cell encapsulation in brain tumour therapy. *Expert Opinion on Biological Therapy* 2003; 3(4):551–561.
93. Bachoud-Levi AC, Deglon N, Nguyen JP, Bloch J, Bourdet C, Winkel L *et al.* Neuroprotective gene therapy for Huntington's disease using a polymer

encapsulated BHK cell line engineered to secrete human CNTF. *Human Gene Therapy* 2000; 11(12):1723–1729.

94. Benoit JP, Faisant N, Venier-Julienne MC, Menei P. Development of microspheres for neurological disorders: from basics to clinical applications. *Journal of Controlled Release* 2000; 65(1–2):285–296.
95. Bloch J, Bachoud-Levi AC, Deglon N, Lefaucheur JP, Winkel L, Palfi S *et al.* Neuroprotective gene therapy for Huntington's disease, using polymer-encapsulated cells engineered to secrete human ciliary neurotrophic factor: Results of a phase I study. *Human Gene Therapy* 2004; 15(10):968–975.
96. Hagihara Y, Saitoh Y, Iwata H, Taki T, Hirano S, Arita N *et al.* Transplantation of xenogeneic cells secreting beta-endorphin for pain treatment: analysis of the ability of components of complement to penetrate through polymer capsules. *Cell Transplantation* 1997; 6(5):527–530.
97. Maysinger D, Berezovskaya O, Fedoroff S. The hematopoietic cytokine colony stimulating factor 1 is also a growth factor in the CNS.2. Microencapsulated CSF-I and LM-10 cells as delivery systems. *Experimental Neurology* 1996; 141(1):47–56.
98. Maysinger D, Krieglstein K, FilipovicGrcic J, Sendtner M, Unsicker K, Richardson P. Microencapsulated ciliary neurotrophic factor: physical properties and biological activities. *Experimental Neurology* 1996; 138(2):177–188.
99. Ponce S, Orive G, Gascon AR, Hernandez RM, Pedraz JL. Microcapsules prepared with different biomaterials to immobilize GDNF secreting 3T3 fibroblasts. *International Journal of Pharmaceutics* 2005; 293(1–2):1–10.
100. Vallbacka JJ, Nobrega JN, Sefton MV. Tissue engineering as a platform for controlled release of therapeutic agents: implantation of microencapsulated dopamine producing cells in the brains of rats. *Journal of Controlled Release* 2001; 72(1–3):93–100.
101. Visted T, Bjerkvig R, Enger PO. Cell encapsulation technology as a therapeutic strategy for CNS malignancies. *Neuro-Oncology* 2001; 3(3):201–210.
102. Winn SR, Emerich DF. Managing chronic pain with encapsulated cell implants releasing catecholamines and endogenous opioids. *Frontiers in Bioscience* 2005; 10:367–378.
103. Chandramouli V, Kailasapathy K, Peiris P, Jones M. An improved method of microencapsulation and its evaluation to protect *Lactobacillus* spp. in simulated gastric conditions. *Journal of Microbiological Methods* 2004; 56(1):27–35.
104. Krasaekoopt W. Evaluation of encapsulation techniques of probiotics for yoghurt. *International Dairy Journal* 2003; 13(1):3–13.
105. Metz T, Jones ML, Chen HM, Halim T, Mirzaei M, Haque T *et al.* A new method for targeted drug delivery using polymeric microcapsules. *Cell Biochemistry and Biophysics* 2005; 43(1):77–85.
106. Garofalo FA, Chang TMS. Effects of mass-transfer and reaction-kinetics on serum-cholesterol depletion rates of free and immobilized *pseudomonas pictorum*. *Applied Biochemistry and Biotechnology* 1991; 27(1):75–91.
107. Babensee JE, Cornelius RM, Brash JL, Sefton MV. Immunoblot analysis of proteins associated with HEMA-MMA microcapsules: Human serum proteins in vitro and rat proteins following implantation. *Biomaterials* 1998; 19(7–9):839–849.
108. Prakash S, Jones ML. Artificial cell therapy: new strategies for the therapeutic delivery of live bacteria. *Journal of Biomedicine and Biotechnology* 2005;(1):44–56.
109. Prakash S, Martoni C. Toward a new generation of therapeutics: artificial cell targeted delivery of live cells for therapy. *Applied Biochemistry and Biotechnology* 2006; 128(1):1–22.

110. Uludag H, de Vos P, Tresco PA. Technology of mammalian cell encapsulation. *Advanced Drug Delivery Reviews* 2000; 42(1–2):29–64.
111. Chen HM, Wei OY, Bisi LY, Martoni C, Prakash S. Reaction of chitosan with genipin and its fluorogenic attributes for potential microcapsule membrane characterization. *Journal of Biomedical Materials Research Part A* 2005; 75A(4):917–927.
112. King A. Evaluation of Alginate Microcapsules for Use in Transplantation of Islets of Langerhans. University of Uppsala, 2001.
113. Gerbsch N, Buchholz R. New processes and actual trends in biotechnology. *FEMS Microbiology Reviews* 1995; 16(2–3):259–269.
114. Gombotz WR, Wee SF. Protein release from alginate matrices. *Advanced Drug Delivery Reviews* 1998; 31(3):267–285.
115. Stokke BT, Smidsrod O, Bruheim P, Skjakbraek G. Distribution of uronate residues in alginate chains in relation to alginate gelling properties. *Macromolecules* 1991; 24(16):4637–4645.
116. Zhang WJ. Purity of alginate affects the viability and fibrotic overgrowth of encapsulated porcine islet xenografts. *Transplantation Proceedings* 2001; 33(7–8):3517–3519.
117. Strand BL, Ryan L, Veld PI, Kulseng B, Rokstad AM, Skjak-Braek G *et al.* Poly-L-lysine induces fibrosis on alginate microcapsules via the induction of cytokines. *Cell Transplantation* 2001; 10(3):263–275.
118. Orive G, Carcaboso AM, Hernandez RM, Gascon AR, Pedraz JL. Biocompatibility evaluation of different alginates and alginate-based microcapsules. *Biomacromolecules* 2005; 6(2):927–931.
119. de Vos P, Hoogmoed CG, Busscher HJ. Chemistry and biocompatibility of alginate–PLL capsules for immunoprotection of mammalian cells. *Journal of Biomedical Materials Research* 2002; 60(2):252–259.
120. de Vos P, van Hoogmoed CG, De Haan BJ, Busscher HJ. Tissue responses against immunoisolating alginate–PLL capsules in the immediate posttransplant period. *Journal of Biomedical Materials Research* 2002; 62(3):430–437.
121. Cui JH, Goh JS, Kim PH, Choi SH, Lee BJ. Survival and stability of bifidobacteria loaded in alginate poly-L-lysine microparticles. *International Journal of Pharmaceutics* 2000; 210(1–2):51–59.
122. Chen H, Ouyang W, Jones M, Haque T, Lawuyi B, Prakash S. *In-vitro* analysis of APA microcapsules for oral delivery of live bacterial cells. *Journal of Microencapsulation* 2005; 22(5):539–547.
123. Tam SK, Dusseault J, Polizu S, Menard M, Halle JP, Yahia L. Physicochemical model of alginate–poly-L-lysine microcapsules defined at the micrometric/nanometric scale using ATR-FTIR, XPS, and ToF-SIMS. *Biomaterials* 2005; 26(34):6950–6961.
124. Bucko M. Immobilization of a whole-cell epoxide-hydrolyzing biocatalyst in sodium alginate–cellulose sulfate–poly(methylene-co-guanidine) capsules using a controlled encapsulation process. *Enzyme and Microbial Technology* 2005; 36(1):118–126.
125. Orive G, Hernandez RM, Gascon AR, Igartua M, Pedraz JL. Development and optimisation of alginate–PMCG–alginate microcapsules for cell immobilisation. *International Journal of Pharmaceutics* 2003; 259(1–2):57–68.
126. Sodini I, Boquien CY, Corrieu G, Lacroix C. Use of an immobilized cell bioreactor for the continuous inoculation of milk in fresh cheese manufacturing. *Journal of Industrial Microbiology & Biotechnology* 1997; 18(1):56–61.

127. Audet P, Paquin C, Lacroix C. Effect of medium and temperature of storage on viability of lactic-acid bacteria immobilized in kappa-carrageenan-locust bean gum gel beads. *Biotechnology Techniques* 1991; 5(4):307–312.
128. Batich C, Vaghefi F. Process for microencapsulating cells. USA patent 6 242 230. 2001 June 2001.
129. Favaro-Trindale CS, Grosso CRF. Microencapsulation of *L. acidophilus* (La-05) and *B. lactis* (Bb-12) and evaluation of their survival at the pH values of the stomach and in bile. *Journal of Microencapsulation* 2002; 19(4):485–494.
130. Canaple L, Rehor A, Hunkeler D. Improving cell encapsulation through size control. *Journal of Biomaterials Science – Polymer Edition* 2002; 13(7):783–796.
131. Lacik I, Anilkumar AV, Wang TG. A two-step process for controlling the surface smoothness of polyelectrolyte-based microcapsules. *Journal of Microencapsulation* 2001; 18(4):479–490.
132. Wang T. An encapsulation system for the immunoisolation of pancreatic islets. *Nature Biotechnology* 1997; 15(4):358–362.
133. Cole ET. *In vitro* and *in vivo* pharmacoscintigraphic evaluation of ibuprofen hypromellose and gelatin capsules. *Pharmaceutical Research* 2004; 21(5):793–798.
134. Esquisabel A, Hernandez RM, Igartua M, Gascon AR, Calvo B, Pedraz JL. Preparation and stability of agarose microcapsules containing BCG. *Journal of Microencapsulation* 2002; 19(2):237–244.
135. Iyer C. Effect of co-encapsulation of probiotics with prebiotics on increasing the viability of encapsulated bacteria under *in vitro* acidic and bile salt conditions and in yogurt. *Journal of Food Science* 2005; 70(1):M18–M23.
136. Ouyang W, Chen H, Jones ML, Metz T, Haque T, Martoni C *et al.* Artificial cell microcapsule for oral delivery of live bacterial cells for therapy: design, preparation, and *in-vitro* characterization. *Journal of Pharmaceutical Pharmacological Science* 2004; 7(3):315–324.
137. Chen HM, Ouyang W, Lawuyi B, Lim T, Prakash S. A new method for microcapsule characterization: use of fluorogenic genipin to characterize polymeric microcapsule membranes. *Applied Biochemistry and Biotechnology* 2006; 134(3):207–222.
138. Chen HM, Ouyang W, Lawuyi B, Prakash S. Genipin cross-linked alginate-chitosan microcapsules: membrane characterization and optimization of cross-linking reaction. *Biomacromolecules* 2006; 7(7):2091–2098.
139. Zimmerman U, Cramer H, Jork A, Thurman F, Zimmerman H, Fuhr G, Hasse C, Rothmund M. Microencapsulation-based cell therapy. In: Reed G, Rehm H J, editors. *Biotechnology*. Wiley-VCH, Weinheim, 2001, pp. 548–571.
140. Gin H, Dupuy B, Baquey C, Ducassou D, Aubertin J. Agarose encapsulation of Islets of Langerhans-reduced toxicity *in vitro*. *Journal of Microencapsulation* 1987; 4(3):239–242.
141. May MH, Sefton MV. Conformal coating of small particles and cell aggregates at a liquid–liquid interface. *Bioartificial Organs II: Technology, Medicine, and Materials* 1999; 875:126–134.
142. Chang TMS, Prakash S. Procedures for microencapsulation of enzymes, cells and genetically engineered microorganisms. *Molecular Biotechnology* 2001; 17(3):249–260.
143. Dusseault J, Leblond FA, Robitaille R, Jourdan G, Tessier J, Menard M *et al.* Microencapsulation of living cells in semi-permeable membranes with covalently cross-linked layers. *Biomaterials* 2005; 26(13):1515–1522.
144. Sakai S, Ono T, Ijima H, Kawakami K. *In vitro* and *in vivo* evaluation of alginate/

sol–gel synthesized aminopropyl-silicate/alginate membrane for bioartificial pancreas. *Biomaterials* 2002; 23(21):4177–4183.
145. Coradin T, Mercey E, Lisnard L, Livage J. Design of silica-coated microcapsules for bioencapsulation. *Chemical Communications* 2001; 23:2496–2497.
146. Rosinski S, Grigorescu G, Lewinska D, Ritzen LG, Viernstein H, Teunou E *et al.* Characterization of microcapsules: recommended methods based on round-robin testing. *Journal of Microencapsulation* 2002; 19(5):641–659.
147. Drury JL, Dennis RG, Mooney DJ. The tensile properties of alginate hydrogels. *Biomaterials* 2004; 25(16):3187–3199.
148. Fery A, Dubreuil F, Mohwald H. Mechanics of artificial microcapsules. *New Journal of Physics* 2004; 6:18.
149. Leblond FA. Quantitative method for the evaluation of biomicrocapsule resistance to mechanical stress. *Biomaterials* 1996; 17(21):2097–2102.
150. Zhao L, Zhang ZB. Mechanical characterization of biocompatible microspheres and microcapsules by direct compression. *Artificial Cells Blood Substitutes and Immobilization Biotechnology* 2004; 32(1):25–40.
151. Gauthier MA, Luo J, Calvet D, Ni C, Zhu XX, Garon M et al. Degree of crosslinking and mechanical properties of crosslinked poly(vinyl alcohol) beads for use in solid-phase organic synthesis. *Polymer* 2004; 45(24):8201–8210.
152. Angelova N. Stability assessment of chitosan-sodium hexametaphosphate capsules. *Journal of Biomaterials Science Polymer Edition* 2001; 12(11):1207–1225.
153. Klemenz A. Investigation of elasto-mechanical properties of alginate microcapsules by scanning acoustic microscopy. *Journal of Biomedical Materials Research* 2003; 65A(2):237–243.
154. Lebedeva OV. Mechanical properties of polyelectrolyte-filled multilayer microcapsules studied by atomic force and confocal microscopy. *Langmuir* 2004; 20(24):10685–10690.
155. Zimmermann H, Hillgartner M, Manz B, Feilen P, Brunnenmeier F, Leinfelder U *et al.* Fabrication of homogeneously cross-linked, functional alginate microcapsules validated by NMR-, CLSM- and AFM-imaging. *Biomaterials* 2003; 24(12):2083–2096.
156. Xu KY, Hercules DM, Lacik I, Wang TG. Atomic force microscopy used for the surface characterization of microcapsule immunoisolation devices. *Journal of Biomedical Materials Research* 1998; 41(3):461–467.
157. Suzuki A, Yamazaki M, Kobiki Y. Direct observation of polymer gel surfaces by atomic force microscopy. *Journal of Chemical Physics* 1996; 104(4):1751–1757.
158. Magonov SN, Qvarnstrom K, Elings V, Cantow HJ. Atomic force microscopy on polymers and polymer-related compounds. 1. Cold-extruded polyethylene. *Polymer Bulletin* 1991; 25(6):689–694.
159. Albrecht TR, Dovek MM, Lang CA, Grutter P, Quate CF, Kuan SWJ *et al.* Imaging and modification of polymers by scanning tunneling and atomic force microscopy. *Journal of Applied Physics* 1988; 64(3):1178–1184.
160. Radmacher M, Fritz M, Hansma PK. Imaging soft samples with the atomic-force microscope – gelatin in water and propanol. *Biophysical Journal* 1995; 69(1):264–270.
161. Baguet J, Sommer F, Duc TM. Imaging surfaces of hydrophilic contact-lenses with the atomic force microscope. *Biomaterials* 1993; 14(4):279–284.
162. Lamprecht A, Schafer UF, Lehr CM. Characterization of microcapsules by confocal laser scanning microscopy: structure, capsule wall composition and encapsulation rate. *European Journal of Pharmaceutics and Biopharmaceutics* 2000; 49(1):1–9.

163. Belton PS, Hills BP, Raimbaud ER. The effects of morphology and exchange on proton NMR relaxation in agarose gels. *Molecular Physics* 1988; 63(5):825–842.
164. Potter K, Balcom BJ, Carpenter TA, Hall LD. The gelation of sodium alginate with calcium-ions studied by magnetic-resonance-imaging (MRI). *Carbohydrate Research* 1994; 257(1):117–126.
165. Potter K, Carpenter TA, Hall LD. Mapping of the spatial variation in alginate concentration in calcium alginate gels by magnetic-resonance-imaging (MRI). *Carbohydrate Research* 1993; 246:43–49.
166. Thu B, Gaserod O, Paus D, Mikkelsen A, Skjak-Braek G, Toffanin R *et al.* Inhomogeneous alginate gel spheres: An assessment of the polymer gradients by synchrotron radiation-induced X-ray emission, magnetic resonance microimaging, and mathematical modeling. *Biopolymers* 2000; 53(1):60–71.
167. Duez JM, Mestdagh M, Demeure R, Goudemant JF, Hills BP, Godward J. NMR studies of calcium-induced alginate gelation. Part I – MRI tests of gelation models. *Magnetic Resonance in Chemistry* 2000; 38(5):324–330.
168. Hills BP, Godward J, Debatty M, Barras L, Saturio CP, Ouwerx C. NMR studies of calcium induced alginate gelation. Part II. The internal bead structure. *Magnetic Resonance in Chemistry* 2000; 38(9):719–728.
169. Simpson NE, Grant SC, Blackband SJ, Constantinidis I. NMR properties of alginate microbeads. *Biomaterials* 2003; 24(27):4941–4948.
170. Shen F, Poncet-Legrand C, Somers S, Slade A, Yip C, Duft AM *et al.* Properties of a novel magnetized alginate for magnetic resonance imaging. *Biotechnology and Bioengineering* 2003; 83(3):282–292.
171. Bodmeier R, McGinity J. Polylactic acid microspheres containing quinidine base and quinidine sulphate prepared by the solvent evaporation technique. I. Methods and morphology. *Journal of Microencapsulation* 1987; 4:279–288.
172. Mathews BR, Nixon JR. Surface characteristics of gelatin microcapsules by scanning electron microscopy. *Journal of Pharmaceutical Pharmacology* 1974; 383–384.
173. Benita S, Benoit JP, Puisieux F, Thies C. Characterization of drug-loaded poly(D,L-lactide) microspheres. *Journal of Pharmaceutical Science* 1984; 73:1721–1724.
174. Benoit JP, Thies C. Microencapsulation – methods and industrial applications. In: Benita S, editor. *Drugs and the Pharmaceutical Sciences*. Marcel Dekker, New York, 1996.
175. Ma XJ, Vacek I, Sun A. Generation of alginate-poly-L-lysine-alginate (APA) Biomicroscopies – the relationship between the membrane strength and the reaction conditions. *Artificial Cells Blood Substitutes and Immobilization Biotechnology* 1994; 22(1):43–69.
176. Ross CJD, Chang PL. Development of small alginate microcapsules for recombinant gene product delivery to the rodent brain. *Journal of Biomaterials Science – Polymer Edition* 2002; 13(8):953–962.
177. Gugerli R. Quantitative study of the production and properties of alginate/poly-L-lysine microcapsules. *Journal of Microencapsulation* 2002; 19(5):571–590.
178. Vandenbossche GMR, Vanoostveldt P, Demeester J, Remon JP. The molecular-weight cutoff of microcapsules is determined by the reaction between alginate and polylysine. *Biotechnology and Bioengineering* 1993; 42(3):381–386.
179. Strand BL, Morch YA, Espevik T, Skjak-Braek G. Visualization of alginate–poly-L-lysine–alginate microcapsules by confocal laser scanning microscopy. *Biotechnology and Bioengineering* 2003; 82(4):386–394.
180. Su JF. Preparation and characterization of double-MF shell microPCMs used in

building materials. *Journal of Applied Polymer Science* 2005; 97(5):1755–1762.
181. Prakash S, Bhathena J. Live bacterial cells as orally delivered therapeutics. *Expert Opinion on Biological Therapy* 2005; 5(10):1281–1301.
182. de Groot M, Schuurs TA, van Schilfgaarde R. Causes of limited survival of microencapsulated pancreatic islet grafts. *Journal of Surgical Research* 2004; 121(1):141–150.
183. Khattak SF, Bhatia SR, Roberts SC. Pluronic F127 as a cell encapsulation material: utilization of membrane-stabilizing agents. *Tissue Engineering* 2005; 11(5-6):974–983.
184. Yang Y, Zhang SF, Jones G, Morgan N, El Haj AJ. Phosphorylcholine-containing polymers for use in cell encapsulation. *Artificial Cells Blood Substitutes and Immobilization Biotechnology* 2004; 32(1):91–104.
185. Gliozzi A. Nanocapsules: coating for living cells and tissues. *Journal of Biotechnology* 2005; 118:S65.
186. Srivastava R, McShane MJ. Application of self-assembled ultra-thin film coatings to stabilize macromolecule encapsulation in alginate microspheres. *Journal of Microencapsulation* 2005; 22(4):397–411.
187. Tao SL, Desai TA. Microfabricated drug delivery systems: from particles to pores. *Advanced Drug Delivery Reviews* 2003; 55(3):315–328.

2
Polymeric hydrophilic polymers in targeted drug delivery

K POON, V CASTELLINO and Y-L CHENG,
University of Toronto, Canada

2.1 Introduction

Rapid progress in therapeutic systems over the past two decades can be attributed to both critical advances in molecular medicine and to the development of new, versatile materials. Foremost among these, polymers and hydrogels have been key components used in the packaging of small molecule drugs, peptide and protein drugs, DNA, viruses and cells in targeted drug delivery and in cell encapsulation applications. In this chapter, we present an overview of the use of hydrophilic polymers in drug targeting, with the goal of illustrating how hydrophilic polymers can be designed for targeting by exploiting physiologic conditions or external stimuli. In Section 2.2, a brief review of hydrophilic polymers is provided, with some detail on natural and synthetic polymers that are prominently featured in targeting applications. Temperature and pH-responsive polymers are also covered. In Section 2.3, mechanisms of drug targeting aimed at sites ranging from macroscopic points of administration to molecular and intracellular level targets are discussed, together with corresponding material design strategies. Illustrative examples are included with references to more thorough reviews.

2.2 Properties and classification of hydrophilic polymers

Hydrophilic polymers have the potential to be biocompatible, and have therefore attracted extensive attention in biomedical and drug delivery applications.[1–5] They can be tailor-designed at both the molecular level and the device level. At the molecular level, they can be synthesized as homopolymers of a single hydrophilic monomer, or as random or block copolymers of a hydrophilic monomer with either hydrophilic or hydrophobic monomers. They can also be grafted onto the backbone of other polymers, or made into branched structures with varying degrees of branching. At the device level, hydrophilic polymers can be fabricated into a variety of physical forms, including crosslinked hydrogel matrices of different geometries and dimensions, crosslinked microparticles and nanoparticles, self-assembled nanoparticulate micelles, reversible sol–gel mixtures, interpenetrating networks, physical blends or composites with other polymers, and grafts on the surfaces of other biomaterials.[5]

Hydrogels are hydrophilic polymers that have been chemically, physically or ionically crosslinked to form a matrix that swells in water.[3,6] Many of the properties that are important in the design of hydrogels as drug delivery vehicles are dictated by the equilibrium degree of swelling. The equilibrium degree of swelling of hydrogels in water is determined by a balance between the free energy of polymer/solvent mixing, ionic interactions and elastic forces, and is influenced by the extent of crosslinking and the chemical nature of the polymer. The degree of swelling in turn determines the mesh size of the hydrogel,[4] which along with the physicochemical nature of both the polymer and the solute, and the morphology and geometry of the device in which the hydrogel is incorporated, determine the solute's diffusion characteristics through a preswollen gel.[4,7–9] Release from swellable hydrogel matrices is most often described by a simple equation $M_t/M_\infty = kt^n$, where the exponent n is an indication of the extent to which release is controlled by diffusion or swelling.[10,11]

Hydrophilic polymers can be derived from either natural or synthetic sources, and can be further classified as anionic, cationic, neutral or amphipathic by the ionic nature of the monomer side group. Hoffman[3] and Peppas *et al.*[4] have listed some of the more commonly used polymers. Stimuli-responsive polymers, particularly pH- and temperature-responsive polymers, have been extensively investigated in targeting applications. A number of naturally derived polymers are also prominently featured in targeting due to their biocompatibility and biodegradability. The basic properties of these classes of polymers are described below.

2.2.1 Temperature-responsive polymers

Similar to small molecules, the solubility of polymers in solvents is usually temperature dependent via the temperature dependence of the Gibb's free energy of mixing ($\Delta G = \Delta H - T\Delta S$). Some polymers show the more common behavior of dissolving when heated to above an upper critical solution temperature (UCST), while others exhibit the reverse behavior and precipitate when heated to above a lower critical solution temperature (LCST). LCST behavior is seen in polymer/solvent systems with negative entropy of mixing, or an increasingly positive $-T\Delta S$ term as temperature increases which would result in a positive ΔG above some critical temperature. Crosslinked LCST-type polymers shrink when heated to above their LCST, corresponding to the precipitation of their uncrosslinked analogs.

Jeong *et al.* have summarized natural and synthetic polymers that show LCST behavior in water between 10 and 125 °C.[12] By far the most thoroughly investigated of these polymers is poly(N-isopropyl acrylamide), or PNIPAAm, primarily because its LCST of ~ 32 °C in water[13] is convenient for biomedical and drug delivery applications. The LCST can be increased or decreased by copolymerization of NIPAAm with either hydrophilic or hydrophobic comonomers, respectively,[14,15] and is also influenced by solution properties including pH and the presence of co-solutes,[15–17] but has been reported to be invariant to crosslinking level.[18]

Block copolymers of hydrophilic and hydrophobic polymers can also show temperature-responsive behavior through the temperature dependence of the hydrophoblic–lipophilic balance. Block copolymers of poly(ethylene oxide) (PEO) with various hydrophobic segments, including poly(propylene oxide) (PPO) and poly(lactide-co-glycolide) (PLGA), are notable examples.[5,12] PEO-b-PPO-b-PEO block copolymers are a family of triblock copolymers available commercially in varying molecular weights and relative copolymer segments lengths (commercial names Pluronic or Poloxamer). The copolymers exist as unimers in aqueous solvents at very low concentrations, self assemble to form micelles with PPO core and PEO corona above the critical micelle concentration (CMC) which are typically in the 10^{-5} to 10^{-6} M range. At high enough concentrations (>20% depending on which Pluronic is used), sol–gel or gel–sol transitions are observed with the transitions occurring in a temperature-dependent fashion. The gelation mechanism has been extensively studied and appears to be related to the formation and packing of micelles, but remains inconclusive.[12,19] Both the micellar formation and the gelation properties have been exploited in targeted delivery. Block copolymers of PEO with other hydrophobic segments such as biodegradable polyesters poly(lactide-co-glycolide) and poly(L-lactide), or PEO-b-PLGA-b-PEO and PEO-b-PLLA–PEO, have also shown similar behavior.[12]

Another intriguing group of temperature-responsive polymers are elastin-like peptides (ELPs). ELPs are peptides with repeating pentapeptide sequences of VPGXG, where X is any amino acid except proline, that aggregate upon heating as the polypeptide conformation changes from extended chains to collapsed β-spirals.[20,21] The transition temperature, T_t, can be controlled by the nature of X, the chain length and the peptide concentration.[22] This polymer has been recently investigated for targeting of tumors in combination with local hyperthermia (see Section 2.3.5).

2.2.2 pH-responsive polymers

A pH-responsive polymer, typically formed from ionic polymer chains and pendant groups, will become protonated or deprotonated depending on the pH of the medium. Anionic polymers, or polyacids, are neutral at low pH, and negatively charged above its pK_a. Similarly, cationic polymers or polybases are positively charged at low pH and neutral above the pK_b of the polymer. As crosslinked hydrogels, such polymers show pH-dependent swelling behavior with higher degrees of equilibrium swelling in their charged forms compared with their neutral forms due to increased electrostatic repulsion.[4] Some commonly studied polyacids are poly(acrylic acid), poly(methacrylic acid), poly(2-ethyl acrylic acid) (PEAAc) and poly(2-propylacrylic acid), while a commonly used polybase is polyl(*N*,*N*′-diethylaminoethyl methacrylate) (PDEAEMA).[5]

2.2.3 Natural polymers

Many naturally occurring polymers generally offer natural hydrophilicity and biocompatibility, which make them suitable candidates in drug delivery or targeting applications. Chitosan and alginate are the most commonly investigated polysaccharides, and gelatin is a commonly studied protein.

Chitosan is a cationic polysaccharide comprising of glucosamine and *N*-acetyl glucosamine residues, and is made by the partial deacetylation of naturally occurring chitin. The extent of deacetylation can range from ~ 60% to 95%, and molecular weight can range from 10 000 to 1 000 000 daltons. It is soluble at acidic pH when the free NH_2 groups exist in protonated NH_3^+ form. It is one of the most attractive and intensely studied materials for a variety of targeting applications because of its biocompatibility, biodegradability, muco-adhesiveness and mucosal transport enhancing properties[221–224] and its degradability by polysaccharidases in the colon microflora. It can also be made into micro- and nano-particles under very mild conditions for use in tumor targeting and mucosal applications including oral, vaginal, nasal and ocular applications.[23] Alginates are anionic polysaccharides most commonly used as the matrix material in cell encapsulation. Alginates are linear, unbranched block copolymers consisting of β-(1-4 linked) D-mannuronic acid (M unit) and α-(1-4 linked) L-guluronic acid (G unit) regions. The polymer blocks consist of either a series of M units, a series of G units, or alternating M-G units, with block ordering dependent on the algae of origin. Gelation occurs when a divalent cation (Mg^{2+}, Ca^{2+}) is introduced, which links G blocks. Physical properties are dependent on composition; a higher G block content increases gel brittleness. Alginate gels, particularly calcium alginate, are popular matrix platforms as a result of their biocompatibility and thermal stability.[24,25] Gelatin is obtained from the thermal denaturation of acid or alkaline processed collagen, consisting mainly of proline and 4-hydroxyproline with every third residue being glycine. Its polypeptide helix fragments can be chemically crosslinked with crosslinkers such as glutaraldehyde to form a hydrogel. Acidic processing yields a cationic gel while alkaline extraction leads to a higher viscosity, anionic gel. Gel strength is closely related to the degree of helix formation within the polypeptide fragments, and can be tailored for a variety of structural functions.[26,27]

2.3 Targeting mechanisms and materials approaches

Drug targeting, or the selective accumulation of administered drug at the site of desired therapeutic action,[28] would lead to increased therapeutic efficacy and reduced adverse side effects, and is therefore a major goal in the development of therapeutics. A number of targeting mechanisms have been identified that give rise to a range of material design strategies.

2.3.1 Targeting through biological interactions

Specific or non-specific interactions between the target biological tissue and the drug delivery vehicle can be exploited for targeting strategies. These targeting mechanisms are often used in combination with other strategies described in Sections 2.3.2–2.3.5, so examples will be deferred until the later sections.

Receptor-mediated targeting

Receptor-mediated targeting is the most specific among this group of strategies. Many targeting ligands are known,[29] including bisphonates for targeting bone tissue,[30] selectin adhesion molecules for targeting vasculature endothelium,[31] and transferrin[32] and folate[33–35] for targeting some classes of cancer cells. Ligands can be attached either directly to the drug, or to nanoparticulate carriers of drugs.[29]

Mucoadhesion

A much less specific biological interaction mechanism is mucoadhesion. Mucoadhesion prolongs contact time of the vehicle with the target tissue, and therefore enhances either absorption of released drug or intracellular uptake of nanoparticulate carriers. A recent issue of *Advanced Drug Delivery Reviews* was devoted entirely to mucoadhesive polymers, including those used in ocular, buccal, vaginal and nasal delivery.[36–43] The major classes of mucoadhesive polymers are thiolated polymers,[41] poly(acrylic acid),[44] and polysaccharides – most notably chitosan.[23] The adhesion process is complex and involves contact, consolidation and the formation of some type of bond between the polymer and the mucus.[45] Thiolated polymers form disulfide bonds with mucus glycoproteins,[41] while bioadhesion of poly(acrylic acid) may be due to ionic interactions.[37,46] The mucoadhesive properties of polycationic chitosan is believed to arise from ionic interaction with sialic acid groups on mucin;[47] furthermore, chitosan is also known to enhance transmucosal absorption by facilitating the transient opening of the tight junction.[48]

Degradative enzymes in microflora

Another biological interaction mechanism is degradative enzymes that can specifically degrade drug delivery carriers of the appropriate material. This mechanism is most prominently used for drug targeting to the colon using degradative enzymes present in the microflora of the colon. Polysaccharidases, for instance, have been exploited to locally degrade polysaccharides such as chitosan, pectin, guar gum, dextran and chondroitin sulfate.[49] Polymers that contain azo-bonds have been designed to exploit the presence of azo reductase in the colon.[50,51] Degradative enzyme-based approaches are typically used in combination with pH-responsive polymers (see pages 54–55).

2.3.2 *In situ* gelation for direct administration

The most obvious targeting approach is the direct administration of drug and vehicle at the macroscopic site of action. The drug vehicle may be simply instilled at a topical site, or surgically implanted or injected. To prolong contact with target tissue, minimize invasiveness and provide control over drug release, recent work has focused on injectable liquid formulations that solidify or gel upon administration, or upon contact with the physiologic environment. *In situ* gelation typically involves ion-, temperature-, and, to a lesser extent, pH-responsive polymers that are soluble under ambient or external conditions, but undergo a phase transition upon injection to form a gel *in situ* and entrap suspended drug particles for sustained control release locally.[12,19,52]

Ionic-gelation

One of the best known materials for *in situ* gelation is Gelrite®, a commercially available deacetylated gellan gum that gels in the presence of monovalent or divalent cations. Cations present in the lacrimal fluid trigger rapid gelation, increase corneal contact time, and thus enhance ocular bioavailability.[53–55] Gelrite® has been used to formulate eyedrops for indomethacin,[56] pilocarpine[57] and timolol maleate.[58,59] Sodium alginate solutions also gel in the eye by interactions with cations in the lacrimal fluid, with the gelation kinetics and gel strengths being dependent on the guluronic acid (G) content in the alginate polymer backbone. Strong gels can be formed rapidly with alginates that have >65% G content, and sustained pilocarpine release and intraocular pressure lowering was demonstrated.[60]

pH-responsive gelation

A pH-responsive gelling system based on Carbopol (poly(acrylic acid)) for timolol maleate was reported by Kumar and Himmelstein who used carboxymethylcellulose as a viscosity thickener to significantly reduce the Carbopol concentration needed for gelation.[61]

Temperature-responsive gelation

Xyloglucan

Xyloglucan is a polysaccharide from tamarind seeds that, when partially degraded by β-galactosidase, shows a sol to gel transition at a transition temperature that is dependent on the degree of galactose elimination that can be designed to be between ambient and physiologic temperatures.[62] It has been explored for rectal,[63] oral[64] and ocular delivery of pilocarpine and timolol[62,65] with promising results, but no gelation mechanism has been hypothesized or reported.

PNIPAAm-based systems

Copolymers based on poly(*N*-isopropylacrylamide) (PNIPAAm) have shown thermoreversible *in situ* gelling properties that can be exploited for targeting. Bae and coworkers[66–68] synthesized high molecular weight (~10^6 Da) copolymers of NIPAAm with 2–5 mol% acrylic acid, and observed four temperature-dependent phases: clear solution, opaque solution, gel and shrunken gel. Gelation is attributed to the collapse of portions of the polymers from expanded coils to globules and the subsequent globule aggregation to form physical crosslinks. The opaque solution to gel transition occurs between 30 and 34 ºC, making these copolymers suitable for injectable gelation applications. Encapsulated islets of Langerhans from Sprague-Dawley rats showed insulin secretion function for a month with shorter lag times and higher permeation rates than alginate matrices more conventionally used for cell encapsulation.

In a different approach, block copolymers of PNIPAAm and poly(ethylene glycol) (PEG) have been shown to exhibit thermoreversible gelation behavior at low concentrations (~4%) in water.[69] The copolymers were designed with PNIPAAm segments as the terminal aggregating blocks connected by central PEG segments as solvating blocks. When heated, the PNIPAAm terminal segments undergo their LCST transitions and form intermolecular aggregates that serve as physical crosslinks connecting the PEG solvation segments. Copolymers of various architectures can be made including two-arm linear (PNIPAAm–PEG–PNIPAAm triblock copolymers), and four-arm and eight-arm branched structures to modulate mechanical and mass transfer properties. Gelation kinetics are quick and appear to be heat conduction limited. These polymers have been used in mixtures with PNIPAAm homopolymers to form cell encapsulating membranes.[70] The presence of PNIPAAm homopolymers enhanced the extent of crosslinking and the resulting membranes were stable in water or phosphate buffered saline (PBS) throughout the 60 day duration of the study. MIN6 cells were not adversely affected by the cell encapsulation process, and remained functional in response to glucose challenge. Wound-healing products are under commercial development (see www.rimontherapeutics.com).

Pluronics (PEO–PPO–PEO block copolymers)

El Kamel[71] combined a viscosity enhancer, methylcellulose, with Pluronic F127 as an ocular temperature-responsive *in situ* gelling formulation and found that ocular availability in albino rabbits was enhanced compared to timolol solution. Wei *et al.*[72] combined Pluronic F127, Pluronic F68 and sodium hyaluronate, a mucoadhesive polysaccharide, to optimize an *in situ* gelling ophthalmic formulation and increase its gelation temperature to well above room temperature. The gel formulation enhanced ocular retention relative to aqueous solutions, but no effect of sodium hyaluronate was seen. The thermosetting properties of Pluronics have also been explored for intratumoral targeting. Paclitaxel/Pluronic F127 solutions injected intratumorally in B16F1 melanoma-bearing mice showed significantly enhanced anti-tumor activity and survival rate.[73] Other *in situ* gelation applications that have been explored include vancomycin delivery at sites at high risk for infections,[74] as well as vaginal and rectal delivery.[75–78]

Poly(acrylic acid)-g-Poloxamer copolymers were synthesized by Hoffman, Bromberg and coworkers[79–87] using two different synthesis methods. The critical gelation concentration was reduced from >20 wt% for Poloxamer F127 to 0.2 to 2.5 wt% for poly(acrylic acid) (PAA)-grafted Poloxamer F127 without changing the gelation temperature, and the thermosetting nature of these polymers, along with the mucoadhesion properties of PAA, were exploited for oral and vaginal applications. Lin and Sung[88] explored the possible synergy between multiple gelation mechanisms for ocular delivery. A mixture of temperature-responsive Pluronic and pH-responsive Carbopol was found to be free flowing at pH 4.0 and 25 °C, and showed more favorable rheological properties, drug release kinetics and *in vivo* pharmacologic response than either Pluronic or Carbopol solutions alone. A similar conclusion was reached with a mixture of cation-responsive alginate and temperature responsive Pluronic.[89]

Biodegradable PEO–b–PLGA–b–PEO and PEO–PLLA–PEO

Kim and coworkers have developed a family of biodegradable and safe block copolymers with thermogelation properties.[90–96] Relatively low molecular weight (~4000 to ~15 000 Da) block copolymers of ABA structures where the terminal A blocks are hydrophilic PEG and the central B blocks are hydrophobic and biodegradable PLLA (poly-L-lactide), PLGA or PCL (poly-ϵ-caprolactone) have been synthesized. PEG–PLLA–PEG block copolymers of 5000 Da PEG segment length show a concentration and PLLA segment length dependent gel to sol transition upon heating – which would require pre-heating for *in situ* gelation applications. Low molecular weight PEG–PLGA–PEG (550–2810–550 Da) copolymers show both a sol to gel transition and a later gel to sol transition when heated through a range of temperatures. The lower temperature sol to gel transition occurs at the convenient temperature of about 30 °C. *In vivo* studies have demonstrated *in situ* gelation into a distinct mass and gel persistence of over a month in rats. Gel formation appears to be driven by micellar growth and increased polymer-polymer interactions with increasing temperature. Intriguingly, low molecular weight block copolymers of the reverse structure, PLGA-b-PEG-b-PLGA also show thermoreversible gelation,[97,98] though the mechanism of gelation for these BAB-type copolymers has not been clearly elucidated. This class of *in situ* gelling polymers have been investigated for the depot delivery of proteins, peptides, water-insoluble drugs,[94,98] and intratumoral paclitaxel injectable formulations are under commercial development (see www.macromed.com).

Chitosan–polyol mixtures

Chenite, Leroux and coworkers[99–105] have developed a novel temperature-sensitive pH-dependent *in situ* gelable material based on chitosan simply by mixing the pH-dependent chitosan with sugar-phosphate salts or polyol-phosphate salts such as glycerol phosphate.[99–105] At physiologic pH, the mixture in water remains liquid at room temperature, but forms a gel at body temperature with some hysteresis observed upon cooling.[99] A decrease in the degree of deacetylation results in an increase in gelation temperature while the molecular weight had little effect. It is believed that gelation is driven by hydrophobic interactions between neutral chitosan molecules, facilitated by the neutralizing effect of polyol salts and the water-structuring effect of polyols at elevated temperatures.[99,100] Proteins and cells could be encapsulated under gentle conditions. Encapsulated bone-inducing growth factor preparation was shown to retain activity, and encapsulated chondrocytes retained >80% cell viability. Intratumoral injections of this hydrogel containing paclitaxel were more effective and less toxic than IV injection of taxol.[104] The formulations could be modified by the entrapment of hydrophilic drug-containing liposomes to further modulate the release of hydrophilic drugs.[102]

2.3.3 Nanoparticles for RES evasion and EPR

The size of the drug delivery vehicle is an important passive targeting parameter. Injectable circulating drug carriers must avoid uptake by the reticuloendothelial system (RES) to have a greater chance of reaching the target site. Reduced macrophage uptake is seen with nanoparticles that have hydrophilic surfaces, especially PEG-modified surfaces, and that are smaller than 200 nm in diameter.[106–108] Nanoparticles of up to 500 nm can also be preferentially localized in tumors due to the enhanced permeability and retention (EPR) effect.[106,109–111] The blood vessel walls in tumors are more permeable than normal tissue, and allow the extravasation of nanoparticles of up to 500 nm in size. Elevated interstitial pressures and lack of lymphatic drainage in tumors result in negligible convection and clearance, and therefore the enhanced retention of extravasated nanoparticles in the interstitium. Much attention has therefore been directed towards designing nanoparticles as drug carriers.[106,111] A wide range of materials have been used including degradable hydrophobic polymers such as PLGA[112] and liposomes.[113,114] Of interest in this chapter, crosslinked hydrophilic polymers such as chitosan, gelatin and alginates, and self-assembled micelles of block copolymers have been extensively studied in tumor targeting; and some illustrative examples are described below.

Hydrophilic nanoparticles

Nanoparticles for RES evasions and EPR-based tumor accumulation have been made by a variety of hydrophilic polymers, including chitosan, gelatin and albumin. Among these polymers, chitosan has received the most attention because in addition to its biocompatibility, biodegradability and mucoadhesive properties, it can be made into drug-loaded nanoparticles by a simple and mild ionic gelation method. The method involves the complexation of polycationic chitosan with one or more polyanions; tripolyphosphate (TPP) is the most commonly used,[115–120] but poly(acrylic acid)[121] and carboxymethylcellulose[122] have also been used. The method has also been adapted to directly form ionic complexes between chitosan and an active agent such as DNA.[123] It has also been used to generate surface-modified fluorescent quantum dots for imaging and paramagnetic gadrinium diethylene triamine pentaacetate for guidance.[124] A number of other methods have also been developed including an emulsion/droplet coalescence method,[132,133,225] a reverse micelle glutaraldehyde cross-linking method that gives ultrafine nanoparticles of narrow distribution,[125,126] and crosslinking via condensation between chitosan amino groups and natural di- or tricarboxylic acids in aqueous media at room temperature.[127] Degree of deacetylation and chitosan molecular weight are among the experimental variables that can be used to control or vary nanoparticle properties in addition to method-specific variables. Chitosan has also been derived with hydrophobic moieties that can self-assemble into nanoparticulate micelles.[128,129]

Crosslinked chitosan nanoparticles designed for tumor targeting via systemic injections include copper (II)-loaded particles of 40–100 nm that were found to show cytotoxic activity in a number of cancer cell lines,[120,130] and dextran–doxorubicin loaded particles of ~100 nm diameter that showed enhanced nanoparticle accumulation in tumor cells, gradual drug release in the tumor, enhanced tumor regression and increased survival in a mouse model.[125,126] Micelles based on hydrophobically modified micelles have been studied for tumor targeting. Chitosan modified with hydrophobic lauryl groups and hydrophilic carboxymethyl groups formed micelles of <100 nm size in solution, solubilized taxol to 1000-fold higher than in buffer, and the micelles showed enhanced cytostatic activity against KB cells *in vitro* than free taxol.[131] Direct conjugation of hydrophobic doxorubicin or fluorescein thiocyanate to glycol chitosan gave micelles that when injected intravenously into tumor-bearing mice, showed enhanced accumulation in tumors and reduced systemic toxicity relative to free doxorubicin.[128] Larger chitosan nanoparticles (~450 nm) containing gadopeneteic acid (Gd-DTPA) were made for direct intratumoral injections, and showed enhanced extracellular and intracellular nanoparticle accumulation in an *in vitro* L929 fibroblast cell study, suggesting both bioadhesion and endocytosis processes were involved in tumor targeting.[225] Thermo-neutron irradiation after intratumoral injection in a melanoma/mice model showed enhance tumor suppression relative to controls.[132,133]

Nanoparticles based on gelatin and albumin protein carriers have also received attention. Au and co-workers formulated paclitaxel into gelatin nanoparticles of 300–900 nm by desolvation/glutaraldehyde crosslinking[134,135] and found that intravenously injected particles preferentially accumulated in the kidney, liver and small intestines of mice in a formulation-dependent fashion, suggesting the possibility of targeting tumors in these organs. Amiji and coworkers used an ethanol precipitation approach to make gelatin and modified gelatin nanoparticles,[136–139] and found enhanced localization of PEGylated nanoparticles in the perinuclear region of BT-20 cells *in vitro*,[136] prolonged circulation and enhanced tumor accumulation in mice after intravenous (IV) injection.[138] As a DNA delivery system, gelatin nanoparticles were injected intratumorally and intravenously to carcinoma-bearing mice, and found to show superior transfection relative to controls, and the transfection efficiency of intravenous injections was 61% of intratumoral injections.[137] Transfection efficiency could be enhanced even further by using thiolated gelatin that form disulfide crosslinks for increased stability during circulation.[139] Langer and coworkers developed a two-step desolvation and glutaraldehyde crosslinking method to make gelatin nanoparticles that were stable in both water and cell medium,[140,141] and prepared human serum albumin (HSA) nanoparticles by a desolvation method.[142] Gelatin nanoparticles tagged with an antibody specific to the human epidermal growth factor receptor (HER2) associated with a number of tumor cell lines was seen to be dependent on HER2 expression level, and increased association results in increased cellular uptake.[143] DTPA (diethylenetriaminepentaacetic acid) coupled to HSA and gelatin nanoparticles were found to be taken up into UKF-NB-3 neuroblastoma cells, and showed increased cytotoxicity to cancer cells.[144]

Micelles

Polymeric micelles are usually prepared by conjugating a hydrophilic polymer to a hydrophobic polymer. The most commonly used hydrophilic polymer is poly(ethylene glycol), while a variety of hydrophobic polymers have been selected, including poly(propylene glycol), poly(L-amino acids), biodegradable polyesters and polyorthoesters, phospholipids, and long chain fatty acids.[145–152] The block copolymers self-assemble in aqueous environments to form a hydrophobic core surrounded by a hydrophilic corona. The hydrophobic core enables the encapsulation of hydrophobic drugs such as paclitaxel, and can be designed to enhance the apparent solubility of drugs to as much as 9000 times the solubility in water.[148] PEGylation of liposome surfaces are known to enhance liposome circulation time by evading the RES,[153] and the PEG corona on PEG-based micelles appear to have a similar effect.[146] Micelles are typically in the range of tens of nanometers, which is appropriate for RES evasion and EPR targeting.[146,154] Since micelles would be significantly diluted upon IV injection, so micelle stabilization is a design concern. For example, Rapoport and coworkers[155,156] have also devised two micelle stabilization methods: reduction of CMC by combining P105 with either PEG2000-diacylphospholipid (PEG-DSPE) or poly(ethylene glycol)-co-poly(β-benzyl-L-aspartate) to form mixed micelles, and physical stabilization of P105 micelles with an interpenetrating network (IPN) of the hydrophobic micelle core and a crosslinked thermo-responsive hydrogel such as PNIPAAm or poly(*N*,*N*-diethylacrylamide). IPN-stabilized micelles were called Plurogel®, and ranged in size from 30 to 400 nm, significantly larger than Pluronic micelles. These micelles were used with ultrasonic activation for tumor targeting and are described in more detail on page 58. Some illustrative passive tumor targeting examples are described below.

Block copolymers of PEG or PEG-derivatives with biodegradable polyesters such as PLGA, PLA and PCL have also been explored as micelle nano-carriers.[157–163] As one example, intraperitoneal (IP) injections of poly(DL-lactide)-b-methoxy polyethylene glycol (PDLLA–MePEG) micelles with entrapped paclitaxel increased survival time in a murine P388 leukemia model compared to a more conventional Cremophor formulation.[162] Although these copolymers were shown to be biocompatible and non-toxic, radiolabeled paclitaxel was shown to be released quickly from the micelles after IV injection,[158] which may compromise the efficacy of these carriers for tumoral targeting. Recently, stereocomplex micelles were prepared by using equimolar mixtures of two block copolymers: PEG-b-poly(D-lactide) and PEG-b-poly(L-lactide). The stereocomplex micelles were shown to be more stable than micelles made of either block copolymer alone, or of the racemic block copolymer PEG-b-poly(DL-lactide). The results suggest that longer circulation times can be achieved with the stereocomplex micelles, which would favor tumor targeting. Drug release and efficacy studies remain to be reported.[163] Using a different biodegradable polymer as the hydrophobic segement, Toncheva *et al.*[149] recently reported a promising pH-dependent micelle made from PEG and polyorthoester block copolymers. Owing to the acid-labile acetal bonds in the polyorthoester segments, the micelles were stable at pH 7.4 for 3 days, but degraded within 2 hours at pH 5.5.

Kataoka and coworkers[164–169] synthesized PEG-poly(α,β-aspartic acid) block copolymer with conjugated doxorubicin (DOX; formerly called adriamycin) (PEG-PAsp(DOX)), and physically entrapped additional unconjugated DOX within the micelles. The conjugated DOX made PAsp sufficiently hydrophobic to drive micelle formation, and interacts with physically trapped free DOX through π–π interactions to reduce leakage of DOX in circulation. Micelles showed prolonged circulation, enhanced accumulation in solid tumors, and release of physically entrapped DOX in the tumors led to complete tumor regression. To avoid chemical conjugation, the same group used poly(ethylene glycol)–poly(β-benzyl-L-aspartate) block copolymers (PEG–PBLA) to physically entrap DOX; the DOX-entrapped micelles showed enhanced stability of both DOX and the micelles possibly due to mutual stabilization through interactions between the benzyl groups in PBLA and DOX. Enhanced blood circulation of DOX, higher antitumor activity after IV injection against C26 tumor in mouse.[165,166,170–172]

Kataoka and coworkers[173–175] also developed a clever concept for using micelles as a DNA delivery vehicle. Block copolymers of PEG–polylysine (PEG-PLys) were mixed with DNA in solution. The polylysine segments complexed with DNA, and the resulting complexed block copolymers formed micelles in solution. Increasing the length of the polylysine segment decreased the rate of exchange between complexed DNA and other polyanions in solution, increased stability of complexed DNA to nuclease attack, and increased transfection efficiency.

2.3.4 pH-responsive targeting

Variations in pH through the GI system, and between intracellular compartments have led to targeting strategies primarily using pH-responsive hydrogels.

Colon targeting

Oral delivery of drugs is the least invasive means of therapy, but the acid and proteolytic environment of the upper GI tract can degrade therapeutic agents before they can reach the absorptive colon surfaces; much effort has therefore been directed towards developing colon-targeting delivery vehicles.[49,176,177] pH-responsive mechanisms have been used to protect encapsulated drugs at low pH, and expose drugs for release in the colon. Methacrylic acid copolymers and derivatives that dissolve at neutral pH, including the commercially available Eudragit® family of polymers, are the most commonly investigated enteric coatings.[178–183] For example, Eudragit was used to encapsulate a drug-containing chitosan core.[183] The Eudragit coatings protected the drug at low pH, then upon dissolution at neutral pH, giving sustained release over 12 h. The mucoadhesiveness of chitosan and its degradability by polysaccharidases in the flora[23,49] were also exploited. In another molecular structure, PMAA with grafted PEG chains were used to make a pH-sensitive oral delivery system for calcitonin; release was found to be swelling-controlled zeroth order.[184,185]

pH-sensitive crosslinked hydrogels that swell at neutral pH to release a drug are also a common physical form for oral delivery,[186] which is sometimes combined with enzymatic degradation to further enhance targeting to the colon. Bajpai and Saxena[187,188] made interpenetrating starch/PAA hydrogels. At pH 2.0, minimal swelling and therefore minimal drug release was observed due to complex formation. At pH 7.4, the hydrogel swells, more rapid drug release is observed which is accelerated in the presence of amylase that enzymatically degrades the starch component of the hydrogel. Hydrogels have also been made by crosslinking with azoaromatic groups that are susceptible to degradation by azoreductase in the colon. pH-responsive hydrogels crosslinked in this way are collapsed in the stomach, swollen in the small intestine, and degraded in the colon to localize drug targeting to the colon.[50,51] To optimize swelling and release characteristics, the concept has been further refined to include interpenetrating networks of an enzymatically degradable network and a non-degradable network.[189] Azo-polysaccharide gels have also been made for colon delivery applications.[190]

Intracellular targeting

An exciting application of pH-responsive polymers is in intracellular targeting, particularly for gene delivery to the nucleus. Endocytosed drug carriers are enclosed in endosomes, which have acidic environments and must escape the endosomes into the cytoplasm before uptake by the nucleus can occur.[191] One approach exploits the pH-dependent protonation of polybasic polymers of the appropriate pK_b; protonation of these polymers at endosomal pH results in enhanced interaction with negatively charged membrane phospholipids and membrane disruption.[192,193] A more extensively investigated approach is based on pH-responsive acidic polymers. Hoffman, Stayton and coworkers[194] have designed two types of pH-responsive acidic polymers that are hydrophobic and insert into and disrupt lipid membranes in the low pH environment of endosomal compartments, but have no membrane disruptive activity in the neutral pH environment of the cytoplasm. The polymers in the first class have both carboxylate and hydrophobic groups in their backbone. Hemolytic studies with red blood cells with a number of such polymers showed that membrane disruption activity and the pH of maximum activity can be controlled by tailoring the molecular structure of the polymer.[195] A physical mixture of poly(propylacrylic acid) (PPAA) along with a cationic lipid dioeyltrimethylammonium propane (DOTAP) and pCMVβ plasmid DNA was shown to significantly enhance gene expression and fraction of transfected NIH3T3 fibroblast cells compared with the DOTAP/pCMVβ plasmid DNA control group.[196] Incubation of Jurkat T-cell lymphoma cell line with a fluorescently labeled ternary complex of biotinylated anti-CD3 antibody, streptavidin and biotinylated poly(propylacrylic acid) showed the presence of the complex in the cytoplasm indicating its endosomal escape, while anti-CD3-streptavidin complexes were retained in the endosome.[197] Potential enhanced therapeutic activity was demonstrated when the presence of PPAA in a green fluorescent protein (GFP)-encoding plasmid formulation was shown to result in increased GFP expression in a wound healing model.[198] The second type of polymers, called 'encrypted polymers', have hydrophobic backbones that are attached to PEG chains via acid-labile bonds. The PEG chains render the polymer conjugates soluble at neutral pH, but upon degradation of the acid-labile bonds the hydrophobic chains are released to exert their membrane-disruptive action. PEG chains were grafted to terpolymers of dimethylaminoethyl methacrylate, butyl methacrylate (BMA) and styrene benzaldehyde via acetal bonds. The degradation kinetics and red blood cell hemolysis kinetics were shown to be pH-dependent and correlated to each other. When the PEG chains were fluorescently labeled, fluorescence was observed throughout the cells indicating endosomal escape.[194] Other polymers of this class have been shown to deliver oligonucleotides and peptides to the cytoplasm of hepatocytes and cultured macrophages.[199,200]

2.3.5 Targeting via external triggers

A growing number of reports are appearing in the literature that exploit an external signal to either guide drug delivery vehicles to the site of action, or to trigger drug release once the vehicle has accumulated at the site of action. Some examples are given below.

Local hyperthermia or hypothermia

Chung *et al.*[201–203] showed that block copolymers of PNIPAAm with a hydrophobic polymer could self-assemble in water to form core-in-shell micelles with hydrophobic cores and PNIPAAm-enriched shells that exhibit LCST at the same temperature as PNIPAAm. Release of adriamycin from PNIPAAm-b-PBMA (polybutyl methacrylate) micelles was found to be minimal below the LCST and dramatically elevated above the LCST in agreement with heating-induced micelle structural changes. *In vitro* cytotoxicity studies showed corresponding temperature dependence. The result suggests the possibility of passive targeting of nanoparticulate PNIPAAm-b-PBMA micelles, followed by the active hyperthermic triggering of drug release locally. Nayak *et al.*[204] combined folate receptor targeting with thermoresponsive drug release. Folic acid was coupled to PNIPAAm hydrogel nanoparticles through amine groups enriched on the shell layer of the nanoparticles. Folate-positive nanoparticles were specifically internalized by KB cells *in vitro*, and cytotoxicity was found to be significantly higher at 37 °C than at 27 °C – a finding hypothesized to be due to particle aggregation above the LCST of PNIPAAm. Higher temperature LCST systems are being developed to allow for the cytotoxic effects to be controlled by local hyperthermia. Deng *et al.*[205] proposed the concept of magnetic targeting followed by hyperthermia triggered local drug release, and reported their initial investigation using core-in-shell nanocapsules of fluorescently labeled magnetic silica core encapsulated in a crosslinked PNIPAAm shell. Nanocapsules ranged in size from ~300 nm to ~550 nm depending on PNIPAAm crosslinking density and temperature. Magnetically directed accumulation in the liver could be seen in rabbit studies. Unfortunately, the *in vitro* release of encapsulated doxorubicin was examined while temperature was ramped up continuously, which made it difficult to analyze the authors' interpretation of the data. Nevertheless, this appears to be a promising concept for targeting.

Chilkoti and co-workers have proposed the combination of local hyperthermia and circulating thermoresponsive polymer carriers for anti-cancer drugs to enhance targeted delivery to tumors.[21,22,206–209] A recursive directional ligation recombinant method to synthesize elastin-like peptides of varying composition and transition temperature T_t, and copolymers of NIPAAm and acrylamide (AAm)[21,209] were designed to have LCSTs above body temperature. Without having optimized either class of thermoresponsive polymers, accumulation of P(NIPAAm-co-AAm) nanoparticles and drug–polymer conjugates were seen to be enhanced in hyperthermia groups compared to unheated groups, but the enhancement was lower than in elastin-like peptides. Rhodamine-conjugated ELPs with a T_t of 40 °C were injected intravenously in nude mice with implanted ovarian tumors in the dorsal skin fold; accumulation of fluorescence in tumors heated to 42 °C was found to be double the accumulation in unheated tumors.[21] *In vitro* studies with rhodamine–ELP conjugates showed uniform distribution of fluorescence in the cytoplasm below, and fluorescent particles above, the transition temperature, suggesting that enhanced accumulation may be due to enhanced endocytosis of ELP aggregates that formed from the ELP conjugates in the heated groups.[206] Furthermore, aggregates were observed to form and attach to vessel walls in the microvasculature of tumors *in vivo* and to grow in size with time – presumably as more circulating ELPs precipitated.[207] Nanoparticles were also investigated[208] using micelles that self-assembled from diblock copolymers containing ELP segments with LCSTs of 90 and 35 °C. The diblock copolymer was found to form micelles of ~40 nm at 40 °C, which transformed to micelles of ~110 nm at 48 °C, and finally micrometer-sized aggregates at 51 °C. It was speculated that the temperature-dependent reversible micelle formation may be coupled with local hypothermia to trigger the disaggregation of micelles that have been accumulated in tumors by EPR, and thus trigger the release of its drug contents. Drug release and efficacy studies remain to be reported, but this is an intriguing approach.

Ultrasound

Rapoport and coworkers[210–218] have proposed and developed the targeting approach that combines the ability of the PEO corona in Pluronic micelles to evade the RES, the passive EPR effect for targeting of tumors, and the ultrasonically triggered destabilization of micelles to localize drug release at the desired sites. Heating of Pluronic micelles that have been stabilized with an interpenetrating network of a thermoresponsive hydrogel to above its LCST resulted in expulsion of a fraction of loaded lipophilic probes, suggesting that lipophilic drugs can be released locally by heating/cooling cycles; and indeed ultrasound triggered release of encapsulated DOX and Rb *in vitro* as well as intracellular drug uptake. For P105 and mixed micelles, negligible release of DOX was seen without ultrasound, but ultrasound irradiation triggered DOX release, which depends on the duration, frequency and power intensity of the ultrasound signal. *In vitro*, release and re-encapsulation of DOX occurred cyclically with the pulsing of the ultrasound signal – by an apparent cavitation related phenomenon, with the release conforming to zero order kinetics in contrast to what would be expected for diffusional release. The re-encapsulation of drug upon turning off ultrasound suggests that drug-containing micelles would circulate without releasing their contents; thus drug release can be targeted by localized ultrasound signals.[216] *In vitro* studies using human leukemia HL-60 cells incubated with DOX-containing micelles show increased cellular uptake of DOX[215] and increased DNA damage[218] for cells exposed to ultrasound relative to controls. *In vivo*, intraperitoneal and intravenous injections in ovarian cancer-bearing nu/nu mice show significantly enhanced accumulation of Pluronic micelles in tumors,[211] and colon cancer-bearing rat models showed significantly reduced tumor sizes.[214]

Magnetic targeting

Zhao *et al.*[219] coated magnetic nanoparticles with *O*-carboxymethylated chitosan, and attached a membrane translocation signal peptide (tat) to improve translocational property and cellular uptake of the nanoparticles. Methotrexate was loaded as a model cancer drug. Ligand attached particles showed greater tumor toxicity in an *in vitro* U-937 tumor cell model than particles with no ligands. Size, magnetic properties and the cell-specific ligands could potentially act synergistically to enhance targeting. Kim *et al.*[220] synthesized magnetite-containing chitosan nanoparticles by spray-coprecipitation. When particles of zeta potential close to zero were injected into mice, increased circulation time and reduced uptake in organs, suggesting the possibility of increased tumour targeting. The authors proposed that the particles may be used with localized hyperthermia for tumor treatment. Tan and Zhang[124] made chitosan nanoparticles containing fluorescent quantum dots and paramagnetic Gd-DTPA (gadrinium diethylene triamine pentaacetate) for immunoassays and MRI contrasting, respectively. Nanoparticles were made by the direct complexation of positively charged chitosan, negatively charged surface modified CdSe/ZnS quantum dots and negatively charged Gd-DTPA. The nanoparticles can be used to target intracellular compartments for enhanced imaging, as well as to elucidate mechanisms of chitosan uptake into cells.

2.4 Conclusions

A number of targeting strategies along with a survey of representative hydrophilic polymeric materials used in targeting have been presented. Targeting is the selective accumulation of drugs at the desired site of therapeutic action, and numerous strategies have been devised for protecting drug and preventing drug release prior to reaching the target site, guiding carriers to the target site, and triggering drug release at the target site.

Even without being thoroughly comprehensive in this survey, it is clear that a wealth of material chemistry and physical properties can be exploited for targeting applications. While many encapsulation and delivery techniques have been successfully implemented using traditional hydrogels, nanoparticles and micelles, emerging targeting strategies are now focused on creating hybrid, multi-component systems. In particular, combinations of new environmentally responsive materials now allow for unprecedented control of sustained drug delivery, and the use of external triggers further enrich the tool chest of stimuli and can be exploited to independently control the guidance and trigger drug release.

2.5 References

1. Chang, S. M. T. Therapeutic applications of polymeric artificial cells. *Nat. Rev. Drug Discov.* **4**, 221–235 (2005).
2. Langer, R. Biomaterials: status, challenges, and perspectives. *AICHE J.* **46**, 1286–1289 (2000).
3. Hoffman, A. S. Hydrogels for biomedical applications. *Adv. Drug Deliv. Rev.* **54**, 3–12 (2002).
4. Peppas, N. A., Bures, P., Leobandung, W. & Ichikawa, H. Hydrogels in pharmaceutical formulations. *Europ. J. Pharmaceutics Biopharmaceutics* **50**, 27–46 (2000).
5. Gil, E. S. & Hudson, S. M. Stimuli-reponsive polymers and their bioconjugates. *Progr. Polym. Sci.* **29**, 1173–1222 (2004).
6. Hennink, W. E. & Van Nostrum, C. F. Novel crosslinking methods to design hydrogels. *Adv. Drug Deliv. Rev.* **54**, 13–36 (2002).
7. Amsden, B. Solute diffusion within hydrogels. Mechanisms and models. *Macromolecules* **31**, 8382–8395 (1998).
8. Hoch, G., Chauhan, A. & Radke, C. J. Permeability and diffusivity for water transport through hydrogel membranes. *J. Membr. Sci.* **214**, 199–209 (2003).
9. Kosto, K. B. & Deen, W. M. Diffusivities of macromolecules in composite hydrogels. *AICHE J.* **50**, 2648 (2004).

10. Ritger, P. L. & Peppas, N. A. A simple equation for description of solute release I. Fickian and non-Fickian release from non-swellable devices in the form of slabs, spheres, cylinders or discs. *J. Control. Rel.* **5**, 23–36 (1987).
11. Ritger, P. L. & Peppas, N. A. A simple equation for description of solute release II. Fickian and anomalous release from swellable devices. *J. Control. Rel.* **5**, 37–42 (1987).
12. Jeong, B., Kim, S. W. & Bae, Y. H. Thermosensitive sol–gel reversible hydrogels. *Adv. Drug Deliv. Rev.* **54**, 37–51 (2002).
13. Schild, H. G. Poly(N-isopropylacrylamide): experiment, theory and application. *Progr. Polym. Sci.* **17**, 163–249 (1992).
14. Chen, G. & Hoffman, A. S. Graft copolymers that exhibit temperature-induced phase transitions over a wide range of pH. *Nature* **373**, 49–52 (1995).
15. Pei, Y. *et al.* The effect of pH on the LCST of poly(*N*-isopropylacrylamide) and poly(*N*-isopropylacrylamide-co-acrylic acid). *J. Biomater. Sci. Polym. Edition* **15**, 585–594 (2004).
16. Garret-Flaudy, F. & Freitag, R. Influence of small uncharged but amphiphilic molecules on the lower critical solution temperature of highly homogeneous *N*-alkylacrylamide oligomers. *J. Polym. Sci. Part A: Polym. Chem.* **38**, 4218 (2000).
17. Lin, H. & Cheng, Y. *In-situ* thermoreversible gelation of block and star copolymers of poly(ethylene glycol) and poly(*N*-isopropylacrylamide) of varying architectures. *Macromolecules* **34**, 3710–3715 (2001).
18. Zhang, X., Wu, D. & Chu, C. Effect of the crosslinking level on the properties of temperature-sensitive poly(*N*-isopropylacrylamide) hydrogels. *J. Polym. Sci. Part B: Polym. Phys.* **41**, 582 (2003).
19. Ruel-Gariépy, E. & Leroux, J. *In situ*-forming hydrogels – review of temperature-sensitive systems. *Europ. J. Pharmaceutics Biopharmaceutics* **58**, 409–426 (2004).
20. Reiersen, H., Clarke, A. R. & Rees, A. R. Short elastin-like peptides exhibit the same temperature-induced structural transitions as elastin polymers: implications for protein engineering. *J. Mol. Biol.* **283**, 255–264 (1998).
21. Meyer, D. E., Shin, B. C., Kong, G. A., Dewhirst, M. W. & Chilkoti, A. Drug targeting using thermally responsive polymers and local hyperthermia. *J. Control. Rel.* **74**, 213–224 (2001).
22. Meyer, D. E. & Chilkoti, A. Quantification of the effects of chain length and concentration on the thermal behavior of elastin-like polypeptides. *Biomacromolecules* **5**, 846–851 (2004).
23. Agnihotri, S. A., Mallikarjuna, N. N. & Aminabhavi, T. M. Recent advances on chitosan-based micro- and nanoparticles in drug delivery. *J. Control. Rel.* **100**, 5–28 (2004).
24. Rowley, J. A., Madlambayan, G. & Mooney, D. J. Alginate hydrogels as synthetic extracellular matrix materials. *Biomaterials* **20**, 45–53 (1999).
25. Kong, H. J., Smith, M. K. & Mooney, D. J. Designing alginate hydrogels to maintain viability of immobilized cells. *Biomaterials* **24**, 4023–4029 (2003).
26. Tabata, Y. & Ikada, Y. Protein release from gelatin matrices. *Adv. Drug Deliv. Rev.* **31**, 287–301 (1998).
27. Ulubayram, K., Eroglu, I. & Hasirci, N. Gelatin microspheres and sponges for delivery of macromolecules. *J. Biomater. Appl.* **16**, 227–241 (2002).
28. Torchilin, V. P. Drug targeting. *Europ. J. Pharmaceut. Sci.* **11, pt. Supplement 2**, S81–S91 (2000).
29. Fahmy, T. M., Fong, P. M., Goyal, A. & Saltzman, W. M. Targeted for drug delivery. *Materials Today* **8**, 18–26 (2005).

30. Gittens, S. A., Bansal, G., Zernicke, R. F. & Uludag, H. Designing proteins for bone targeting. *Adv. Drug Del. Rev.* **57**, 1011–1036 (2005).
31. Omolola Eniola, A. & Hammer, D. A. *In vitro* characterization of leukocyte mimetic for targeting therapeutics to the endothelium using two receptors. *Biomaterials* **26**, 7136–7144 (2005).
32. Lopez-Barcons, L. A., Polo, D., Llorens, A., Reig, F. & Fabra, A. Targeted adriamycin delivery to MXT-B2 metastatic mammary carcinoma cells by transferrin liposomes: effect of adriamycin ADR-to-lipid ratio. *Oncol. Rep.* **14**, 1337–1343 (2005).
33. Hattori, Y. & Maitani, Y. Folate-linked lipid-based nanoparticle for targeted gene delivery. *Curr. Drug Deliv.* **2**, 243–252 (2005).
34. Hilgenbrink, A. R. & Low, P. S. Folate receptor-mediated drug targeting: from therapeutics to diagnostics. *J. Pharmaceut. Sci.* **94**, 2135–2146 (2005).
35. Kim, S. H., Jeong, J. H., Chun, K. W. & Park, T. G. Target-specific cellular uptake of PLGA nanoparticles coated with poly(L-lysine)–poly(ethylene glycol)–folate conjugate. *Langmuir 21*, 8852–8857 (2005).
36. Ludwig, A. The use of mucoadhesive polymers in ocular drug delivery. *Adv. Drug Delivery Rev.* **57**, 1595–639. Epub: 2005 Sep 28 (2005).
37. Grabovac, V., Guggi, D. & Bernkop-Schnurch, A. Comparison of the mucoadhesive properties of various polymers. *Adv. Drug Delivery Rev.* **57**, 1713–1723. Epub: 2005 Sep 23 (2005).
38. Salamat-Miller, N., Chittchang, M. & Johnston, T. P. The use of mucoadhesive polymers in buccal drug delivery. *Adv. Drug Delivery Rev.* **57**, 1666—1691. Epub: 2005 Sep 23 (2005).
39. Valenta, C. The use of mucoadhesive polymers in vaginal delivery. *Adv. Drug Delivery Rev.* **57**, 1692–1712. Epub: 2005 Sep 22 (2005).
40. Ugwoke, M. I., Agu, R. U., Verbeke, N. & Kinget, R. Nasal mucoadhesive drug delivery: background, applications, trends and future perspectives. *Adv. Drug Delivery Rev.* **57**, 1640–1665. Epub: 2005 Sep 21 (2005).
41. Bernkop-Schnurch, A. Thiomers: a new generation of mucoadhesive polymers. *Adv. Drug Delivery Rev.* **57**, 1569–1582. Epub: 2005 Sep 19 (2005).
42. Bernkop-Schnurch, A. Mucoadhesive polymers: strategies, achievements and future challenges. *Adv. Drug Delivery Rev.* **57**, 1553–1555. Epub: 2005 Sep 19 (2005).
43. Takeuchi, H. *et al.* Novel mucoadhesion tests for polymers and polymer-coated particles to design optimal mucoadhesive drug delivery systems. *Adv. Drug Delivery Rev.* **57**, 1583–1594. Epub: 2005 Sep 16 (2005).
44. Huang, Y. B., Leobandung, W., Foss, A. & Peppas, N. A. Molecular aspects of muco- and bioadhesion: Tethered structures and site-specific surfaces. *J. Control. Rel.* **65**, 63–71 (2000).
45. Smart, J. D. The basics and underlying mechanisms of mucoadhesion. *Adv. Drug Delivery Rev.* **57**, 1556–1568 (2005).
46. Mahrag Tur, K. & Ch'ng, H. Evaluation of possible mechanism(s) of bioadhesion. *Int. J. Pharm.* **160**, 61–74 (1998).
47. He, P., Davis, S. S. & Illum, L. *In vitro* evaluation of the mucoadhesive properties of chitosan microspheres. *Int. J. Pharm.* **166**, 75–88 (1998).
48. Artursson, P., Lindmark, T., Davis, S. S. & Illum, L. Effect of chitosan on the permeability of monolayers of intestinal epithelial cells (Caco-2). *Pharm. Res.* **11**, 1358–1361 (1994).
49. Chourasia, M. & Jain, S. Polysaccharides for colon targeted drug delivery. *Drug Delivery* **11**, 129 (2004).

50. Ghandehari, H., Kopečková, P. & Kopeček, J. *In vitro* degradation of pH-sensitive hydrogels containing aromatic azo bonds. *Biomaterials* **18**, 861–872 (1997).
51. Wang, D., Dušek, K., Kopečková, P., Dušková-Smrčkovč, M. & Kopeček, J. Novel aromatic azo-containing pH-sensitive hydrogels: synthesis and characterization. *Macromolecules* **35**, 7791–7803 (2002).
52. Hatefi, A. & Amsden, B. Biodegradable injectable *in situ* forming drug delivery systems. *J. Control. Rel.* **80**, 9–28 (2002).
53. Paulsson, M., Hägerström, H. & Edsman, K. Rheological studies of the gelation of deacetylated gellan gum (Gelrite®) in physiological conditions. *Europ. J. Pharmaceut. Sci.* **9**, 99–105 (1999).
54. Carlfors, J., Edsman, K., Petersson, R. & Jörnving, K. Rheological evaluation of Gelrite® *in situ* gels for ophthalmic use. *Europ. J. Pharmaceut. Sci.* **6**, 113–119 (1998).
55. Rozier, A., Mazuel, C., Grove, J. & Plazonnet, B. Functionality testing of gellan gum, a polymeric excipient material for ophthalmic dosage forms. *Int. J. Pharm.* **153**, 191–198 (1997).
56. Grove, J., Chastaing, G., Rozier, A. & Plazonnet, B. Ophthalmic GELRITE increases ocular bioavailability of indomethacin. *Exp. Eye Res.* **55, pt. Supplement 1**, 54 (1992).
57. Meseguer, G., Buri, P., Plazonnet, B., Rozier, A. & Gurny, R. Gamma scintigraphic comparison of eyedrops containing pilocarpine in healthy volunteers. *J. Ocular Pharmacol. Therapeutics: Off. J. Assoc. Ocular Pharmacol. Therapeutics* **12**, 481–488 (1996).
58. Nelson, M. D., Bartlett, J. D., Corliss, D., Karkkainen, T. & Voce, M. Ocular tolerability of timolol in Gelrite in young glaucoma patients. *J. Am. Optom. Assoc.* **67**, 659–663 (1996).
59. Rozier, A., Mazuel, C., Grove, J. & Plazonnet, B. Gelrite®: a novel, ion-activated, *in-situ* gelling polymer for ophthalmic vehicles. Effect on bioavailability of timolol. *Int. J. Pharm.* **57**, 163–168 (1989).
60. Cohen, S., Lobel, E., Trevgoda, A. & Peled, Y. A novel *in situ*-forming ophthalmic drug delivery system from alginates undergoing gelation in the eye. *J. Control. Rel.* **44**, 201–208 (1997).
61. Kumar, S. & Himmelstein, K. J. Modification of *in situ* gelling behavior of carbopol solutions by hydroxypropyl methylcellulose. *J. Pharmaceut. Sci.* **84**, 344–348 (1995).
62. Miyazaki, S. *et al. In situ* gelling xyloglucan formulations for sustained release ocular delivery of pilocarpine hydrochloride. *Int. J. Pharm.* **229**, 29 (2001).
63. Miyazaki, S. *et al.* Thermally reversible xyloglucan gels as vehicles for rectal drug delivery. *J. Control. Rel.* **56**, 75–83 (1998).
64. Miyazaki, S. *et al.* Oral sustained delivery of paracetamol from *in situ* gelling xyloglucan formulations. *Drug Dev. Ind. Pharm.* **29**, 113–119 (2003).
65. Burgalassi, S., Chetoni, P., Panichi, L., Boldrini, E. & Saettone, M. F. Xyloglucan as a novel vehicle for timolol: pharmacokinetics and pressure lowering activity in rabbits. *J. Ocular Pharmacol. Therapeutics: Off. J. Assoc. Ocular Pharmacol. Therapeutics* **16**, 497–509 (2000).
66. Han, C. K. & Bae, Y. H. Inverse thermally-reversible gelation of aqueous *N*-isopropylacrylamide copolymer solutions. *Polymer* **39**, 2809–2814 (1998).
67. Bae, Y. H., Vernon, B., Han, C. K. & Kim, S. W. Extracellular matrix for a rechargeable cell delivery system. *J. Control. Rel.* **53**, 249–258 (1998).
68. Vernon, B., Kim, S. W. & Bae, Y. H. Thermoreversible copolymer gels for

extracellular matrix. *J. Biomed. Mater. Res.* **51**, 69 (2000).
69. Ista, L. K. & López, G. P. Lower critical solubility temperature materials as biofouling release agents. *J. Ind. Microbiol. Biotechnol.* **20**, 121 (1998).
70. Lu, H-F., Targonsky, E. D., Wheeler, M. B. & Cheng, Y-L. Thermally induced gelable polymer networks for living cell encapsulation. *Biotechnology and Bioengineering 96*(1), 146–155 (2007).
71. El-Kamel, A. H. *In vitro* and *in vivo* evaluation of Pluronic F127-based ocular delivery system for timolol maleate. *Int. J. Pharm.* **241**, 47 (2002).
72. Wei, G., Xu, H., Ding, P. T., Li, S. M. & Zheng, J. M. Thermosetting gels with modulated gelation temperature for ophthalmic use: the rheological and gamma scintigraphic studies. *J. Control. Rel.* **83**, 65–74 (2002).
73. Bonhomme-Faivre, L., Mathieu, M. C., Depraetere, P., Grossiord, J. L. & Seiller, M. Formulation of a charcoal suspension for intratumoral injection: influence of the Pluronic F68® concentration. *Int. J. Pharm.* **152**, 251–255 (1997).
74. Veyries, M. L. *et al.* Controlled release of vancomycin from Poloxamer 407 gels. *Int. J. Pharm.* **192**, 183 (1999).
75. Choi, H. G. *et al.* Development of *in situ* gelling and mucoadhesive acetaminophen liquid suppository. *Int. J. Pharm.* **165**, 33–44 (1998).
76. Ryu, J. M., Chung, S. J., Lee, M. H., Kim, C. K. & Shim, C. K. Increased bioavailability of propranolol in rats by retaining thermally gelling liquid suppositories in the rectum. *J. Control. Rel.* **59**, 163–172 (1999).
77. Chang, J. Y., Oh, Y. K., Choi, H. G., Kim, Y. B. & Kim, C. K. Rheological evaluation of thermosensitive and mucoadhesive vaginal gels in physiological conditions. *Int. J. Pharm.* **241**, 155–163 (2002).
78. Chang, J. Y. *et al.* Prolonged antifungal effects of clotrimazole-containing mucoadhesive thermosensitive gels on vaginitis. *J. Control. Rel.* **82**, 39–50 (2002).
79. Chen, G. & Hoffman, A. S. Block and graft copolymers and methods relating thereto. United States Patent No. 6486213.
80. Bromberg, L. Polyether-modified poly(acrylic acid): Synthesis and applications. *Ind. Eng. Chem. Res.* **37**, 4267–4274 (1998).
81. Bromberg, L. Novel family of thermogelling materials via C–C bonding between poly(acrylic acid) and poly(ethylene oxide)-b-poly(propylene oxide)-b-poly(ethylene oxide). *J. Phys. Chem. B* **102**, 1956–1963 (1998).
82. Bromberg, L. Properties of aqueous solutions and gels of poly(ethylene oxide)-b-poly(propylene oxide)-b-poly(ethylene oxide)-g-poly(acrylic acid). *J. Phys. Chem. B* **102**, 10736–10744 (1998).
83. Bromberg, L. Scaling of rheological properties of hydrogels from associating polymers. *Macromolecules* **31**, 6148–6156 (1998).
84. Bromberg, L. Self-assembly in aqueous solutions of polyether-modified poly(acrylic acid). *Langmuir* **14**, 5806–5812 (1998).
85. Bromberg, L. & Magner, E. Release of hydrophobic compounds from micellar solutions of hydrophobically modified polyelectrolytes. *Langmuir* **15**, 6792–6798 (1999).
86. Hoffman, A. S. *et al.* in *Graft Copolymers of PEO-PPO-PEO Triblock Polyethers on Bioadhesive Polymer Backbones: Synthesis and Properties* 525 (ACS, Washington, DC, 1997).
87. Bromberg, L. *et al.* Responsive polymer networks and methods of their use. United States Patent No. 5939485.
88. Lin, H. & Sung, K. C. Carbopol/Pluronic phase change solutions for ophthalmic drug delivery. *J. Control. Rel.* **69**, 379–388 (2000).

89. Lin, H. R., Sung, K. C. & Vong, W. J. *In situ* gelling of alginate/pluronic solutions for ophthalmic delivery of pilocarpine. *Biomacromolecules* **5**, 2358–2365 (2004).
90. Jeong, B., Bae, Y. H., Lee, D. S. & Kim, S. W. Biodegradable block copolymers as injectable drug-delivery systems. *Nature* **388**, 860–862 (1997).
91. Jeong, B., Han Bae, Y. & Kim, S. W. Biodegradable thermosensitive micelles of PEG–PLGA–PEG triblock copolymers. *Colloids and Surfaces B: Biointerfaces* **16**, 185 (1999).
92. Jeong, B., Bae, Y. H. & Kim, S. W. Thermoreversible gelation of PEG–PLGA–PEG triblock copolymer aqueous solutions. *Macromolecules* **32**, 7064–7069 (1999).
93. Jeong, B., Bae, Y. H. & Kim, S. W. *In situ* gelation of PEG–PLGA–PEG triblock copolymer aqueous solutions and degradation thereof. *J. Biomed. Mater. Res.* **50**, 171 (2000).
94. Jeong, B., Bae, Y. H. & Kim, S. W. Drug release from biodegradable injectable thermosensitive hydrogel of PEG–PLGA–PEG triblock copolymers. *J. Control. Rel.* **63**, 155–163 (2000).
95. Shim, M. S. *et al.* Poly(D,L-lactic acid-co-glycolic acid)-b-poly(ethylene glycol)-b-poly (D,L-lactic acid-co-glycolic acid) triblock copolymer and thermoreversible phase transition in water. *J. Biomed. Mater. Res.* **61**, 188 (2002).
96. Kwon, Y. M. & Kim, S. W. Biodegradable triblock copolymer microspheres based on thermosensitive sol–gel transition. *Pharm. Res.* **21**, 339–343 (2004).
97. Lee, D. *et al.* Novel thermoreversible gelation of biodegradable PLGA-block-PEO-block-PLGA triblock copolymers in aqueous solution. *Macromolecular Rapid Commun.* **22**, 587 (2001).
98. Zentner, G. M. *et al.* Biodegradable block copolymers for delivery of proteins and water-insoluble drugs. *J. Control. Rel.* **72**, 203–215 (2001).
99. Chenite, A. *et al.* Novel injectable neutral solutions of chitosan form biodegradable gels *in situ*. *Biomaterials* **21**, 2155–2161 (2000).
100. Chenite, A., Buschmann, M., Wang, D., Chaput, C. & Kandani, N. Rheological characterisation of thermogelling chitosan/glycerol-phosphate solutions. *Carbohydr. Polym.* **46**, 39–47 (2001).
101. Ruel-Gariépy, E., Chenite, A., Chaput, C., Guirguis, S. & Leroux, J. Characterization of thermosensitive chitosan gels for the sustained delivery of drugs. *Int. J. Pharm.* **203**, 89 (2000).
102. Ruel-Gariépy, E., Leclair, G., Hildgen, P., Gupta, A. & Leroux, J. Thermosensitive chitosan-based hydrogel containing liposomes for the delivery of hydrophilic molecules. *J. Control. Rel.* **82**, 373–383 (2002).
103. Molinaro, G., Leroux, J., Damas, J. & Adam, A. Biocompatibility of thermosensitive chitosan-based hydrogels: an *in vivo* experimental approach to injectable biomaterials. *Biomaterials* **23**, 2717–2722 (2002).
104. Ruel-Gariépy, E. *et al.* A thermosensitive chitosan-based hydrogel for the local delivery of paclitaxel. *Europ. J. Pharmaceutics Biopharmaceutics* **57**, 53–63 (2004).
105. Berger, J. *et al.* Pseudo-thermosetting chitosan hydrogels for biomedical application. *Int. J. Pharm.* **288**, 197–206 (2005).
106. Brannon-Peppas, L. & Blanchette, J. O. Nanoparticle and targeted systems for cancer therapy. *Adv. Drug Delivery Rev.* **56**, 1649–1659 (2004).
107. Gaur, U. *et al.* Biodistribution of fluoresceinated dextran using novel nanoparticles evading reticuloendothelial system. *Int. J. Pharm.* **202**, 1 (2000).
108. Storm, G., Belliot, S. O., Daemen, T. & Lasic, D. D. Surface modification of nanoparticles to oppose uptake by the mononuclear phagocyte system. *Adv. Drug*

Delivery Rev. **17**, 31–48 (1995).
109. Maeda, H., Seymour, L. W. & Miyamoto, Y. Conjugates of anticancer agents and polymers – advantages of macromolecular therapeutics *in vivo*. *Bioconjug. Chem.* **3**, 351–362 (1992).
110. Cassidy, J. & Schatzlein, A. G. Tumour-targeted drug and gene delivery: principles and concepts. *Expert Rev. Mol. Med.* **6**, 1–17 (2004).
111. Feng, S. & Chien, S. Chemotherapeutic engineering: Application and further development of chemical engineering principles for chemotherapy of cancer and other diseases. *Chem. Eng. Sci.* **58**, 4087–4114 (2003).
112. Lutsiak, M. E. C., Robinson, D. R., Coester, C., Kwon, G. S. & Samuel, J. Analysis of poly(D,L-lactic-co-glycolic acid) nanosphere uptake by human dendritic cells and macrophages *in vitro*. *Pharm. Res.* **19**, 1480–1487 (2002).
113. Managit, C., Kawakami, S., Nishikawa, M., Yamashita, F. & Hashida, M. Targeted and sustained drug delivery using PEGylated galactosylated liposomes. *Int. J. Pharmaceutics* **266**, 77–84 (2003).
114. Lian, T. & Ho, R. J. Y. Trends and developments in liposome drug delivery systems. *J. Pharm. Sci.* **90**, 667–680 (2001).
115. Huang, M., Ma, Z. S., Khor, E. & Lim, L. Y. Uptake of FITC-chitosan nanoparticles by a549 cells. *Pharm. Res.* **19**, 1488–1494 (2002).
116. Calvo, P., RemunanLopez, C., VilaJato, J. L. & Alonso, M. J. Chitosan and chitosan ethylene oxide propylene oxide block copolymer nanoparticles as novel carriers for proteins and vaccines. *Pharm. Res.* **14**, 1431–1436 (1997).
117. Fernandez-Urrusuno, R., Calvo, P., Remunan-Lopez, C., Vila-Jato, J. L. & Alonso, M. J. Enhancement of nasal absorption of insulin using chitosan nanoparticles. *Pharm. Res.* **16**, 1576–1581 (1999).
118. De Campos, A. M., Sanchez, A. & Alonso, M. J. Chitosan nanoparticles: a new vehicle for the improvement of the delivery of drugs to the ocular surface. Application to cyclosporin A. *Int. J. Pharm.* **224**, 159–168 (2001).
119. Dyer, A. M. *et al.* Nasal delivery of insulin using novel chitosan based formulations: A comparative study in two animal models between simple chitosan formulations and chitosan nanoparticles. *Pharm. Res.* **19**, 998–1008 (2002).
120. Qi, L. F., Xu, Z. R., Jiang, X., Li, Y. & Wang, M. Q. Cytotoxic activities of chitosan nanoparticles and copper-loaded nanoparticles. *Bioorg. Med. Chem. Lett.* **15**, 1397–1399 (2005).
121. Hu, Y. *et al.* Synthesis and characterization of chitosan-poly(acrylic acid) nanoparticles. *Biomaterials* **23**, 3193–3201 (2002).
122. Cui, Z. R. & Mumper, R. J. Chitosan-based nanoparticles for topical genetic immunization. *J. Control. Rel.* **75**, 409–419 (2001).
123. Roy, K., Mao, H. Q., Huang, S. K. & Leong, K. W. Oral gene delivery with chitosan-DNA nanoparticles generates immunologic protection in a murine model of peanut allergy. *Nat. Med.* **5**, 387–391 (1999).
124. Tan, W. B. & Zhang, Y. Multifunctional quantum-dot-based magnetic chitosan nanobeads. *Adv. Mater.* **17**, 2375–2380 (2005).
125. Banerjee, T., Mitra, S., Singh, A. K., Sharma, R. K. & Maitra, A. Preparation, characterization and biodistribution of ultrafine chitosan nanoparticles. *Int. J. Pharm.* **243**, 93–105 (2002).
126. Mitra, S., Gaur, U., Ghosh, P. C. & Maitra, A. N. Tumour targeted delivery of encapsulated dextran-doxorubicin conjugate using chitosan nanoparticles as carrier. *J. Control. Rel.* **74**, 317–323 (2001).
127. Bodnar, M., Hartmann, J. F. & Borbely, J. Preparation and characterization of

chitosan-based nanoparticles. *Biomacromolecules* **6**, 2521–2527 (2005).
128. Park, J. H. *et al.* Self-assembled nanoparticles based on glycol chitosan bearing hydrophobic moieties as carriers for doxorubicin: *in vivo* biodistribution and anti-tumor activity. *Biomaterials* **27**, 119–126 (2006).
129. Miwa, A. *et al.* Development of novel chitosan derivatives as micellar carriers of taxol. *Pharm. Res.* **15**, 1844–1850 (1998).
130. Qi, L., Xu, Z., Jiang, X., Hu, C. & Zou, X. Preparation and antibacterial activity of chitosan nanoparticles. *Carbohydr. Res.* **339**, 2693–2700 (2004).
131. Aoyagi, T. & Okano, T. Targeting of anticancer drug using intelligent polymers. *Nippon Rinsho. Jap. J. Clin. Med.* **56**, 644–648 (1998).
132. Tokumitsu, H. *et al.* Gadolinium neutron-capture therapy using novel gadopentetic acid–chitosan complex nanoparticles: *in vivo* growth suppression of experimental melanoma solid tumor. *Cancer Lett.* **150**, 177–182 (2000).
133. Tokumitsu, H., Ichikawa, H. & Fukumori, Y. Chitosan–gadopentetic acid complex nanoparticles for gadolinium neutron-capture therapy of cancer: Preparation by novel emulsion-droplet coalescence technique and characterization. *Pharm. Res.* **16**, 1830–1835 (1999).
134. Lu, Z., Yeh, T. K., Tsai, M., Au, J. L. S. & Wientjes, M. G. Paclitaxel-loaded gelatin nanoparticles for intravesical bladder cancer therapy. *Clin. Cancer Res.* **10**, 7677–7684 (2004).
135. Copland, J. A. *et al.* Bioconjugated gold nanoparticles as a molecular based contrast agent: Implications for imaging of deep tumors using optoacoustic tomography. *Mol. Imaging. Biol.* **6**, 341–349 (2004).
136. Kaul, G. & Amiji, M. Long-circulating poly(ethylene glycol)-modified gelatin nanoparticles for intracellular delivery. *Pharm. Res.* **19**, 1061–1067 (2002).
137. Kaul, G. & Amiji, M. Biodistribution and targeting potential of poly(ethylene glycol)-modified gelatin nanoparticles in subcutaneous murine tumor model. *J. Drug Target.* **12**, 585–591 (2004).
138. Kaul, G. & Amiji, M. Tumor-targeted gene delivery using poly(ethylene glycol)-modified gelatin nanoparticles: *in vitro* and *in vivo* studies. *Pharm. Res.* **22**, 951–961 (2005).
139. Kommareddy, S. & Amiji, M. Preparation and evaluation of thiol-modified gelatin nanoparticles for intracellular DNA delivery in response to glutathione. *Bioconjug. Chem.* **16**, 1423–1432 (2005).
140. Coester, C. J., Langer, K., Von Briesen, H. & Kreuter, J. Gelatin nanoparticles by two step desolvation – a new preparation method, surface modifications and cell uptake. *J. Microencapsul.* **17**, 187–193 (2000).
141. Coester, C., Kreuter, J., von Briesen, H. & Langer, K. Preparation of avidin-labelled gelatin nanoparticles as carriers for biotinylated peptide nucleic acid (PNA). *Int. J. Pharm.* **196**, 147–149 (2000).
142. Weber, C., Coester, C., Kreuter, J. & Langer, K. Desolvation process and surface characterisation of protein nanoparticles. *Int. J. Pharm.* **194**, 91 (2000).
143. Wartlick, H. *et al.* Highly specific HER2-mediated cellular uptake of antibody-modified nanoparticles in tumour cells. *J. Drug Target.* **12**, 461–471 (2004).
144. Michaelis, M. *et al.* Pharmacological activity of DTPA linked to protein-based drug carrier systems. *Biochem. Biophys. Res. Commun.* **323**, 1236–1240 (2004).
145. Lavasanifar, A., Samuel, J. & Kwon, G. S. Poly(ethylene oxide)-block-poly(l-amino acid) micelles for drug delivery. *Adv. Drug Delivery Rev.* **54**, 169–190 (2002).
146. Lukyanov, A. N. & Torchilin, V. P. Micelles from lipid derivatives of water-soluble

polymers as delivery systems for poorly soluble drugs. *Adv. Drug Delivery Rev.* **56**, 1273–1289 (2004).
147. Jie, P., Venkatraman, S. S., Min, F., Freddy, B. Y. & Huat, G. L. Micelle-like nanoparticles of star-branched PEO-PLA copolymers as chemotherapeutic carrier. *J. Control. Rel.* **110**, 20–33. Epub: 2005 Nov 11 (2005).
148. Lee, H., Zeng, F., Dunne, M. & Allen, C. Methoxy poly(ethylene glycol)-block-poly(delta-valerolactone) copolymer micelles for formulation of hydrophobic drugs. *Biomacromolecules* **6**, 3119–3128 (2005).
149. Toncheva, V., Schacht, E., Ng, S. Y., Barr, J. & Heller, J. Use of block copolymers of poly(ortho esters) and poly(ethylene glycol) micellar carriers as potential tumour targeting systems. *J. Drug Target.* **11**, 345 (2003).
150. Croy, S. R. & Kwon, G. S. The effects of Pluronic block copolymers on the aggregation state of nystatin. *J. Control. Rel.* **95**, 161–171 (2004).
151. Kabanov, A. V. & Alakhov, V. Y. Pluronic block copolymers in drug delivery: from micellar nanocontainers to biological response modifiers. *Crit. Rev. Therapeut. Drug Carrier Syst.* **19**, 1–72 (2002).
152. Kabanov, A. V. *et al.* A new class of drug carriers: micelles of poly(oxyethylene)–poly(oxypropylene) block copolymers as microcontainers for drug targeting from blood in brain. *J. Control. Rel.* **22**, 141–157 (1992).
153. Klibanov, A. L., Maruyama, K., Torchilin, V. P. & Huang, L. Amphipathic polyethyleneglycols effectively prolong the circulation time of liposomes. *FEBS Lett.* **268**, 235–237 (1990).
154. Kataoka, K., Harada, A. & Nagasaki, Y. Block copolymer micelles for drug delivery: design, characterization and biological significance. *Adv. Drug Delivery Rev.* **47**, 113–131 (2001).
155. Rapoport, N. Stabilization and activation of Pluronic micelles for tumor-targeted drug delivery. *Colloids Surfaces B: Biointerfaces* **16**, 93 (1999).
156. Husseini, G. A., Christensen, D. A., Rapoport, N. Y. & Pitt, W. G. Ultrasonic release of doxorubicin from Pluronic P105 micelles stabilized with an interpenetrating network of *N,N*-diethylacrylamide. *J. Control. Rel.* **83**, 303–305 (2002).
157. Letchford, K., Zastre, J., Liggins, R. & Burt, H. Synthesis and micellar characterization of short block length methoxy poly(ethylene glycol)-block-poly(caprolactone) diblock copolymers. *Colloids Surfaces B: Biointerfaces* **35**, 81–91 (2004).
158. Burt, H. M., Zhang, X., Toleikis, P., Embree, L. & Hunter, W. L. Development of copolymers of poly(D,L-lactide) and methoxypolyethylene glycol as micellar carriers of paclitaxel. *Colloids Surfaces B: Biointerfaces* **16**, 161 (1999).
159. Kim, S. Y., Shin, I. G. & Lee, Y. M. Preparation and characterization of biodegradable nanospheres composed of methoxy poly(ethylene glycol) and DL-lactide block copolymer as novel drug carriers. *J. Control. Rel.* **56**, 197–208 (1998).
160. Xichen, Z., Jackson, J. K. & Burt, H. M. Development of amphiphilic diblock copolymers as micellar carriers of taxol. *Int. J. Pharm.* **132**, 195–206 (1996).
161. Zhang, X. *et al.* An investigation of the antitumour activity and biodistribution of polymeric micellar paclitaxel. *Cancer Chemother. Pharmacol.* **40**, 81 (1997).
162. Ramaswamy, M., Zhang, X., Burt, H. M. & Wasan, K. M. Human plasma distribution of free paclitaxel and paclitaxel associated with diblock copolymers. *J. Pharm. Sci.* **86**, 460 (1997).
163. Kang, N. *et al.* Stereocomplex block copolymer micelles: core-shell nanostructures with enhanced stability. *Nano Lett.* **5**, 315–319 (2005).

164. Masayuki, Y. *et al.* Polymer micelles as novel drug carrier: adriamycin-conjugated poly(ethylene glycol)-poly(aspartic acid) block copolymer. *J. Control. Rel.* **11**, 269–278 (1990).
165. Yokoyama, M., Okano, T., Sakurai, Y. & Kataoka, K. Improved synthesis of adriamycin-conjugated poly (ethylene oxide)-poly (aspartic acid) block copolymer and formation of unimodal micellar structure with controlled amount of physically entrapped adriamycin. *J. Control. Rel.* **32**, 269–277 (1994).
166. Yokoyama, M. *et al.* Characterization of physical entrapment and chemical conjugation of adriamycin in polymeric micelles and their design for *in vivo* delivery to a solid tumor. *J. Control. Rel.* **50**, 79–92 (1998).
167. Yokoyama, M. *et al.* Selective delivery of adiramycin to a solid tumor using a polymeric micelle carrier system. *J. Drug Targeting* **7**, 171–186 (1999).
168. Yokoyama, M. *et al.* Toxicity and antitumor activity against solid tumors of micelle-forming polymeric anticancer drug and its extremely long circulation in blood. *Cancer Res.* **51**, 3229–3236 (1991).
169. Masayuki, Y. *et al.* Influencing factors on *in vitro* micelle stability of adriamycin-block copolymer conjugates. *J. Control. Rel.* **28**, Special Issue, 59–65 (1994).
170. Kwon, G. S. *et al.* Physical entrapment of adriamycin in AB block copolymer micelles. *Pharm. Res.* **12**, 192–195 (1995).
171. Kwon, G. *et al.* Block copolymer micelles for drug delivery: loading and release of doxorubicin. *J. Control. Rel.* **48**, 195–201 (1997).
172. Kataoka, K. *et al.* Doxorubicin-loaded poly(ethylene glycol)–poly(β-benzyl-L-aspartate) copolymer micelles: their pharmaceutical characteristics and biological significance. *J. Control. Rel.* **64**, 143–153 (2000).
173. Yamasaki, Y., Katayose, S., Kataoka, K. & Yoshikawa, K. PEG–PLL block copolymers induce reversible large discrete coil-globule transition in a single DNA molecule through cooperative complex formation. *Macromolecules* **36**, 6276–6279 (2003).
174. Katayose, S. & Kataoka, K. Remarkable increase in nuclease resistance of plasmid DNA through supramolecular assembly with poly(ethylene glycol)–poly(L-lysine) block copolymer. *J. Pharm. Sci.* **87**, 160 (1998).
175. Katayose, S. & Kataoka, K. Water-soluble polyion complex associates of DNA and poly(ethylene glycol)–poly(L-lysine) block copolymer. *Bioconjug. Chem.* **8**, 702–707 (1997).
176. Yang, L., Chu, J. S. & Fix, J. A. Colon-specific drug delivery: new approaches and in vitro/in vivo evaluation. *Int. J. Pharm.* **235**, 1 (2002).
177. Vandamme, T., Lenourry, A., Charrueau, C. & Chaumeil, J. The use of polysaccharides to target drugs to the colon. *Carbohydr. Polym.* **48**, 219–231 (2002).
178. Dai, J. *et al.* pH-sensitive nanoparticles for improving the oral bioavailability of cyclosporine A. *Int. J. Pharm.* **280**, 229–240 (2004).
179. Di Colo, G., Falchi, S. & Zambito, Y. *In vitro* evaluation of a system for pH-controlled peroral delivery of metformin. *J. Control. Rel.* **80**, 119–128 (2002).
180. Carelli, V., Di Colo, G., Nannipieri, E., Poli, B. & Serafini, M. F. Polyoxyethylene–poly(methacrylic acid-co-methyl methacrylate) compounds for site-specific peroral delivery. *Int. J. Pharm.* **202**, 103 (2000).
181. Sinha, V. R. & Kumria, R. Binders for colon specific drug delivery: an *in vitro* evaluation. *Int. J. Pharm.* **249**, 23–31 (2002).
182. Khan, M. Z. I., Prebeg, Z. & Kurjaković, N. A pH-dependent colon targeted oral drug delivery system using methacrylic acid copolymers. *J. Control. Rel.* **58**, 215–

222 (1999).

183. Lorenzo-Lamosa, M. L., Remuñán-López, C., Vila-Jato, J. L. & Alonso, M. J. Design of microencapsulated chitosan microspheres for colonic drug delivery. *J. Control. Rel.* **52**, 109–118 (1998).
184. Torres-Lugo, M. & Peppas, N. A. Molecular design and *in vitro* studies of novel pH-sensitive hydrogels for the oral delivery of calcitonin. *Macromolecules* **32**, 6646–6651 (1999).
185. Torres-Lugo, M., Garcia, M., Record, R. & Peppas, N. A. pH-sensitive hydrogels as gastrointestinal tract absorption enhancers: transport mechanisms of salmon calcitonin and other model molecules using the caco-2 cell model. *Biotechnol. Prog.* **18**, 612–616 (2002).
186. Mahkam, M. Using pH-sensitive hydrogels containing cubane as a crosslinking agent for oral delivery of insulin. *J. Biomed. Mater. Res. Part B, Appl. Biomater.* **75**, 108–112 (2005).
187. Bajpai, S. K. & Saxena, S. Enzymatically degradable and pH-sensitive hydrogels for colon-targeted oral drug delivery. I. Synthesis and characterization. *J. Appl. Polym. Sci.* **92**, 3630 (2004).
188. Bajpai, S. K. & Saxena, S. Dynamic release of riboflavin from a starch-based semi IPN via partial enzymatic degradation: part II. *React. Funct. Polym.* **61**, 115–129 (2004).
189. Chivukula, P. *et al.* Synthesis and characterization of novel aromatic azo bond-containing pH-sensitive and hydrolytically cleavable IPN hydrogels. *Biomaterials* **27**, 1140–1151. Epub: 2005 Aug 11 (2006).
190. Stubbe, B., Maris, B., Van den Mooter, G., De Smedt, S. C. & Demeester, J. The *in vitro* evaluation of 'azo containing polysaccharide gels' for colon delivery. *J. Control. Rel.* **75**, 103–114 (2001).
191. Stayton, P. S. *et al.* Molecular engineering of proteins and polymers for targeting and intracellular delivery of therapeutics. *J. Control. Rel.* **65**, 203–220 (2000).
192. Asayama, S., Nogawa, M., Takei, Y., Akaike, T. & Maruyama, A. Synthesis of novel polyampholyte comb-type copolymers consisting of a poly(L-lysine) backbone and hyaluronic acid side chains for a DNA carrier. *Bioconjug. Chem.* **9**, 476–481 (1998).
193. Asayama, S., Maruyama, A., Cho, C. & Akaike, T. Design of comb-type polyamine copolymers for a novel pH-sensitive DNA carrier. *Bioconjug. Chem.* **8**, 833–838 (1997).
194. Hoffman, A. S. *et al.* Design of 'Smart' polymers that can direct intracellular drug delivery. *Polym. Adv. Technol.* **13**, 992 (2002).
195. Murthy, N., Robichaud, J. R., Tirrell, D. A., Stayton, P. S. & Hoffman, A. S. The design and synthesis of polymers for eukaryotic membrane disruption. *J. Control. Rel.* **61**, 137–143 (1999).
196. Cheung, C. Y., Murthy, N., Stayton, P. S. & Hoffman, A. S. A pH-sensitive polymer that enhances cationic lipid-mediated gene transfer. *Bioconjug. Chem.* **12**, 906–910 (2001).
197. Lackey, C. A., Press, O. W., Hoffman, A. S. & Stayton, P. S. A biomimetic pH-responsive polymer directs endosomal release and intracellular delivery of an endocytosed antibody complex. *Bioconj. Chem.* **13**, 996–1001 (2002).
198. Kyriakides, T. R. *et al.* pH-sensitive polymers that enhance intracellular drug delivery *in vivo*. *J. Control. Rel.* **78**, 295–303 (2002).
199. Murthy, N., Campbell, J., Fausto, N., Hoffman, A. S. & Stayton, P. S. Design and synthesis of pH-responsive polymeric carriers that target uptake and enhance the

intracellular delivery of oligonucleotides. *J. Control. Rel.* **89**, 365–374 (2003).
200. Murthy, N., Campbell, J., Fausto, N., Hoffman, A. S. & Stayton, P. S. Bioinspired pH-responsive polymers for the intracellular delivery of biomolecular drugs. *Bioconjug. Chem.* **14**, 412–419 (2003).
201. Chung, J. E., Yokoyama, M., Aoyagi, T., Sakurai, Y. & Okano, T. Effect of molecular architecture of hydrophobically modified poly(*N*-isopropylacrylamide) on the formation of thermoresponsive core-shell micellar drug carriers. *J. Control. Rel.* **53**, 119–130 (1998).
202. Chung, J. E. *et al.* Thermo-responsive drug delivery from polymeric micelles constructed using block copolymers of poly(*N*-isopropylacrylamide) and poly(butylmethacrylate). *J. Control. Rel.* **62**, 115–127 (1999).
203. Chung, J. E., Yokoyama, M. & Okano, T. Inner core segment design for drug delivery control of thermo-responsive polymeric micelles. *J. Control. Rel.* **65**, 93–103 (2000).
204. Nayak, S., Lee, H., Chmielewski, J. & Lyon, L. A. Folate-mediated cell targeting and cytotoxicity using thermoresponsive microgels. *J. Am. Chem. Soc.* **126**, 10258–10259 (2004).
205. Deng, Y. *et al.* Preparation, characterization, and application of multistimuli-responsive microspheres with fluorescence-labeled magnetic cores and thermoresponsive shells. *Chemistry (Weinheim an der Bergstrasse, Germany)* **11**, 6006–6013 (2005).
206. Raucher, D. & Chilkoti, A. Enhanced uptake of a thermally responsive polypeptide by tumor cells in response to its hyperthermia-mediated phase transition. *Cancer Res.* **61**, 7163–7170 (2001).
207. Meyer, D. E., Kong, G. A., Dewhirst, M. W., Zalutsky, M. R. & Chilkoti, A. Targeting a genetically engineered elastin-like polypeptide to solid tumors by local hyperthermia. *Cancer Res.* **61**, 1548–1554 (2001).
208. Chilkoti, A., Dreher, M. R. & Meyer, D. E. Design of thermally responsive, recombinant polypeptide carriers for targeted drug delivery. *Adv. Drug Delivery Rev.* **54**, 1093–1111 (2002).
209. Chilkoti, A., Dreher, M. R., Meyer, D. E. & Raucher, D. Targeted drug delivery by thermally responsive polymers. *Adv. Drug Delivery Rev.* **54**, 613–630 (2002).
210. Gao, Z., Fain, H. D. & Rapoport, N. Controlled and targeted tumor chemotherapy by micellar-encapsulated drug and ultrasound. *J. Control. Rel.* **102**, 203–222 (2005).
211. Gao, Z., Fain, H. D. & Rapoport, N. Ultrasound-enhanced tumor targeting of polymeric micellar drug carriers. *Molecular Pharmaceutics* **1**, 317–330 (2004).
212. Rapoport, N. Combined cancer therapy by micellar-encapsulated drug and ultrasound. *Int. J. Pharm.* **277**, 155–162 (2004).
213. Rapoport, N. Y., Christensen, D. A., Fain, H. D., Barrows, L. & Gao, Z. Ultrasound-triggered drug targeting of tumors *in vitro* and *in vivo*. *Ultrasonics* **42**, 943–950 (2004).
214. Rapoport, N., Pitt, W. G., Sun, H. & Nelson, J. L. Drug delivery in polymeric micelles: from *in vitro* to *in vivo*. *J. Control. Rel.* **91**, 85–95 (2003).
215. Marin, A. *et al.* Drug delivery in Pluronic micelles: effect of high-frequency ultrasound on drug release from micelles and intracellular uptake. *J. Control. Rel.* **84**, 39–47 (2002).
216. Husseini, G. A., Rapoport, N. Y., Christensen, D. A., Pruitt, J. D. & Pitt, W. G. Kinetics of ultrasonic release of doxorubicin from Pluronic P105 micelles. *Colloids Surfaces B: Biointerfaces* **24**, 253 (2002).

217. Husseini, G. A., Myrup, G. D., Pitt, W. G., Christensen, D. A. & Rapoport, N. Y. Factors affecting acoustically triggered release of drugs from polymeric micelles. *J. Control. Rel.* **69**, 43–52 (2000).
218. Husseini, G. A., El-Fayoumi, R. I., O'Neill, K. L., Rapoport, N. Y. & Pitt, W. G. DNA damage induced by micellar-delivered doxorubicin and ultrasound: comet assay study. *Cancer Lett.* **154**, 211–216 (2000).
219. Zhao, A. J. *et al.* Synthesis and characterization of tat-mediated O-CMC magnetic nanoparticles having anticancer function. *J. Magn. Magn. Mater.* **295**, 37–43 (2005).
220. Kim, D. H. *et al.* Biodistribution of chitosan-based magnetite suspensions for targeted hyperthermia in ICR mice. *IEEE Trans. Magn.* **41**, 4158–4160 (2005).
221. Illum, L. Chitosan and its use as a pharmaceutical excipient. *Pharm. Res.* **15**, 1326–1331 (1998).
222. Aspden, T. J., Illum, L. & Skaugrud, Ø. Chitosan as a nasal delivery system: evaluation of insulin absorption enhancement and effect on nasal membrane integrity using rat models. *Europ. J. Pharmaceut. Sci.* **4**, 23–31 (1996).
223. Dodane, V., Amin Khan, M. & Merwin, J. R. Effect of chitosan on epithelial permeability and structure. *Int. J. Pharm.* **182**, 21 (1999).
224. Schipper, N. G. M. *et al.* Chitosans as absorption enhancers for poorly absorbable drugs 2: mechanism of absorption enhancement. *Pharm. Res.* **14**, 923–929 (1997).
225. Shikata, F., Tokumitsu, H., Ichikawa, H. & Fukumori, Y. *In vitro* cellular accumulation of gadolinium incorporated into chitosan nanoparticles designed for neutron-capture therapy of cancer. *Eur. J. Pharm. Biopharm.* **53**, 57–63 (2002).

3

The artificial cell design: liposomes

G BARRATT, Centre d'Etudes Pharmaceutiques, France

3.1 Introduction

Liposomes were first described in the 1960s as models of biological membranes (Bangham *et al.*, 1965). Soon afterwards, their potential as carriers for biologically active material was pointed out by Gregoriadis *et al.* (1971). The role of a drug carrier is to control the fate of a drug after administration. This is normally determined by a combination of several processes: distribution, metabolism and elimination when given intravenously; absorption, distribution, metabolism and elimination when an extravascular route is used. Regardless of the mechanisms involved, the result depends mainly on the physicochemical properties of the drug and therefore on its chemical structure. Although drug carriers usually modify drug distribution within the organism, they may also affect absorption, metabolism and elimination. Drug delivery systems can be classified either according to their physical form or according to their functional properties. In the latter case, a division into first, second and third generations has been proposed (Barratt *et al.*, 2002).

The so-called first-generation systems are capable of delivering the active substance specifically to the intended target but cannot be considered as 'carriers', because they have to be implanted as closely as possible to the site of action. Microcapsules and microspheres for chemoembolization belong to this group, as do similar systems used for the controlled release of proteins and peptides or for drug delivery within the brain.

In contrast, 'second-generation' systems are true carriers and are usually soluble or particulates less than a micrometre in diameter. They are capable not only of releasing an active product at the intended target but also of carrying it there after administration by a general route. This group includes so-called passive colloidal carriers such as liposomes, nanocapsules and nanospheres and certain 'active' carriers such as temperature-sensitive liposomes and magnetic nanospheres, which release their contents after a specific signal. It should be noted, however, that after intravenous administration most colloidal carriers are rapidly removed from the circulation by phagocytic cells in the liver and spleen.

This limits their potential to deliver their contents to specific sites. Therefore, systems whose surface properties have been modified to reduce the deposition of plasma proteins and show diminished recognition by phagocytes have been developed. These are known as sterically stabilized carriers (or 'Stealth^{TM}' carriers according to a registered trade mark of Liposome Technology Inc.) and may remain in the blood compartment for a considerable time. Although such colloidal particles cannot cross normal continuous capillary endothelium, they have been shown to extravasate into sites where the endothelium is more permeable, such as solid tumors or regions of inflammation and infection.

The systems referred to as 'third generation' are also true carriers and, furthermore, are capable of specific recognition of the target. For example, monoclonal antibodies belong to this group, as do certain second-generation particulate systems (liposomes, nanocapsules, nanospheres) piloted by monoclonal antibodies or other ligands. Of course, targeted colloidal carriers will be much more effective if they are also sterically stabilized.

As far as the physical form of drug delivery systems is concerned, these can be molecular or particulate. Molecular carriers include soluble polymers to which drug molecules have been covalently attached, sometimes with targeting moieties coupled to the same molecule, drug–antibody conjugates such as immunotoxins as well as conjugates with other naturally occurring macromolecules, and lipophilic prodrugs. Drugs trapped within the central cavity of water-soluble cyclodextrins can also be considered as belonging to this category. Although molecular carriers allow a wide distribution of the associated drug, one limitation to their use is the payload of drug which can be carried by each molecule. In contrast, particulate delivery systems can carry a large number of drug molecules in one entity. The most 'natural' particulate carriers are cells from the patient which have been loaded *ex vivo*; these could be resealed erythrocytes or lymphocytes that have been loaded by electroporation. The disadvantage of this approach is the complexity of the procedures involved, not least the regulatory aspect, since each preparation can be considered as a separate 'batch'.

The potential of synthetic particulate drug delivery systems depends on their size. As explained above, polymeric microspheres with a diameter of more than 1 μm cannot be given by a general route and have to be implanted close to the desired site of action. On the other hand, submicronic particles, often referred to as 'colloidal drug carriers', can be given by parenteral routes, including intravenous, and may be able to deliver the drug to a site distant from the site of administration, subject to some limitations which will be discussed below. Liposomes were the first type of colloidal carrier proposed for this application and over the past 35 years a large body of results has been acquired.

3.2 Liposome structure and preparation

Liposomes consist of one or more phospholipid bilayers enclosing an aqueous phase. They can be classified as large multilamellar liposomes (MLV), small unilamellar vesicles (SUV) or large unilamellar vesicles (LUV) depending on their size and the number of lipid bilayers (Fig. 3.1). Water-soluble drugs can be included within the aqueous compartments, while lipophilic or amphiphilic compounds can be associated with the lipid bilayers. In some cases, the resulting objects are better described as lipid complexes than liposomes, since they do not contain an internal aqueous phase. Non-ionic surfactant vesicles or niosomes are similar systems obtained from synthetic surfactants and cholesterol (Uchegbu and Florence, 1995).

Methods for the preparation of liposomes are reviewed in Gregoriadis (1993) and summarized in Fig. 3.2. All these methods require that phospholipids, which have very low critical micellar concentrations, are dispersed in an aqueous phase. This can be done either by hydration of a lipid film, or by formation of an emulsion and subsequent removal of organic solvent or by use of a detergent which is later removed. Size reduction of liposomes can be achieved by sonication or by extrusion through membranes with calibrated pore size. The latter method is preferable because ultrasound may degrade lipids or the material which is to be entrapped. An important parameter to consider during liposome formulation is the phase transition temperature of the principal phospholipids. Phospholipids which are in the gel state at physiological temperatures will yield vesicles which are more stable in the presence of biological fluids, but liposomes must be formed above the transition temperature, which may be incompatible with the substance to be encapsulated. Cholesterol may be added to phospholipid bilayers to increase *in-vivo* stability (Senior, 1987).

Water-soluble substances can be added to the aqueous phase during liposome preparation. In this case, the percentage encapsulation will depend on the proportion of the aqueous phase which is included within the vesicles and may be quite low for some types of preparation, such as small unilamellar vesicles prepared by sonication. Higher entrapment yields for water-soluble compounds can be obtained by the freeze–thaw or dehydration–rehydration techniques. When the drug is a weak acid or base, it is also possible to load pre-formed

Multilamellar vesicle
Several micrometres
Several bilayers
Small encapsulation volume

Small unilamellar vesicle
25–50 nm Single bilayer
Small encapsulation volume

Large unilamellar vesicle
100 nm–1 μm Single bilayer
Large encapsulation volume

3.1 Different types of liposomes.

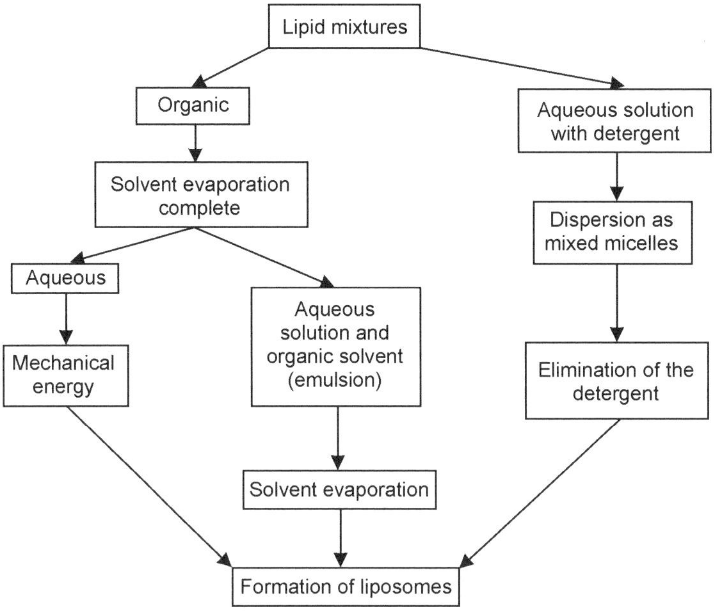

3.2 Outline of methods for liposome preparation.

liposomes by means of a pH gradient, as has been shown for doxorubicin (Harrigan *et al.*, 1993).

In the rest of this chapter we will look first at conventional liposome formulations. Their biological fate and distribution after administration by different routes will be explained and some successful and some less successful applications will be discussed. The development of liposomal formulations which persist longer in the circulation will then be described with some potential therapeutic applications for these systems. The next section will deal with liposomes which are able to deliver their contents to the cytoplasm and with liposomes targeted to specific cell sites. Finally, some future trends will be suggested.

3.3 Conventional liposome formulations

3.3.1 Distribution and fate *in vivo*

Intravenous administration

The fate of liposomes after intravenous administration has been discussed by Poste *et al.* (1984). The authors describe three different types of capillary endothelium and stress the fact that colloidal particles cannot extravasate except in tissues with a discontinuous capillary endothelium; i.e. the liver, spleen and bone marrow. In these particular capillaries, called sinusoids in the liver, the basement membrane is discontinuous or absent and the endothelial cells are perforated by pores (fenestrations) of about 100 nm in diameter. However, liposome membranes are flexible and, depending on the composition, vesicles as large as 400 nm can penetrate the capillary endothelium of the liver and reach the hepatocytes (Daemen *et al.*, 1997a; Romero *et al.*, 1999). It has also been shown that colloidal carriers can extravasate into solid tumours and into inflamed or infected sites, where the capillary endothelium is defective. However, for 'conventional' liposomes, the usual fate is opsonization by plasma proteins followed by uptake by phagocytic cells: either polymorphonuclear leucocytes in the blood or fixed macrophages, particularly the Kupffer cells in the liver (Senior, 1987; Juliano 1988). Liposomes which penetrate the endothelial barrier can also be taken up by hepatocytes, and since these cells are more numerous than Kupffer cells, this can account for a considerable proportion of the dose (Daemen *et al.*, 1997a). Complement activation by the alternative pathway is an important component of carrier recognition and uptake and depends on liposome composition (Scherphof *et al.*, 1981); but opsonization by other plasma proteins, for example non-specific adsorption of IgG, also intervenes. Liposomes are also destabilized by interactions with circulating lipoproteins, especially high-density lipoproteins (Scherphof *et al.*, 1981). The stability of liposomes in plasma is determined by the phase transition temperature of the phospholipids and is increased when the bilayer is stabilized by cholesterol (Senior, 1987).

After uptake by phagocytic cells, the drug carrier systems will be localized in the acidic environment of endosomes and lysosomes and will be degraded by lysosomal enzymes. If the associated drug is not able to escape from this compartment, it may also be degraded and will not reach its site of action unless this is within the lysosomes and the drug is stable in this environment.

Oral administration

The most convenient route of drug administration is the oral one. However, this route presents a number of barriers to the use of colloidal carriers, since conditions within the gastrointestinal tract can disrupt many of them. It has been shown that the concerted action of duodenal enzymes and bile salts destroys the lipid bilayers of most types of liposome, thus releasing the drug (Woodley, 1986). Multilamellar liposomes prepared from phospholipids with phase transition temperatures above 37 °C and which contain cholesterol in their bilayers are the most resistant to degradation (Ouadahi *et al.*, 1998). Lipids that contain a diacetylene moiety in the acyl chain can be polymerized by UV irradiation and thus form a more stable bilayer structure which can resist the conditions of oral administration (Alonso-Romanowski *et al.*, 2003).

Even if the carrier is stable, anatomical considerations mean that only a small proportion of the administered drug-carrier systems can be absorbed intact across the intestinal mucosa into the circulation or the lymphatics. Passage across enterocytes by diffusion is restricted to small, lipophilic molecules, and transcytosis, which is rare, to particles less than 200 nm in diameter. Passage by the paracellular pathway is impossible if the tight junctions are intact. Nevertheless, a number of studies have reported the appearance of particles in the circulation after oral dosing (reviewed in Shakweh *et al.*, 2004).

Other routes of administration

When colloidal drug carriers are administered by other routes, e.g. subcutaneous or intramuscular injection or topical application, they are generally retained at the site of administration longer than free drug. When a liposome-associated drug is applied to the skin, the amount penetrating into the superficial layers may be increased compared with free drug, while its passage to the systemic circulation may be reduced (Mezei, 1988). After subcutaneous or intraperitoneal administration, small liposomes may be taken up by regional lymph nodes (Hawley *et al.*, 1995). Liposomes instilled into the eye are also retained at the site of administration, leading to important therapeutic potential in this area (Kaur *et al.*, 2004).

3.3.2 Therapeutic potential of conventional liposomes

The therapeutic potential of liposome-based systems is influenced by their distribution, as described above. This section describes the possible spheres of application of 'conventional' liposomes by different routes of administration. Structural modifications designed to modify their distribution will be discussed in the following section.

Administration by the intravenous route

The potential applications by the intravenous route can be summarized as concentrating drugs in accessible sites, re-routing drugs away from sites of toxicity and increasing the circulation time of labile or rapidly eliminated drugs (e.g. peptides and proteins).

Delivery to macrophages

Since colloidal drug carriers are naturally concentrated within macrophages, it is logical to use them to deliver drugs to these cells. A good example is the delivery of muramyldipeptide (MDP) and chemically related compounds to stimulate the antimicrobial and antitumoral activity of macrophages. MDP is a low molecular weight, soluble, synthetic compound based on the structure of peptidoglycan from mycobacteria, and, although it acts on intracellular receptors, it penetrates poorly into macrophages. Furthermore, it is eliminated rapidly after intravenous (IV) administration. These problems can be overcome by encapsulation within liposomes or nanocapsules. Early work using soluble MDP within liposomes showed activity against pulmonary (Fidler *et al.*, 1981) and liver (Daemen *et al.*, 1990) metastases in mice. Later, lipophilic derivatives such as muramyltripeptide-cholesterol (Barratt *et al.*, 1989) or muramyltripeptide-phosphatidylethanolamine (Asano and Kleinerman, 1993) were developed to increase the efficiency of encapsulation. *In vitro* studies have shown increased intracellular penetration of muramyl peptides into macrophages when these compounds are associated with liposomes. Activity against hepatic metastases has been observed in a number of mouse models; however, this treatment is only curative when the tumour burden is low (Asano and Kleinerman, 1993). Liposomes containing muramyltripeptide-phosphatidylethanolamine have also been shown to be effective against bacterial infections, for example, by *Klebsiella pneumoniae* (Melissen *et al.*, 1994).

Liposomes can also be used to concentrate antibiotics at the site of infection for direct treatment of bacterial and parasitic infections, particularly when the microorganism is within the lysosomes (reviewed in Pinto-Alphandary *et al.*, 2000). A liposomal formulation of amikacin (MiKasome®) has been tested in clinical trials against complicated bacterial infections and is reputed to be better tolerated than the free antibiotic (Fielding *et al.*, 1999).

The potential of liposomes as immunological adjuvants (reviewed by Kersten and Crommelin, 2003) was recognized as early as 1974. In the case of protein antigens, encapsulation increases capture by antigen-presenting cells such as macrophages (Gregoriadis, 1994). In an alternative strategy, immunogenic peptides have been coupled to the surface in order to activate B- and T-cell clones directly (Boeckler *et al.*, 1999). Liposomes have also been used as carriers for DNA vaccines (Gregoriadis *et al.*, 1996).

Reduction of toxicity

Colloidal carriers can be used to divert drugs from sites of toxicity after intravenous administration. For example, the anticancer drug doxorubicin (adriamycin) is active against a wide spectrum of tumours, but provokes dose-limiting cardiotoxicity. Encapsulation within liposomes reduces this toxicity, by reducing the amount of drug which reaches the myocardium (Gabizon, 1992; Orditura *et al.*, 2004). This is the rationale behind the development of commercially available liposome-based anthracyclin formulations (reviewed by Theodoulou and Hudis, 2004). Small unilamellar liposomes containing daunorubicin, marketed as DaunoXome® by Astellas Pharm US Inc., are indicated for AIDS-linked Kaposi sarcoma. Doxorubicin is available both as conventional liposomes (Myocet, Elan Pharmaceuticals) and liposomes with poly(ethylene glycol) (PEG) at the surface (Caelyx® or Doxil®, Ortho Biotech Products). As well as Kaposi sarcoma, these formulations are indicated for ovarian and breast cancer, as well as being tested for other types of cancer, often in combination with other agents. All reduce cardiac toxicity compared with free drug; however, the 'pegylated' formulation with a much longer circulating half-life is able to concentrate in tumours by the 'enhanced permeability and retention' (EPR) effect and seems to have the highest potential.

A corollary is that concentrations of doxorubicin in the liver increase considerably. A temporary depletion of Kupffer cells, and hence the ability to clear bacteria, in rats has been observed after treatment with doxorubicin-containing liposomes; however, this was less marked when long-circulating liposomes were used (Daemen *et al.*, 1997b). Thus, altered distribution may generate new types of toxicity and this must be borne in mind when developing carrier systems.

Amphotericin B (AmB), an antifungal drug used in deep-seated systemic disease, is another drug with specific dose-limiting side effects, in this case to the kidneys. In 1984 it was noted that association with colloidal lipid systems could increase the maximum tolerated dose of AmB and thereby enhance activity (reviewed by Hartsel and Bolard, 1996). Many different colloidal delivery systems for AmB have since been developed: emulsions, nanoparticles, liposomes and lipid complexes, some of which are now on the market. AmBisome® (Astellas Pharm US Inc.) is a true liposomal formulation of small, unilamellar vesicles with a low AmB/lipid ratio which is marketed by Gilead Sciences. Higher AmB/lipid ratios yield complexes: Amphocil® (Liposome Technology Inc.) consists of small discoid particles of the antibiotic with cholesterol sulphate, while Abelcet® (Enzon Pharmaceuticals) is an association of AmB with two phospholipids which forms ribbon-like structures of several micrometres in length.

Reduced toxicity of these systems is probably the result of both an altered distribution and the physico-chemical state of the AmB in the carrier systems. It has been shown that lipid or polymer association maintains AmB in its monomeric form which is less disruptive to mammalian cell membranes than the self-associated form while maintaining its antifungal activity towards fungal cell walls (Hartsel and Bolard, 1996). Association of AmB with lipid carriers also reduces its transfer to low-density lipoprotein, thereby lowering its renal toxicity (Wasan and Lopez-Berestein, 1997). Colloidal preparations of AmB also have an increased therapeutic index against *Leishmania* infections (Yardley and Croft, 1997); this is an intracellular infection and the use of carriers increases the concentration of drug within macrophages. On the other hand, lipid association also reduces some immunostimulating effects of AmB (such as nitric oxide and tumour necrosis factor-α production) compared with free AmB at the same dose, which may contribute to the reduced toxicity (Larabi *et al.*, 2001). However, the different lipid carrier systems for AmB are not equivalent: their behaviour depends on the size and composition of the particles and on the AmB : lipid ratio.

Prolongation of circulating half-life

Although 'conventional' colloidal carriers are cleared from the circulation quickly, they are still capable of increasing the plasma half-lives of some labile or rapidly eliminated molecules. These could be proteins, peptides or small molecules. For example, the circulating half-lives of cytokines such as gamma interferon (Hockertz *et al.*, 1989) and interleukin-2 (IL-2, Kedar *et al.*, 2000) have been shown to be prolonged by encapsulation within liposomes. As an example of a small molecule, the encapsulation of ATP in liposomes increases its circulating half-life and allows it to protect rats against repeated episodes of cerebral ischæmia (Chapat *et al.*, 1992).

Administration by the oral route

In the 1970s and 1980s many studies with insulin encapsulated in liposomes gave controversial results, which may have been due to differences in liposome composition affecting their stability in digestive fluids (reviewed by Devissaguet *et al.*, 1992). Polymerizable liposomes (Alonso-Romanowski *et al.*, 2003) may have potential for oral immunization.

Administration by other routes

Colloidal drug carrier systems have been used to concentrate interferon-gamma in the skin for the treatment of cutaneous herpes. The cytokine accumulated in the stratum corneum, rather than remaining on the surface as occurred after administration of a simple solution (Weiner *et al.*, 1989). The effect was particularly marked when skin lipids were used to prepare the liposomes. Recently, cholesterol-based surfactants which allow liposomes to deliver their content to keratinocytes have been developed (Barragan-Montero *et al.*, 2005).

Application of carrier formulations to the eye retards elimination of drug from the corneal surface. Drug delivery to the eye by means of vesicles has been reviewed recently (Kaur *et al.*, 2004). In particular, the use of liposomes with or without a gel to retard clearance even further can greatly increase the persistence of antisense oligonucleotides in the vitreous humour, thus opening the way for antisense therapy of viral infections such as herpes simplex or cytomegalovirus (Fattal *et al.*, 2004).

Subcutaneous (Konno *et al.*, 1990) or intra-peritoneal (Parker *et al.*, 1982) administration of anti-cancer agents in liposomes has been shown to deliver the drug to lymphatic metastases. The biotin–avidin system has been used to aggregate liposomes within lymph nodes and thus increase targeting (Phillips *et al.*, 2000).

Active carriers

As far as so-called 'active' carriers are concerned, a comprehensive account is outside the scope of this chapter. Recently, thermosensitive liposomes have been described which are able to accumulate in tumours because of their long-circulating properties (see below) and to release doxorubicin in response to local heating (Ishida *et al.*, 2000; Chen *et al.*, 2004). Release of doxorubicin from liposomes has also been achieved by high-intensity focused ultrasound (Yuh *et al.*, 2005).

Another type of active carrier which is attracting much attention at the moment is liposomes containing magnetic iron particles, or magnetoliposomes (reviewed by Ito *et al.*, 2005). These carrier systems have potential both as contrast agents for magnetic resonance imaging and for guided drug delivery to tumours, especially when long-circulating formulations are used to take advantage of the so-called EPR effect (Martina *et al.*, 2005). Since these systems can be destabilized by hyperthermia, they could be used for site-specific release of an encapsulated drug. The attachment of a tumour-specific ligand at the surface would further enhance the efficacy of this approach (Kullberg *et al.*, 2005).

3.3.4 Long-circulating liposomes

Despite encouraging results with these 'conventional' carrier systems, much research has been devoted towards designing carriers with modified distribution and new therapeutic applications. One major axis is the development of sterically stabilized, or 'Stealth™', carriers which undergo greatly reduced opsonization and uptake by the mononuclear phagocyte system and therefore open up new perspectives for applications by the intravenous route. Secondly, since internalization of colloidal carriers usually leads to the lysosomal compartment, it may be necessary to modify the intracellular distribution of the carrier. This is particularly true when the encapsulated drug is a nucleic acid. Delivery systems are necessary for this type of molecule because they are susceptible to nuclease-mediated degradation in the circulation and penetrate poorly through membranes. However, they are also susceptible to nuclease attack within the lysosomes and their site of action is either in the cytoplasm in the case of an antisense or small interfering RNA strategy or in the nucleus in the case of gene replacement or anti-gene therapy. Thus, systems have been developed which either fuse with the plasma membrane or have a pH-sensitive configuration which changes conformation in the lysosomes and allows escape of the encapsulated material into the cytoplasm. Finally, the ultimate goal would be to be able to direct the drug carrier system to a specific cell type; that is, to develop a third-generation carrier.

Formulation of long-circulating liposomes

Early work with liposomes defined some factors which led to increased persistence in the circulation: small size, inclusion of cholesterol and/or phospholipids with a high phase transition temperature, use of some negatively charged lipids such as the ganglioside GM1 (Allen and Chonn, 1987). However, a major breakthrough in the liposome field consisted in the use of phospholipids grafted with PEG chains of molecular weight from 1 to 5 kDa (Woodle, 1998). This provides a 'cloud' of hydrophilic chains at the particle surface which repels plasma proteins, as discussed theoretically by Jeon *et al.* (1991). These 'sterically stabilized' liposomes have circulating half-lives of about 20 h in rodents and up to 45 h in humans, as opposed to a few hours or even minutes observed for conventional liposomes in rodent models. Their ultimate fate is, however, the same as that of conventional liposomes and the majority will eventually be taken up by the liver and spleen. They have been shown to function as reservoir systems and can penetrate into sites such as solid tumours (Gabizon *et al.*, 1994; Allen *et al.*, 1998; Harrington *et al.*, 2000). These carriers have been termed 'Stealth™' particles because they are 'invisible' to macrophages.

Another strategy for preparing long-circulating colloidal systems can be considered as biomimetic in that it seeks to imitate cells or pathogens which avoid phagocytosis by reducing or inhibiting complement activation. One example is the development of liposomes with a membrane composition similar to that of erythrocytes: e.g. liposomes containing GM_1 (Allen and Chonn, 1987) and those coated with polysialic acids (Gregoriadis *et al.*, 1998). These systems may show circulating half-lives as long as those of liposomes bearing PEG.

Applications of long-circulating liposomes

Long-circulating drug carrier systems can be used as circulating reservoirs of drug and to deliver drugs to intravascular targets. They can also pass through the vascular endothelium in some circumstances, one of the most important of which being the case of solid tumours (Fig. 3.3). The rapidly expanding tumour vasculature often has a discontinuous endothelium with gaps between the cells which may be as large as several hundred nanometres (Jain, 1987). Combined with the fact that tumours often lack an effective lymphatic drainage, this means that particulate matter can be trapped within this zone by the so-called EPR effect. Other pathological situations in which the vascular endothelium is more permeable are infection and inflammation, when the endothelial cells have been activated by chemokines and cytokines to allow extravasation of leucocytes, and hypoxic areas after myocardial infarction.

Prolongation of circulating half-life

Prolongation of circulating half-life by encapsulation in liposomes has been shown for both a small peptide (vasopressin, Woodle *et al.*, 1992) and for a

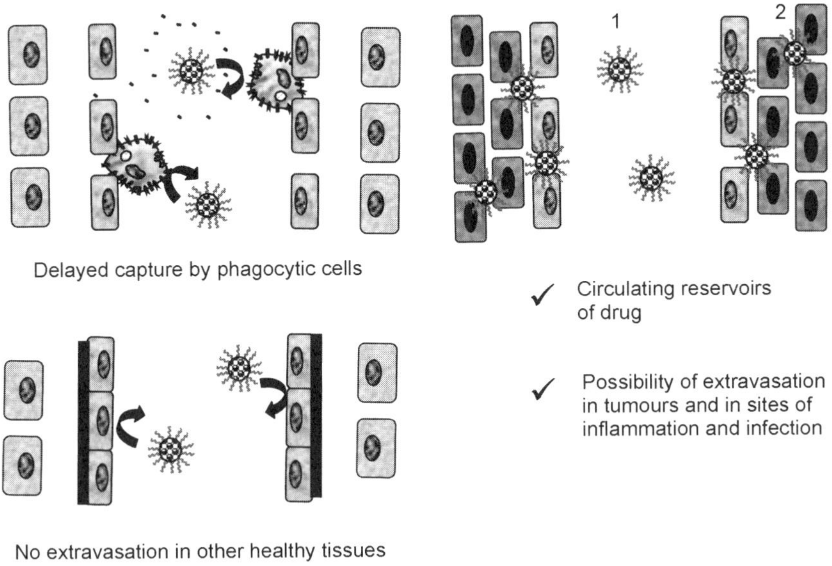

3.3 Potential applications of long-circulating liposomes.

cytokine (IL-2, Kedar *et al.*, 2000). In the first case, the anti-diuretic activity of the peptide administered in PEG-containing liposomes appeared after a 2-day lag period and was prolonged for 4 days. This was interpreted as sustained release from liposomes taken up by a compartment other than phagocytic cells (Woodle *et al.*, 1992). In the study of IL-2-loaded liposomes, the biological activity of the cytokine was evaluated at the same time as its circulation time. Although long-circulating liposomes increased the plasma half-life of the protein longer than conventional liposomes, the latter were more effective in increasing leucocyte levels in the blood and in providing adjuvant activity for sub-unit vaccines (Kedar *et al.*, 2000). This illustrates that the circulation time of the drug carrier is not the only important factor; its ability to interact with the target cells and to deliver the encapsulated material is also important. Long-circulating liposomes loaded with haemoglobin have been proposed as blood substitutes (Awasthi *et al.*, 2004).

Treatment of intravascular disease

An intravascular application of long-circulating drug carriers would be the treatment of leukaemia. This has been demonstrated by Lopes de Menezes *et al.* (1998) with a human B-cell lymphoma in nude mice.

Treatment of solid tumours

The main application to date for long-circulating liposomes has undoubtedly been the treatment of solid tumours, because of the EPR effect mentioned above.

Cytosine arabinoside, vincristine, epirubicin and doxorubicin are among the drugs which have been formulated in this way (Gabizon, 1992; Gabizon *et al.*, 1994; Allen *et al.*, 1998). As mentioned above, a doxorubicin-containing formulation based on 'Stealth^{TM}' is commercially available.

Applications in inflammation and infection

As well as accumulating in solid tumours, long-circulating liposomes can extravasate into sites of inflammation and infection (Oyen *et al.*, 1996). This provides possibilities for delivering antibiotic and anti-inflammatory agents to these sites (Schiffelers *et al.*, 2001; Bakker-Woudenberg *et al.*, 2005), as well as the possibility of imaging using liposomes labelled with gamma emitters (Phillips *et al.*, 2001).

Extravascular applications

Long-circulating liposomes may be useful for prolonging the residence time of drugs administered by the subcutaneous (Oussoren and Storm, 2001) and intraocular routes (Bochot *et al.*, 2000).

3.4 Targeted drug delivery using liposomes

3.4.1 Liposomes avoiding the lysosomal compartment

It may be sufficient for a carrier system to concentrate the drug in the tissue of interest. However, in the case of hydrophilic molecules that cross the plasma membrane with difficulty (e.g. nucleic acids; Lasic, 1997), intracellular delivery is required. If the carrier is taken up by endocytosis, its ultimate destination will be the lysosomes, in which hydrolytic enzymes will degrade both the carrier and its contents. Therefore, a number of liposome-based systems, have been developed to avoid the lysosomal compartment either by fusing directly with the plasma membrane with the help of fusogenic proteins or peptides (Fattal *et al.*, 1994; Pecheur *et al.*, 1997) or by destabilizing the endosome. Endosome disruption can be achieved by the use of pH-sensitive liposomes, in which the lipid undergoes a phase change at acid pH (De Oliveira *et al.*, 1997), or by cationic liposomes (Mönkkönen and Urrti, 1998). In this way, the encapsulated material can be delivered to the cytoplasm. These systems are particularly appropriate for the delivery of genes and oligonucleotides, as a non-immunogenic alternative to viral vectors.

For example, an antisense oligonucleotide against the Friend leukaemia virus was found to have a better antiviral effect *in vitro* when encapsulated in pH-sensitive liposomes than in conventional ones (De Oliveira *et al.*, 1997). More recently, liposomes rendered pH-sensitive by the incorporation of cholesterol hemisuccinate (CHEMS) have been used to carry a phosphothiolate antisense oligonucleotide to tumour necrosis factor alpha (TNF-α) in ethanol-fed rats (Ponnappa *et al.*, 2005). The use of the carrier greatly reduced the quantity of oligonucleotide necessary to reduce serum TNF-α and protect against liver injury. Similar pH-sensitive liposomes have been used to delivery a trypanocidal drug, etanidazole, to the cytoplasm of infected macrophages. *In vivo*, this treatment provoked a significant decrease in parasitaemia in *Trypanosoma cruzi*-infected mice (Morilla *et al.*, 2005). However, nucleic acid delivery can also be achieved by conventional liposome formulations. For example, small interfering RNA against the oncoprotein EphA2 has been delivered to intraperitoneal ovarian cancer cells in mice by means of neutral liposomes (Landen *et al.*, 2005).

The use of lipid-based systems for nucleic acid delivery has been reviewed by Liu and Huang (2002) and by Hashida *et al.* (2005).

3.4.2 Liposomes targeted to specific cell populations

An ideal drug carrier system would contain a specific 'homing group' capable of being recognized by the target cells. Much work has been devoted to coupling specific ligands to the surface of liposomes. Monoclonal antibodies or fragments thereof have often been used because of their specificity (reviewed by Barbet, 1995). Other targeting systems that have been investigated are sugar–lectin interactions, e.g. the mannose/fucose receptor of macrophages and the galactose receptor of hepatocytes, hormone and growth factor receptors and receptors for cell nutrients such as transferrin and folic acid, which are over-expressed in some tumours. Impressive results have been obtained *in vitro* (Barbet, 1995) but, with the exception of targeting to the liver, these often have not been confirmed *in vivo*, since the use of specific ligands cannot overcome physiological constraints. Firstly, even targeted systems will be recognized by macrophages if their surfaces are not modified to reduce opsonization. On the other hand, PEG chains can mask a ligand attached directly to the liposome surface. This has been overcome by attaching the targeting group to the end of the PEG chain and, ideally, using a longer PEG chain to carry the ligand than the PEG chains used to confer 'Stealth^{TM}' properties (Mercadal *et al.*, 1999). It is necessary to limit the degree of substitution to avoid recreating a surface on which opsonization can occur easily (Zalipsky *et al.*, 1998). Secondly, the permeability of the vascular endothelium still has to be taken into account. Thirdly, if intracellular delivery is required, the targeting ligand must not simply be bound to the surface, but also internalized after binding, and the carrier system must be small enough to be taken up by receptor-mediated endocytosis in non-phagocytic cells, that is, 200 nm or less (Barbet, 1995). Furthermore, if this internalization occurs, it will lead to the lysosomal compartment unless an endosome-disrupting element is present.

In the light of such constraints and considerations, it seems that the most suitable targets for drug carriers would be cells in accessible sites such as liver metastases, circulating cells (e.g. leukaemia) and cells in sites were the endothelium is leaky (tumours, inflammation, infection, including within the blood–brain barrier). Another strategy is to target the carrier to a particular region of capillary endothelium, to concentrate the drug within a particular organ and allow it to diffuse from the carrier to the target tissue. A refinement of this approach is to choose a receptor which mediates transcytosis across the endothelium.

Targeting to macrophages and liver cells

Although 'conventional' carriers are naturally captured by phagocytic cells, the extent and rate of capture can be increased by the presence of a ligand for a receptor expressed by these cells. The receptor most often exploited for this purpose is the mannose/fucose receptor of macrophages (for a review see Barratt and Schuber, 1993). For example, an immunomodulator can be delivered efficiently in mannose-grafted liposomes, in order to stimulate the anti-tumoral properties of macrophages (Barratt *et al.*, 1987). More recently, this strategy has been applied to the delivery of Amphotericin B (Vyas *et al.*, 2000) and another fungally derived antibiotic (Mitra *et al.*, 2005) to macrophages for treatment of leishmanias. Another targeting ligand which has been used in a similar application is the tetrapeptide tuftsin (Thr-Lys-Pro-Arg; Agrawal *et al.*, 2002). This peptide has the advantage of being both a targeting element and a macrophage activator. The anti-leishmanial activity of the drug is thus reinforced by macrophage-mediated effects.

Liposomes coated with mannan or pullanan, a glucose-containing polysaccharide, loaded with AmB, were delivered to rats as an aerosol to target alveolar macrophages. Drug concentrations were three-fold higher than unmodified liposomes after 6 h and remained high for 24 h (Vyas *et al.*, 2005).

Another type of carbohydrate receptor which has been employed for targeting to the liver is that which recognizes galactose. Proteins able to bind this sugar are expressed on both macrophages and hepatocytes. Although targeting to hepatocytes can be achieved *in vitro* (Cho *et al.*, 2001), capture by Kupffer cells seems to predominate over that of hepatocytes (Shimada *et al.*, 1997). Nevertheless, liposomes coated with 1-amino-lactose have been proposed for targeting to hepatoma cells (Yamamoto *et al.*, 2000). A monogalactosyl lipid incorporated in liposomes through a cholesterol anchor was able to promote doxorubicin uptake by the liver in mice, with 88% of the accumulated drug in parenchymal cells (Wang *et al.*, 2005). Apolipoprotein E, which mediates lipoprotein uptake by hepatocytes, has also been shown to give some specificity for hepatocytes (Scherphof *et al.*, 1997).

Targeting to tumour cells: antibody-targeted systems

For the treatment of leukaemia, effective targeting of sterically stabilized liposomes containing doxorubicin and bearing anti-CD19 to malignant B cells has been observed *in vitro* and *in vivo* in mice (Lopes de Menezes *et al.*, 1998). The same liposome system has also been used to target circulating myeloma cells, in order to prevent relapse after bone marrow transplant (Lopes de Menezes *et al.*, 2000). The release rate of the drug from the liposomes has to be controlled to obtain maximal efficacy and minimal toxicity (Allen *et al.*, 2005).

Another cell-surface receptor which has been targeted by such 'immuno-liposomes' is the her2 (ErbB2) antigen (Park *et al.*, 2001). In this case, a phospholipid bearing a PEG chain terminated by an anti-her2 antibody fragment was inserted into preformed commercially available doxorubicin-loaded liposomes (Doxil^{TM}). Binding to this receptor is followed by internalization and in this way the drug concentrations can be dramatically increased in her2-overexpressing tumour cells. This targeted system has shown increased efficacy against human breast cancer cells growing in nude mice (Nielsen *et al.*, 2002). Another growth factor receptor which has been targeted in the same way is vascular endothelial growth factor receptor 2 (VEGFR-2, Rubio Demirovic *et al.*, 2005).

The diganglioside GD_2 is another potential tumour-associated antigen. Brignole *et al.* (2005) have directed liposomes containing anti-c-myb antisense oligonucleotides to this glycolipid with a surface-coupled monoclonal antibody. This construct inhibits the growth of neuroblastoma cell lines *in vitro* and and prolongs the survival of tumour-bearing mice.

However, antibody-targeted systems do not always show an advantage over simple sterically stabilized liposomes in solid tumours; as observed by Allen and Moase (1996) in a human ovarian tumour growing in nude mice, probably because the larger antibody-decorated particles diffuse less easily within the tumours, where the hydrostatic pressure is increased due to the lack of lymphatic drainage (Jain, 1987). Furthermore, it has been shown that such immuno-liposomes can in some cases be opsonized and cleared rapidly from the circulation (Laverman *et al.*, 2001) and may even give rise to immunological and pseudoallergic reactions (Phillips *et al.*, 1994; Szebeni *et al.*, 2001). The use of Fab′ fragments could reduce these problems, even if avidity is reduced. Thus, Tuffin *et al.* (2005) have used Fab′ fragments of the OX7 monoclonal antibody directed against the Thy1.1 antigen to target liposomes to mesangial cells.

An approach that avoids coupling a large protein to the carrier surface is a two- or three-step procedure using bispecific antibodies, which combine two immunological recognition functions in the same molecule. The antibody can therefore act as a bridge between the target cell and a carrier bearing a low molecular weight ligand. For example, in the protocol described by Cao and Suresh (2000), a bispecific antibody was engineered to recognize a tumour-specific antigen and biotin. This was given intravenously and after allowing a time for distribution, conventional multilamellar liposomes bearing biotin were given. These bound excess antibody in the circulation and carried it to the liver, but could not reach the antibody already bound to tumour cells. A few hours later, small liposomes bearing PEG chains and biotin were administered; these could circulate, extravasate and bind to the antibody present on the surface of the tumour cells. This approach can concentrate the carrier within the tumour, but will not necessarily promote internalization, since in order for the procedure to function, the bispecific antibody must remain at the cell surface. However, this may be sufficient in many cases; for example, if the carrier is loaded with a radioisotope to provide local irradiation. An example of this is the use of a bispecific antibody to the tumour antigen CA 125 and biotin to deliver biotin-decorated immunoliposomes loaded with the radionuclide ^{90}Yo to human ovarian cancer cells growing in nude mice (McQuarrie *et al.*, 2004).

One strategy for avoiding the necessity of an internalizing antigen would be to deliver a lipophilic prodrug which could diffuse passively into the target cell after being released in the vicinity. Thus, Koning *et al.* (2002) synthesized a dipalmitoylated derivative of 5-fluorodeoxyuridine and encapsulated it in liposomes bearing an antibody specific for a rat colon carcinoma. Although only a small proportion of the liposomes were internalized, the majority of the prodrug was metabolized to the active drug, probably at the level of the cell membrane. This approach allows a bystander effect whereby neighbouring cells which do not express the surface antigen are also exposed to the drug.

Another means of achieving site-specific local high concentrations of drug is the ADEPT (antibody-directed enzyme prodrug therapy) strategy. This requires that an inactive prodrug is cleaved to the active agent by an enzyme not normally present in the organism which is directed to the target cell. In the original concept (Bagshawe *et al.*, 1988), the enzyme was coupled directly to an antibody either chemically or as a fusion protein. However, the amount of enzyme delivered can be improved by using immunoliposomes which have several enzyme molecules coupled to the surface as well as the targeting antibody (Fonseca *et al.*, 1999). In a variant of this approach, Huysmans *et al.* (2005) have encapsulated a nucleoside hydrolase within liposomes containing porins in their membrane, thus allowing specific transport of the substrate to the aqueous core, while avoiding a host immune response to the exogenous enzyme. A second step would be to direct these 'nanoreactors' to the target cells by means of specific ligands.

Targeting to tumour cells: transferrin receptors

Many proteins and peptides other than antibodies have been used to target carriers to specific cell types. The transferrin receptor system has been widely studied because it is over-expressed on many tumours. Transferrin, an iron-transporting protein of 80 kDa, can be coupled to the surface of liposomes by a number of different methods (Anabousi *et al.*, 2005). Long-circulating liposomes bearing transferrin have been shown to be taken up by receptor-mediated endocytosis in Colon 26 cells (Ishida *et al.*, 2001) and to deliver doxorubicin to

C6 glioma cells *in vitro* (Eavarone *et al.*, 2000). However, although transferrin-bearing PEG-liposomes accumulated more in HeLa cells than non-targeted ones, they were less efficient at delivering a photosensitizer, hypericin, because of leakage of the drug from the liposomes (Derycke and De Witte, 2002). On the other hand, pH-sensitive pegylated liposomes have been shown to deliver their contents to the cytoplasm of human T-leukaemia cells (Fonseca *et al.*, 2005).

Targeting to tumour cells: low-molecular-weight ligands

Another receptor which is over-expressed on many tumour cells is the folate receptor. Folic acid has some advantages over transferrin or antibodies as a ligand for long-circulating carriers because it is a much smaller molecule which is unlikely to interact with opsonins and can be coupled easily to a PEG chain without loss of receptor-binding activity. PEG-liposomes bearing folic acid, have been shown to be effective carriers for intracellular delivery of nucleic acids and anti-cancer drugs to tumour cells *in vitro* (Wang and Low, 1998; Shi *et al.*, 2002; Hilgenbrink and Low, 2005). Folate has also been coupled to cationic liposomes in order to deliver a plasmid coding for the tumour suppressor p53, in order to sensitize the cells to chemotherapy and radiotherapy (Xu *et al.*, 2001).

Another relatively small molecule which has been used to target long-circulating liposomes is antagonist G, a hexapeptide analogue of the neuro-transmitter substance P, which blocks the action of several neuropeptides by binding to their receptors. This ligand promotes liposome binding to and internalization by a human small cell lung cancer cell line, H69, and improves the nuclear delivery of encapsulated doxorubicin (Moreira *et al.*, 2001).

The RGD (Arg-Gly-Asp) peptide is a sequence recognized by several integrins (cell adhesion molecules) present on the surface of both tumour cells and their associated endothelial cells. A number of authors have coupled this peptide to the surface of drug carrier systems in order to promote uptake by tumour cells. For example, Xiong *et al.* (2005) have achieved improved delivery of doxorubicin in sterically stabilized liposomes to tumour cells both *in vitro* and *in vivo*. Thompson *et al.* (2005) have developed colloidal particles for gene delivery in which the nucleic acid is complexed by cationic lipids in association with other lipids which produce a net neutral particle. PEG–cholesterol can be modified after particle formation to add an RGD motif.

Exploiting tumour cell properties

An alternate approach to specific targeting would be to exploit particular characteristics of tumour cells to bring about site-specific release of drugs from liposomes. One example of this takes advantage of the high levels of phospholipase A_2 (PLA_2) in tumour tissue. The incorporation of short-chain phospholipids, which are preferred substrates, renders the liposomes susceptible to degradation by PLA_2, even when PEG chains are present, thus promoting drug release (Davidsen *et al.*, 2003, Andresen *et al.*, 2005). The system can be made thermosensitive, for even better control of release. In a further refinement, a PLA_2-sensitive prodrug, a derivative of an anticancer ether lipid, can be entrapped in the liposomes (Andresen *et al.*, 2005).

Targeting to endothelium

An example of targeting to endothelium is the use of sterically stabilized liposomes containing Amphotericin B bearing an antibody specific for pulmonary endothelium at the end of the PEG chains (Otsubo *et al.*, 1998). Accumulation of antibiotic in the lungs was observed, as opposed to its remaining in the blood in the case of non-targeted PEG-bearing liposomes or accumulating in the liver in the case of conventional liposomes. This was accompanied by increased efficacy against experimental aspergillosis in mice.

The adhesion molecules selectively expressed on endothelial cells during infection or inflammation, which promote leucocyte arrest and extravasation, can be considered as useful targets for drug carriers. As hypothesized by Muro and Muzykantov (2005), targeting of antioxidant and antithrombotic drugs to these exposed adhesion molecules could give a considerable therapeutic advantage. Immunoliposomes directed against ICAM-1 are bound and internalized by interferon-gamma activated bronchial epithelial cells (Mastrobattista *et al.*, 1999). Liposomes have been targeted to selectins on activated endothelium by means of monoclonal antibodies (Kessner *et al.*, 2001). Similarly, the coupling of antibodies to VCAM-1 onto the surface of liposomes promotes binding to activated human endothelial cells. A large proportion of these bound liposomes are internalized by receptor-mediated endocytosis and a small percentage cross the cells by transcytosis, as observed by fluorescence microscopy (Voinea *et al.*, 2005).

Kamps *et al.* (1997) have developed liposomes with a high specificity for hepatic endothelial cells by modifying their surface with anionized human serum albumin. The majority of the available free amino groups of the protein were derivatized with *cis*-aconitic anhydride before coupling to conventional liposomes of about 100 nm in diameter. The high uptake by endothelial cells was attributed to their high level of expression of scavenger receptors.

Carbohydrate-coated systems similar to those described above as biomimetic might also be useful for targeting to endothelial cells. Liposomes coated with modified dextrans interact with human endothelial cells (Cansell *et al.*, 1999) and with vascular smooth muscle cells (Letourner *et al.*, 2000). However, Lestini *et al.* (2002) saw no targeting advantage with oligodextrans expressed on liposomes. The same authors described liposomes conjugated with the RGD peptide described above to target the GPIIb-IIIa integrin expressed on activated platelets. This should direct the liposomes to regions of damaged endothelium.

A similar concept has been developed by the group of Torchilin, in which liposomes bearing antibodies to the cytoskeleton protein myosin accumulate in damaged regions of the myocardium, where this protein is exposed. The presence of the liposomes is sufficient to 'plug' the lesions, and protection of cardiocytes from hypoxic injury has been demonstrated *in vitro* (Khaw *et al.*, 1995).

The vascular endothelium of tumours would be another valid target for drug carrier systems. The RGD peptide has also been used in this respect to deliver liposomes containing the antivascular drug combretastatin (Patillo *et al.*, 2005). Growth delay in irradiated melanoma cells in mice was achieved by this means. Another small peptide recognized by angiogenic vessels in tumours, APRPG, has been used to direct liposomes containing a photosensitizer, to allow photodynamic therapy of tumours (Ichikawa *et al.*, 2005). Benzinger *et al.* (2000) designed immunoliposomes directed towards an accessible domain (KDR) of the vascular endothelial growth factor receptor with the aim of delivering anti-angiogenic drugs.

Targeting to the brain

The group of Pardridge has pioneered the concept of using receptor-mediated transcytosis to carry drug across the blood–brain barrier (Pardridge, 1999). A number of hydrophilic and/or high molecular weight substrates that cannot diffuse across this barrier are transported in this way. One such receptor is that for transferrin. Thus, Shi *et al.* (2001) have used sterically stabilized liposomes with 1% of the PEG chains modified with a monoclonal antibody against the transferrin receptor to introduce a plasmid encoding for the beta-galactosidase gene into the brain. Widespread expression of the transgene in the brain, but also in peripheral tissues such as liver and spleen, was observed after IV injection. Daunomycin has been encapsulated in liposomes conjugated to anti-transferrin receptor antibodies via the biotin–streptavidin system (Schnyder *et al.*, 2005). This allows the drug to bypass the P-glycoprotein in multidrug-resistant brain capillary endothelial cells.

Liposomes coupled directly to transferrin itself have been used to deliver 5-fluorouracil to brain endothelium (Soni *et al.*, 2005). Horseradish peroxidase, as a model of a protein drug, has also been encapsulated in transferrin-modifed liposomes (Visser *et al.*, 2005). In this case, although internalization by brain capillary endothelial cells was observed, the liposomes were directed to the lysosomes and the protein was degraded.

The receptor for low-density lipoprotein is also expressed on brain capillary endothelial cells. One ligand for this receptor is apolipoprotein E (apoE). A peptide derived from the receptor-binding domain of this protein has been coupled to PEG-bearing liposomes and has been shown to promote binding to and uptake by brain capillary endothelial cells (Sauer *et al.*, 2005).

Targeting by extravascular routes

Liposomes have been designed to target particular cell types within lymph nodes after subcutaneous administration. Thus, vesicles containing a bacterial antigen have been modified with mannose or with phosphatidylserine to direct them to dendritic cells (Arigita *et al.*, 2003). Targeting enhanced both liposome uptake and interleukin 12 production. Liposomes coupled with a monoclonal antibody against the CD134 surface antigen bound to a $CD4^+$ T cell subset thought to be responsible for adjuvant arthritis (Boot *et al.*, 2005). Encapsulation of a lipophilic prodrug of 5-fluorodeoxyuridine inhibited proliferation of the target cells and reversed symptoms of arthritis in rats.

3.5 Future trends

Colloidal drug delivery vehicles have been studied in the laboratory for more than 30 years, but the few liposome-based formulations already on the market are mainly concerned with reducing the side effects of the encapsulated drugs. Now that the interactions between particles and biological milieux are better understood, 'Stealth™' liposomes and nanoparticles which show diminished phagocytosis have been developed and the range of sites which can be reached has been extended. Even without specific targeting technologies, it has already been shown that sites of inflammation and infection and solid tumours can be reached, as well as intravascular sites. If specificity for a particular cell type is required, ligands such as monoclonal antibodies, sugars, lectins or growth factors can be coupled to these long-circulating systems.

The use of bi-specific antibodies for targeting, in a two-step protocol, would reduce the problems of coupling large targeting ligands directly to liposomes. It could also give a more flexible approach, allowing multiple combinations of drug-loaded liposomes and targeting antibodies. A further improvement in specificity could be achieved by using two different targeting antibodies recognizing two different epitopes on the target cell surface.

Site-specific release of drug is another method by which the specific of effect of carrier systems could be increased. This release could be triggered by physical means such as heat or electromagnetic irradiation, particularly when magnetic or other metallic particles are included within the liposomes. It could also be promoted by enzymes, either exogenous, as in the ADEPT strategy, or endogenous, for example phospholipase A_2.

Colloidal drug carriers such as liposomes will be particularly useful for formulating new drugs derived from biotechnology (peptides, proteins, genes, oligonucleotides) because they can provide protection from degradation in biological fluids and promote their penetration into cells. However, liposome encapsulation also provides an ultradispersed form of small hydrophobic molecules without the use of irritating solvents and allows rapid drug dissolution. The ability of liposomes to retain a drug at the site of administration and their bioadhesive properties means that their use is not restricted to the intravenous route.

Therefore, it is to be expected that commercial products able to improve the efficacy of both established drugs and new molecules will soon be available. As well as therapeutic systems, liposomes associated with contrast agents, ferrofluids or radioisotopes could be useful diagnostic agents. Finally, specifically targeted systems could be appropriate tools for *ex-vivo* cellular therapy.

3.6 Sources of further information and advice

Liposome Technology, Third Edition (2007). Edited by G Gregoriadis, Informa Healthcare US Inc., New York. A three-volume book covering all aspects of liposome formulation in a practical approach.

Methods in Enzymology, Volume 391 (2005) *Liposomes*, edited by N Duzgunes. 25 articles written by experts in the field.

Andresen T L, Jensen S S and Jorgensen K (2005), 'Advanced strategies in liposomal cancer therapy: problems and prospects of active and tumor specific drug release', *Progress in Lipid Research*, vol. 44, pp. 68–97. A review that concentrates on liposome-mediated drug delivery to tumour cells.

3.7 References

Agrawal A J, Agrawal A, Pal A, Guru P Y and Gupta C M (2002), 'Superior chemotherapeutic efficacy of Amphotericin B in tuftsin-bearing liposomes against *Leishmania donovani* infection in hamsters', *J Drug Target*, 10, 41–45.

Allen T M and Chonn A (1987), 'Large unilamellar liposomes with low uptake into the reticulo-endothelial system', *FEBS Lett*, 223, 42–46.

Allen T M and Moase E H (1996), 'Therapeutic opportunities for targeted liposomal drug delivery', *Adv Drug Deliv Rev*, 21, 117–133.

Allen T M, Lopes de Menezes D, Hansen C B and Moase E H (1998), 'Stealth™ liposomes for the targeting of drugs in cancer', in Gregoriadis G and McCormack B, *Targeting of Drugs 6: Strategies for Stealth Therapeutic Systems*, New York, Plenum Press, 61–75.

Allen T M, Mumbengegwi D R and Charrois G J R (2005), 'Anti-CD19-targeted liposomal doxorubicin improves the therapeutic efficacy in murine B-cell lymphoma and ameliorates the toxicity of liposomes with varying drug release rates', *Clin Cancer Res*, 11 (9), 3567–3573.

Alonso-Romanowski S, Chiaramoni N S, Lioy V S, Gargini R A, Viera L I and Taira M C (2003), 'Characterization of diacetylenic liposomes as carriers for oral vaccines', *Chem Phys Lipids*, 122, 191–203.

Anabousi S, Laue M, Lehr C-M, Bakowsky U and Ehrhardt C (2005), 'Assessing transferrin modification of liposomes by atomic force microscopy and transmission electron microscopy', *Eur J Pharm Biopharm*, 60, 295–303.

Andresen T L, Jensen S S and Jorgensen K (2005), 'Advanced strategies in liposomal cancer therapy: problems and prospects of active and tumor specific drug release',

Prog Lipid Res, 44, 68–97.

Arigita C, Bevaart L, Everse L A , Koning G A, Hennink W E, Crommelin D J A, van de Winkel J G J, van Vugt M J, Kersten G F A and Jiskoot W (2003), 'Liposomal meningococcal B vaccination: role of dendritic cell targeting in the development of a protective response', *Infect Immun*, 71, 5210–5218.

Asano T and Kleinerman E S (1993), 'Liposome-entrapped MTP-PE, a novel biologic agent for cancer therapy', *J Immunother*, 14, 286–292.

Awasthi V D, Garcia D, Klipper K, Phillips W T and Goins B A (2004), 'Kinetics of liposome-encapsulated haemoglobin after 25% hypovolemic exchange transfusion', *Int J Pharm*, 283, 53–62.

Bagshawe K D, Springer C J, Searle F, Antoniw P, Sharma S K, Melton R G and Sherwood R F (1988), 'A cytotoxic agent can be generated selectively at cancer sites', *Br J Cancer*, 58, 700–703.

Bakker-Woudenberg I A, Schiffelers R M, Storm G, Becker M J and Guo L (2005), 'Long-circulating sterically stabilized liposomes in the treatment of infections', *Methods Enzymol*, 391, 228–260.

Bangham A D, Standish M M and Watkins J C (1965), 'Diffusion of univalent ions across the lamellae of swollen phospholipids', *J Mol Biol*, 13, 238–252.

Barbet J (1995), 'Immunoliposomes', in Puisieux F, Couvreur P, Delattre J and Devissaguet J-Ph, *Liposomes: New Systems and New Trends in their Applications*, Paris, Editions de Santé, 159–191.

Barragan-Montero V, Winum J-Y, Molès J-P, Juan E, Clavel C and Montero J-L (2005), 'Synthesis and properties of isocannabinoid and cholesterol derivatized rhamnosurfactant: application to liposomal targeting of keratinocytes and skin', *Eur J Med Chem*, 40, 1022–1029.

Barratt G and Schuber F (1993), 'Targeting of liposomes with mannose terminated ligands', in Gregoriadis G, *Liposome Technology*, Vol. III, Second Edition, Boca Raton, CRC Press Inc., 199–218.

Barratt G M, Nolibé D, Yapo A, Petit J-F and Tenu J-P (1987), 'Use of mannosylated liposomes for *in vivo* targeting of a macrophage activator and control of artificial pulmonary metastases', *Ann Inst Pasteur (Immunol)*, 138, 437–450.

Barratt G M, Yu W P, Fessi H, Devissaguet J Ph, Petit J F, Tenu J P, Israel L, Morère J F and Puisieux F (1989), 'Delivery of MDP-L-alanyl-cholesterol to macrophages: comparison of liposomes and nanocapsules', *Cancer J*, 2, 439–443.

Barratt G, Couarraze G, Couvreur P, Dubernet C, Fattal E, Gref R. Labarre D, Legrand P. Ponchel G and Vauthier C (2002), 'Polymeric micro and nanoparticles as drug carriers', in Dumitriu S, *Polymeric Biomaterials*, Second Edition, New York, Marcel Dekker Inc., 753–782.

Benzinger P, Martiny-Baron G, Reusch P, Siemeister G, Kley J T, Marmé D, Unger C and Massing U (2000), 'Targeting of endothelial KDR receptors with 3G2 immunoliposomes *in vitro*', *Biochim Biophys Acta*, 1466, 71–78.

Bochot A, Couvreur P and Fattal E (2000), 'Intravitreal administration of antisense oligonucleotides: potential of liposomal delivery,' *Prog Ret Eye Res*, 19, 131–147.

Boeckler C, Dautel D, Schelté P, Frisch B, Wachsmann D, Klein J P and Schuber F (1999), 'Design of highly immunogenic liposomal constructs combining structurally independent B cell and T helper cell peptide epitopes', *Eur J Immunol* 29, 2297–2308.

Boot E P J, Koning G A, Storm G, Wagenaar-Hilbers J P A, van Eden W, Everse L A and Wauben M H M (2005), 'CD134 as target for specific drug delivery to auto-aggressive $CD4^+$ T cells in adjuvant arthritis', *Arthritis Res Ther*, 7, R604–R615.

Brignole C, Marimpietri D, Pagnan G, Di Paolo D, Zancolli M, Pistoia V, Ponzoni M and Pastorino F (2005), 'Neuroblastoma targeting by c-myb-selective antisense oligonucleotides entrapped in anti-GD_2 immunoliposome: immune cell-mediated anti-tumor activities', *Cancer Lett*, 228, 181–186.

Cansell M, Parisel C, Jozefonvicz J and Letourneur D (1999), 'Liposomes coated with chemically modified dextrans interact with human endothelial cells', *J Biomed Mater Res*, 44, 140–148.

Cao Y and Suresh M R (2000), 'Bispecific MAb aided liposomal drug delivery', *J Drug Target*, 8, 257–266.

Chapat S, Frey V, Claperon N, Bouchaud C, Puisieux F, Couvreur P, Rossignol P and Delattre J (1992), 'Efficiency of liposomal ATP in cerebral ischemia: bioavailability features', *Brain Res Bull*, 26, 339–342.

Chen Q, Tong S, Dewhirst M W and Yuan F (2004), 'Targeting tumor microvessels using doxorubicin encapsulated in a novel thermosensitive liposome', *Mol Cancer Ther*, 3 (10), 1311–1317.

Cho C-S, Kobayashi A, Takei R, Ishihara T, Maruyama A and Akaike T (2001), 'Receptor-mediated cell modulator delivery to hepatocyte using nanoparticles coated with carbohydrate-carrying polymers', *Biomaterials*, 22, 45–51.

Daemen T, Velinova M J, Regts J, de Jager M, Kalicharan R, Donga J, van der Want J J and Scherphof G L (1997a), 'Different intrahepatic distribution of phosphatidylglycerol and phosphatidylserine liposomes in the rat', *Hepatology*, 26, 416–423.

Daemen T, Regts J, Meesters M, Ten Kate M T, Bakker-Woudenberg I A J M and Scherphof G (1997b), 'Toxicity of doxorubicin entrapped within long-circulating liposomes', *J Control Res*, 44, 1–9.

Daemen T, Dontje B H, Veninga A, Scherphof G L and Oosterhuis W L (1990), 'Therapy of murine liver metastases by administration of MDP encapsulated in liposomes', *Select Cancer Ther*, 6, 63–71.

Davidsen J, Jorgensen K, Andresen T L and Mouritsen O G (2003), 'Secreted phospholipase A_2 as a new enzymatic trigger mechanism for localised liposomal drug release and absorption in diseased tissue', *Biochim Biophys Acta* 1609, 95–101.

De Oliveira M C, Fattal E, Ropert C, Malvy C and Couvreur P (1997), 'Delivery of antisense oligonucleotides by means of pH-sensitive liposomes', *J Control Rel*, 48, 179–184.

Derycke A S L and De Witte P A M (2002), 'Transferrin-mediated targeting of hypericin embedded in sterically stabilized PEG-liposomes', *Int J Oncol*, 20, 181–187.

Devissaguet J-Ph, Fessi H, Ammoury N and Barratt G (1992), 'Colloidal drug delivery systems for gastrointestinal applications', in Junginger H E, *Drug Targeting and Delivery – Concepts in Dosage Form Design*, Chichester, Ellis Horwood Ltd., 71–91.

Eavarone D A, Yu X and Bellamkonda R V (2000), 'Targeted drug delivery to C6 glioma by transferrin-coupled liposomes', *J Biomed Mat Res*, 51, 10–14.

Fattal E, Nir S, Parente R A and Szoka F C Jr (1994), 'Pore-forming peptides induce rapid phospholipid flip-flop in membranes', *Biochemistry*, 31, 6721–6731.

Fattal E, De Rosa G and Bochot A (2004), 'Gel and solid matrix systems for the controlled delivery of drug-carrier associated nucleic acids', *Int J Pharm* 277, 25–30.

Fidler I J, Sone S, Fogler W E and Barnes Z L (1981), 'Eradication of spontaneous metastases and activation of alveolar macrophages by intravenous injection of liposomes containing muramyl dipeptide', *Proc Natl Acad Sci USA*, 78, 1680–1684.

Fielding R M, Moon-McDermott L and Lewis R O (1999), 'Bioavailability of a small unilamellar low-clearance liposomal amikacin formulation after extravascular administration', *J Drug Target*, 6 (6), 415–426.

Fonseca M J, Haisma H J, Klaassen S, Vingerhoeds M H and Storm G (1999), 'Design of immuno-enzymosomes with maximum enzyme targeting capability: effect of the enzyme density on the enzyme targeting capability and cell binding properties', *Biochim Biophys Acta*, 1419, 272–282.

Fonseca C, Moreira J N, Cuidad C J, Pedrosa de Lima M C and Simoes S (2005), 'Targeting of sterically stabilised pH-sensitive liposomes to human T-leukemia cells', *Eur J Pharm Biopharm*, 59, 359–366.

Gabizon A (1992), 'Selective tumour localization and improved therapeutic index of anthracyclines encapsulated in long-circulating liposomes', *Cancer Res*, 52, 891–896.

Gabizon A, Catane R, Uziely B, Kaufman B and Barenholz Y (1994), 'Prolonged circulation time and enhanced accumulation in malignant exudates of doxorubucin encapsulated in polyethylene-glycol-coated liposomes', *Cancer Res*, 54, 987–992.

Gregoriadis G (1993), *Liposome Technology*, Second Edition, Vol. I, Boca Raton, CRC Press.

Gregoriadis G (1994), 'The immunological adjuvant and vaccine carrier properties of liposomes', *J Drug Target*, 2, 351–356.

Gregoriadis G, Leathwood P D and Ryman B E (1971), 'Enzyme entrapment in liposomes', *FEBS Lett*, 14, 95–99.

Gregoriadis G, Saffie R and Hart S L (1996), 'High yield incorporation of plasmid DNA within liposomes: effect of DNA integrity and transfection efficiency', *J Drug Target*, 3, 469–475.

Gregoriadis G, Fernandes A, McCormack B, Mital M and Zhang X (1998), 'Polysialic acids: potential for long circulating drug, protein, liposome and other microparticle constructs', in Gregoriadis G and McCormack B, *Targeting of Drugs 6: Strategies for Stealth Therapeutic Systems*, New York, Plenum Press, 193–205.

Harrigan P R, Wong K F, Redelmeier T E, Wheeler J J and Cullis P R (1993), 'Accumulation of doxorubicin and other lipophilic amines into large unilamellar vesicles in response to transmembrane pH gradients', *Biochim Biophys Acta*, 1149 (2), 329–338.

Harrington K J , Rowlinson-Busza G, Syrigos K N, Uster P S, Abra R M and Stewart J S (2000), 'Biodistribution and pharmacokinetics of ^{111}In-DTPA-labelled pegylated liposomes in a human tumour xenograft model: implications for novel targeting strategies', *Br J Cancer*, 83, 232–238.

Hartsel S and Bolard J (1996), 'Amphotericin B: new life for an old drug', *Trends Pharmacol Sci*, 17, 445–449.

Hashida M, Kawakami S and Yamashita F (2005), 'Lipid carrier systems for targeted drug and gene delivery', *Chem Pharm Bull*, 53 (8), 871–880.

Hawley A E, Davis S S and Illum L (1995), 'Targeting of colloids to lymph nodes: influence of lymphatic physiology and colloidal characteristics', *Adv Drug Deliv Rev*, 17, 129–148.

Hilgenbrink A R and Low P S (2005), 'Folate receptor-mediated drug targeting: from therapeutics to diagnostics', *J Pharm Sci*, 94 (10), 2135–2146.

Hockertz S, Franke G, Kniep E and Lohman-Matthes M L (1989), 'Mouse interferon-γ in liposomes. Pharmacokinetics, organ distribution and activation of spleen and liver macrophages *in vivo*', *J Interferon Res*, 9, 591–602.

Huysmans G, Ranquin A, Wyns L, Steyaert J and van Gelder P (2005), 'Encapsulation of therapeutic nucleosidase hydrolase in functionalised nanocapsules', *J Control Rel*, 102, 171–179.

Ichikawa K, Hikita T, Maeda N, Yonezawa S, Takeuchi Y, Asai T, Namba Y and Oku N (2005), 'Antiangiogenic photodynamic therapy (PDT) by using long-circulating liposomes modified with peptide specific to angiogenic vessels', *Biochim Biophys Acta*, 1669, 69–74.

Ishida O, Maruyama K, Yanagie H, Eriguchi M and Iwatsuru M (2000), 'Targeting chemotherapy to solid tumors with long-circulating thermosensitive liposomes and local hyperthermia', *Jpn J Cancer Res*, 91, 118–126.

Ishida O, Maruyama K, Tanahashi H, Iwatsuru M, Sasaki K, Eriguchi M and Yanagie H (2001), 'Liposomes bearing polyethyleneglycol-coupled transferrin with intracellular targeting property to the solid tumors *in vivo*', *Pharm Res*, 18, 1042–1048.

Ito A, Shinkai M, Honda H and Kobayashi T (2005), 'Medical application of functionalized magnetic nanoparticles', *J Biosci Bioeng*, 100 (1), 1–11.

Jain R K (1987), 'Transport of molecules across tumor vasculature', *Cancer Metastasis Rev*, 6, 559–593.

Jeon S I, Lee J H, Andrade J D and De Gennes P G (1991), 'Protein-surface interactions in the presence of polyethylene oxide; 1. Simplified theory', *J Colloid Interf Sci*, 142, 149–166.

Juliano R L (1988), 'Factors affecting the clearance kinetics and tissue distribution of liposomes, microspheres and emulsions', *Adv Drug Deliv Rev*, 2, 31–54.

Kamps J A A M, Morselt H W M, Swart P J, Meijer D K F and Scherphof G L (1997), 'Massive targeting of liposomes, surface modified with anionized albumins, to hepatic endothelial cells', *Proc Natl Acad Sci USA*, 94, 11681–11685.

Kaur I P, Garg A, Singla A K and Aggarwal D (2004), 'Vesicular systems in ocular drug delivery: an overview', *Int J Pharm* 269, 1–14.

Kedar E, Gur H, Babai I, Samira S, Even-Chen S and Barenholz Y (2000), 'Delivery of cytokines by liposomes. Hematopoietic and immunomodulatory activity of interleukin-2 encapsulated in conventional liposomes and long-circulating liposomes', *J Immunother*, 23, 131–145.

Kersten G F A and Crommelin D J A (2003), 'Liposomes and ISCOMs', *Vaccine*, 21, 915–920.

Kessner S, Krause A, Rothe U and Bendas G (2001), 'Investigation of the cellular uptake of E-selectin-targeted immunoliposomes by activated human endothelial cells', *Biochim Biophys Acta*, 1514, 177–190.

Khaw B A, Torchilin V P, Vural I and Narula J (1995), 'Plug and seal: prevention of hypoxic cardiocyte death by sealing membrane lesions with antimyosin-liposomes', *Nature Med*, 1, 1195–1198.

Koning G A, Kamps J A and Scherphof G L (2002), 'Efficent intracellular delivery of 5-fluorodeoxyuridine into colon cancer cells by targeted immunoliposomes', *Cancer Detect Prevent*, 26 (4), 299–307.

Konno H, Tadakuma T, Kumai K, Takahashi T, Ishibiki K, Abe O and Sagaguchi S (1990), 'The antitumor effects of adriamycin entrapped in liposomes on lymph node metastases', *Jpn J Surg*, 20, 424–428.

Kullberg M, Mann K and Owens J L (2005), 'Improved drug delivery to cancer cells: a method using magnetoliposomes that target epidermal growth factor receptors', *Medical Hypotheses*, 64, 468–470.

Landen C N Jr, Chavez-Reyes A, Bucana C, Schmandt R, Deavers M T, Lopez-Berestein G and Sood A K (2005), 'Therapeutic EphA2 gene targeting *in vivo* using neutral liposomal small interfering RNA delivery', *Cancer Res*, 65 (15), 6910–6918.

Larabi M, Legrand P, Appel M, Gil S, Lepoivre M, Devissaguet J P, Puisieux F and Barratt G (2001), 'Reduction of NO synthase expression and TNF alpha production in macrophages by Amphotericin B lipid carriers', *Antimicrob Agents Chemother*, 45, 553–562.

Lasic D D (1997), *Liposomes in Gene Therapy*, Boca Raton, CRC Press.

Laverman P, Boerman O C, Oyen W J G, Corstens F H M and Storm G (2001), '*In vivo* applications of PEG liposomes: unexpected observations', *Crit Rev Ther Drug Carr Syst*, 18, 551–566.

Lestini B J, Sagnella S M, Xu Z, Shive M S, Richter N J, Jayaseharan J, Case A J, Kottke-Marchant K, Anderson J M and Marchant R E (2002), 'Surface modification of liposomes for selective cell targeting in cardiovascular drug delivery', *J Control Rel*, 78, 235–247.

Letourneur D, Parisel C, Prigent-Richard S and Cansell M (2000), 'Interactions of functionalized dextran-coated liposomes with vascular smooth muscle cells', *J Control Rel*, 65, 83–91.

Liu F and Huang L (2002), 'Development of non-viral vectors for systemic gene delivery', *J Control Rel*, 78, 259–266.

Lopes de Menezes D, Pilarski L and Allen T M (1998), '*In vitro* and *in vivo* targeting of immunoliposomal doxorubicin to human B-cell lymphoma', *Cancer Res*, 58, 3320–3330.

Lopes de Menezes D E, Pilarski L M, Belch A R and Allen T M (2000), 'Selective targeting of immunoliposomal doxorubicin against human multiple myeloma *in vitro* and *ex vivo*', *Biochim Biophys Acta*, 1466, 205–220.

Martina M-S, Fortin J-P, Ménager C, Clément O, Barratt G, Grabielle-Madelmont C, Gazeau F, Cabuil V and Lesieur S (2005), 'Generation of superparamagnetic liposomes revealed as highly efficient MRI contrast agents for *in vivo* imaging', *J Am Chem Soc*, 127, 10676–10685.

Mastrobattista E, Storm G, van Bloois L, Reszka R, Bloemen P G M, Crommelin D J A and Hanricks P A (1999), 'Cellular uptake of liposomes targeted to intercellular adhesion molecule-1 (ICAM-1) on bronchial epithelial cells', *Biochim Biophys Acta*, 1419, 353–363.

McQuarrie S A, Mercer J R, Syme A, Suresh M R and Miller G G (2004), 'Preliminary results of nanopharmaceuticals used in the radioimmunotherapy of ovarian cancer', *J Pharm Pharmaceut Sci*, 7 (4), 29–34.

Melissen P M B, van Vianen W and Bakker-Woudenberg, I A J M (1994), 'Treatment of Klebsiella *pneumoniae* septicemia in normal and leukopenic mice by liposome-encapsulated muramyl tripeptide phosphatidylethanolamine', *Antimicrob Agents Chemother*, 38, 147–150.

Mercadal M, Domingo J C, Petriz J C, Garcia J. and De Madariaga M M (1999), 'A novel strategy affords high-yield coupling of antibody to extremities of liposomal surface grafted PEG chains', *Biochim Biophys Acta*, 1418, 232–238.

Mezei M (1988), 'Liposomes in the topical applications of drugs: a review', in Gregoriadis G, *Liposomes as Drug Carriers*, New York, John Wiley and Sons, 663–677.

Mitra M, Mandal A K, Chatterjee T K and Das N (2005), 'Targeting of mannosylated liposome incorporated benzyl derivative of *Penicillium nigricans* derived compound MT81 to reticuloendothelial systems for the treatment of visceral leishmanias', *J Drug Target*, 13 (5), 285–293.

Mönkkönen J and Urtti A (1998), 'Lipid fusion in oligonucleotide and gene delivery with cationic lipids', *Adv Drug Del Rev*, 34, 37–49.

Moreira J N, Hansen C B, Gaspar R and Allen T M (2001), 'A growth factor antagonist as a targeting agent for sterically stabilized liposomes in human small cell lung cancer', *Biochim Biophys Acta*, 1514, 303–317.

Morilla M J, Montanari J, Frank F, Malchiodi E, Corral R, Petray P and Romero E L (2005), 'Etanidazole in pH-sensitive liposomes: design, characterization and *in vitro/in vivo* anti-*Trypanosoma cruzi* activity', *J Control Rel*, 103, 599–607.

Muro S and Muzykantov V R (2005), 'Targeting of antioxidant and anti-thrombotic drugs to endothelial cell adhesion molecules', *Curr Pharm Des*, 11 (18), 2383–2401.

Nielsen U B, Kirpotin D B, Pickering E M, Hong K, Park J W, Shalaby M R, Shao Y, Benz C C and Marks J D (2002), 'Therapeutic efficacy of anti-ErbB2 immunolipsomes targeted by a phage antibody selected for cellular endocytosis', *Biochim Biophys Acta*, 1591, 109–118.

Oriditura M, Quaglia F, Morgillo F, Martinelli E, Lieto E, De Rosa G, Comunale D, Diadema M R, Ciardello F, Catalano F and De Vita F (2004), 'Pegylated liposomal doxorubicin: pharmacologic and clinical evidence of potent antitumor activity with reduced antracycline-induced cardiotoxicity (review)', *Oncol Rep*, 12 (3) 549–556.

Otsubo T, Maruyama K, Maesaki S, Miyazaki Y, Tanaka E, Takizawa T, Moribe K, Tomono K and Tashiro T (1998), 'Long-circulating immunoliposomal amphotericin B against invasive pulmonary aspergillosis in mice', *Antimicrob Agents Chemother*, 42, 40–44.

Ouadahi S, Paternostre M, André C, Genin I, Xuan Thao T, Devissaguet J Ph and Barratt G (1998), 'Liposomal formulations for oral immunotherapy: *in-vitro* stability in synthetic intestinal media and *in-vivo* efficacy in the mouse', *J Drug Target*, 5, 365–378.

Oussoren C and Storm G (2001), 'Liposomes to target the lymphatics by subcutaneous administration', *Adv Drug Deliv Rev*, 50, 143–156.

Oyen W J, Boerman O C, Storm G, van Bloois L, Koenders E B, Crommelin D J, van der Meer J W and Corstens F H (1996), 'Labeled Stealth™ liposomes in experimental infection: an alternative to leukocyte scintigraphy?', *Nucl Med Commun*, 17, 742–748.

Pardridge W M (1999), 'Vector-mediated drug delivery to the brain', *Adv Drug Deliv Rev*, 36, 299–321.

Park J W, Kirpotin D B, Hong K, Shalaby R, Shao Y, Nielsen U B, Marks J D, Papahadjopoulos D and Benz C C (2001), 'Tumor targeting using anti-her2 immunoliposomes', *J Control Rel*, 74, 95–113.

Parker R J, Priester E R and Sieber S M (1982), 'Comparison of lymphatic uptake, metabolism, excretion and biodistribution of free and liposome-entrapped [^{14}C] cytosine β-D-arabinofuranoside following intraperitoneal administration to rats', *Drug Met Dis*, 10, 40–46.

Pattillo C B, Sari-Sarraf F, Nallamothu R, Moore B M, Wood G C and Kiani M F (2005), 'Targeting of the antivascular drug combretastatin to irradiated tumors results in tumor growth delay', *Pharm Res*, 22, 1117–1120.

Pecheur E I, Hoekstra D, Sainte-Marie J, Maurin L, Bienvenüe A and Philippot J R (1997), 'Membrane anchorage brings about fusogenic properties in a short synthetic peptide', *Biochemistry*, 36, 3773–3781.

Phillips N C, Gagné L, Tsoukas C and Dahman J (1994), 'Immunoliposome targeting to murine CD4$^+$ leucocytes is dependent on immune status', *J Immunol*, 152, 3168–3174.

Phillips W T, Klipper R and Goins B (2000), 'Novel method of greatly enhanced delivery of liposomes to lymph nodes', *J Pharm Exp Ther*, 295, 309–313.

Phillips W T, Andrews T, Liu H-L, Klipper R, Landry A J Jr, Blumhardt R and Goins B (2001), 'Evaluation of [^{99m}Tc] liposomes as lymphoscintigraphic agents: comparison with [^{99m}Tc] sulfur colloid and [^{99m}Tc] human serum albumin', *Nucl Med Biol* 28, 435–444.

Pinto-Alphandary H, Andremont A and Couvreur P (2000), 'Targeted delivery of antibiotics using liposomes and nanoparticles: research and applications', *Int J Antimicrob Agents* 13, 155–168.

Ponnappa B C, Israel Y, Aini M, Zhou F, Russ R, Cao Q, Hu Y and Rubin R (2005), 'Inhibition of tumor necrosis factor alpha secretion and prevention of liver injury in ethanol-fed rats by antisense oligonucleotides', *Biochem Pharmacol*, 69, 569–577.

Poste G, Kirsh R and Koestler T (1984), 'The challenge of liposome targeting *in vivo*', in Gregoriadis G, *Liposome Technology*, First Edition, Vol. III, Boca Raton, CRC Press, 1–28.

Romero E L, Morilla M J, Regts J, Koning G A and Scherphof G L (1999), 'On the mechanism of hepatic transendothelial passage of large liposomes', *FEBS Lett*, 448, 193–196.

Rubio Demirovic A, Marty C, Console S, Zeisberger S M, Ruch C, Jaussi R, Schwendener R A and Ballmer-Hofer K (2005), 'Targeting human cancer cells with VEGF receptor-2-directed liposomes', *Oncol Rep*, 13 (2), 319–324.

Sauer I, Dunay I R, Weisgraber K, Bienert M and Dathe M (2005), 'An apolipoprotein E-derived peptide mediates uptake of sterically stabilized liposomes into brain capillary endothelial cells', *Biochemistry*, 44, 2021–2029.

Scherphof G, Damen J and Hoekstra D (1981), 'Interactions of liposomes with plasma proteins and components of the immune system', in Knight, G, *Liposomes: From Physical Structure to Therapeutic Applications*, Amsterdam, Elsevier, 299–322.

Scherphof G, Velinova M, Kamps J, Donga J, van der Want H, Kuipers F, Havekes L and Daemen T (1997), 'Modulation of pharmacokinetic behavior of liposomes', *Adv Drug Deliv Rev*, 24, 179–191.

Schiffelers R M, Storm G, ten Kate M T and Bakker-Woudenberg I A (2001), 'Therapeutic efficacy of liposome-encapsulated gentamicin in rat *Klebsiella pneumoniae* pneumonia in relation to impaired host defense and low bacterial susceptibility to gentamicin', *Antimicrob Agents Chemother*, 45 (2), 464–470.

Schnyder A, Krahenbuhl S, Drewe J and Huwyler J (2005), 'Targeting of daunomycin using biotinylated immunoliposomes: pharmacokinetics, tissue distribution and *in vitro* pharmacological effects', *J Drug Target*, 13 (5), 325–335.

Senior J H (1987), 'Fate and behaviour of liposomes *in vivo*: a review of controlling factors', *Crit Rev Drug Carr Syst*, 3, 123–193.

Shakweh M, Ponchel G and Fattal E (2004), 'Particle uptake by Peyer's patches: a pathway for drug and vaccine delivery', *Expert Opin Drug Deliv*, 1 (1), 141–163.

Shi G, Guo W, Stephenson S M and Lee R J (2002), 'Efficient intracellular drug and gene delivery using folate receptor-targeted pH-sensitive liposomes composed of cationic/anionic lipid combinations', *J Control Rel*, 80 (1–3), 309–319.

Shi N, Boado R J and Pardridge W M (2001), 'Receptor-mediated gene targeting to tissues *in vivo* following intravenous administration of pegylated immunoliposomes', *Pharm Res*, 18, 1091–1095.

Shimada K, Kamps J A, Regts J, Ikeda K, Shiozawa T, Hirota S and Scherphof G L (1997), 'Biodistribution of liposomes containing synthetic galactose-terminated diacylglyceryl-poly(ethyleneglycol)', *Biochim Biophys Acta*, 1326, 329–341.

Soni V, Kohli D V and Jain S K (2005), 'Transferrin couple liposomes as drug delivery carriers for brain targeting of 5-fluorouracil', *J Drug Target*, 13 (4), 245–250.

Szebeni J (2001), 'Complement activation-related pseudoallergy caused by liposomes, micellar carriers of intravenous drugs, and radiocontrast agents', *Crit Rev Ther Drug Carr Syst*, 18, 567–606.

Theodoulou M and Hudis C (2004), 'Cardiac profiles of liposomal antracyclines – greater cardiac safety versus conventional doxorubicin?', *Cancer*, 100 (10) 2052–2063.

Thompson B, Mignet N, Hofland H, Lalons D, Seguin J, Nicolazzi C, de la Figuera N, Kuen R L, Meng X Y, Scherman D and Bessodes M (2005), 'Neutral postgrafted colloidal particles for gene delivery', *Bioconjugate Chem*, 16, 608–614.

Tuffin G, Waelti E, Huwyler J, Hammer C and Marti H P (2005), 'Immunoliposome targeting to mesangial cells: a promising strategy for specific drug delivery to kidney', *J Am Soc Nephrol*, 16, 3295–3305.

Uchegbu I F and Florence A T (1995), 'Non-ionic surfactant vesicles (niosomes): physical and pharmaceutical chemistry', *Adv Colloid Interf Sci*, 58, 1–55.

Visser C C, Stevanovic S, Voorwinden L H, van Bloois L, Gaillard P J, Danhof M, Crommelin D J A and de Boer A G (2005), 'Targeting liposomes with protein drugs to the blood-brain barrier *in vitro*', *Eur J Pharm Sci*, 25, 299–305.

Voinea M, Manduteanu I, Dragomir E, Capraru M and Simionescu M (2005), 'Immunoliposomes directed toward VCAM-1 interact specifically with activated endothelial cells – a potential tool for specific drug delivery', *Pharm Res*, 22, 1906–1917.

Vyas, S P, Katare Y K, Mishra V and Sihorkar V (2000), 'Ligand directed macrophage targeting of amphotericin B loaded liposomes', *Int J Pharm*, 210, 1–14.

Vyas S P, Quraishi S, Gupta S and Jaganathan K S (2005), 'Aerosolized liposome-based delivery of amphotericin B to alveolar macrophages', *Int J Pharm*, 296, 12–25.

Wang S and Low P S (1998), 'Folate-mediated targeting of antineoplastic drugs, imaging agents and nucleic acids to cancer cells', *J Control Rel*, 53, 39–48.

Wang S, Deng Y, Xu H, Wu H, Qiu Y and Chen D (2006), 'Synthesis of a novel galactosylated lipid and its application to the hepatocyte-selective targeting of liposomal doxorubicin', *Eur J Pharm Biopharm*, 62, 32–38.

Wasan K M and Lopez-Berestein G (1997), 'Diversity of lipid-based polyene formulations and their behavior in biological systems', *Eur J Clin Microb Infect Dis*, 16, 81–92.

Weiner N, Williams N, Birch G, Ramachandran C, Shipman C Jr and Flynn G (1989), 'Topical delivery of liposomally encapsulated interferon evaluated in a cutaneous herpes guinea-pig model', *Antimicrob Agents Chemother*, 33, 1217–1221.

Woodle M C (1998), 'Controlling liposome blood clearance by surface-grafted polymers', *Adv Drug Deliv Rev*, 32, 139–152.

Woodle MC, Storm G, Newman M S, Jekot J J, Collins L R, Martin F J and Szoka F C Jr (1992), 'Prolonged systemic delivery of peptide drugs by long-circulating liposomes: illustration with vasopressin in the Brattleboro rat', *Pharm Res*, 9, 260–265.

Woodley J F (1986), 'Liposomes for oral administration of drugs', *Crit Rev Drug Carr Syst*, 2, 1–18.

Xiong X-B, Huang Y, Lu W-L, Zhang X, Zhang H, Nagai T and Zhang Q (2005), 'Intracellular delivery of doxorubicin with RGD-modified sterically stabilized liposomes for an improved antitumor efficacy: *in vitro* and *in vivo*', *J Pharm Sci*, 94, 1782–1793.

Xu L, Pirollo K F and Chang E H (2001), 'Tumor-targeted p53-gene therapy enhances the efficacy of conventional chemo/radiotherapy', *J Control Rel*, 74, 115–128.

Yamamoto M, Ichinose K, Ishii N, Khoji T, Akiyoshi K, Moriguchi N, Sunamoto J and

Kanematsu T (2000) 'Utility of liposomes coated with polysaccharide bearing 1-amino-lactose as targeting chemotherapy for AH66 hepatoma cells', *Oncol Rep*, 7, 107–111.

Yardley V and Croft S L (1997), 'Activity of liposomal amphotericin B against experimental cutaneous leishmaniasis', *Antimicrob Agents Chemother*, 41, 752–756.

Yuh E L, Shulman S G, Mehta S A, Xie J, Chen L, Frenkel V, Bednarski M D and Li K C (2005), 'Delivery of systemic chemotherapeutic agent to tumors by using focused ultrasound: study in a murine model', *Radiology*, 234 (2), 431–437.

Zalipsky S, Gittelman J, Mullah N, Qazen M M and Harding J (1998), 'Biologically active ligand-bearing polymer-grafted liposomes', in Gregoriadis G. and McCormack B, *Targeting of Drugs 6: Strategies for Stealth Therapeutic Systems*, New York, Plenum Press, 131–138.

4

The artificial cell design: nanoparticles

W. YU and D. SU, Celsion Corporation, USA

4.1 Introduction

Nanoparticles have been studied as carriers to deliver different types of drug for many years. A variety of techniques have been developed. Recently, great efforts have been made to develop nanoparticulate drug carriers such as nanoparticles, nanocapsules, micelles, liposomes and conjugates (Torchilin and Trubetskoy, 1995; Jones and Leroux, 1999; Panyam and Labhasetwar, 2003). Nanoparticles have unique physical and chemical properties. Advances in nanotechnology enable drugs to be delivered in ways that preserve their efficacy and to precise therapeutic targets. This creates opportunities for drug delivery. A variety of nanostructures are being investigated as functional drug carriers for a wide range of therapies, most notably cardiovascular medicine, autoimmune diseases, blood substitutes, diagnostic applications and cancer therapy. One of the advantages of nanoparticles is that they can reach certain organs, cells or sites in the body. It has been shown that nanoparticles enhanced the activity of actinomycin D in an experimental subcutaneous sarcoma (Brasseur *et al.*, 1980). Nanocapsules or nanoparticles also have the capacity to increase the ocular therapeutic effect of associated drugs (Zimmer *et al.*, 1994; Losa *et al.*, 1991). Treatment with muramyldipeptide or its lipophilic derivative included in nanocapsules stimulates the macrophages' cytotoxic effect towards tumor cells (Barratt *et al.*, 1994). The capabilities that evolving technologies possess will lead to a dramatic increase in the number of therapies delivered via nanoparticles. Although there were obstacles of nanoparticulate systems for drug delivery, such as extensive uptake of the particles by the reticuloendothelial systems (RES) (Alakhov *et al.*, 1999; Jones and Leroux, 1999; Kataoka *et al.*, 2000), a solution for avoiding the RES uptake has been developed already. The surface of the nanoparticles can be stabilized by hydrophilic poly(ethylene glycol) (PEG), and form so-called 'Stealth^{TM} particles' (Barratt, 2000). These stabilized nanoparticulate drug carriers provide attractive characteristics in that they can avoid uptake of the drug by the RES *in vivo*, and hence can circulate in the blood for a long period of time.

Nanoparticle development involves multidisciplinary scientific approaches, derived from engineering, biology, physics and chemistry. Currently, many of these drug delivery approaches have been approved or are in clinical development. Doxil, a liposomal doxorubicin, is one of the first approved nanoparticulate drug delivery system. This drug delivery system can significantly reduce the doxorubicin toxicity.

This chapter focuses on the potential of the nanotechnology as a platform for drug delivery.

4.2 *In vivo* properties of nanoparticles

Nanocapsules make up a colloidal system. After intravenous injection, the particles without surface modification are rapidly captured by the RES especially by macrophages. We studied the interaction of the nanoparticles containing a fluorescent marker with macrophages. In our earlier study (Yu, 1990), we prepared nanoparticles containing perylene, a fluorescent marker. These nanoparticles were then incubated with macrophages. As a control, a free fluorescent marker was incubated with the macrophages. The study showed that nanoparticles containing the fluorescent marker were easily captured by macrophage. This property can be used to deliver the drug to the organ rich in macrophages such as the liver, spleen and lung.

More recently, a solution for avoiding the reticuloendothelial system uptake has been developed. As already stated, the surface of the nanoparticles can be stabilized by hydrophilic polymer such as PEG, and form so-called 'Stealth™ particles' (Barratt, 2000). In this way, the nanoparticles can be used to target organs other than the liver, spleen and lung. Chang *et al.* (2003) reported that nanoparticles can be used as hemoglobin carrier to prepare artificial blood. The nanoparticles prepared by this approach have a much longer circulatory half-life than the nanoparticles without a stabilized surface.

4.3 Design of polymeric nanoparticles

Nanoparticles can be made using varieties of techniques and different polymers. Natural and synthetic polymers including albumin, fibrinogen, alginate, chitosan, collagen, cyanoacrylate acid and polylactic acid have been used. However, polylactic acid, lactic-glycolic acid copolymers and other biodegradable synthetic polymers are the most frequently employed materials owing to their biocompatibility and biodegradability.

There are many reports on the nanoparticle preparation method. One of the earlier methods used in nanoparticle preparation was interfacial polymerization method. Al Kouri (1984) reported this method in 1980s. Briefly, an alcoholic solution containing the monomer, such as isobutyl-cyanoacrylate, drug, benzyl benzoate and phospholipids, was slowly injected through a syringe with a needle into an aqueous solution containing Poloxamer and under magnetic agitation. The nanocapsules were formed by interfacial polymerization immediately. The suspension was concentrated by rotary evaporation under vacuum. The nanoparticles formed were then filtered through a sintered glass filter.

We studied the nanoparticles' morphology under an electronic microscope. The polymer forms the membrane around the particles. The particles have a mean size of about 200 nm. Magnetic polybutylcyanoacrylate (PBCA) nanospheres encapsulated with acacinomycin A (MPNS-ACM) were prepared by interfacial polymerization (Gao *et al.*, 2004). Insulin-loaded poly (isobutylcanoacrylate) nanocapsules were made and found to reduce blood glucose level after oral administration to diabetic rats and dogs (Aboubakar *et al.*, 1999).

Yu *et al.* (1999) reported an interfacial polymer deposition method to prepare nanoparticles. The polymer, such as polylactic acid, was dissolved in an acetone and ethanol mixed solution containing phospholipid. The polymer solution was then slowly injected into water. The nanoparticles were immediately formed under magnetic stirring. The organic solvents were removed by rotary evaporation. The nanoparticles formed have a particle size around 200 nm.

Another method is by layer-by-layer self-assembly. This technique is based on alternate adsorption of oppositely charged materials, mostly linear polyelectrolytes, via electrostatic interactions. Multilayer ultra-thin films can be developed by 'molecular architecture' design with precise control of thickness and molecular composition. Layer-by-layer stepwise self-assembly of the polyelectrolytes poly(allyamine hydrochloride) and poly(styrenesulfonate) was used to create a macromolecular nanoshell around drug nanoparticles (approximately 150 nm in diameter) (Zahr *et al.*, 2005).

Velinova *et al.* (2004) reported layer-by-layer coating. In this method, the nanocapsules were prepared by hydrating dry lipid film (1.2 μmol) consisting of mixtures of lipids in molar ratios as indicated with the drug (e.g. 5 mM Cisplatin) in water at 37 °C for 30 min, followed by 10 freeze–thaw cycles using ethanol/dry-ice (−70 °C). The suspension was centrifuged at low speed in a tube to collect the nanocapsules in pellet.

Muller *et al.* (2000) used a spray-drying process to prepare nanoparticles. The drug is solubilized or dispersed in an organic solution of the polymer to be nebulized in a hot air flow. The solvent is almost instantly evaporated and dried nanocapsules are formed.

Elamanchili *et al.* (2004) prepared poly(D,L-lactic acid-co-glycolic acid) (PLGA) copolymer nanoparticles by using multiple emulsion process. This biodegradable nanoparticle enhanced delivery of antigens to murine marrow derived dentritic cells (DCs) *in vitro*.

Hamaguchi *et al.* (2005) reported a self-association process to prepare nanoparticles. Molecular self-association in liquids is a physical process that can dominate cohesion (interfacial tension) and miscibility. In water, self-association is a powerful organizational force leading to a three-dimensional hydrogen-bonded network (water structure). Paclitaxel (PTX), a micellar nanoparticle formulation, was developed to overcome neurotoxicity and to enhance the antitumor activity of PTX. Via the self-association process, PTX was incorporated into the inner core of the micelle system by the physical entrapment through hydrophobic interactions between the drug and the well-designed block copolymers for PTX.

Shenoy and Amiji (2005) prepared Tamoxifen nanocapsules in poly(ethylene oxide)-modified poly(varepsion-caprolactone) (PEO-PCL). The nanoparticles were prepared by a solvent displacement process that allowed *in situ* surface modification via physical adsorption of poly(ethylene oxide)–poly(propylene oxide)–poly(ethylene oxide) (PEO–PPO–PEO) triblock polymeric stabilizer. Xanthone-loaded nanospheres (nanomatrix systems) and nanocapsules (nanoreservoir systems), made of poly(DL-lactide-co-glycolide) (PLGA), were also prepared by the solvent displacement technique (Teixerira *et al.*, 2005).

Feng *et al.* (2004) developed a solvent extraction/evaporation process. The nanoparticles of PLGA were made by a modified solvent extraction/evaporation technique, in which natural emulsifiers, such as phospholipids, cholesterol and vitamin E were used to achieve high drug encapsulation efficiency, desired drug released kinetics and high cell intake.

Table 4.1 summarizes nanoparticle preparation methods.

Table 4.1 Nanoparticles preparation method

Polymer	Process method	Nature of particle	References
Polyelectrolyte	Layer-by-layer self-assembly	Nanocapsules	Ai and Gao (2004)
Poly-*n*-butylcyanoacrylate (PNBCA)	Interfacial polymerization	Nanocapsules	Miyazaki *et al.* (2003)
Zwitterionic phosphatidylcholine and negatively charged phosphatidylcholine	Layer-by-layer coating	Nanocapsules	Velinova *et al.* (2004)
Poly(ethylene oxide)-modified poly(varepsion-caprolactone) (PEO-PCL)	Solvent displacement	Nanoparticles	Shenoy and Amiji (2005)
Poly(lactic-co-glycolic acid) (PLGA)	Solvent extraction/ evaporation technique	Nanoparticles	Feng *et al.* (2004)
Pegylated poly(lactide) and poly(lactide-co-glycolide)	Emulsification-solvent evaporation	Nanocapsules	Avgoustakis (2004)
Poly-ϵ-caprolactone and Eudragit S90®	Spray-drying	Nanocapsules	Muller *et al.* (2000)

4.4 Medical applications of nanoparticles

The development of drug delivery systems has improved the therapeutic and toxicological properties of existing chemotherapies and facilitated the implementation of new ones. Since the drugs are used in technologically optimized drug delivery systems or conjugated with different polymers, it is useful to have a brief presentation of these techniques.

Paclitaxel is a widely used drug developed since the mid-1980s, with significant antitumor activity against ovarian, head and neck, bladder, breast and lung cancers (Rowinsky *et al.*, 1993). Paclitaxel promotes microtubule assembly and stabilizes microtubule dynamics, resulting in inhibition of cell proliferation and induction of apoptosis (Jordon *et al.*, 1993). Paclitaxel is a water-insoluble drug. Currently, it is formulated in a mixture of Cremophor EL (polyoxyethylated castor oil) and ethanol (1:1, w/w) (C/E formulation). Intravenous administration of this formulation resulted in peak paclitaxel concentration in plasma at the first sampling time, followed by biphasic declines over time (Wiernik *et al.*, 1987). Paclitaxel disposition is nonlinear, in part due to the presence of Cremophor (Gianni *et al.*, 1995; Kearns *et al.*, 1995). Hepatic metabolism and biliary excretion are the main elimination pathways for the C/E formulation (Monsarrat *et al.*, 1993; Walle *et al.*, 1995). The dose-limiting toxicity of paclitaxel is neutropenia, which is related to the duration of time that plasma paclitaxel concentrations are at or above a threshold value. The use of Cremophor is also associated with hypersensitivity reactions (Weiss *et al.*, 1990). Many approaches have been focused on developing safer formulation.

Teng *et al.* (2005) formulated paclitaxel in nanoparticles and evaluated its tissue distribution by comparing the blood/plasma and tissue pharmacokinetics of paclitaxel-loaded gelatin nanoparticles and the C/E paclitaxel formulation. The paclitaxel–gelatin nanoparticle can be used to target tumors in the kidneys, liver and small intestines.

In a new formulation, ABI-007, paclitaxel molecules are bound to and/or coated with human serum albumin, resulting in particles with a mean particle diameter of 120–150 nm (Damascelli *et al.*, 2001; Kolodgie *et al.*, 2002). ABI-007 shows a higher activity in breast cancer patients compared with the C/E formulation (O'Shaughnessy *et al.*, 2003). Its plasma pharmacokinetics in patients has been reported (Ibrahim *et al.*, 2002). A preliminary preclinical study reported suggested different disposition and/or tissue distribution as mechanisms of the different activity of the C/E and nanoparticles formulations (Desai *et al.*, 2000).

Cisplatin is another one of the most important anticancer drugs. Cisplatin nanocapsules represent a novel lipid formulation of the anticancer drug *cis*-diamminedichoroplatinum (II) (cisplatin), in which nanoprecipitate of cisplatin are coated by a phospholipid bilayer consisting of 1:1 mixture of zwitterionic phosphatidylcholine and negatively charged phosphatidylserine. Cisplatin nanocapsules exhibit increased *in vitro* cytotoxicity compared with the free drug (Burger *et al.*, 2002; Velinova *et al.*, 2004).

Cisplatin nanocapsules are prepared by the repeated freezing and thawing of an equimolar dispersion of phosphatidylserine and phosphatidylcholine in a concentrated aqueous solution of cisplatin (Chupin *et al.*, 2005). Briefly, cisplatin nanocapsules were made by hydrating dry lipid mixture films (1.2 μmol) with 1.25 mM cisplatin in water at 37 °C for 30 min, and followed by 10 freeze–thaw cycles using ethanol/dry-ice (−70 °C) and a water bath at 37 °C. The resulting suspension was centrifuged at low speed (4 min at 2100 rpm) in an Eppendorf centrifuge. The nanocapsules were collected in the pellet. The MLV and free cisplatin were removed. The nanocapsules were washed with 1 ml of water. For a typical formulation containing a lipid mixture of dioleoylphosphatidylcholine/dioleoylphosphatidylserine (DOPC/DOPS), 30% of cisplatin and 15% of phospholipids were recovered in the nanocapsule pellet (Velinova *et al.*, 2004).

Doxorubicin is another class of the potent drug used to treat various types of cancer. It is a cytotoxic anthracycline antibiotic isolated from cultures of *Streptomyces peucetius* var. *caesius*. Doxorubicin binds to nucleic acids, presumably by specific intercalation of the planar anthracycline nucleus with the DNA double helix. The toxic effects of doxorubicin limited its potential application. The nanoparticles have been developed for target delivery of the doxorubicin.

Yi *et al.* (2005) prepared a doxorubicin-loaded polymeric nanoparticle (Dox-PNP) by using a solvent evaporation method. Briefly, the Dox-PNP was prepared as follows: 10 mg (0.017 mmol) doxorubicin hydrochloride was dissolved in 5 ml of ethanol–water (9:1 v/v) in a round-bottomed flask. Then 190 mg (0.17 mmol) of the biodegradable polyester polylaurylmethacrylate (PLMA-COONa) was added thereto and completely dissolved with a rotary evaporator at an elevated temperature of 50 °C for 30 min to give a clear solution. Some 890 mg (0.21 mmol) of the amphiphilic block copolymer poly(lactic acid)tocopherol (mPEG-PLA-Toco) was added and dissolved in the solution. The solvent was evaporated at an elevated temperature under vacuum with a rotary evaporator. Three milliliters of an aqueous solution of lactose (20% by weight) was added, and the flask was rotated at 100 rpm to form the Dox-PNP colloidal solution in aqueous medium. The solution was filtered using 0.22 μm polyvinylidene fluoride (PVDF) membrane filter. The filtered solution was freeze-dried and stored in a refrigerator until use. The particle size of Dox-PNP was measured by the dynamic light scattering method. The doxorubicin content was determined by high-performance liquid chromatography (HPLC) using daunorubicin as the internal standard. The entrapment efficiency, size distribution and *in vitro* release profile at various pH conditions were characterized. *In vitro* cellular uptake was investigated by confocal microscopy, flow cytometry and MTT assay using drug-sensitive and drug-resistant cell lines.

Pharmacokinetics and biodistribution were evaluated in rats and tumor-bearing mice. Doxorubicin was efficiently loaded into the PNP (higher than 95%

of entrapment efficiency), and the diameter of Dox-PNP was in the range 20–25 nm with a narrow size distribution. The *in-vitro* study showed that Dox-PNP exhibited higher cellular uptake into both human breast cancer cell (MCF-7) and human uterine cancer cell (MES-SA) than free doxorubicin solution (free-Dox), especially into drug-resistant cells (MCF-7/ADR and MES-SA/Dx-5). In pharmacokinetics and tissue distribution study, the bioavailability of Dox-PNP calculated from the area under the blood concentration time curve (AUC) was 69.8 times higher than that of free-Dox in rats, and Dox-PNP exhibited 2 times higher bioavailability in tumor tissue of tumor-bearing mice.

Kaul and Amiji (2005) reported tumor-targeted gene delivery using PEG-modified gelatin nanoparticles. The *in vitro* and *in vivo* results of their study clearly showed that a long-circulating, biocompatible and biodegradable, DNA-encapsulating nanoparticle system would be highly desirable for systemic delivery of genetic constructs to solid tumors.

A novel drug-carrier system for photodynamic therapy was reported by Roy *et al.* (2003). The carrier system can provide stable aqueous dispersion of hydrophobic photo-sensitizer, and preserve the key step of photo-generation of singlet oxygen, necessary for photodynamic action.

Lboutounne *et al.* (2004) reported nanocapsules as a drug system for topical application. In their study, the transport of chlorhexidine-loaded poly(epsilon-caprolactone) nanocapsules through full-thickness and stripped hairless rat skin was investigated in a static-diffusion cell. The behavior of nanocapsules at the skin interface was investigated by contact angle and surface tension measurements.

4.5 Commercial development of nanoparticles

Abraxane is the first nanoparticle drug formulation to have been approved. It consists of albumin-bound paclitaxel nanoparticles, and is free of toxic solvents. This delivery system demonstrated a superior response rate with an almost doubling of the reconciled target lesion response when compared with the solvent-based Taxol. This nanoparticle formulation was approved in January 2005 for breast cancer in women.

Table 4.2 summarizes current commercial nanoparticle products.

4.6 Future trends

An ideal drug delivery system requires a non-protein adsorbing surface, a predictable drug release profile and an interaction with only the disease site. Nano-artificial cell approaches may be a solution for the future of drug delivery system. Reconstruction of biomembranes and the transport of drug, protein, enzyme and gene over a polymeric capsule may find a place in the design of artificial cells.

Table 4.2 Commercializing nanoparticles for medical applications (Salata, 2004)

Company	Major area of activity	Technology
Abraxis Bioscience, Inc.	Drug delivery	Commercialized the first product using nanotechnology (Abraxane)
Alnis Biosciences, Inc.	NanoGel treatment	Biodegradable polymeric nanoparticles for drug delivery
Biophan Technologies, Inc.	MRI shielding	Nanomagnetic/carbon composite materials to shield medical devices from radiofrequency fields
Capsulution Nanoscience AG	Pharmaceutical coatings to improve solubility of drugs	Layer-by-layer poly-electrolyte coatings, 8–50 nm
Effel Technologies	Drug delivery	Reducing size of the drug particles to 50–100 nm
Envirosystems, Inc.	Surface disinfectant	Nanoemulsions
Evident Technologies	Luminescent biomarkers	Semiconductor quantum dots with amine or carboxyl groups on the surface, emission from 350–2500 nm
NanoBio Corp.	Pharmaceutical	Antimicrobial nano-emulsions
Nanocarrier Co., Ltd	Drug delivery	Micellar nanoparticles for encapsulation of drugs, proteins and DNA
NanoPharm AG	Drug delivery	Polybutylcyanoacrylate nanoparticles are coated with drugs and then with surfactant, can go across the blood–brain barrier
NanoMed Pharmaceutical, Inc.	Drug delivery	Nanoparticles for drug delivery
Celsion Corp	Drug delivery	Heat-activated lipsome drug for anticancer

Surface modification of nanoparticles with hydrophilic non-immunogenic polymers has been successfully applied to obtain drug carriers capable of circulating for a long time. Long-lived drug-entrapped liposomes have become the first clinically approved drug delivery system. Initial clinical trials showed encouraging results in terms of reduced toxicity and drug targeting to the tumor site. Some of the problems encountered with liposomal systems may be overcome with the use of more stable biodegradable and polymeric nanoparticles with a hydrophilic surface.

The medical application of nanotechnology has led to the emergence of a new discipline known as nanomedicine. As part of nanomedicine, nanocapsules and nanoparticles may overcome current drug delivery limitations.

4.7 Sources of further information and advice

http://www.azonano.com

Gunter Schmid (2004), *Nanoparticles: From Theory to Applications*, Wiley-VCH, Weinheim.

4.8 References

Aboubakar M., Puisieux F., Couvreur P., Deyme M. and Vauthier C. (1999), 'Study of the mechanism of insulin encapsulation in poly(isobutylcanoacrylate) nanocapsules obtained by interficial polymerization', *J Biomed. Mater. Res.*, 47(4): 568–576.

Ai H. and Gao J. (2004), 'Size-controlled polyelectrolyte nanocapsules via layer-by-layer self-assembly', *J. Mater. Sci.*, 39: 1429–1432.

Alakhov V., Klinski E., Li S., Pietrzynski G., Venne A., Batrakova E., Bronitch T. and Kabanov A. (1999), 'Block-copolymer-based formulation of doxorubicin. From cell screen to clinical trials', *Colloids Surf. B Biointerfaces* 16: 113–134.

Al-Khouri N. (1984), 'Nanocapsules de polyalkylcyanoacrylate. Nouveaux vecteurs de medicaments. Etude galenique et pharmacocinetique'. These de Doctorat d'Etat, Universite Paris Sud.

Avgoustakis K. (2004), 'Pegylated poly(lactide) and poly(lactide-co-glycolide) nanoparticles: preparation, properties and possible applications', *Drug Delivery*, 1(4): 321–333.

Barratt G.M. (2000), 'Therapeutic applications of colloidal drug carriers', *Pharm. Sci. Technol. Today* 3: 163–171.

Barratt G., Puisieux F., Yu W.P., Foucher C.H. and Devissaguet J.P. (1994), 'Anti-metastatic activity of MDP-L-alanyl-choloesterol incorporated into various types of nanocapsules', *Int. J. Immunopharm.*, 16: 457–461.

Brasseur E.L., Couvreur P., Kanta B., Deckers-Passau L., Roland M., Deckers C. and Speiser P. (1980), 'Actinomycin D adsorbed on polymethylcyanacrylate nanoparticles: increase efficiency against an experimental tumor', *Eur. J. Cancer*, 16: 1441–1445.

Burger K.N.J., Staffhorst R.W.H.M., de Vijlder H.C., Velinova M.J., Bomans P.H., Frederik P.M. and de Kruijff B. (2002), 'Nanocapsules: lipid-coated aggregates of cisplatin with high cytotoxicity', *Nature Medicine*, 8: 81–84.

Chang T.M., Powanda D. and Yu W.P. (2003), 'Analysis of polyethylene-glycol-polylactide nano-dimension artificial red blood cells in maintaining systemic hemoglobin levels and prevention of methemoglobin formation', *Artif. Cells Blood Substit. Immobil. Biotechnol.* 31: 3231–3247.

Chupin V., de Kroon A.I. and de Kruijff B. (2005), 'Molecular architecture of nanocapsules, bilayer-enclosed solid particles of Cisplatin', *Cancer Therapy*, 3: 131–138.

Damascelli B., Cantu G., Mattavelli F., Tamplenizza P., Bidoli P., Leo E., Dosio F., Cerrotta A.M., Di Tolla G., Frigerio L.F., Garbagnati F., Lanocita R., Marchiano A., Patelli G., Spreafico C., Ticha V., Vespro V. and Zunino F. (2001), 'Intraarterial chemotherapy with polyoxyethylated castor oil freepaclitaxel, incorporated in albumin nanoparticles (ABI-007): phase II study of patients with squamous cell carcinoma of the head and neck and anal canal: preliminary evidence of clinical activity', *Cancer*, 92: 2592–2602.

Desai N.P., Louie L., Ron N., Magdassi S. and Soo-Shiong P. (2000), 'Protein-based nanoparticles for drug delivery of paclitaxel', *Trans. World Biomater. Congr.* 1: 199.

Elamanchili P., Diwan M., Cao M. and Samuel J. (2004), 'Characterization of poly(D,L-lactic-co-glycolic acid) based nanoparticulate system for enhanced delivery of antigens to dendritic cells', *Vaccine*, 22(19): 2406–12.

Feng S.S., Mu L., Win K.Y. and Huang G. (2004), 'Nanoparticles of biodegradable polymers for clinical administration of paclitaxel', *Curr. Med. Chem.*, 11(4): 413–424.

Gao H., Wang J.Y., Shen X.Z., Deng Y.H. and Zhang W. (2004), 'Preparation of magnetic polybutylcyanoacrylate nanospheres encapsulated with aclacinomycin A and its effect on gastric tumor', *World J. Gastroenterol.*, 10(14): 2010–2013.

Gianni L., Kearns C.M., Giani A., Capri G., Vigano L., Lacatelli A., Bonadonna G. and Egorin M.J. (1995), 'Nonlinear pharmacokinetics and metabolism of paclitaxel and its pharmacokinetic/pharmacodynamic relationships in humans', *J. Clin. Oncol.*, 13: 180–190.

Hamaguchi T., Matsumura Y., Suzuki M., Shimizu K., Goda R., Nakatomi I., Yokoyama M., Kataoka K. and Kakizoe T. (2005), 'NK105, a paclitaxel-incorporating micellar nanoparticles formulation, can extend *in vivo* antitumour activity and reduce the neurotoxicity of paclitaxel', *Br. J. Cancer*, 92(7): 1240–1246.

Ibrahim N.K., Desai N., Legha S., Soon-Shiong P., Theriault P.R.L., Rivera E., Esmaeli B., Ring S.E., Bedikian A., Hortobagyi G.N. and Ellerhorst A. (2002), 'Phase I and pharmacokinetic study of ABI-007, a Cremophor-free, protein-stabilized, nanoparticle formulation of paclitaxel', *Clin. Cancer Res.*, 8: 1038–1044.

Jones M.C. and Leroux J.C. (1999), 'Polymeric micelles – a new generation of colloidal drug carriers', *Eur. J. Pharm. Biopharm.*, 48: 101–111.

Jordan M.A., Toso R.J., Thrower D. and Wilson L. (1993), 'Mechanism of mitotic block and inhibition of cell proliferation by taxol at low concentrations', *Proc. Natl Acad. Sci. USA*, 90: 9552–9556.

Kataoka K., Matsumoto T., Yokoyama M., Okano T., Sakurai Y., Fukushima S., Okamoto K. and Kwon G. S. (2000), 'Doxorubicin-loaded poly(ethylene glycol)-poly(β-benzyl-L-aspartate) copolymer micelles: their pharmaceutical characteristics and biological significance', *J. Control. Rel.* 64: 143–153.

Kaul G. and Amiji M. (2005), 'Tumor-targeted gene delivery using ploy (ethylene glycol)-modified gelatin nanoparticles: *in vitro* and *in vivo* studies', *Pharm. Res.*, 22(6): 951–961.

Kearns C.M., Gianni L. and Egorin M.J. (1995), 'Paclitaxel pharmacokinetics and pharmacodynamics', *Semin. Oncol.*, 22: 16–23.

Kolodgie F.D., John M., Khurana C., Farb A., Wilson P.S., Acampado E., Desai N., Soon-Shiong P. and Virmani R. (2002), 'Sustained reduction of in-stent neointimal growth with the use of a novel systemic nanoparticle paclitaxel', *Circulation*, 106: 1195–1198.

Lboutounne H., Faivre V., Falson F. and Pirot F. (2004), 'Characterization of transport of chlorexidine-loaded nanocapsules through hairless and wistar rat skin', *Skin Pharmacol. Physiol.*, 17(4): 176–182.

Losa C., Calvo P., Castro E., Vila-Jato L. and Alonso M. (1991), 'Improvement of ocular penetration of amikacin sulphate by association to poly(butylcyanoacrylate) nanoparticles', *J. Pharm. Pharmacol.*, 43: 548–552.

Miyazaki S., Takahashi A. and Kubo W. (2003), 'Poly *n*-butylcyanoacrylate (PNBCA) nanocapsules as a carrier for NSAIDs: *in vitro* release and *in vivo* skin penetration', *J. Pharm. Pharmaceut. Sci.*, 6(2): 240–245.

Monsarrat B., Alvinerie P., Wright M., Dubois J., Gueritte-Voegelein F., Guenard D., Donehower R.C. and Rowinsky E.K. (1993), 'Hepatic metabolism and biliary excretion of Taxol in rats and humans', *J. Natl. Cancer Inst. Monogr.*, 15: 39–46.

Muller C.R, Bassani V.L., Pohlmann A.R., Michalowski C.B., Petrovick P.R. and Guterres S.S. (2000), 'Preparation and characterization of spray-dried polymeric nanocapsules', *Drug Develop. Ind. Pharm.*, 26(3): 343–347.

O'Shaughnessy J., Tjulandin S., Davidson N., Shaw H., Desai N., Hawkins M.J. and Gradisha W.J. (2003), 'ABI-007 (ABRAXANEi), a nanoparticle albumin-bound (nab) paclitaxel demonstrates superior efficacy vs taxol in MBC: a phase III trial', *Breast Cancer Res. Treat.*, 82(Suppl 1): 44.

Panyam J. and Labhasetwar V. (2003), 'Biodegradable nanoparticles for drug and gene delivery to cells and tissue', *Adv. Drug Deliv. Rev.* 55: 329–347.

Rowinsky E.K., Wright M., Monsarrat B., Lesser G.J. and Donehower R.C. (1993), 'Taxol: pharmacology, metabolism and clinical implications', *Cancer Surv.*, 17: 283–304.

Roy I., Ohulchanskyy T.Y., Pudavar H.E., Bergey E.J., Oseroff A.R., Morgan J., Dougherty T.J. and Prasad P.N. (2003), 'Ceramic-based nanoparticles entrapping water-insoluble photosensitizing anticancer drugs: a novel drug-carrier system for photodynamic therapy', *J. Am. Chem. Soc.*, 125(26): 7860–7865.

Salata O.V. (2004) 'Applications of nanoparticles in biology and medicine', *J. Nanotechnol.*, 2: 3.

Shenoy D.B. and Amiji M.M. (2005), 'Poly(ethylene oxide)-modified poly(varepsilon-caprolactone) nanoparticles for targeted delivery of tamoxifen in breast cancer', *Int. J. Pharm.*, 293(1–2): 261–270.

Teixerira M., Alonso M.J., Pinto M.M., Barbosa C.M. (2005), 'Development and characterization of PLGA nanospheres and nanocapsules containing xanthone and 3-methoxyxanthone', *Eur. J. Pharm. Biopharm.*, 59(3): 491–500.

Teng Y.K., Lu Z., Wientjes M.G. and Au J.L. (2005), 'Formulating paclitaxel in nanoparticles alters its disposition', *Pharm. Res.*, 22(6): 867–874.

Torchilin V.P. and Trubetskoy V.S. (1995), 'Which polymers can make nanoparticulate drug carriers long-circulating?', *Adv. Drug Deliv. Rev.* 16: 141–155.

Velinova M.J., Staffhoorst R.W.H.M., Mulder W.J.M., Dries A.S., Jansen B.A.J., de Kruijff B. and de Kroon A.I.P.M. (2004), 'Preparation and stability of lipid-coated nanocapsules of cisplatin: anionic phospholipid specificity', *Biochim. et Biophys. Acta*, 1663: 135–142.

Walle T., Walle U.K., Kumar G.N. and Bhalla K.N. (1995), 'Taxol metabolism and disposition in cancer patients', *Drug Metab. Dispos.*, 23: 506–512.

Weiss B., Donehower R.C., Wiernik P.H., Ohnuma T., Gralla R.J., Trump D.L., Baker J.R. Jr, Van Echo D.A., Von Hoff D.D. and Leyland Jones B. (1990), 'Hypersensitivity reactions from taxol', *J. Clin. Oncol.*, 8: 1263–1268

Wiernik P.H., Schwartz E.L., Strauman J.J., Dutcher J.P., Lipton R.B. and Paietta E. (1987), 'Phase I clinical and pharmacokinetic study of taxol', *Cancer Res.*, 47: 2486–2493.

Yi Y., Kim J.H., Kang H.-W., Oh H.S., Kim S.W. and Seo M.H. (2005), 'A polymeric nanoparticle consisting of mPEG-PLA-Toco and PLMA-COONa as a drug carrier: improvements in cellular uptake and biodistribution', *Pharm. Res.*, 22 (2): 200–208.

Yu W.P. (1990), 'Etude pharmacotechnique et biologique de vecteurs colloidaux d'un immunomodulateur lipophile: le muramyltripeptide-cholestérol (MTP-chol)'. Thèse doctorale, No. 160, Université Paris Sud.

Yu W.P., Wong J.P. and Chang T.M.S. (1999), 'Biodegradable polylactic acid nanocapsules containing ciprofloxacin: preparation and characterization', *Art. Cells Blood Subs. Immob. Biotech.*, 27(3): 263–278.

Zahr A.S., de Villiers M. and Pishko M.V. (2005), 'Encapsulation of drug nanoparticles in self-assembled macroshells', *Langmuir*, 21(1): 403–410.

Zimmer A., Mutshler E., Lambrecht G., Mayer D. and Kreuter J. (1994), 'Pharmacokinetic and pharmacodynamic aspects of an ophthalmic pilcarpine nanoparticle delivery system', *Pharm. Res.*, 11(10): 1435–1442.

Part II

Cell engineering

5

The cutting edge: apoptosis and therapeutic opportunity

C GRIFFIN, D GUEORGUIEVA, A McLACHLAN-BURGESS, M SOMAYAJULU-NITU and S PANDEY,
University of Windsor, Canada

5.1 Apoptosis

Programmed cell death is an important physiological process that plays a significant role during the development and maintenance of the living organisms. Two distinct categories of cell death processes, namely apoptosis and necrosis, have been defined (Kerr *et al.*, 1972). Necrosis represents pathological cell death characterized by the swelling of the cells and organelles and lysis of the plasma membrane. Apoptosis, on the other hand, is characterized by plasma membrane blebbing, cell shrinkage, condensation of chromatin and DNA fragmentation into high molecular weight and/or oligonucleosomal fragments. Thus, cells undergoing apoptosis systematically fragment their DNA, shrink from within and get disposed of quietly by phagocytosis without causing any inflammation or damage to their surroundings. While apoptosis is a normal cellular event during development, it can also be induced by pathological conditions and injuries. Apoptosis is involved in a wide range of pathologies including viral diseases, neurodegeneration, cancer and aging.

Apoptosis is a genetically regulated cellular process that can be switched on or off depending on the ratio of pro-apoptotic to anti-apoptotic factors. In the past two decades significant progress has been made to elucidate the biochemical mechanism of apoptosis. Various proteins and other stimuli actively participating in the initiation and execution of apoptosis have been identified. Two major pathways, namely the extrinsic pathway involving a cell death receptor-induced mechanism and the intrinsic pathway involving mitochondrial permeabilization-apoptosome mediated pathway, have been proposed.

In the receptor-mediated apoptosis pathway (Fig. 5.1), interaction of ligands such as Fas, TNF and TRAIL with their specific cell surface receptor is followed by oligomerization of these receptors and their association with and recruitment of various proteins on the cytosolic side of the plasma membrane. This complex is referred to as death inducing signalling complex (DISC) (Ashkenzai and Dixit, 1998). One of the main consequences of DISC formation is the activation

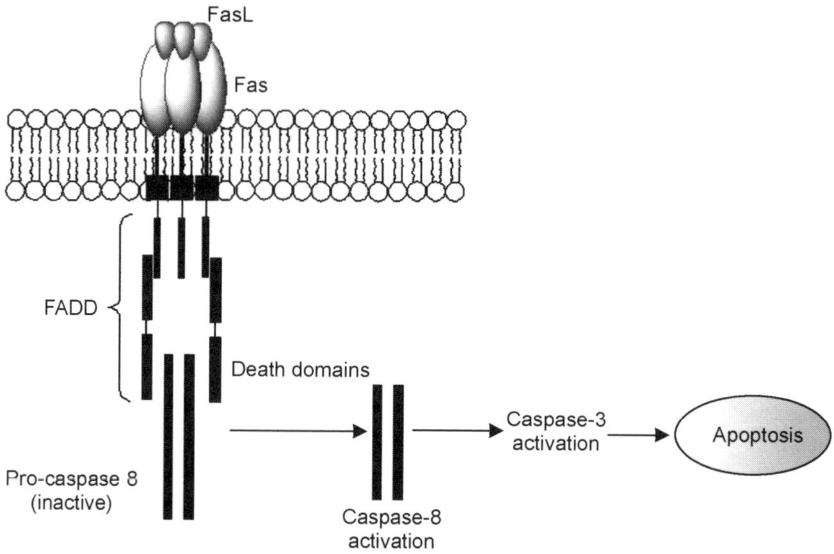

5.1 Receptor-mediated induction of apoptosis resulting from FasL binding to Fas receptor and subsequent activation of initiator caspase-8 and effector caspase-3. An alternative result of caspase-8 activation is the activation of proaptotic protein (tBid), which also ends in apoptosis.

of the cysteine-specific aspartase caspase-8, which is subsequently recruited to this complex. Active caspase-8 further activates executioner caspases, such as caspase-3, by proteolysis. Once caspase-3 is activated it causes proteolysis of various key proteins including inhibitor of caspase-activated DNase (ICAD). This results in the activation of caspase-activated DNase (CAD) followed by its translocation to the nucleus where it proceeds to fragment the DNA and ultimately results in apoptosis.

The Fas-receptor mediated extrinsic pathway is also linked to mitochondria-apoptosome mediated intrinsic pathway since the activated caspase-8 cleaves pro-apoptotic Bid protein, activating Bax and mediating its association with mitochondria, resulting in cytochrome *c* release (Korsmeyer *et al.*, 2000).

The intrinsic apoptotic pathway is centered around the permeabilization of the mitochondrial outer membrane (MOM) and the release of pro-apoptotic proteins from the inner mitochondrial space (IMS). Currently there are several proposed mechanisms for MOM opening including the Bax pore (Fig. 5.2(a)) and permeability transition pore (PTP) formation (Fig. 5.2(b)). The former hypothesis suggests that pro-apoptotic proteins such as Bax and Bak form a pore in the MOM either by homo or heterodimerization or activation by other BH3 proteins such as Bid (Kuwana, 2002). However, since these finding are based on experimentation on cell-free systems, the hypothesis of the PTP formation is useful in understanding the events that occur within the cell which may lead to mitochondrial destabilization.

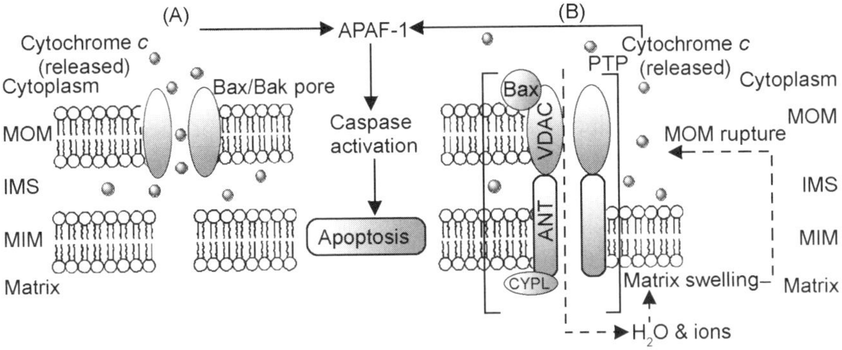

5.2 Events leading to mitochondrial membrane potential collapse in the extrinsic apoptotic pathway.

There are several components to the PTP, which span from the inner to the outer membrane of the mitochondria. The voltage-dependent anion channel (VDAC) protein is found to transverse the MOM while two other components, cyclophilin D (cyp D) a soluble matrix protein and the adenine nucleotide translocase (ANT) protein are found on the mitochondrial inner membrane (MIM). Pro-apoptotic Bcl-2 family proteins are understood to be closely associated with the PTP and play a role in its activation, leading to the collapse of the mitochondria membrane potential (Murphy *et al.*, 2000).

Despite the debate over the mechanism behind mitochondrial membrane potential collapse this event in evidently results in the release of cytochrome *c* and/or apoptosis-inducing factor (AIF) from the inner membrane space of the mitochondria (Fig. 5.3). Subsequently, cytochrome *c* binds to apoptosis activating factor (APAF) and caspase-9 (together this complex is referred as the apoptosome), leading to the activation of caspase 3 that in turn activates itself and/or other executioner caspases such as caspase-6 and 7.

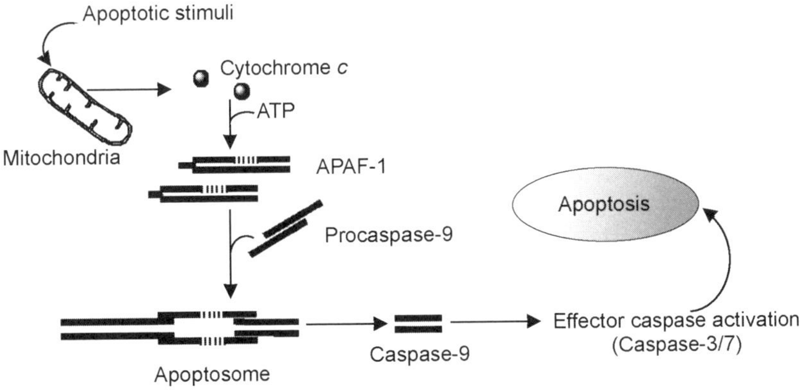

5.3 Mitochondrial membrane permealization leads to release of pro-apoptotic proteins.

Cell death caused by radiation, chemotherapy, oxidative stress and DNA-damaging agents induce apoptosis through the mitochondrial permeabilization mediated intrinsic pathway. Anti-apoptotic proteins of the Bcl-2 family, such as Bcl-2 and Bcl-X_L, stabilize mitochondria and prevent its permeabilization by neutralizing or titrating-out pro-apoptotic proteins like Bax, also a member of the Bcl-2 family.

Identification of the key players of apoptosis, including several death receptors, initiator and executioner caspases, pro- and anti-apoptotic Bcl-2 family proteins, cytochrome c and certain nucleases, provides a great opportunity to design modulators for diseases that involve apoptosis (Hu and Kavanagh, 2003). Currently there is an unprecedented amount of interest in this relatively young field of research. Development of modulators against the aforementioned targets in order to activate apoptosis selectively in cancer cells and inhibit apoptosis in post-mitotic cells to prevent progression of neurodegenerative diseases and immune cell depletion in HIV patients (Reed, 2002), has become a major focus in research today. A brief description of some of these diseases in light of involvement of apoptosis and the therapeutic opportunities is described in the sections below.

5.2 Apoptosis and neurodegeneration

Throughout the development of the central nervous system, induction of cell death is not uncommon as many excess cells and those developing improperly are eliminated by apoptosis (Lindsten *et al.*, 2003). However, serious problems arise when apoptosis occurs after development has been completed, resulting in the loss of vital post-mitotic cells such as neurons and cardiac muscle cells as is the case for multiple neurodegenerative as well as cardiovascular disease, including stroke and cardiac arrest.

5.3 Cell death during stroke

Though the biochemical makeup of brain and heart muscle cells is quite different, the effects of ischemia as it occurs in both organs during stroke and cardiac arrest respectively are comparable. In cases of stroke, neurons directly devoid of blood flow, found in the core region, are killed within minutes of this effect, while neurons in the surrounding region termed the penumbra retain their potential for survival for hours or even days. Specifically, this latter region can be saved by reperfusion when blood flow is returned, however, this is certainly not without consequences (Cuzzocrea *et al.*, 2001). As blood supply is returned, this reperfusion stage is linked to a massive increase in reactive oxygen species (ROS) as a result of the altered function of the mitochondria's electron transport chain which can directly and indirectly trigger apoptotic cell death (Adhihetty and Hood, 2003). The ROS-mediated pathway is involved in apoptosis in several ways, including the alteration of the mitochondrial permeability transition pores (mPTP), allowing for the release of pro-apoptotic proteins, promoting release of cytochrome *c* from the inner mitochondrial membrane, as well as the activation of transcription factors of various pro- and anti-apoptotic factors (Nomura *et al.*, 2000).

Numerous other pathways have also been identified as leading to apoptosis in post-mitotic cells such as receptor mediated, as well as caspase dependent and independent; however in cases of neurodegenerative diseases such as stroke, an increase in ROS after injury is the primary trigger of neuronal death because of the brain's strong susceptibility to damage from oxidative stress. Specifically, there are several reasons for the detrimental effects of ROS on the brain: primarily, the brain makes up a small percentage of the overall body weight (2%) but consumes nearly 20% of oxygen entering the body. Secondly, the brain contains relatively low levels of antioxidant species while levels of lipids sensitive to oxidation remain high. In addition, some regions are found to be rich in iron, which can catalyze formation of free radicals (Cherubini *et al.*, 2005).

5.4 Stroke management

5.4.1 Stroke management: current treatment strategies

Presently, treatment of stroke involves several drug therapies including anticoagulants such as warfarin and thrombolytic recombinant tissue plasminogen activator (r-TPA) (Margaill *et al.*, 2005). Anticoagulant drugs therapies are used to treat stroke or cardiac arrest patients; however this is more of a preventive measure. For example, acetylsalicylic acid (ASA) and warfarin work by decreasing the ability of clot-forming components (platelet cells in the blood) from sticking together, thus limiting the possibility of clot formation.

Thrombolytic drugs are used in cases of ischemic stroke; that is when a clot is blocking the blood supply to a particular part of the brain. In these cases, this therapy must be applied within 3 h of stroke onset, which is severely limiting since many patients are unaware of their symptoms and take hours or days to seek treatment. However, when taken within the correct time period this therapy can successfully allow reperfusion of the oxygen-deprived region. In some cases, if the affected part of the brain is small enough, rTPA can lead to complete reversal of symptoms. Unfortunately, it is not uncommon for patients to experience weakness or paralysis in various body parts as well as difficulty in speaking owing to irreversible damage to the core and penumbra regions of the brain, discussed above.

5.4.2 Stroke management: recent progress

Although reperfusion can restore proper blood flow to affected regions in the brain, as previously discussed, a consequence to this treatment is the generation of free radicals as cells take up high amounts of oxygen at this stage. As a result, current research involving quenching of free radicals has become an important area of study in order to prevent the damaging side effects of strokes. One pharmacological area of study involves the prevention of free radical production through cyclo-oxygenase 2 (COX-2) inhibitors such as nimesulide and NS-398. Candelario-Jalil *et al.* (2004) have reported on the neuroprotective potential of these drugs in ischemic conditions of rat brains, showing both histological and functional recovery even after administering the drugs 24 h after stroke onset. These results are quite exciting since there are no current treatments which are able to treat the later phase of this condition. It is suggested that COX-2 inhibition may delay the progression of the ischemic legion, limiting the size of the penumbra region although the exact mechanism of COX-2 activation in cases of ischemia in humans is still unclear (Candelario-Jalil *et al.*, 2004).

Other antioxidant therapies involve searching out and quenching free radicals released in ischemic areas. One such compound is the glutathione precursor *N*-acetylcysteine (NAC) which, as reported by Khan *et al.* (2004), is able to limit infarct size in rat stroke models and improve neurologic scores while limiting apoptotic cell death and inflammation. Such optimistic results were observed even 6 h after the onset of stroke and because of the previously demonstrated safety and efficacy of NAC in other neurodegenerative diseases, this treatment offers strong potential for preventing the effects of ischemic stroke.

Research in our laboratory focuses on preventing apoptosis in postmitotic cells such as neurons by limiting ROS production, which results from mitochondrial permealization by the Bax protein. Because of the central role which Bax plays in inducing apoptosis we are currently attempting to block the pro-apoptotic action of this protein through the use of specific anti-Bax llama single domain antibodies (sdAb).

Unlike conventional human antibodies which contain several components (variable light (VL) and heavy (VH) chains and constant light and heavy chains) llama and camelid antibodies do not contain any of the light chains, significantly reducing their overall size. Typically, the smallest functional units of human antibodies (the antigen binding site) are the fragments Fvs (VH and VL) or scFvs (VH and VL connected by a peptide linker) (Fig. 5.4(a)). In contrast, the antigen-binding sites of the llama (and camelid) species comprise only the VH region (Fig. 5.4(b)). As a result these smaller fragments, termed single domain antibodies (sdAbs), which contain the specificity of the parent antibody, are significantly smaller and offer greater stability when working intra-cellularly (Gueorguieva *et al.*, 2006).

A sdAb to any target antigen can be isolated from a naive llama phage display library, a process known as panning. Using Bax as an antigen of interest, in our panning assay results we were able to identify of six unique anti-Bax sdAbs, with which we have been able to successfully quench the activity of this pro-apoptotic protein, both *in vitro* and *in vivo*. Specifically, we have been able to

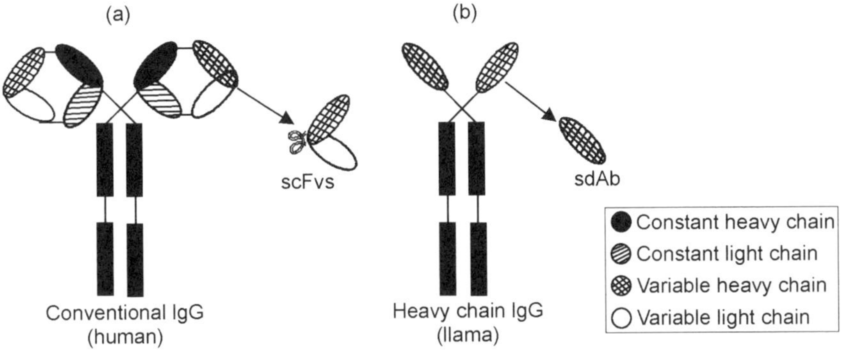

5.4 Comparing conventional and single domain antibodies (SdAb).

show a significant decrease in ROS production when mitochondria were incubated with various anti-Bax sdAb and Bax protein compared to mitochondria in presence of Bax alone.

Furthermore, by transfecting neuroblastoma (SHSY-5Y) cells with each of the six anti-Bax sdAb and developing six unique stable cell lines, our *in vivo* studies have shown that these antibodies are non-toxic to the cell and can protect cells from high levels of oxidative stress (200 μM of H_2O_2 for 1–3 h), resulting in cell viability nearly equivalent to control of non-transfected, non-treated cells. In addition, treated cells show no limitation in division and proliferation.

Although gene therapy is one possible application for these antibodies in prevention of oxidative stress, we hope to use a chemical combinatorial library to select a low molecular weight compound which can mimic our most efficient anti-Bax sdAb. This compound will be put under further scrutiny for its ability to cross the blood–brain barrier and prevent spread of oxidative stress which occurs after reperfusion in ischemic stroke conditions (Gueorguieva *et al.*, 2006).

5.5 Neuronal cell death and Parkinson's disease

The significance of apoptosis in maintaining normal functioning of the body is clearly demonstrated when it malfunctions. Excessive death of cells by apoptosis has been linked to neurodegenerative diseases such as Parkinson's disease (PD) and Alzheimer's disease (AD).

PD is caused by the depletion of dopaminergic neurons of the substantia nigra. Classical symptoms of PD such as bradykinesia, resting tremors, impaired postural reflexes and rigidity are generally exhibited only after a loss of about 75–80% of dopamine in the striatum (Fearnley *et al.*, 1991; Marsden, 1994; Steece-Collier *et al.*, 2002). The exact cause of the disease is unknown but most cases are sporadic in contrast to familial cases, which constitute only about 1–2%.

Several mutations in genes such as α-synuclein and parkin have been linked to PD (Shimura *et al.*, 2000; Lotharius *et al.*, 2002; Fernandez-Espejo, 2004), while exposure to certain chemicals and pesticides has been shown to increase the risk of PD (Semchuk *et al.*, 1991; Tipton *et al.*, 1993; Seidler *et al.*, 1996; Koller *et al.*, 1990). In addition, there is strong evidence that oxidative stress from a variety of reactive oxygen species plays an important role in the demise of dopaminergic neurons (Jenner, 2003).

Generally, the neurons in the substantia nigra are exposed to higher oxidative stress because dopamine metabolism gives rise to dopamine-quinone species as well as superoxide anion radical and hydrogen peroxide. It has also been shown that levels of iron are elevated in the nigral region of PD patients (Jenner, 1994; Koller and Cersosimo, 2004) and this is of particular concern since iron can catalyze the Fenton reaction where hydrogen peroxide is converted into hydroxyl radicals. Also, a decline in the amount of GSH, a tripeptide that neutralizes excess ROS in the glia of nigral tissue of PD patients, has been observed (Sian *et al.*, 1994). Furthermore, an increase in nitric oxide synthase (NOS) activity through glial cell activation has also been observed in PD cases, which leads to elevated levels of nitric oxide and peroxynitrite (Hunot *et al.*, 1996). Increased levels of lipid peroxidation, protein carbonyls and oxidized nucleic acids – markers of oxidative stress – have also been reported in the substantia nigra of the PD patients.

The cause of sporadic PD which accounts for more than 95% of all cases is not yet fully understood. However, recent findings show that 1-methyl-4-phenyl-1,2,3,6-tetrahydropyridine (MPTP), a neurotoxin, selectively damages the nigrostriatal dopaminergic neurons causing a Parkinsonian syndrome in humans and other mammals (Langston *et al.*, 1983; Date *et al.*, 1990; Kurlan *et al.*, 1991). As MPTP is a well-known inhibitor of complex-I of electron transport chain, mitochondrial dysfunction might have an important role in PD. Several researchers (Parker *et al.*, 1989; Krige *et al.*, 1992; Benecke *et al.*, 1993) have found a selective decrease in the complex-1 activity in the subtantia nigra region of PD patients. These results suggest that one of the causes of sporadic PD could be exposure to environmental toxins that affect mitochondria or induce oxidative stress. It was first reported that humans exposed to the herbicide paraquat (structurally similar to MPTP), had an increased risk of PD (Liou *et al.*, 1997). *In-vitro* results from our laboratory definitely show that paraquat also causes mitochondrial dysfunction in differentiated human neuroblastoma (SHSY-5Y) cells (McCarthy *et al.*, 2004). Other studies have demonstrated that intra-peritoneal injections of paraquat or maneb cause selective degeneration of nigral dopaminergic neurons in mice (McCormack *et al.*, 2002; Thiruchelvam *et al.*, 2002; Chicchetti *et al.*, 2005) and rats, accompanied by severely reduced overall general motor activity (Thiruchelvam *et al.*, 2002; Chicchetti *et al.*, 2005). Mice suffer greater deficits when exposed to these herbicides at 18 months, rather than at 6 weeks or at 5 months of age.

There is mounting evidence that the dopaminergic neurons in the substantia nigra are under immense oxidative stress; leading to the multiple misfolding of proteins such as parkin and α-synuclein, which form fibrils constituting the lewy body, a pathological hallmark of PD (Schapira *et al.*, 1990; Bucciantini *et al.*, 2002).

5.6 Therapeutics

Currently, dopamine antagonists, monoamine oxidase inhibitors, anti-excitotoxic drugs are being studied and developed to treat PD patients. A lot of research is also focused on the use of antioxidants, which will help abate the deleterious effects of oxidative stress on dopaminergic neurons. Cell transplantation therapy is also one of the approaches to ameliorate the symptoms of neuro-degenerative diseases in animal models (Björklund and Stenevi, 1979; Date *et al.*, 1996; Borlongan *et al.*, 1998; Saporta *et al.*, 1999). The treatment proposes the use of fetal neuronal cells, which is a huge difficulty from the ethical point of view as well as the availability of fetus. However, recent studies have indicated the use of more readily available cells such as genetically engineered cells, encapsulated cells and different cell lines (Bankiwicz *et al.*, 1993; Freed, 1993; Hammang *et al.*, 1995; Date *et al.*, 1996; Raymon *et al.*, 1997; Schinstine *et al.*, 1997). Neurons are post-mitotic tissues. It is difficult to study the biochemical mechanisms of neurodegenerative diseases due to the lack of availability of human neurons. Human teratocarcinoma (NT2N) and human neuroblastoma (SH-SY5Y) cells have been used as neuronal models to study neuronal functions. They resemble human primary neurons and like them, elaborate processes that differentiate into dendrites (Pleasure *et al.*, 1992, 1993).

Differentiated human teratocarcinoma (NT2-N) cells have been shown to survive and integrate within the host brain after transplantation and help in function recovery in animal models of stroke, PD and Huntington's disease (Borlongan *et al.*, 1998; Hartley *et al.*, 1999; Muir *et al.*, 1999; Sandhu *et al.*, 2003).

5.7 Advances in development of neuro-protective agents

Coenzyme Q_{10} (Fig. 5.5) is an essential electron and proton carrier that functions in the production of biochemical energy in aerobic organisms. Coenzyme Q_{10} also has antioxidant and membrane-stabilizing properties that serve to prevent the cellular damage that results from normal metabolic processes. It is a highly hydrophobic, lipid-soluble molecule with a quinone ring attached to an isoprene side chain.

Coenzyme Q_{10} has been shown to have protective effects in lymphocytes, skin cells and heart diseases (Singh *et al.*, 1999; Tomasetti *et al.*, 1999; Hoppe *et*

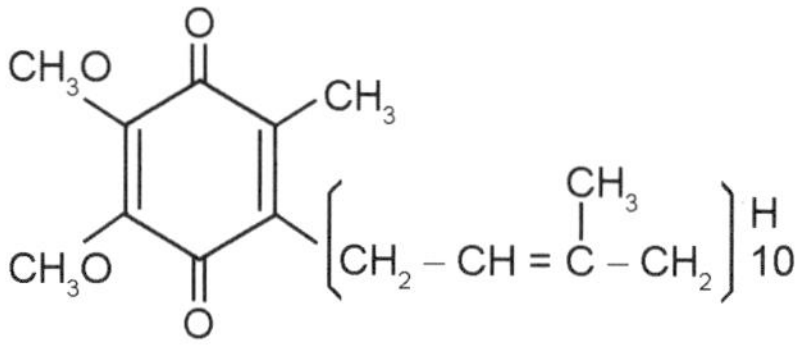

5.5 Coenzyme Q_{10}.

al., 1999). Recently, CoQ_{10} supplementation in PD patients has been shown to cause symptomatic benefit (Muller *et al.*, 2003; Shults *et al.*, 2004). One of the major problems in using Coenzyme Q_{10} in cell culture models and *in vivo* studies is its insolubility in aqueous media. Dr Sikorska's laboratory at NRC Ottawa has developed a novel solubilization technology to solubilize lipophilic compounds in aqueous media. The water-soluble formulation of CoQ_{10} developed in NRC Ottawa has been shown to be absorbed five times better than the regular oil soluble formulae in mice (Borowy-Borowski *et al.*, 1999). Significantly increased levels of Coenzyme Q_{10} have been detected in the brains of the animals fed with this formula. We have used water-soluble Coenzyme Q_{10} to assess its protective effects on neuronal cells challenged with glutamate excitotoxicity and direct oxidative stress (Sandhu *et al.*, 2003; Somayajulu *et al.*, 2005). Differentiated human teratocarcinoma (NT2N) cells were subjected to oxidative stress in the absence and presence of CoQ_{10} and cellular morphology was studied. Changes in morphology such as beading in the neuronal processes (indicated by arrows in Fig. 5.6(b)) were evident in cells under oxidative stress when compared with untreated cells (Fig. 5.6(a)). However, the cells treated with CoQ_{10} prior to oxidative stress appeared healthier and apoptotic features were considerably inhibited (Fig. 5.6(c)).

Our findings have indicated that a water-soluble formulation of Coenzyme Q_{10} (Borowy-Borowski *et al.*, 1999; Pandey *et al.*, 2006) shows unprecedented near-complete protection against oxidative stress-induced cell death of cultured neurons (Sandhu *et al.*, 2003; McCarthy *et al.*, 2004; Somayajulu *et al.*, 2005). These results prompted us to test the neuro-protective abilities of coenzyme Q_{10} *in vivo* in the rat. The preliminary results indicate that water-soluble CoQ_{10} protects dopamine neurons of the substantia nigra from paraquat-induced neurotoxicity *in vivo* (Somayajulu-Nitu *et al.*, unpublished work).

5.8 Neuronal cell death and Alzheimer's disease

Another disease caused by excessive loss of neurons is Alzheimer's disease. Alzheimer's disease leads to a progressive worsening of cognitive function and loss of memory. The disease has two pathological hallmarks, which comprise the plaques of amyloid–beta (Aβ) peptide aggregates and the neurofibrillary tangles that are made of phosphorylated tau protein (Chong *et al.*, 2005).

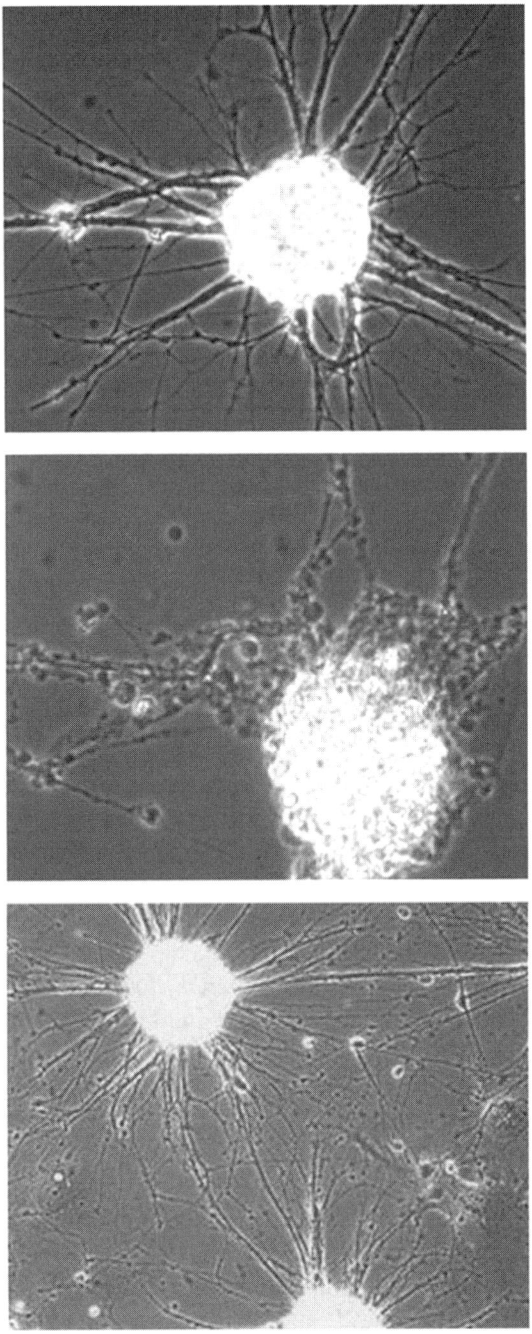

(A)

(B)

(C)

5.6 Morphology of differentiated neuronal cells after oxidative stress: (A) normal cells; (B) under direct oxidative stress, leading to apoptosis; (C) protected by coenzyme Q_{10}.

Unfortunately, no cure for this disease has been discovered yet; some of the factors and cellular mechanisms involved in AD are discussed below.

Research on the peripheral brain from AD patients, as well as animal models, have indicated that oxidative stress plays an important role in the onset and progression of AD (Butterfield *et al.*, 2002; Choi *et al.*, 2003). Increased lipid peroxidation (Lovell *et al.*, 1999), protein oxidation (Smith *et al.*, 1998) and DNA oxidation (Flaherty *et al.*, 2000) have been reported during progression of Alzheimer's disease, and correlation has been suggested between oxidative stress and Aβ deposition (Monji *et al.*, 2001). Specifically, Aβ has been found to lead to ROS production and cause damage to the neurons (Combs *et al.*, 2001) and this free radical production ability is caused by aggregation of Aβ peptides (Zhang *et al.*, 2002). Furthermore, Aβ can also elicit the expression of inducible nitric oxide synthase, leading to production of peroxynitrite (Nakagawa *et al.*, 2000). Overall, oxidative stress induced apoptosis appears to be the reason for the demise of the neurons in Alzheimer's disease. Studies have indicated that up-regulation of the anti-apoptotic protein Bcl-2 can enhance neuronal survival during Alzheimer's disease (Troy *et al.*, 2001).

The involvement of caspase-mediated apoptotic pathway during Alzheimer's disease has also been proposed (Pompl *et al.*, 2003; Gastard *et al.*, 2003). Post-mortem studies of brain tissue from Alzheimer's patients revealed an elevation of caspase genes (Banik *et al.*, 1997). Moreover, it is proposed that caspase-3 activation may contribute to the formation of neurofibrillary tangles (Hardy, 1997). Specifically, caspase-3 has been reported to lead to cleavage of the tau protein at the C-terminus, thereby enhancing its polymerization (Zhu *et al.*, 2004). Caspase-3 mediated cleavage of amyloid precursor protein (APP) has also been observed in brains of AD patients. Genetic mutations in APP and presenilins-1 and -2 are associated with early-onset PD. The presenilins are required for the proteolytic processing of APP (Janicki and Monteiro, 1999). Over-expression of both presenilins-1 and -2 can lead to cell cycle arrest in the G1 phase (Janicki *et al.*, 2000). Current research focuses on the development of anti-apoptotic and anti-inflammatory agents to prevent the death of neurons in AD.

5.9 Apoptosis and cancer

Cancer is an ever-evolving disease and, despite research efforts, remains largely unconquered to date. The challenge in finding a treatment is that cancer cells are our own body's cells, and unlike bacteria or viruses, which can be easily distinguished from normal cells and targeted for death, cancer cells divide and live using the same proteins and signals as healthy cells – with only slight and sometimes hidden changes. These changes are a result of genetic mutations in key regulatory genes that can be either oncogenes or tumor suppressor genes. If over-expressed, oncogenes lead to uncontrolled proliferation and immunity to cell death signals. Mutations of tumor suppressor genes result in cancer cells without functional apoptotic machinery and a prolonged lifespan (Chabner and Roberts, 2005). This damage is often due to a combination of inherited mutations and those acquired from environmental and physiological factors; thus each cancer is as unique as the person afflicted by it.

Successful cancer therapy must include several specific treatments in order to eliminate each cancerous cell. Current chemotherapy, radiation and surgery techniques may be able to either cure the disease, cause remission and relapse, simply slow progression or have no effect on the cancer, depending on type and phase. Exciting new methods of cancer treatment aim to induce apoptosis (programmed cell suicide) specifically in cancer cells by exploiting specific proteins or vulnerable membranes, while leaving normal healthy cells unscathed (Hu and Kavanagh, 2003). Current chemotherapies and cutting edge treatments for blood cancers and breast carcinoma, two of the major types of cancer, will be discussed here in detail.

5.9.1 Blood cancers: lymphoma and leukemia

Lymphoma is the general term for cancers originating in the lymphatic system, and is categorized as either Hodgkin or non-Hodgkin lymphoma. Lymphomas are not inherited, but are caused by DNA damage to lymphocytes of the lymphatic system. Non-Hodgkin lymphoma is the sixth most common type of cancer in North America after skin, lung, colon-rectal, breast and prostate carcinoma. Cancer of immune system cells is known as leukemia, and is divided into lymphocytic, myeloid or monocytic leukemia, depending on the cell type involved. There are both acute and chronic forms of lymphocytic and myeloid leukemia; the cells of acute cancers grow more rapidly, and are less functional, whereas chronic leukemia progresses slowly and allows for more developed cells to grow in great numbers (Schattner, 2002).

Burkitt's lymphoma is a non-Hodgkin B-cell lymphoma with an endemic form, associated with contraction of the Epstein–Barr virus, and a sporadic form more common in North America (Takada, 2001). Most non-Hodgkin lymphomas have a gene translocation that results in rearrangement and subsequent over-expression of the Bcl-2 gene, discovered through the study of Burkitt's cell lymphoma (Reed, 1996). Over-expression of Bcl-2, an anti-apoptotic protein, blocks the inherent cell death mechanism, allowing cancer cells to grow uncontrolled. Up-regulation of the Bcl-2 gene is found in approximately half of all human malignancies (Kim, 2005).

Chronic myeloid leukemia (CML) is a unique type of leukemia; it is caused by a specific genetic mutation leading to DNA rearrangement and formation of the *Philadelphia chromosome*, which translates into the cancer-promoting protein Bcr-Abl. This non-inherited mutation occurs in a stem cell of the bone marrow, typically in adults over 35 years old (Sharifi and Steinman, 2002).

5.9.2 Therapy options

With chronic myeloid leukemia, the cure comes from a revolutionary designer drug, Imatinib (Gleevec^{TM}), which inactivates the onco-protein unique to cells of this type of leukemia. By blocking the active site of Bcr-Abl kinase, this drug disables further growth and proliferation of cancerous cells (Sharifi and Steinman, 2002). Kinase inhibitors in general are revolutionary new drug concepts, which target the 'kinase crutch' found in many cancers (Mendoza *et al.*, 2005).

Classical chemotherapy treatments, such as paclitaxel and etoposide, have genomic targets that are found in both cancerous and healthy cells (Thatte *et al.*, 2000). Blood cells regularly undergo mitosis and apoptosis, making them vulnerable to drugs that target cell cycle proteins. The common method of intravenous chemotherapy delivery means direct exposure of these toxic agents to blood cells, which can drastically compromise immune system function and, worse, cause irreversible DNA damage to blood cells that may spawn secondary cancers. Radiation therapy, used to treat lymphoid tumors, causes massive DNA damage and ultimately apoptosis of both cancer cells and healthy neighboring cells (Martin *et al.*, 2001). Alternate therapy options include bone marrow transplantation or blood transfusion, usually coupled with chemotherapy and/or radiation therapy.

Current research efforts focus on the development of modulators of apoptosis in cancer cells (Hu and Kavanagh, 2003). Various biochemical pathways involved in apoptosis are being targeted, such as monoclonal antibodies against TRAILR1, inhibitors of Bcl_2, and proteasome inhibitors (Srivastava, 2001; Ciardiello and Tortora, 2001; LeBlanc *et al.*, 2002). New drugs are categorized by the target they identify, whether its cell death machinery, protein kinases, transcription factors or cell-surface antigens (Kim, 2005).

Monoclonal antibody therapy involves removal of the body's own immune cells and stimulation with cancer cell-surface antigens. These immune cells raise specific antibodies and are transplanted back into the body, where they attack cancer cells that display the specific cell-surface protein. Combination therapy links the new antibody with chemotherapy agents, or makes use of radioactive antibodies to deliver radiation therapy directly to the cancer cells (Chabner and Roberts, 2005).

Drugs that target cell death machinery, for example CD95 membrane receptors common in lymphoma cells, may exploit the observation that cancerous cells tend to over-express these receptors. Laboratories including ours have found targeting CD95 to activate caspase-3 at the membrane level, rapidly inducing apoptosis in an unorthodox mechanism (Aouad *et al.*, 2004; Kekre *et al.*, 2005). Preliminary results indicate that the natural compound pancratistatin is able to specifically induce apoptosis in both Jurkat cells (human B-cell lymphoma cell line) and types of leukemia obtained from newly diagnosed and relapsed patients (Kekre *et al.*, 2005; Griffin *et al.*, 2006). This compound, which was derived from the Hawaiian spider lily (*Pancratium littorale*) (Pettit *et al.*, 1986), has proven to be non-genotoxic to normal healthy cells with no effect on their proliferation following treatment (McLachlan *et al.*, 2005).

5.9.3 Breast cancer

Breast cancer is one of the most prominent cancers in women today (falling only second to lung cancer) and by far one of the deadliest: one out of every three patients will ultimately die from the disease (Simstein *et al.*, 2003). Although great strides have been taken in the search for the cure, many avenues have yet to be explored.

Breast cancer, like other cancers, is a result of genetic mutations that lead to over-expression of oncogenes or the decreased expression/function of tumor suppressor genes. Although breast cancer can be linked to several inherited gene mutations (such as *BRCA* and *p53*), the presence or absence of these genes does not guarantee the fate of the individual, as many gene mutations can be acquired throughout an individual's lifetime.

A number of receptors have been found to be over-expressed in breast cancer cells; breast cancer can be estrogen receptor (ER) positive or progesterone receptor (PR) positive. Cancers with over-expression of one of these receptors (or both), tend to respond well to hormone therapy. Hetero-dimerization of the HER-2 protein (also known as c-erbB-2 or neu) with other HER family members forms complexes that initiate intracellular signaling (Olayioye *et al.*, 2000). Amplification of the *HER-2* gene can be found in 15–30% of primary invasive tumors. Where normal cells have two copies of the gene, some breast cancers may have up to 100 copies, which can lead to a 100-fold increase in the number of receptors per cell (Duffy, 2005).

5.9.4 Current therapies

Current anti-cancer treatments for breast cancer can be divided into two major categories: local therapy (surgery, radiation therapy) and systematic therapy (chemotherapy, hormone therapy and immuno-therapy).

Surgery and radiation therapy work by targeting the affected area, and are often used in combination with other treatments. While surgery works by attempting to remove all of the cancerous cells, this is often not enough; approximately 70% of diagnosed cancer patients will receive radiation therapy (Martin, 2001). Radiation therapy induces apoptosis in cancer and normal cells by focusing a beam of ionizing radiation (IR) at the specific tumor and interferes with cell division of the cells by introducing oxidative stress and injury to DNA. Cancer cells have been shown to be more susceptible to apoptosis and irreparable DNA damage following IR treatment than normal cells (McLachlan-Burgess *et al.*, 2006), but the reason for this remains unknown. Radiation therapy often leads to harsh side effects due to the large number of normal cells that acquire DNA damage and oxidative stress. Additionally, DNA damage that goes unrepaired in normal cells may lead to the formation of secondary cancers.

Chemotherapeutic agents such as DNA-targeting agents (such as anti-metabolites, alkylating agents, topoisomerase inhibitors) or microtubule targeting agents (e.g. vincas and taxanes) induce apoptosis in cancer cells by targeting fast-dividing cells. On average, there is a substantial decrease in apoptosis by 3 months after the start of chemotherapy (Dowsett *et al.,* 1999). Changes in the apoptotic machinery of cancer cells decrease the ability of cancer cells to undergo apoptosis, which may play a role in chemo-resistance (Simstein *et al.*, 2003). Additionally these treatments are often genotoxic to our normal cells causing harsh side effects and secondary cancers.

Current hormone therapies include: anti-estrogens (tamoxifen), aromatase inhibitors (anastrozole and letrozole), and luteinizing hormone-releasing hormone agonists (goserelin). Anti-estrogen therapy is widely used in breast cancer therapy. Tamoxifen and other anti-estrogens compete with estrogens for binding to the ER on breast cancer cells and thereby block estrogen's ability to activate transcription in these cells. Additionally, researchers have found that tamoxifen triggers apoptosis of breast cancer cells by two different pathways: ER-dependent and ER-independent. While both pathways lead to mitochondrial membrane permeabilization, one pathway leads to the activation of caspase-9 and caspase-3, while the other does not (Obrero *et al.*, 2002). There are several serious side effects associated with long-term use of tamoxifen including increased incidence of endometrial cancer, thromboembolic events and the development of resistance (Baum, 2005). By preventing estrogen production, ER positive cancers are less likely to grow. After menopause, other steroid hormones (primarily androgens) are converted to estrogen by peripheral aromatase. Aromatase inhibitors work by blocking the production of estrogen in post-menopausal women thereby limiting the growth of estrogen receptor tumors (Jones and Buzdar, 2004). Luteinizing hormone-releasing hormone agonists work by temporarily stopping the functioning of the ovaries. If the individual has not yet gone through menopause and their cancer is ER-positive, ovarian ablation stops the ovaries from producing the estrogen that enables further growth (Jones and Buzdar, 2004).

The therapeutic monoclonal antibody trastuzumab (Herceptin) binds to the growth promoting protein HER-2 that is overly-expressed on the surface of many breast cancers. Herceptin induces inhibition of growth in addition to causing down-modulation of HER-2, inhibition of cell cycle progression as a result of p27 induction, inhibition of angiogenesis, and induction of an immune response (Duffy, 2005). However, the benefits of Herceptin therapy seem to have limited application; even in the selected group of patients with very high levels of HER-2 over-expression the response rate from this highly specific targeted agent is limited in magnitude and duration (Albanell and Baselga, 2001).

Although some success has come from targeting the DNA or over-expressed receptors in cancer cells, resistance to treatment often develops. Additionally, the harsh side effects that result from damage to normal cells, and the appearance of secondary cancers make these treatments risky and often unbearable.

Tumor necrosis factor, TNF-α, is a cytokine that is naturally secreted by cells of the immune system, and has been found to be selectively cyto-toxic to cancerous cells. Although harsh side effects result from the systematic use of TNF-α as chemotherapy, TNF family members such as TRAIL have been found to minimize toxicity and augment chemotherapy-induced apoptosis. Current *in vitro* research has shown these drugs to be promising prospective therapy for breast cancer, particularly in combination with classic chemotherapeutic drugs (Simstein *et al.*, 2003).

5.9.5 Mitochondrial proteins as novel differential targets to develop anti-cancer therapy

Mitochondria play a significant role in apoptosis by releasing caspase activators such as cytochrome *c*, releasing caspase-independent death effectors such as AIF and causing the loss of essential mitochondrial functions like ATP production (Green and Kroemer, 2004). Recent studies have found differences between the mitochondria of cancerous and non-cancerous cells (Don and Hogg, 2004; Fantin and Leder, 2004) and researchers are on the search for anti-cancer agents that can target these differences and specifically induce apoptosis in cancerous cells, including breast cancer.

Uptake of the lipophilic cation Rh123 by mitochondria has been found to be significantly higher in cancerous cell lines including MCF-7 (human breast carcinoma) than in normal cell lines. Researchers have found that the increased membrane potential of cancerous mitochondria may be the reason for the differential uptake of Rh123, and also that selective mitochondrial toxicity and apoptosis result from this uptake (Modica-Napolitano and Singh, 2002). These findings have led to the development of delocalized lipophilic cations as a new class of potential anti-cancer agents.

Research work in our laboratory focuses on exploiting the differential mitochondrial sensitivities of cancerous cells to potential anti-cancer agents. As described above, pancratistatin (PST) is a natural compound that has shown to exhibit anti-neoplastic activity (Luduena *et al.*, 1992). We have shown that PST can induce apoptosis specifically in a variety of cancerous cells, including breast cancer, while normal cells are not harmed (McLachlan *et al.*, 2005; Kekre *et al.*, 2005; Pandey *et al.*, 2005). Interestingly, our recent findings indicate that PST causes mitochondrial membrane potential collapse and permeabilization

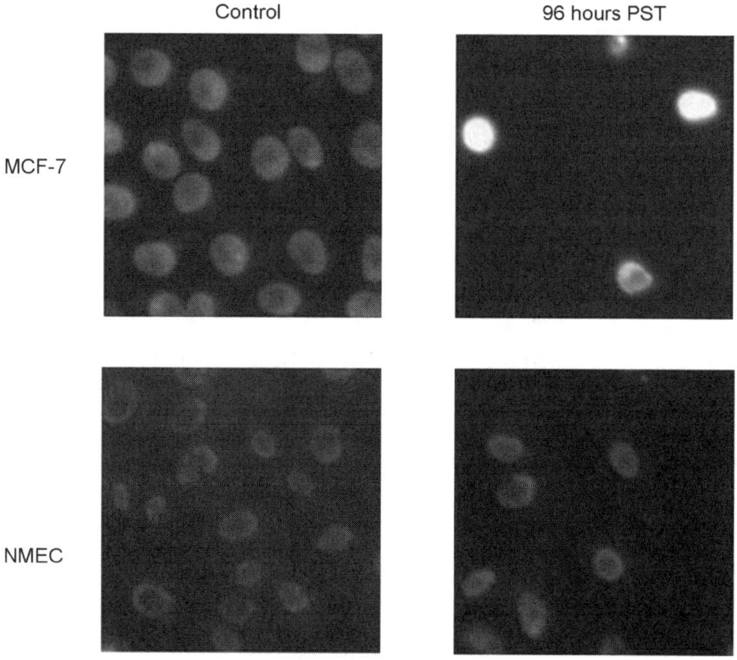

5.7 Pancratistatin induces apoptosis specifically in breast cancer (MCF-7) cells; normal mammary epithelial (NMEC) cells remain unaffected. MCF-7 and NMEC cells were treated with 1 μm PST Hoechst 33342 dye. Fluorescent pictures were taken after 96 h at 400× magnification. Nuclear condensation is evident in cells with bright, condensed and rounded nuclei.

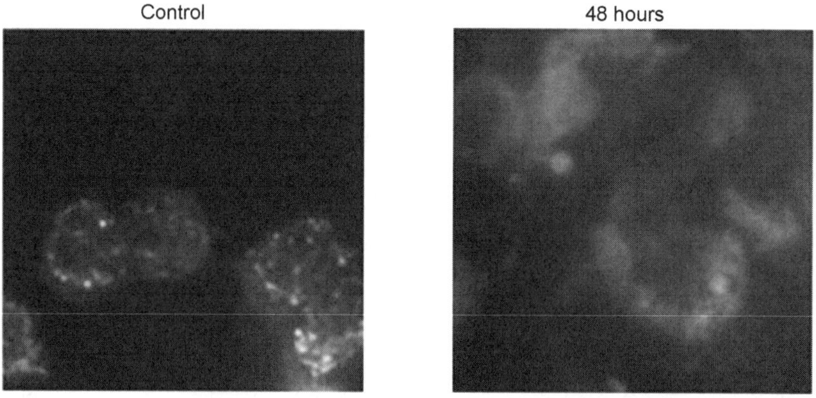

5.8 Pancratistatin-induced mitochondrial membrane potential collapse in breast cancer (MCF-7) cells. Fluorescent pictures were taken 48 h after treatment at 400× magnification. Bright punctuate fluorescence staining is indicative of intact healthy mitochondria; dull and diffuse staining is indicative of a loss in mitochondrial membrane potential.

selectively in breast cancer cells. More specifically, when breast cancer cells (MCF-7) and their non-cancerous counterpart, normal mammary epithelia cells, were treated with PST for 96 h, the treatment induced apoptosis in only the cancerous MCF-7 cells (over 90% apoptotic) (McLachlan-Burgess *et al.*, 2006). The normal mammary epithelia cells were unaffected after identical treatments (Fig. 5.7). Upon examination, we found that mitochondrial membrane potential collapse was the initial event in the induction of apoptosis by PST in breast cancer cells. Figure 5.8 shows breast cancer (MCF-7) cells stained with a membrane potential sensitive dye (JC-1). This experiment indicates that while untreated cells have healthy, intact mitochondria, the breast cancer cells treated with PST have lost their mitochondrial potential (McLachlan-Burgess *et al.*, 2006). Most importantly, we have found that PST causes mitochondrial dysfunction in isolated breast cancer cell mitochondria, while PST treatment has no effect on isolated non-cancerous mitochondria (McLachlan-Burgess *et al.*, 2006).

5.10 Future trends

As we learn more about the mechanism of apoptotic cell death, the important role of apoptosis in various diseases is being uncovered. More and more opportunities are available for developing novel therapeutic agents that can either block cell death (AIDS, neurodegenerative and cardiovascular diseases) or those that induce cell death (cancer). In particular, the recent development in the elucidation of the central role that the mitochondria and mitochondrial proteins play in the initiation and execution of apoptosis has opened a new area of research to modulate cell death. The application of new approaches will result in the discovery of new therapeutics for the aforementioned diseases. These techniques may include identifying specific blockers of apoptosis using a phage-display library of single domain antibodies for specific target proteins involved in apoptosis, or screening combinatorial pharmacophore libraries for potential compounds that can block or induce apoptosis without causing toxicity. Further work with small natural compounds such a water-soluble CoQ_{10}, which can stabilize mitochondria and prevent neuronal cell death under oxidative stress, will supply novel preventive strategies for neurodegenerative diseases. The recent finding that the mitochondria of cancer cells are vulnerable to certain natural compounds presents an opportunity to develop a new battery of anti-cancer compounds. We are currently at a transition stage in the era of cellular biology and disease research: the application of knowledge regarding the biochemical mechanism of cell death will transform therapeutic strategies and development of therapeutic agents for cancer, cardiovascular and neuro-degenerative diseases.

5.11 Acknowledgements

This work was supported by research grants to Dr Siyaram Pandey from the Heart and Stroke Foundation of Ontario, Canada, by the Lotte and John Hecht Memorial Foundation of Vancouver, BC, Canada, and through funds raised by the Knights of Columbus, St Clair Beach, Council 9671, ON, Canada.

5.12 References and further reading

Adhihetty P, Hood D. Mechanisms of apoptosis in skeletal muscle. *Basic Applied Myology*. 2003; 13(4): 171–179.

Albanell J, Baselga J. Unravelling resistance to trastuzumab (herceptin): insulin-like growth factor-1 receptor, a new suspect. *Journal of the National Cancer Institute*. 2001; 93: 1830–1832.

Andrews PW. Retinoic acid induces neuronal differentiation of a cloned human embryonal carcinoma cell line *in-vitro*. *Developmental Biology*. 1987; 103: 285–293.

Aouad SM, Cohen LY, Sharif-Askari F, Haddad EK, Alam A, Sekaly RP. Caspase-3 is a component of Fas death-inducing signaling complex in lipid rafts and its activity is required for complete caspase-8 activation during Fas-mediated cell death. *Journal of Immunology*. 2004; 172(4): 2316–2323.

Ashkenazi A, Dixit VM. Death receptors: signaling and modulation. *Science*. 1998; 281: 1305–1308.

Banik NL, Matzelle D, Gantt-Wilford G, Hogan EL. Role of calpain and its inhibitors in tissue degeneration and neuroprotection in spinal cord injury. *Annals of the New York Academy of Science*. 1997; 825: 120–127.

Bankiewicz K, Mandel RJ, Sofroniew MV. Trophism, transplantation, and animal models of Parkinson's disease. *Experimental Neurology*. 1993; 24(1): 140–149.

Baum M. Adjuvant endocrine therapy in postmenopausal women with early breast cancer: where are we now? *European Journal of Cancer*. 2005; 41: 1667–1677.

Benecke R, Strumper P, Weiss H. Electron transfer complexes I and IV of platelets are abnormal in Parkinson's disease but normal in Parkinson-plus syndromes. *Brain*. 1993; 116: 1451–1463.

Björklund A, Stenevi U. Reconstruction of the nigrostriatial dopamine pathway by intracerebral nigral transplants. *Brain Research*. 1979; 177: 555–560.

Borlongan CV, Tajima Y, Trojanowski JQ, Lee VM, Sanberg PR. Transplantation of cryopreserved human embryonal carcinoma-derived neurons (NT2N cells) promotes functional recovery in ischemic rats. *Experimental Neurology*. 1998; 149: 310–321.

Borowy-Borowski H, Sikorska M, Walker PR. Water soluble composition of bioactive lipophilic compounds. US Patent No. 6,045,826 (1999).

Bucciantini M, Giannoni E, Chiti F, Baroni F, Formigli L, Zurdo J, Taddei N, Ramponi G, Dobson CM, Stefani M. Inherent toxicity of aggregates implies a common mechanism for protein misfolding diseases. *Nature*. 2002; 416: 507–511.

Butterfield DA, Castegna A, Lauderback CM, Drake J. Evidence that amyloid beta-peptide-induced lipid peroxidation and its sequelae in Alzheimer's disease brain contribute to neuronal death. *Neurobiology of Aging*. 2002; 23: 655–664.

Candelario-Jalil EG, Garcia-Cabrera M, Leon OS, Fiebich BL. Wide therapeutic time window for nimesulide neuroprotection in a model of transient focal cerebral ischemia in the rat. *Brain Research*. 2004; 1007: 98–108.

Chabner BA, Roberts TG Jr. Timeline: chemotherapy and the war on cancer. *Nature Reviews Cancer*. 2005; 5(1): 65–72.

Chanan-Khan A. Bcl-2 antisense therapy in hematologic malignancies. *Current Opinion Oncology*. 2004; 16(6): 581–585.

Cherubini A, Polidori MC, Mecocci P. Potential markers of oxidative stress in stroke. *Free Radical Biology & Medicine*. 2005; 39: 841–852.

Chicchetti F, Lapointe N, Roberge-Tremblay A, Saint-Pierre M, Jimenez L, Ficke BW, Gross RF. Systematic exposure to paraquat and maneb models early Parkinson's disease in young adult rats. *Neurobiology of Disease*. 2005; 30: 360–371.

Choi J, Malakowsky CA, Talent JM, Conrad CC, Carroll CA, Weintraub ST, Gracy RW. Anti-apoptotic proteins are oxidized by A-beta25-35 in Alzheimer's fibroblasts. *Biochimica et Biophysica Acta*. 2003; 1637: 135–141.

Chong ZZ, Li F, Maiese K. Stress in the brain: novel cellular mechanisms of injury linked to Alzheimer's disease. *Brain Research Reviews*. 2005; 49: 1–21.

Ciardiello F, Tortora G. Inhibition of Bcl-2 as cancer therapy. *Annual Oncology*. 2001; 13: 501–502.

Combs CK, Karlo JC, Kao SC, Landreth GE. β-Amyloid stimulation of microglia and monocytes results in TNF-alpha dependent expression of inducible nitric oxide synthase and neuronal apoptosis. *Journal of Neuroscience*. 2001; 21: 1179–1188.

Cuzzocrea SR, Caputi AP, Salvemini D. Antioxidant therapy: a new pharmacological approach in shock, inflammation, and ischemia/reperfusion injury. *Pharmacological Reviews*. 2001; *53*: 135–159.

Date I, Felten DL, Felten SY. Long-term effect of MPTP in the mouse brain in relation to aging: neurochemical and immunocytochemical analysis. *Brain Research*. 1990; 519: 266–276.

Date I, Ohmoto T, Imaoka T, Ono T, Hammang JP, Francis J, Greco C, Emerich DF. Co-grafting with polymer-encapsulated human nerve growth factor-secreting cells and chromaffin cell survival and behavioral recovery in hemiparkinsonian rats. *Journal of Neurosurgery*. 1996; 84: 1006–1012.

Don A, Hogg P. Mitochondria as cancer drug targets. *Trends in Molecular Medicine*. 2004; 10: 8.

Dowsett M, Archer C, Assersohn L, Gregory RK, Ellis PA, Salter J, Chang J, Mainwaring P, Boeddinghaus I, Johnston SR, Powles TJ, Smith IE. Clinical studies of apoptosis and proliferation in breast cancer. *Endocrine-related Cancer*. 1999; 6(1): 25–28.

Duffy MJ. Predictive markers in breast and other cancers: a review. *Clinical Chemistry*. 2005; 51: 494–503.

Fantin VR, Leder P. F16, a mitochondriotoxic compound, trigger apoptosis or necrosis depending on the genetic background of the target carcinoma cell. *Cancer Research*. 2004; 64: 329–336.

Fearnley JM, Lees AJ. Ageing and Parkinson's disease: substantia nigra regional selectivity. *Brain*. 1991; 114: 2283–2301.

Fernandez-Espejo E. Pathogenesis of Parkinson's disease: prospects of neuroprotective and restorative therapies. *Molecular Neurobiology*. 2004; 29: 15–30.

Flaherty DB, Soria JP, Tomasiewicz HG, Wood JG. Phosphorylation of human tau protein by microtubule-associated kinases: GSK3beta and cdk5 are key participants. *Journal of Neuroscience Research*. 2000; 62: 463–472.

Freed WJ. Neural transplantation: prospects for clinical use. *Cell Transplant*. 1993; 2: 13–31.

Gastard MC, Troncoso JC, Koliatsos VE. Caspase activation in the limbic cortex of subjects with early Alzheimer's disease. *Annals of Neurology*. 2003; 54: 393–398.

Gibson GE, Huang HM. Oxidative stress in Alzheimer's disease. *Neurobiology of Aging*. 2005; 26: 575–578.

Green D, Kroemer G. The pathology of mitochondrial cell death. *Science*. 2004; 305: 626–629.

Griffin C, McNulty J, Hamm C, Pandey S. Pancratistatin: a novel highly selective anticancer agent that induces apoptosis by the activation of membrane-Fas-receptor associated caspase-3. *Trends in Cell Apoptosis Research*, Nova Science Publishers Inc., in press.

Gueorguieva D, Li S, Walsh N, Mukerji A, Tanha J, Pandey S. Identification of single-domain, Bax-specific intrabodies that confer resistance to mammalian cells against oxidative-stress induced apoptosis. *FASEB Journal*. E-pub ahead of print 23 October 2006.

Hammang JP, Emerich DF, Winn SR, Lee A, Lindner MD, Gentile FT, Doherty EJ, Kordower JH, Baetge EE. Delivery of neurotrophic factors to the CNS using encapsulated cells: developing treatments for neurodegenerative diseases. *Cell Transplant*. 1995; 4, (Suppl 1): S27–S28.

Hardy J. Amyloid, the presenilins and Alzheimer's disease. *Trends in Neuroscience*. 1997; 20: 154–159.

Hartley RS, Margulis M, Fishman PS, Lee VM, Tang CM. Functional synapses are formed between human Ntera 2 (NT2N, hNT) neurons grown on astrocytes. *The Journal of Comparative Neurology*. 1999; 407: 1–10.

Hoppe U, Bergemann J, Diembeck W, Ennen J, Gohla S, Harris I, Jacob J, Kielholz J, Mei W, Pollet D, Schachtschabel D, Sauermann G, Schreiner V, Stab F, Steckel F. Coenzyme Q10, a cutaneous antioxidant and energizer. *Biofactors*. 1999; 9(2–4): 371–378.

Hu W, Kavanagh J. Anticancer therapy targeting the apoptotic pathway. *The Lancet Oncology*. 2003; 4: 721–729.

Hunot S, Boissière F, Faucheux B, Brugg B, Mouatt-Prigent A, Agid Y, Hirsch EC. Nitric oxide synthase and neuronal vulnerability in Parkinson's disease. *Neuroscience*. 1996; 72: 355–363.

Janicki SM, Monteiro MJ. Presenilin over-expression arrests cells in the G1 phase of the cell cycle: arrest potentiated by the Alzheimer's disease PS2 (N141I) mutant. *American Journal of Pathology*. 1999; 155: 135–144.

Janicki SM, Stabler SM, Monteiro MJ. Familial Alzheimer's disease presenilin-1 mutants potentiate cell cycle arrest. *Neurobiology of Aging*. 2000; 21: 829–836.

Jenner P. Oxidative stress in neurodegenerative disease. *The Lancet*. 1994; 344: 796–798.

Jenner P. Oxidative stress in Parkinson's disease. *Annals of Neurology*. 2003; 53: 526–537.

Jones KL, Buzdar AU. A review of adjuvant hormonal therapy in breast cancer. *Endocrine-Related Breast Cancer*. 2004; 11(3): 391–406.

Kekre N, Griffin C, McNulty J, Pandey S. Pancratistatin causes early activation of caspase-3 and the flipping of phosphatidyl serine followed by rapid apoptosis specifically in human lymphoma cells. *Cancer Chemotherapy & Pharmacology*. 2005; 56(1): 29–38.

Kerr JF, Wyllie AH, Currie AR. Apoptosis: A basic biological phenomenon with wide ranging implications in tissue kinetics. *British Journal of Cancer*. 1972; 26: 239–257.

Khan M, Sekhon B, Jatana M, Giri S, Gilg AG, Sekhon C, Singh I, Singh AK. Administration of *N*-acetylcysteine after focal cerebral ischemia protects brain and reduces inflammation in a rat model of experimental stroke. *Journal of Neuroscience Research*. 2004; 76: 519–527.

Kim R. Recent advances in understanding the cell death pathways activated by anticancer therapy. *Cancer*. 2005; 103(8): 1551–1560.

Koller WC, Cersosimo MG. Neuroprotection in Parkinson's disease: elusive goal. *Current Neurology and Neuroscience Reports*. 2004; 4: 277–283.

Koller W, Vetere-Overfield B, Gray C, Alexander C, Chin T, Dolezal J, Hassanein R, Tanner C. Environment risk factors in Parkinson's disease. *Neurology*. 1990; 40(8): 1218–1221.

Korsmeyer SJ, Wei MC, Saito M. Proapoptotic cascade activates BID which oligomerizes BAK or BAX into pores that result in the release of cytochrome *c*. *Cell Death and Differentiation*. 2000; 7: 1166–1173.

Krige D, Carrol MT, Cooper JM, Marsden CD, Schapira AHV. Platelet mitochondrial function in Parkinson's disease. *Annals of Neurology*. 1992; 32: 782–788.

Kurlan R, Kim MH, Gash DM. The time course and magnitude of spontaneous recovery of Parkinsonism produced by intracarotid administration of 1-methyl-4-phenyl-1,2,3,6-tetrahydropyridine to monkeys. *Annals of Neurology*. 1991; 29: 677–679.

Kuwana T, Mackey MR, Perkins G, Ellisman MH, Latterich M, Schneiter R, Green DR, Newmeyer DD. Bid, Bax and lipids cooperate to form supramolecular openings in the outer mitochondrial membrane. *Cell*. 2002; 111(3): 331–342.

Langston JW, Ballard P, Tetrud JW, Irwin I. Chronic Parkinsonism in humans due to a produce of meperidine-analog synthesis. *Science*. 1983; 219: 979–980.

LeBlanc R, Catley LP, Hideshima T, Lentzsch S, Mitsiades CS, Mitsiades N, Neuberg D, Goloubeva O, Pien CS, Adams J, Gupta D, Richardson PG, Munshi NC, Anderson KC. Proteasome inhibitor PS-341 inhibits human myeloma cell growth *in vivo* and prolongs survival in a murine model. *Cancer Research*. 2002; 62(17): 4996–5000.

Lindsten T, Golden JA, Zong W, Minarcik J, Harris M, Thompson C. The pro-apoptotic activities of Bax and Bak limit the size of the neural stem cell pool. *Journal of Neuroscience*. 2003; 23(35): 11112–11119.

Liou HH, Tsai MC, Chen CJ, Jeng JS, Chang YC, Chen SY, Chen RC. Environmental risk factors and Parkinson's disease: a case control study in Taiwan. *Neurology*. 1997; 48: 1583–1588.

Lotharius J, Barg S, Wiekop P, Lundberg C, Raymon HK, Brundin P. Effect of mutant alpha-synuclein on dopamine homeostasis in a new human mesencephalic cell line. *Journal of Biological Chemistry*. 2002; 277: 38884–38894.

Lovell MA, Gabbita SP, Markesbery WR. Increased DNA oxidation and decreased levels of repair products in Alzheimer's disease ventricular CSF. *Journal of Neurochemistry*. 1999; 72: 771–776.

Luduena RF, Roach MC, Prasad V, Pettit GR. *Biochemical Pharmacology*. 1992; 43: 539.

Margaill I, Plotkine M, Lerouet D. Antioxidant strategies in the treatment of stroke. *Free Radical Biology & Medicine*. 2005; 39(4): 429–443.

Marsden CD. Problems with long-term levodopa therapy for Parkinson's disease. *Clinical Neuropharmacology*. 1994; 17(2): S32–S44.

Martin NMB. DNA repair inhibition and cancer therapy. *Journal of Photochemistry & Photobiology. B, Biology*. 2001; 63: 162–170.

McCarthy S, Somayajulu M, Hung M, Sikorska M, Borowy-Borowski H, Pandey S. Paraquat induces oxidative stress and neuronal cell death; neuro-protection by water-soluble coenzyme Q_{10}. *Toxicology & Applied Pharmacology*. 2004; 201: 21–31.

McCormack AL, Thiruchelvam M, Manning-Bog AB. Environmental risk factors and Parkinson's disease: selective degeneration of nigral dopaminergic neurons caused by the herbicide paraquat. *Neurobiology of Disease*. 2002; 10: 119–127.

McLachlan A, Kekre N, McNulty J, Pandey S. Pancratistatin: a natural anti-cancer compound that targets mitochondria specifically in cancer cells to induce apoptosis. *Apoptosis*. 2005; 10(3): 619–630.

McLachlan-Burgess A, McCarthy S, Griffin C, Richer J, Pandey S. The differential response induced by exposure to low-dose ionizing radiation in SHSY-5Y and normal human fibroblast cells. *Applied Biochemistry & Biotechnology* 2006; 135(2): 159–178.

Mendoza FJ, Espino PS, Cann KL, Bristow N, McCrea K, Los M. Anti-tumor chemotherapy utilizing peptide-based approaches – apoptotic pathways, kinases, and proteasome as targets. *Archivum Immunologiae et Therapiae Experimentalis*. 2005; 53(1): 47–60.

Modica-Napolitano J, Singh K. Mitochondria as targets for detection and treatment of cancer. *Expert Reviews in Molecular Medicine*. 11 April, 2002; http://www-ermm.cbcu.cam.ac.uk/02004453h.htm

Monji A, Utsumi H, Ueda T, Imoto T, Yoshida I, Hashioka S, Tashiro K, Tashiro N. The relationship between the aggregational state of the amyloid-beta peptides and free radical generation by the peptides. *Journal of Neurochemistry*. 2001; 77: 1425–1432.

Muir JK, Raghupathi R, Saatman KE, Wilson CA, Lee VM, Trojanowski JQ, Philips MF, McIntosh TK. Terminally differentiated human neurons survive and integrate following transplantation into the traumatically injured rat brain. *Journal of Neurotrauma*. 1999; 16(5): 403–414.

Muller T, Buttner T, Gholipour AF, Kuhn W. Coenzyme Q10 supplementation provides mild symptomatic benefit in patients with Parkinson's disease. *Neuroscience Letters*. 2003; 341(3): 201–204.

Murphy KM, Ranganathan V, Farnsworth ML, Kavallaris M, Lock RB. Bcl-2 inhibits Bax translocation from cytosol to mitochondria during drug-induced apoptosis of human tumor cells. *Cell Death & Differentiation*. 2000; 7: 102–111.

Nakagawa T, Zhu H, Morishima N, Li E, Xu J, Yankner BA, Yuan J. Caspase-12 mediates endoplasmic-reticulum-specific apoptosis and cytotoxicity by amyloid-beta. *Nature*. 2000; 403: 98–103.

Nomura K, Koumura T, Kobayashi T, Nakagawa Y. Mitochondrial phospholipids hydroperoxide glutathione peroxidase inhibits the release of cytochrome *c* from mitochondria by suppressing the peroxidation of cardiolipin in hypoglycaemia induced apoptosis. *Biochemistry Journal*. 2000; 351: 183–193.

Obrero M, Yu DV, Shapiro DJ. Estrogen receptor-dependent and estrogen receptor-independent pathways for tamoxifen and 4-hydroxytamoxifen-induced programmed cell death. *Journal of Biological Chemistry*. 2002; 277: 45695–45703.

Olayioye MA, Neve RM, Lane HA, Hynes NE. The erbB signalling network: heterodimerization in development and cancer. *EMBO Journal*. 2000; 19: 3159–3167.

Pandey S, Kekre N, Naderi J, McNulty J. Induction of apoptotic cell death specifically in rat and human cancer cells by pancratistatin. *Artificial Cells, Blood Substitutes, and Biotechnology*. 2005; 33: 279–295.

Pandey S, Somayajulu M, Vergel de Dios J, Matei A, Parameswarann V, Cohen J, Sandhu J, Borowy-Borowski H, Sikorska M. Paraquat induces oxidative stress, neuronal loss in substantia-nigra region and Parkinsonism in rats: neuro-protection and amelioration of symptoms by water-soluble CoQ_{10}. *Abstract accepted for platform presentation at the Society of Toxicology Annual General Meeting, San Diego, CA, USA*. 4–9 March 2006.

Parker WD Jr., Boyson SJ, Parks JK. Abnormalities of the electron transport chain in idiopathic Parkinson's disease. *Annals of Neurology*. 1989; 26: 719–723.

Pettit GR, Gaddamidi V, Herald DL, Singh SB, Cragg GM, Schmidt JM, Boettner FE, Williams M, Sagawa Y. Antineoplastic agents, 120. Pancratium littorale. *Journal of Natural Products*. 1986; 49(6): 995–1002.

Pleasure SJ, Page C, Lee VMY. Pure, postmitotic, polarized human neurons derived from NT2 cells provide a system for expressing exogenous proteins in terminally differentiated neurons. *Journal of Neuroscience Research*. 1992; 12: 1802–1815.

Pleasure SJ, Page C, Lee VMY. Ntera 2 cells: a human cell line which displays characteristics expected of a human committed neuronal progenitor cell. *Journal of Neuroscience Research*. 1993; 35: 585–602.

Pompl PN, Yemul S, Xiang Z, Ho L, Haroutunian V, Purohit D, Mohs R, Pasinetti GM. Caspase gene expression in the brain as a function of the clinical progression of Alzheimer disease. *Archives of Neurology*. 2003; 60: 369–376.

Raymon HK, Thode S, Gage FH. Application of *ex-vivo* gene therapy in the treatment of Parkinson's disease. *Experimental Neurology*. 1997; 144: 82–91.

Reed JC. Mechanisms of Bcl-2 family protein function and dysfunction in health and disease. *Behring Institute Mitteilungen*. 1996; 97: 72–100.

Reed JC. Mechanisms of apoptosis. *American Journal of Pathology*. 2000; 157(5): 1415–1426.

Reed JC. Apoptosis-based therapies. *Nature Reviews. Drug Discovery*. 2002; 1: 111–121.

Sandhu JK, Pandey S, Ribecco M, Monette R, Borowy-Borowski H, Walker PR, Sikorska M. Molecular mechanism of glutamate neurotoxicity in mixed cultures of NT-2-derived neurons and astrocytes: protective effects of CoQ_{10}. *Journal of Neuroscience Research*. 2003; 72: 691–703.

Saporta S, Borlongan CV, Sanberg PR. Neural transplantation of human neuroteratocarcinoma (hNT) neurons into ischemic rats: a quantitative dose–response analysis of cell survival and behavioural recovery. *Neuroscience*. 1999; 91(2): 519–525.

Schapira AH, Cooper JM, Dexter D, Clark JB, Jenner P, Marsden CD. Mitochondrial complex I deficiency in Parkinson's disease. *Journal of Neurochemistry*. 1990; 54(3): 823 827.

Schattner EJ. Apoptosis in lymphocytic leukemias and lymphomas. *Cancer Investigation*. 2002; 20(5–6): 737–748.

Schinstine M, Ray J, Gage FH. Potential effect of cytokines on transgene expression in primary fibroblasts implanted into the rat brain. *Molecular Brain Research*. 1997; 47: 195–201.

Seidler A, Hellenbrand W, Robra BP, Vieregge P, Nischan P, Joerg J, Oertel WH, Ulm G, Schneider E. Possible environmental, occupational, and other etiologic factors for Parkinson's disease: a case-control study in Germany. *Neurology*. 1996; 46(5): 1275–1284.

Semchuk KM, Love EJ, Lee RG. Parkinson's disease and exposure to rural environmental factors: a population based case-control study. *Canadian Journal of Neurological Sciences*. 1991; 18(3): 279–286.

Sharifi N, Steinman RA. Targeted chemotherapy: chronic myelogenous leukemia as a model. *Journal of Molecular Medicine*. 2002; 80(4): 219–232.

Shimura H, Hattori N, Kubo S, Mizuno Y, Asakawa S, Minoshima S, Shimizu N, Iwai K, Chiba T, Tanaka K, Suzuki T. Familial Parkinson disease gene product, parkin, is a ubiquitin-protein ligase. *Nature Genetics*. 2000; 25(3): 302–305.

Shults CW, Flint BM, Song D, Fontaine D. Pilot trial of high dosages of coenzyme Q10 in patients with Parkinson's disease. *Experimental Neurology*. 2004; 188(2): 491–494.

Sian J, Dexter DT, Lees AJ, Daniel S, Jenner P, Marsden CD. Glutathione-related enzymes in brains of Parkinson's disease. *Annals of Neurology*. 1994; 36(3): 356–361.

Simstein R, Burow M, Parker A, Weldon C, Beckman B. Apoptosis, chemoresistance, and breast cancer: insights from the MCF-7 cell model system. *Experimental Biological Medicine (Maywood)*. 2003; 228(9): 995–1003.

Singh RB, Niaz MA, Rastogi SS, Shukla PK, Thakur AS. Effect of hydrosoluble coenzyme Q_{10} on blood pressures and insulin resistance in hypertensive patients with coronary artery disease. *Journal of Human Hypertension*. 1999; 13(3): 203–208.

Smith MA, Hirai K, Hsiao K, Pappolla MA, Harris PL, Siedlak SL, Tabaton M, Perry G. Amyloid-beta deposition in Alzheimer transgenic mice is associated with oxidative stress. *Journal of Neurochemistry*. 1998; 70: 2212–2215.

Somayajulu M, McCarthy S, Hung M, Sikorska M, Borowy-Borowsi H, Pandey S. Role of mitochondria in neuronal cell death induced by oxidative stress; neuroprotection by coenzyme Q_{10}. *Neurobiology of Disease*. 2005; 18: 618–627.

Srivastava RK. TRAIL/Apo-21: mechanism and clinical application in cancer. *Neoplasia*. 2001; 3: 535–546.

Steece-Collier K, Maries E, Kordower JH. Etiology of Parkinson's disease: genetics and environment revisited. *Proceedings of the National Academy of the Sciences of the United States of America*. 2002; 99: 13972–13974.

Takada K. Role of Epstein-Barr virus in Burkitt's lymphoma. *Current Topics in Microbiology & Immunology*. 2001; 258: 141–151.

Thatte U, Bagadey S, Dahanukar S. Modulation of programmed cell death by medicinal plants. *Cellular Molecular Biology*. 2000; 46(1): 199–214.

Thiruchelvam M, Richfield, EK, Goodman BM, Baggs RB, Cory-Slechta DA. Developmental exposure to the pesticides paraquat and maneb and the Parkinson's disease phenotype. *Neurotoxicology*. 2002; 23(4–5): 621–633.

Thiruchelvam M, Prokopenko O, Cory-Slechta DA, Richfield EK, Buckley B, Mirochnitchenko O. Overexpression of superoxide dismutase or glutathione peroxidase protects against the paraquat + maneb-induced Parkinson disease phenotype. *Journal of Biological Chemistry*. 2005; 280: 22530–22539.

Tipton KF, Singer TP. Advances in our understanding of the mechanisms of the neurotoxicity of MPTP and related compounds. *Journal of Neurochemistry*. 1993; 61(4): 1191–1206.

Tomasetti M, Littarru GP, Stocker R, Alleva R. Coenzyme Q10 enrichment decreases oxidative DNA damage in human lymphocytes. *Free Radical Biology & Medicine*. 1999; 27(9–10): 1027–1032.

Troy CM, Rabacchi SA, Xu Z, Maroney AC, Connors TJ, Shelanski ML, Greene LA. beta-Amyloid-induced neuronal apoptosis requires c-Jun N-terminal kinase activation. *Journal of Neurochemistry*. 2001; 77: 157–164.

Zhang Y, McLaughlin R, Goodyer C, LeBlanc A. Selective cytotoxicity of intracellular amyloid beta peptide142 through p53 and Bax in cultured primary human neurons. *Journal of Cellular Biology*. 2002; 156: 519–529.

Zhu X, Raina AK, Perry G, Smith MA. Alzheimer's disease: the two-hit hypothesis. *The Lancet Neurology*. 2004; 3: 219–226.

6

Bone marrow stromal cells in myocardial regeneration and the role of cell signaling

R ATOUI, R C J CHIU and D SHUM-TIM,
McGill University Health Center, Canada

6.1 Introduction

Myocardial infarction remains a widespread and important cause of morbidity and mortality among adults, accounting for more than 15 million new cases worldwide each year.[1] The loss of cardiomyocytes that results, combined with the limited endogenous repair mechanism, sets into play the remodeling process that ultimately leads to progressive heart failure. End-stage heart failure still has a grave prognosis with an estimated 5-year mortality rate of 70%.[2]

In addition to medical therapy, the management of heart failure currently includes the use of mechanical ventricular assist devices, pacing for ventricular synchronization, and other surgical techniques such as ventricular resection and mitral valve repair. Heart transplantation has been successful, but benefits only a few patients due to limited donor supply. A novel approach currently under intensive investigation is cellular transplantation which is directly aimed to overcome the problem of myocardial cell loss. We first introduced the term 'cellular cardiomyoplasty' in 1995 to indicate this new therapeutic strategy consisting of replacing dead cardiomyocytes with newly functional contracting cells.[3] Since then several infarct models performed on rodents, sheep, dogs, swine or monkeys have shown that the transplantation of a wide range of stem cells and progenitor cells is possible and contributes to the improvement in the ventricular function. Notable among the donor cells are the satellite cells[3]/ myoblasts[4] derived from the skeletal muscle, embryonic stem cells,[5] adult marrow stem cells[6] (MSC), dermal fibroblasts, fetal or neonatal cardiomyocytes,[7] other bone marrow-derived cells (CD34+) and proendothelial cells.[8]

In this chapter we will first briefly introduce the concept of cellular cardiomyoplasty, and focus on the evidence available so far regarding the cardiomyocytic differentiation of MSC. We will review the most updated information regarding the cellular and molecular signaling mechanisms for these MSC to be recruited and home in to the injury site and then undergo 'milieu-dependent' or *in situ* differentiation.[9] Finally, we will discuss the clinical and therapeutic implications of these studies.

6.2 Cellular cardiomyoplasty and myocardial repair

The ideal candidate donor cell would be a cell that can be relatively easily obtained, and expanded. Once implanted, it can home to the injury site, proliferate and differentiate into morphologically and functionally normal cardiomyocyte. As mentioned above, various types of cells have been administered to an ischemic myocardium and studies by different groups have repeatedly documented the successful engraftment of these cells in adult myocardium. Furthermore, most of these studies have also shown an improvement in the ventricular function after transplantation and are summarized in several good review articles.[10,11]

One clear division of the stem cell family is between those found in the embryo, known as embryonic stem cells (ES), and those found in adult somatic tissue. Skeletal myoblasts can be isolated from adult muscle and expanded in culture.[12] In the early 1990s, our laboratory reported the first successful transplantation of satellite cells into the injured myocardium.[3] Since then, several other groups have reproduced this finding, both on animal models and since 2001, in clinical trials on humans.[13] Although it still remains controversial, their long-term clinical utility may be limited by the finding that these cells are capable of differentiating only into mature skeletal myofibers and are unable to form functional gap junctions with the host cardiomyofibers.[7,14,15] Fetal cardiomyocytes and ES cell-derived cardiomyocytes have also been transplanted.[5,16] However, major ethical, moral and legal limitations as well as shortage of donors and the issue of chronic rejection hinder their clinical use. Thus in this chapter, we will focus our discussion on the MSC as the donor cells for cellular cardiomyoplasty.

In contrast to other cell types, MSC appear to possess some unique properties. They can be harvested and handled relatively easily, multiplied in culture and implanted without encountering immuno-rejection, as will be further discussed later. Furthermore, they have been shown to differentiate into several lineages, including the cardiomyocytic phenotype.[17] With this apparent plasticity, MSC could be an ideal cellular source for cell therapy.

6.3 Mesenchymal stromal cells as adult stem cells

A stem cell is generally defined as a primitive cell capable of self-renewal, and able to undergo pluripotent differentiation when exposed to the appropriate conditions. Bone marrow stromal cells, also called 'stromal stem cells', 'marrow progenitor cells', 'marrow mononuclear cells', 'mesenchymal stem cells' and 'marrow-derived adult stem cells' essentially represent a heterogeneous population of fibroblast-like cells, which can be found in the bone marrow (BM) stroma. There is evidence that at least some 'adult stem cells' isolated from muscle, skin, adipose tissue and peripheral blood originated from the bone marrow. Furthermore, the pluripotent stem cells derived from the amniotic fluid, placenta, and umbilical cord blood show some characteristics similar to those of MSCs.

The MSCs residing in the bone marrow were previously believed to play only supportive roles for hematopoiesis by expressing various cytokines, growth factors and adhesion molecules. Cohnheim in the 19th century[18] first implied the presence of these cells in the blood and their possible role in wound repair. Friedenstein and his group were the first in the early 1970s to better describe these MSC in a number of species, including mice, rats, rabbits, guinea pigs, hamsters and humans, showing their differentiation potential into cells of mesenchymal lineage including chrondrocytes, osteoblasts, myocytes and adipocytes.[19–21] Because these cells appeared clonal in nature, they were initially termed colony-forming unit fibroblasts (CFU-F). Isolation of MSC was then undertaken by Caplan who described a technique still used today by isolating the cells that adhered to the bottom of the plates when the bone marrow cells are cultured *in vitro*.[22] Furthermore, several *in vivo* and *in vitro* studies have confirmed the pluripotent potential of these cells and have observed the presence of injected MSC in host adipose tissue, lung, cartilage, central nervous system, liver, spleen, thymus and skeletal muscle.[23–28] In the last few years, studies have also shown the capacity of these MSC to differentiate into cells of lineages other than mesenchymal, such as hepatocytes,[23] kidney and even early astrocytes.[29]

6.3.1 MSC characteristics and subpopulations

Although MSCs' pluripotent potential has been demonstrated in many studies, controversy still exists as to what proportion of these cells is truly pluripotent. Thus, although they are collectively called marrow-stromal cells, not all stromal cells are pluripotent.[30] In fact, it was reported that up to one-third of the initial adherent stromal colonies are truly polypotent.[31] Plating studies confirm the rarity of MSC in the adult bone marrow, representing approximately 0.01% to 0.05% of the nucleated cells, being much less abundant than their hematopoietic counterpart.[32] Nonetheless both cell types appear to contribute to myocardial repair.

The human MSC can be cloned and expanded to greater than 1 million-fold and still retain the ability to differentiate into several mesenchymal lineages. After isolating human MSC from over 600 patients, Pittenger and his coworkers have shown that these cells behaved as a homogeneous population, and retained their multilineage potential for several passages, although not indefinitely.[31]

Unlike hematopoietic cells, MSC are CD34− and CD45−. Although still not fully identified, some other characteristic MSC surface markers include CD29, CD44, CD71, CD90, CD106, CD120a, CD124, SH2, SH3 and SH4-69 (Table 6.1). It is important to keep in mind that this is an incomplete list, and as

Table 6.1 Phenotype characterization of human MSC

Antigens	Expression/References
Adhesion molecules	
ICAM-1,2,3	+ [31,35] or − [171]
VCAM	+ [31,35]
LFA-3	+ [32,35] or − [171]
LFA-1	− [31]
L-selectin	+ [35]
P-selectin	− [31,32,35]
Growth factors and cytokine receptors	
IL-1,3,4,6,7 R	+ [35]
INFγ-R	+ [35]
TNF-α-R	+ [35]
FGFR	+ [35]
PDGFR	+ [35]
Integrins	
VLA-α1,2,3,5,6	+ [32,35]
B4-integrins	+ [35]
LFA-1,2	+ [32,35]
Additional markers	
B2 microglobulin	+ [32]
Nestin	+ [35]
Endoglin, SH2	+ [31,32,35,171]
HLA ABC	+ [161,171]
HLA DR	− [161,171]
B7.1,2	− [161,172]
Transferrin receptor	+ [31,35] or − [171]

mentioned above, some variation has been seen from laboratory to laboratory.[30] In fact there is currently considerable confusion regarding the definition and composition of such cells. For this reason, no unique phenotype has been identified that allows the reproducible isolation of MSC with predictable lineage differentiation. The reasons behind such uncertainty lie primarily in the experimental conditions used such as the heterogeneity of culture conditions, cell separation techniques and different molecular cell markers used by various investigators. Thus, while the principle of clonal homogeneity is used by some experts to define these cells,[31] others use a different combination of molecular cell markers such as c-kit+/Lin− cells,[32] Sac-1+ Lin−/cKit+ cells,[33] c-kit+/CD34− cells,[34] etc. Furthermore, early studies suggested a common precursor between the hematopoietic and mesenchymal lineages,[36] identified as CD34+, CD38−, HLA DR−. Waller and his group further subdivided the two lineages based on the CD50 marker; thus defining MSC precursors as CD50−, CD34+ cells.[37]

Because of such differences, it is often difficult to compare the findings among different studies.[35] Standardization of such classifications is of paramount importance as it will be very helpful in the further exploration of the mechanisms of MSC differentiation. One possible reason behind this confusion might be because only fully mature cells can be characterized by a defined set of specific markers. In fact, because of their undifferentiated state, a constantly changing set of markers may be continuously defining the 'labile' phenotype of MSC.

6.3.2 Pathophysiological role of MSC in cardiac injury

In contrast to the ES cells, whose goal is to develop a new organism, cumulative information gathered during the past several years suggests that adult stem cells participate in tissue growth and repair throughout postnatal life.[38,39] In fact, there is currently ample evidence suggesting that MSC can be recruited from the BM to various tissues to participate in tissue repair and regeneration in response to either apoptosis or tissue injury.[40,41] In fact, progenitor stem cells have been shown to be recruited from the BM to contribute to angiogenesis in wound healing, vascularization post-myocardial ischemia, and even growth of certain tumors.[42]

A hypothesis driven by our laboratory to explain the role of MSC in the bone marrow is that they serve as 'reserve' cells to participate in tissue repair when needed.[43] Indeed, several studies have shown that MSC differentiation occurs almost exclusively in organs that have been damaged. For instance, differentiation to endothelial cells, hepatocytes and myoblasts is seen in cases of ischemia, cirrhosis and muscular dystrophy.[26,44,45] In this case, it is hypothesized that upon injury, stem cells can proliferate *in vivo* and are then recruited to the injured environment. There, they will differentiate in response to local cues[38,46] (see below).

Several pieces of evidence published in the last several years has confirmed this pathophysiological role of marrow-derived adult stem cells. Orlic and his group have shown that labeled MSC can be mobilized within hours of myocardial infarction to home to the injured myocardium.[47] Furthermore, by using a coronary artery ligation model, Bittira *et al.* from our laboratory,[38] demonstrated that in response to a myocardial injury, labeled MSC are recruited from the bone marrow, and travel through the circulation to home to the peri-infarct area within hours to days. In the following weeks, these MSC underwent 'milieu-dependent' differentiation and expressed various phenotypes including cardiomyocytes, myofibroblasts, endothelial and smooth muscle cells. Our hypothesis is that each type of cell is somehow involved in the pathophysiological process following myocardial infarction. For instance, newly formed endothelial cells can participate in the process of angiogenesis; cardiomyocytes can functionally integrate into the myocardium; and myofibroblasts can contribute to scar maturation, which favorably alters the remodeling process.

It is important to keep in mind that this hypothesis, although appearing very credible, can only explain part of the mechanism since it is clear that these cells do not always fully repair the damaged myocardium, as evidenced by the clinical consequences of a myocardial infarct. Further questions remain to be answered. For instance, it is known that immediately after birth, a low level of quiescent progenitor cells, including stem cell precursors, are released into the peripheral circulation.[48] Although it has been shown that these circulating stem cells can repopulate areas of damaged bone marrow and thymus, the exact physiological role of these cells and their fate are currently unknown.

6.4 Plasticity of adult MSC: milieu-dependent differentiation

Plasticity describes a property that allows adult stem cells, assumed until recently to be committed to a fixed lineage, to switch to produce other specialized sets of cells appropriate to their new microenvironment.[49] To explain this, Verfaillie proposed a hypothesis in which these MSC can proliferate and differentiate in response to local cues provided by the environment they are recruited to.[46]

Stem cells have been identified in most organ tissues. The best studied so far is the hematopoietic stem cell (HSC).[47] Several studies have shown that HSC can repopulate the hematopoietic cell pool when transplanted into lethally irradiated animals or humans.[48,49] Many studies later confirmed the differentiation potential of the MSC with respect to the mesenchymal lineage, in particular bone and cartilage.[35] For instance, it was found that human MSC can express genes characteristic of both the osteoblastic and adipocytic lineages, thus clearly indicating their progenitor phenotype.[50] Furthermore, it was also clear that differentiated human adventicular reticular cells can mature into adipocytes upon pharmacological myelosuppression *in vivo*. These cells are thus able to switch phenotypes among two terminal stages within the progeny of the MSC.[51] This finding may highlight the plasticity of the bone marrow stroma and distinguishes it from the hematopoietic system in which such shifts among differentiated cells do not occur.

Historically, the connection between the bone marrow and osteogenesis was first observed in the 19th century, and later revived by Friedenstein and his group.[21] It was clear from these series of experiments that the extent of bone formation varies broadly depending on the transplantation conditions. Placement of the cells into diffusion chambers allowed the flow of nutrients, but not the movement of host cells. The production of mesenchymal lineage following transplantation confirmed that the differentiation capacity lies with the donor MSC.[52] This finding was later confirmed by Owen and his group on rabbit bone marrow cells.[53] Since these initial observations, more definitive evidence for the multipotential differentiation of the MSC have been reported by other investigators showing the ability of the MSC to repopulate several non-hematopoietic tissues, such as skeletal myoblasts,[25,26] neuronal cells,[27,54] cardiomyocytes,[32,55] endothelial cells,[32] hepatocytes,[23] and lung, gut, kidney, pancreas and skin epithelia.[28] It is, however, important to keep in mind that most of these studies did not conclusively demonstrate that a single cell could differentiate into different cell lineages. Although some studies have shown that the involved cell populations were rich in hematopoietic stem cells,[26,44] they did not identify the exact phenotype of the cell capable of differentiating.

6.5 Cardiomyocytic differentiation of MSC

Data from a number of laboratories have shown that MSC, once exposed to a variety of physiological or non-physiological stimuli, differentiate into cells with a cardiomyocytic phenotype. These exhibit a myotube-like structure with typical sarcomeres, are positive for markers specific for cardiomyocytes, express multiple contractile proteins and display sinus node-like and ventricular cell-like action potentials.[56–60]

6.5.1 Animal *in vitro* studies

Wakitani and his group were the first to show that a hypomethylating agent, 5-azacytidine, can convert rat MSC to multinucleated myotubes that contracted when exposed to acetylcholine and stained positively for skeletal muscle-specific myosin.[59] Another landmark study by Makino *et al.* established a cardiomyocytic cell lineage after treating MSC with 5-azacytidine, expressing cardiomyocyte-specific genes, with evidence of ventricular-like action potentials[56] and expressing β-adrenergic and muscarinic receptors.[61] However, 5-azacytidine is known to be toxic *in vivo*.

In order to determine the nature of the possible *in vivo* signals involved, Tomita and his group used a co-culture system and found that when the labeled-MSC were co-cultured with cardiomyocytes, only with direct cell-to-cell contact could they induce cardiomyocytic differentiation. Separating the two populations with a filter shield, hence allowing the passage of macromolecules but preventing direct cell-to-cell contact, failed to induce such differentiation.[62] These results are consistent with the hypothesis, described above, that cell-to-cell contact may play a crucial role in the milieu-dependent differentiation of MSC, relaying cardiac environmental signals. Furthermore, other studies have shown the presence of specific gap junctions allowing direct cell-to-cell contact between the implanted human MSC and ventricular myocytes,[63] as well as the remarkable cellular and molecular similarity between 'true' cardiac cells in culture and the cardiomyocytic-like cell that differentiated from MSC.[64]

6.5.2 Animal *in vivo* studies

To confirm the *in vitro* studies mentioned above, many laboratories, including ours, looked at the potential of the MSC to differentiate *in vivo* into functional cardiomyocytes. Tomita and his coworkers were the first to report the differentiation of rat MSC into myogenic cells expressing cardiac-specific genes.[65] After creating ischemic rat myocardium, 5-azacytidine-pretreated MSC were observed in the transplanted area but not in the control scars. Furthermore, a higher degree of angiogenesis, a smaller transmural scar as well as an improved ventricular function was observed in the transplanted group. It should be noted that bone marrow, as opposed to a purified population of MSC, was used in this study.

In our laboratory, we explored the hypothesis that MSC, when implanted, will choose to express a specific phenotype based on the principle of milieu-dependent differentiation.[66] If this is the case, we would not need to pre-treat the MSC with 5-azacytidine to induce the cardiomyocytic phenotype. In our experiment, we implanted labeled MSC near the peri-infarct area in rats. It was noted that the stem cells surrounded by scar tissue appear to differentiate into fibroblasts, whereas those in direct contact with native cardiomyocytes expressed phenotypic molecular markers specific to cardiomyocytes such as connexin-43 and troponin I-C.[66] This finding supported our hypothesis that these cells received signals from neighboring cells to express phenotypes specific to their microenvironment. Thus, depending on the surrounding milieu, our cells appear to have differentiated into cardiomyocytes, fibroblasts, endothelial cells or adipocytes. From an evolution point of view, one can postulate that by obtaining such signals from the surrounding neighborhood, MSCs may avoid undergoing heterotopic differentiation.[43]

It is of interest to note that if one injects 5-azacytidine pre-treated MSC into the scar tissue of an injured myocardium, cardiomyogenic differentiation of these cells can still occur within the scar. One can thus suggest that such a pretreatment *in vitro* can alter MSC gene expression such that these cells will no longer respond to microenvironmental signals, but rather undergo lineage-specific differentiation.[58,65]

In another series of experiments, Wang *et al.* in our laboratory injected male rats with labeled rat MSC and showed that these cells will develop into cells morphologically similar to cardiomyocytes, exhibiting organized contractile fibers.[6] This view appeared consistent with our subsequent study,[67] whereby MSC were injected directly into the coronary arteries of an ischemic rat myocardium. These cells were subsequently found to migrate out of the coronary vasculature and differentiate into cells of multiple lineages, depending on their microenvironment. These studies further supported the hypothesis that the fate of the implanted MSC is defined by its cardiac microenvironment, thus consolidated the concept of 'milieu-dependent' differentiation, a term that was originally suggested by Edelman in relation to embryogenesis.[9]

These findings were subsequently confirmed by many laboratories around the world. Pittenger and his group used a swine myocardial infarction model and demonstrated the differentiation of MSC toward a myogenic lineage with expression of α-actinin, troponin-T and tropomyosin[68] resulting in an improved overall left ventricular function. Furthermore, Orlic *et al.* injected labeled Lin−/ckit+ cells from male mice into an ischemic female mouse myocardium[47] and found newly formed Y-containing myocytes occupying up to 2/3 of the infarct area. Similar findings were obtained by Toma *et al.* using adult mice,[17] and by Davani and his group.[69] In a swine model, investigators have used magnetic resonance (MR) fluoroscopy to identify target sites on the myocardium in order to guide their injections. In these studies, not only were MSC shown to engraft and express several cardiac markers, but a significant improvement in the ventricular function was also noted accompanied with a reduction in wall thinning.[68,70] Moreover, Kawada *et al.* transplanted labeled MSC into the BM of a mouse ischemic model that was treated with G-CSF. They were then able to demonstrate the presence of labeled cells in the ischemic myocardium, suggesting that most of the labeled cardiomyocytes originated from the implanted MSC.[71]

Other studies focusing on gender mismatched human heart transplants have found a wide difference in the estimate of the levels of Y chromosome-containing cells ranging from 0.04% to 18%.[72,73] The discrepancies in the amount of chimerism among different groups are probably due to technical differences. Although these results are still controversial,[74] they highlighted the repair function of extracardiac stem cells and their potential of regenerating the injured myocardium.

In addition, recent studies evaluating the effect of MSC on myocardial perfusion have also shown the ability of these cells to enhance neovascularization. Implanted MSC were shown to express von Willebrand factor (vWF), vascular endothelial growth factor (VEGF) and other proteins indication ongoing angiogenesis.[32,75] Kinnaird *et al.* recently showed that MSC secreted a variety of angiogenic cytokines such as fibroblast growth factor (FGF), VEGF, insulin-like growth factor (IGF), hepatocyte growth factor (HGF), matrix metalloproteinases (MMP), platelet-derived growth factor (PDGF), IL-1, angiopoetin, TGF-β, TNF-α and many others, most of which are upregulated following a myocardial infarction and probably contribute to stimulating neovascularization following a myocardial infarction.[76,77] Among all these factors, VEGF seems to be the key regulatory cytokine orchestrating endogenous neovascularization by modulating stem cell migration and proliferation.[76] Not only does it stimulate the development of microvessels, but it also contributes to endothelial cell survival through VEGF-mediated phosphorylation of protein kinase B and nitric oxide synthase proteins.[77]

In a recent study, Tang *et al.* highlighted the paracrine action of the engrafted MSC in the ischemic myocardium and the resulting stimulation of neovascularization.[173] They also showed that the release of bFGF, VEGF and SDF-1 not only leads to efficient vascular regeneration but also attenuates the apoptotic pathway by downregulating the prosurvival protein Bax. Finally, it was shown that local injection of MSC-derived conditioned media alone containing several arteriogenic cytokines can enhance collateral perfusion in a murine model of hindlimb ischemia, hence highlighting the important role of paracrine signaling.[76]

6.5.3 Trans-differentiation *vs* fusion

The studies reviewed above support the idea of MSC differentiation. However, this concept has been challenged recently with the demonstration of cell fusion whereby a new cell is derived from the fusion of the implanted cell and a native host cell. Several *in vitro* and *in vivo* studies published in the last few years[78–80] showed that cell fusion can be responsible for a certain percentage of phenotypic changes observed following transplantation. Terada and his group demonstrated the presence of polyploidy DNA content when female BM cells were co-cultured with male ES cells.[78] Although still highly controversial, it is important to keep in mind that the frequency of this phenomenon varied widely among different studies and cannot by itself explain all the significant regeneration observed in previous studies. In any case, future studies must examine this mechanism with rigor in order to better understand the mechanisms of cellular transplantation.

Furthermore, in the last several years, compelling evidence has emerged showing the potential of cardiomyocytes to re-enter the cell cycle and undergo mitosis.[81,82] To add more to this confusion, the concept of resident cardiac stem cells was recently introduced as well.[83–85] Although they may indeed exist, their physiological role appears so far minimal.

6.5.4 Improvement in ventricular function

Most of these previous studies have noted an improvement in the ventricular function in the transplanted group. It is of interest to note that even the experiments that did not show extensive myocytic differentiation nonetheless showed an improvement in the global functioning of the heart. How can these cells, without apparent connection to the native myocardium contribute to the improvement in ventricular function remains perplexing. A number of mechanisms have been proposed[86] and include contraction of the implanted cells, changes in the extracellular matrix by an autocrine mechanism, with an improvement in the elastic property of the transplanted region, thus limiting the remodeling process; and an enhancement of the angiogenesis in order to rescue hibernating myocardium.

6.6 Stem cell niche

Recently, it has been realized that stem cells exist in various adult tissues, not just in constitutively renewing organs. The concept of stem cell niches containing adult stem cells was first introduced by Schofield in the end of 1970 and refers to a subset of tissue cells such as fibroblasts, adipocytes, osteoblasts and endothelial cells and their extracellular substrates, which *in vivo* regulates stem cell fate.[87] These niches are believed to be present across many species, from the primitive *Drosophila* to humans. They are composed of not only the stem cells themselves but also of a framework of other cell types within a rich dynamic microenvironment that allows these cells to maintain their unique intrinsic abilities of self-renewal and differentiation.[88] The complex regulatory mechanisms governing stem cell fate must be able to maintain a balance between a self-renewal phenotype and that of a differentiated specialized cell. This involves many signal transduction pathways, not yet well defined. A detailed comprehensive coverage of this topic can be found in a recent review by Fuchs *et al.*[88]

6.6.1 Molecular regulation of stem cell fate

Several mechanisms, both temporal and spatial, have been demonstrated to play key roles in the regulation of a stem cell fate. It is now believed that extrinsic control of stem cell fate relies on highly conserved signaling pathways[89] that interact with each other and include Wingless (Wnt), Notch/Delta, transforming growth factor-β/bone morphogenic protein (BMP-2), cyclin-dependent kinase inhibitor p21, proto-oncogene Bmi-1, transcription factor HoxB4, JAK/STAT, Rho and Sonic Hedgehog pathways.[88,90–92] Furthermore, it has been suggested that inputs from distinct signaling pathways must be integrated by the stem cell in order to produce discrete biological output.[91]

For instance, Wnt signaling pathway was shown to play an active role not only in multiple processes in animal development, but also as a major regulator in the maintenance and self-renewal of HSC.[91] Briefly, after binding to its receptor, GSK-3β-mediated phosphorylation of β-catenin is inhibited, which subsequently leads to the interaction of catenin with members of the LEF-1/TCF family of transcription factors, thus activating expression of target genes previously implicated in HSC renewal such as HoxB4 and Notch1. Furthermore, Duncan *et al.* have recently shown how Notch signaling is critical for the maintenance of an undifferentiated state by HSC and may hence act as a 'gate-keeper' between self-renewal and differentiation.[93] This mechanism might actually be activated by the Wnt pathway, highlighting once more the integration of several pathways within a stem cell.

In addition to the signal identity, signal strength, which can be achieved by altering either receptor or ligand concentration, has also been shown to have an important effect on stem cell fate. This mechanism is highlighted in endothelial cell differentiation and maintenance which is highly dependent on VEGF.[94] Furthermore, temporal and spatial mechanisms involving the timing and duration of the stimuli, as well as the content of the microenvironment surrounding the cell do also influence stem cell fate. In fact, various studies have shown the importance of direct cell-to-cell contact, as well as the stem cell size and the paracrine and autocrine signaling events that surround stem cell fate.[89,90]

6.7 Cell signaling and mechanisms of differentiation

Numerous studies analyzed the signaling mechanisms involving stem cells' regulation and proliferation. Most of these reports focused on HSC signaling. However, as we will see, and perhaps not surprisingly, many of the factors and pathways involved have also been shown to be implicated in MSC differentiation.

A number of studies have tried to analyze the endogenous and environmental factors that are involved in the regulation of stem cells, including inflammatory cytokines, growth factors, surface receptors, proteases, transcription factors, telomerase activities, hypoxia-responding proteins and stem cell–matrix interaction. Furthermore, Lapidot and Petit even suggested the existence of a dynamic situation in which there is a constant turnover, proliferation, migration and homing of stem cells as part of their developmental program, a process that may even be linked to the dynamic interaction between osteoblasts/osteoclasts in BM remodeling.[48] Interestingly, G-CSF stimulation induced both MSC mobilization and osteoclast-mediated BM resorption.[95]

6.7.1 Stem cell mobilization

Several previous studies focusing on knock-out embryos revealed the critical roles of SDF-1α, a member of the CXC chemokine family, which was shown to bind to its 7-transmembrane-spanning G protein-coupled receptor CXCR4. Its constitutive expression in various tissues as well as its highly conserved amino acid sequences between different species highlights its important biological role, namely in cardiogenesis, stem cell hematopoiesis, vasculogenesis and cerebral development.[48] These studies confirmed the role of SDF-1 as the key regulator of stem cell trafficking between the BM and the peripheral circulation. In fact, Peled *et al.* have demonstrated in a series of studies that SDF-1/CXCR4 interactions not only tightly regulate stem cells homing but are also involved in transendothelial migration by mobilizing progenitor stem cells and activation of major integrins such as LFA-1, VLA-4 and VLA-5 mediating cell-to-cell and cell-to-matrix interactions in response to tissue stress or injury.[96,97]

The mobilization process whereby stem cells are released from the BM was first documented in 1970s and has been shown to be induced both in animals and humans by a wide number of molecules, including cytokines such as G-CSF, GM-CSF, interleukin IL-7, IL-3, IL-12, stem cell factor (SCF) and flt-3 ligand, chemokines such as IL-8, Mip-1α, Groβ, or SDF-1 and a variety of chemotherapeutic agents.[98] For instance, IL-8, which is secreted in response to SDF-1 stimulation, is believed to stimulate stem cell mobilization by activating MMP-9 and the integrin LFA-1.[99–102] Similarly, it was found that both SDF-1 and steel factor act cooperatively to attract progenitor stem cells from the bone marrow.[35] Furthermore, Sweeney *et al.* found that sulfated polysaccharides can increase the levels of SDF-1 which can ultimately result in an up-regulation of the MSC mobilization.[103] In this study, they demonstrated that these polysaccharides compete for SDF-1 binding to the BM endothelium. Furthermore, Yamaguchi *et al.* confirmed the hypothesis that locally administered SDF-1 can, by augmenting the levels of endothelial progenitor cells to the site of ischemia, enhance the efficacy of neovascularization after systemic EPC transplantation.[104]

Although the exact mechanism of mobilization remains not fully understood, it is believed to be a multi-step process whereby a key process involves the disruption of the adhesion interactions between the stem cells and the BM when stimulated by specific signals such as an ischemic injury or stress. In fact, Papayannopoulou *et al.* demonstrated the critical role of VLA-4 in the mobilization process[105] and several previous studies have shown the role of proteolytic enzymes such as elastases, peptidases, cathepsins G, MMP-2 and MMP-9 in inactivating SDF-1 by cleaving part of its N-terminus. Furthermore, a recent study showed that the increase in the level of cathepsin G and elastase correlate with stem cell mobilization.[106] In accordance with these studies, Petit and her group demonstrated the proteolytic degradation of SDF-1 by elastase induced by G-CSF, accompanied by a gradual increase in CXCR4 expression on bone marrow cells.[107] Studies by Moore and coworkers confirmed further the critical role of MMP-2 and MMP-9, as well as their natural tissue inhibitors (TIMP) in allowing SDF-1 and G-CSF mediated stem cell mobilization.[108] The importance of these metalloproteinases is also highlighted by their role in maintaining low levels of surface CXCR4 to keep the stem cells in the circulation.[109] Furthermore, Heissig and his group[110] showed that the activation of MMP-9 is followed by the release of SCF into the circulation, which is essential for SDF-1 mediated stem cell mobilization and proliferation. Finally, Flores *et al.*[111] recently highlighted the role of telomerase Tert and telomere length as critical determinants in the mobilization and proliferation of epidermal stem cells for their niches.

6.7.2 BM: niche for HSC

Engraftment of the stem cells in the bone marrow can be seen as the end of a complex series of events in which circulating HSC are first recruited by the BM vasculature followed by their transendothelial migration into the extravascular hematopoietic cords of the marrow.[112] Once in the BM, the regulation of HSC proliferation and differentiation occurs by a complex interplay of cells, growth factors, adhesion molecules and other signals, not yet fully understood. This is then followed by lodgment whereby cells selectively migrate to a suitable niche in the extravascular compartment of the BM.

Previous studies suggested key roles for P-selectin,[113] E-selectin,[114] the $\beta 1$ integrin very late antigen-4 (VLA-4),[115] SDF-1 and CXCR4,[116] in the homing of HSC to the bone marrow. Thus, to home to the BM, stem cells must first roll on E and P selectins, which are expressed on the BM vascular cells. After their adhesion to the vessel wall via the major integrins (VLA-4, -5 and LFA-1) and their vascular ligands (VCAM-1 and ICAM-1), they extravasate into the hematopoietic compartment. Almost every one of these previous steps has been shown to be activated by SDF-1.[117]

Furthermore, recent data demonstrate that flt-3 ligand plays an important role in the proliferation of HSC in tightly regulating the actions of VLA-4 and VLA-5.[118] Also, Driessen and coworkers recently confirmed the transmembrane isoform of SCF as important in the lodgment of HSC in their niche.[119]

In another recent study, Mohle and his group showed that other non-peptide mediators such as cysteinyl-leukotriene receptor CysLT1 similarly stimulate HSC migration.[120] Also, Netelenbos *et al.* introduced the role of proteoglycans such as heparin and dermatan sulfate in HSC homing by showing their attachment to SDF-1.[121] The glycosaminoglycan hyaluronic acid (HA) recently found to be synthesized by HSC, was also found to have a key role on their migration and engraftment.[122] Hence, it seems that the processes of mobilization and homing are 'mirror images' involving the same molecules in 'opposite directions'.[123] Thus, by activating adhesion molecules, SDF-1 plays an important role in homing of the stem cells and engraftment in the BM. On the other hand, desensitization of the SDF-1/CXCR4 pathway is required for the successful mobilization of the stem cells from the BM.

Other studies have focused on the impact of cytokine exposure on the homing mechanism. Ahmed *et al.*[124] reported that cytokine-activated CD34+ cells (with IL-3, IL-6 and SCF) showed irreversible impaired homing ability, possibly through the induction of pro-apoptotic genes such as Fas/CD95.[125] Furthermore, Zheng *et al.* reported significant up-regulation in the concentrations of homing-related signals such as CD49, CD54, CXCR4, MMP-4 and MMP-2 when stem cells were shortly exposed to SCF.[126]

Although these findings can give us some insight into the mechanism of stem cell mobilization, it is clear that many more studies are needed to fully understand this complex event that involves the interplay between several adhesion molecules, chemo-cytokines, proteolytic enzymes and the BM. Clinically, this is very relevant since it is possible that the manipulation of SDF-1/CXCR-4 interactions, as well as the simultaneous infusion of stromal cells with the hematopoietic component, could improve the outcome of human BM transplantations.[127] In fact, several studies are currently underway taking advantage of the MSC in autologous and allogeneic transplantation.[48]

6.7.3 Interactions between MSC and HSC

As we described previously, the interactions between MSC, HSC as well as other mediators in the BM are important in the homing and proliferation process. In fact, several studies have identified numerous receptors on MSC important for cell adhesion with HSC and the rest of the extracellular matrix such as ICAM, VCAM, platelet endothelial cell adhesion molecule PECAM, L-and P-selectins (Table 6.1). It is also likely that the adhesive interactions that occur between HSC and MSC help not only in the homing and engraftment of the HSC in the BM, but are also involved in the proliferation and differentiation process of progenitor cells.[112] For instance, several studies have identified a number of growth factors, expressed in MSC cultures which are associated with hematopoietic support such as IL-6, IL-11, LIF, CSF, G-CSF, GM-CSF and SCF.[128] Furthermore, Calvi and his group recently highlighted the role of osteoblasts, present within the endosteal region, as key cellular elements in influencing stem cell differentiation through Notch activation.[90] Further pathways involving both MSC and HSC have been shown to have an important impact on stem cell differentiation and proliferation such as the Wnt signaling pathway[129] as well as the bone morphogenic protein receptor type 1A activation of specific osteoblastic cells.[130]

6.8 Homing of the MSC to the infarcted site

One of the most intriguing properties of MSC is their ability to home to sites of inflammation or tissue damage. Although the steps responsible for this migration have yet to be fully elucidated, it entails a two-step process whereby stem cells first bind to their adhesive complexes around the injury zone, followed by local chemotaxis to the site of engraftment.[131] This phenomenon has been demonstrated in various settings including infarcted hearts,[132] cerebral ischemia[133] and bone fractures.[134] In fact, Saito *et al.* from our laboratory were the first to demonstrate that MSC administered intravenously engraft within the infarcted myocardium, whereas those injected in healthy rats home to the bone marrow.[132] In another study, Sorger *et al.*[135] showed the remarkable specificity with which MSC can home to infarcted regions. Although the specific factors responsible for this migration have not yet been defined clearly, further complexity is added by a recent finding suggesting that expansion of murine MSC in culture may actually diminish their homing ability.[136]

As we saw previously, SDF-1 and its receptor CXCR4 are required for stem cells to home to the BM. Their role in coronary artery disease is less clear. Previous studies have shown the expression of SDF-1 in atherosclerotic plaques, its up-regulation in the heart early after MI as well as the increase in neovascularization following its exogenous expression.[131] Askari *et al.* further reinforced the role of SDF-1 in stem cell homing in a study whereby cardiac fibroblasts expressing SDF-1 were transplanted into the infarcted regions of rat hearts.[137] After using G-CSF to mobilize stem cells, a significant homing of c-kit cells to the injured area as well as an improved cardiac function was found in treated animals. Orlic and his group have also demonstrated the up-regulation of MSC homing and differentiation with the use of G-CSF.[47] In this study, a 250-fold increase in the levels of Lin−/c-kit+ cells as well as an improvement in the ventricular function were found in rats that were pretreated with G-CSF and SCF. A similar finding was obtained when granulocyte-macrophage stimulating factor (GM-CSF) was used.[138] Although the exact mechanism is yet to be understood, Harada *et al.* recently showed that this G-CSF-mediated stem cell mobilization and improvement in cardiac function occur through the activation of the Jak/Stat pathway in the cardiomyocytes, hence inducing a number of antiapoptotic proteins and angiogenic factors.[139]

Other than SDF-1, SCF is also involved in the regulation of stem cells migration by binding to its tyrosine kinase receptor, c-kit, which is expressed on a variety of stem cell lines.[1,140] This is confirmed by further studies showing the role of SCF in the induction of the expression of CXCR4 on human CD34+ cells, resulting in an increase in their migration in response to SDF-1.[116] A wide variety of chemokines have actually been shown to modulate such migration; however the largest response was seen with α and β SDF-1.[141] Furthermore, it is important to realize that although SDF-1 is required in stem cell mobilization to the injured site, it is not singularly sufficient, hence reflecting the need for additional factors. In fact, patients with acute coronary syndromes have elevated levels of many factors other than SDF-1 including MMP-2, MMP-9, ICAM and VCAM.[131]

As we saw previously, cell-to-cell interactions as well as other environmental factors involving a combination of paracrine growth factors promote stem cell migration and differentiation. Eghbali-Webb has recently reviewed the role of cardiac fibroblasts in regulating myocardial regeneration by the release of various soluble factors within the extracellular matrix such as VEGF, FGF, TGF-β1, PDGF and MMPs, highlighting the coordinated cell-to-cell and cell-to-environment interactions.[142] It is also possible that the hypoxia following an ischemic insult can enhance the expression of some adhesion molecules and thus facilitates MSC migration. For instance, the increase in MMP-9 level following the use of mobilizing agents such as SDF-1, VEGF and G-CSF or after a myocardial infarction, leads to an up-regulation of soluble kit, which ultimately results in an increase in MSC mobilization and proliferation.[143,144]

It is of interest to note that MacDonald *et al.* in our laboratory have observed that MSC which can migrate to the acutely injured site can lose this ability in chronic scar tissue, when the inflammatory mediators have probably subsided.[145]

The clinical application of modulating the SF1-CXCR4 axis to improve stem cell homing using monoclonal antibodies or antagonists against CXCR4 remains to be determined.[144] Furthermore, strategies to improve the engraftment of BM stem cells into the ischemic areas by the local administration of SDF-1 remains to be fully investigated. Finally, it was proposed that in order for the implanted stem cells to survive in the ischemic myocardium, one must also control the different factors that influence apoptosis, including cytokines and growth factors (such as HGF, GATA-4), expression of apoptosis-regulating genes (such as Fas, p53 and caspases) mitochondrial dysfunction, telomerase activities and hypoxia-responding proteins (such as hypoxia-inducible factor HIF-1 and erythropoietin).[146]

6.9 Therapeutic use of MSC

The recognition of the broad growth and differentiation potential of MSC and their relative ease of handling has opened the door to several clinical applications.

In the hematological field, clinical studies have progressed farthest in the use of human MSC in repopulating the BM stroma after myeloablative therapy, in conjunction with the reconstitution of the hematopoietic system with BM transplantation.[128]

Furthermore, the ability of MSC to proliferate makes it an excellent target for retroviral gene therapy.[147] In several studies, it was shown that stromal cells can be efficiently transduced with a variety of growth factors and hematological factors such as VEGF, VWF, factor VIII or XI.[35,148–150] Schwartz *et al.* were able to engineer MSC expressing L-DOPA when implanted into the brain in a rat model of Parkinson's disease.[151]

In the last few years, nine clinical trials around the world[152–155] have been completed and several are ongoing to assess the effect of autologous BM cells transplantation after acute myocardial infarction. With the exception of one,[156] all others showed encouraging results, with a significant increase in the ejection fraction and myocardial viability and a decrease in end-systolic LV volumes. It is of interest to note that in most of these trials, a heterogeneous fraction of BM mononuclear cells was used, containing B and T lymphocytes, myeloid cells, endothelial cells and a low number of hematopoietic and MSC. However, some of these trials involved purified fractions of cells such as CD34+ or CD133+ progenitors, as well as skeletal myoblasts. Furthermore, different delivery techniques were used, concomitant with medical therapy, angioplasty or coronary artery bypass surgery. In a recent randomized trial, Chen *et al.* injected MSC directly into the coronaries of patients post-myocardial infarctions.[157] It was found that treated patients had a decrease in the proportion of hypokinetic and akinetic segments, as well as a significant improvement in the ventricular function and wall motion.

Finally in addition to all these trials mentioned earlier, MSC and allogeneic bone marrow transplantation have also been used, although with limited success so far, in various mitochondrial defects and inborn metabolic diseases.[158]

6.10 Conclusions

This chapter reviews the 'state of the art'[159] in stem cell research and highlights the finding that undifferentiated adult stem cells are not determined progenitor cells with limited differentiation potential. Rather, these cells seem to possess a much broader capacity for cellular differentiation that is dependent on the microenvironment of the engrafted site. As we saw, a number of developmental regulatory pathways appear, perhaps not surprisingly, to be redundantly involved in regulating cell fates, engraftment, migration, lodgment, proliferation and differentiation. Although the basic mechanisms may be conserved among different stem cell lines, it is important to note that the cellular input of self-renewal or differentiation may be unique and confined to this particular stem cell and its microenvironmental niche. Furthermore, the multilineage potential of MSC, their ability to elude detection by the host's immune system, and their relative ease of expansion in culture make MSC transplantation a fascinating new approach for the management of heart disease. Ideally, MSC can be harvested, expanded and cryopreserved, ready for injection into patients following myocardial infarction.

Recent clinical trials have shown the feasibility of adult autologous cell therapy in patients following a myocardial infarction. However, interventions aimed at enhancing donor cell retention, survival, homing and proliferation are still definitely required to achieve a better level of cardiomyocyte engraftment.[160] Moreover, several unresolved questions are still open for future research. In addition to defining which stem cell is most suitable, we must also define which patient groups are suitable for this therapy, and what is the optimum timing, the optimal angiogenic milieu, the dosage and the method of delivery. Furthermore, long-term side effects and arrhythmogenic potential are still unknown as most of the clinical studies are very recent. Further fundamental questions relating to the biology of MSC are still unresolved. What are the specific signals and mechanisms involved in their homing, engraftment and differentiation? What are the exact mechanisms behind the improvement in ventricular function? What are the potential benefits of such therapy in non-ischemic heart failure? In addition, rigorous criteria are needed to better assess the efficacy of cellular cardiomyoplasty, as well as the stability and function of the transplanted cells.

6.11 Future trends

Another fascinating aspect of stem cell therapy involves the recent findings that MSC may have a unique immunological capacity to induce tolerance in immunocompetent allotransplants or even xenotransplant recipients.[132,161]

Liechty *et al.* reported the successful engraftment and site-specific differentiation of human MSC into sheep fetus,[162] even beyond the period of what is considered as 'immature' immune system of the fetus. In our laboratory, Saito *et al.*[132] transplanted labeled mice MSC into adult fully immunocompetent rats, thus producing stable cardiac chimeras for at least 12 weeks without any immunosuppression and with no evidence of rejection. Furthermore, Macdonald *et al.* from our group have shown that not only stable chimeras are formed, but that the overall ventricular function is improved.[163] These findings were once again replicated by Luo *et al.* from our lab (personal communication) using pig MSC into rat myocardium and were confirmed by several *in-vitro* studies using allogeneic and xenogeneic mixed lymphocyte reactions.[164,165]

Although the exact underlying immunological mechanism is not yet well understood, the 'Danger Model' theory of Matzinger[166] was invoked in our laboratory to try to explain the unexpected findings. In addition to the action of veto cells and to the proposed role of tryptophan catabolizing enzyme indoleamine 2,3-dioxygenase (IDO)-mediated tryptophan degradation in MSC-mediated immunosuppression,[167] Aggarwal and Pittenger recently reported that human MSC can secrete PGE2, hence altering the cytokine secretion profile of the T lymphocytes, NK and antigen-presenting cells,[168] namely by inhibiting TNFα and IFN-γ and by stimulating IL-10 to modulate the immune cell response. Clearly, additional studies are required to confirm such claims. If true, this finding would have an incredible implication in the clinical field, whereby 'universal donor cells' would be available, ready to be transplanted.

Whether the results obtained could be optimized by the more specific use of the multipotent adult mesenchymal stem cells (MAPC) isolated by the group of Verfaillie[169] remains to be demonstrated. Finally, it is of interest to note the recent first report of a clonally expanded subpopulation of human MSC that exhibits the capacity for pluripotent differentiation, and showing the ability of such homogeneous cell line to induce *in vivo* not only therapeutic neovascularization but also both exogenous and endogenous cardiogenesis; thus enhancing myocardial regeneration.[170]

It is therefore clear that, in spite of the great promise of stem-cell cardiomyoplasty, many challenges remain to be solved. Investigations at both experimental and clinical levels are being pursued, and it is hoped that this fascinating therapeutic approach could prove to be a potent therapeutic tool aimed at dealing with the failing human heart.

6.12 References

1. Orlic D, Hill J, Arai A. Stem cells for myocardial regeneration. *Circ Res* 2002; 91: 1092–102.
2. Nir S, David R, Zaruba M, Franz W, Eldor J. Human embryonic stem cells for cardiovascular repair. *Cardiovasc Res* 2003; 58: 313–23.

3. Chiu RCJ, Zibaitis A, Kao RL. Cellular cardiomyoplasty: myocardial regeneration with satellite cell implantation. *Ann Thorac Surg* 1995; 60: 12–18.
4. Taylor DA, Atkins BZ, Hungspreugs P *et al.* Regenerating functional myocardium: improved performance after skeletal myoblast transplantation. *Nat Med* 1998; 4: 929–33.
5. Leor J, Patterson M, Quinones MJ *et al.* Transplantation of fetal myocardial tissue into the infarcted myocardium of rat. A potential method for repair of infarcted myocardium? *Circulation* 1996; 94: II-332–6.
6. Wang JS, Shum-Tim D, Galipeau J *et al.* Marrow stromal cells for cellular cardiomyoplasty: feasibility and clinical advantages. *J Thorac Cardiovasc Surg* 2000; 120: 999–1006.
7. Scorsin M, Hagege A, Vilquin JT, Fiszman M, Marotte F, Samuel JL, Rappaport L, Schwartz K, Menasche P. Comparison of the effects of fetal cardiomyocyte and skeletal myoblast transplantation on postinfarction left ventricular function. *J Thorac Cardiovasc Surg* 2000; 119: 1169–75.
8. Fuchs S, Baffour R, Zhou YF, Shou M, Pierre A, Tio FO, Weissman NJ, Leon MB, Epstein SE, Kornowski R. Transendocardial delievery of autologous bone marrow enhances collateral perfusion and regional function in pigs with chronic experimental myocardial ischemia. *J Am Coll Cardiol* 2001; 37: 1726–32.
9. Edelman GM. *Topobiology: an introduction to molecular embryology*. London: HarperCollins Basic Books, 1988: 18, 21.
10. Tang G, Fedak P, Yau T, Weisel R, Kulik A, Mickle D, Li R. Cell transplantation to improve ventricular function in the failing heart. *Eur J Cardiothorac Surg* 2003; 23: 907–16.
11. Lovell M, Mathur A. The role of stem cells for treatment of cardiovascular disease. *Cell Prolif* 2004; 37(1): 67–89.
12. Campion DR. The muscle satellite cell: a review. *Int Rev Cytol* 1984; 87: 225–51.
13. Menasche P, Hagege AA, Vilquin JT, Desnos M, Abergel E, Pouzet B, Bel A, Sarateanu S, Scorsin M, Schwartz K, Bruneval P, Benbunan M, Marolleau JP, Duboc D. Autologous skeletal myoblast transplantation for severe postinfarction left ventricular dysfunction. *J Am Coll Cardiol* 2003; 41: 1078–83.
14. Kao RL, Chiu RC. *Cellular cardiomyoplasty: myocardial repair with cell implantation*. Austin, TX: Landes Bioscience; 1997.
15. Murry CE, Wiseman RW, Schwartz SM, Hauschka SD. Skeletal myoblast transplantation for repair of myocardial necrosis. *J Clin Invest* 1996; 98: 2512–23.
16. Kehat I, Kenyagin-Karsenti D, Snir M *et al.* Human embryonic stem cells can differentiate into myocytes with structural and functional properties of cardiomyocytes. *J Clin Invest* 2001; 108: 407–14.
17. Toma C, Pittenger MF, Cahill KS *et al.* Human mesenchymal stem cells differentiate to a cardiomyocyte phenotype in the adult murine heart. *Circulation* 2002; 105: 93–8.
18. Cohnheim J. *Arch Path Anato Physiol Klin Med* 1867; 40: 1.
19. Friedenstein AJ, Gorskoja V, Kulagina NN. *Exp Hematol* 1976; 4: 276.
20. Friedenstein AJ. Stromal mechanisms of bone marrow: cloning *in vitro* and retransplantation *in vivo*. *Hamatol Bluttransfus* 1980; 25: 19–29.
21. Friedenstein AJ, Petrakova KV, Kurolesova AI, Frolova GP. Heterotopic of bone marrow. Analysis of precursor cells for osteogenic and hematopoietic tissues. *Transplantation* 1968; 6: 230–47.
22. Caplan AI. Mesenchymal stem cells. *J Ortho Res* 1991; 9: 641–50.
23. Peterson BE, Bowen WC, Patrene KD *et al.* Bone marrow as a potential source of

hepatic oval cells. *Science* 1999; 284: 1168–70.
24. Pereira RF, O'Hara MD, Laptev AV *et al.* Marrow stromal cells as a source of progenitor cells for non hematopoietic tissues in transgenic mice with a phenotype of osteogenesis imperfecta. *Proc Natl Acad Sci USA* 1998; 95: 1142–47.
25. Ferrari G, Cusella De Angelis G, Coulta M *et al.* Muscle regeneration by bone marrow derived myogenic progenitors. *Science* 1998; 279: 1528–30.
26. Gussoni E, Soneoka Y, Strickland CD *et al.* Dystrophin expression in the MDX mouse restored by stem cell transplantation. *Nature* 1999; 401: 390–4.
27. Brazelton TR, Rossi FM, Keshet GI *et al.* From marrow to brain: expression of neuronal phenotypes in adult mice. *Science* 2000; 290: 1775–79.
28. Krause DS, Theise ND, Collector MI *et al.* Multi-organ, multi-lineage engraftment by a single bone marrow-derived stem cell. *Cell* 2001; 105: 369–77.
29. McKay R. Stem cell in the central nervous system. *Science* 1997; 274: 69–71.
30. Pittenger MF, Mosca JD, McIntosh KR. Human mesenchymal stem cells: progenitor cells for cartilage, bone, fat and stroma. *Curr Top Microbiol Immunol* 2000; 251: 3–11.
31. Pittenger MF, MacKay AM, Beck SC *et al.* Multilineage potential of adult human mesenchymal stem cells. *Science* 1999; 284: 143–7.
32. Pittenger M, Martin B. Mesenchymal stem cells and their potential as cardiac therapeutics. *Circ Res* 2004; 95: 9–20.
33. Ringes-Lichtenberg SM, Jaeger MD, Fuchs M *et al.* Sustained improvement of left ventricular function after myocardial infarction by intravenous application of Sca-1^+ Lin^-/c-kit^+ bone marrow derived stem cells. *Circulation* 2002; 106(Suppl.): II-14.
34. Mangi AA, Kong D, He H. Genetically modified mesenchymal stem cells perform *in vivo* repair of damaged myocardium. *Circulation* 2002; 106(Suppl.): II-131.
35. Deans R, Moseley A. Mesenchymal stem cells: biology and potential clinical uses. *Exp Hem* 2000; 28: 875–84.
36. Huang S, Terstappen LWMM. Formation of hematopoietic microenvironment and hematopoietic stem cells from single human bone marrow stem cell. *Nature* 1992; 360: 745.
37. Waller EK, Olweus J, Lund-Johansen F *et al.* The 'common stem cell' hypothesis reevaluated: human fetal bone marrow contains separate populations of hematopoietic and stromal progenitors. *Blood* 1995; 85: 2422.
38. Bittira B, Kuang JQ, Piquer S, Shum-Tim D, Al-Khaldi A, Chiu RCJ. The pathophysiological roles of bone marrow stromal cells in myocardial infarction. *Circulation* 2001; 104(Suppl.): II-523.
39. Hoecht E, Kahnert H, Guan K *et al.* Cardiac regeneration by stem cells after myocardial infarction in transplanted human hearts. *Circulation* 2002; 106(Suppl.): II-132.
40. Pereira RF, Halford KW, O'Hara MD *et al.* Cultured adherent cells from marrow can serve as long-lasting precursor cells for bone, cartilage, and lung in irradiated mice. *Proc Natl Acad Sci. USA* 1995; 92: 4857–61.
41. Bittira B, Wang JS, Shum-Tim D, Chiu RCJ. Marrow stromal cells as the autologous adult stem source for cardiac myogenesis. *Card Vasc Regeneration* 2000; 1(3): 205–10.
42. Rafii S, Avecilla S, Shmelkov S, Shido K, Tejada R, Moore M, Heissig B, Hattori K. Angiogenic factors reconstitute hematopoiesis by recruiting stem cells from bone marrow microenvironment. *Ann NY Acad Sci* 2003; 996: 49–60.
43. Chiu RCJ. Bone marrow stem cells as a source for cell therapy. *Heart Failure Rev* 2003; 8:247-251.

44. Lagasse E, Connors H, Al-Dhalimy M *et al.* Purified hematopoietic stem cells can differentiate into hepatocytes *in vivo*. *Nat Med* 2000; 6: 1229–34.
45. Theise ND, Nimmakayalu M, Gardner R *et al.* Liver from bone marrow in humans. *Hepatology* 2000; 32: 11–16.
46. Verfaillie C. Stem cell plasticity. *Graft* 2000; 3(6): 296–9.
47. Orlic D, Kajstura J, Chimenti S. *et al.* Mobilized bone marrow cells repair the infarcted heart, improving function and survival. *Proc Natl Acad Sci USA* 2001; 98: 10344–9.
48. Lapidot T, Petit I. Current understanding of stem cell mobilization: the roles of chemokines, proteolytic enzymes, adhesion molecules, cytokines and stromal cells. *Exp Hem* 2002; 30: 973–81.
49. Lovell MJ, Mathur A. The role of stem cells for treatment of cardiovascular disease. *Cell Prolif* 2004; 37: 67–87.
50. Richard DJ, Kassem M, Hefferan TE, Sarkar G, Spelsberg TC, Riggs BL. Isolation and characterization of osteoblast precursor cells from human bone marrow. *J Bone Miner Res* 1996; 11: 312.
51. Bianco P, Costantini M, Dearden LC, Bonucci E. Alkaline phosphatase positive precursors of adipocytes in human bone marrow. *Br J Haematol* 68: 401–3.
52. Kuznetsov S, Robey P. A look at the history of bone marrow stromal cells. *Graft* 2000; 3(6): 278–83.
53. Ashton BA, Allen TD, Howlett CR *et al.* Formation of bone and cartilage by marrow stromal cells in diffusion chambers *in vivo*. *Clin Ortho Rel Res* 1980; 151: 294–307.
54. Mezey E, Chandross KJ, Harta G *et al.* Turning blood into brain: cells bearing neuronal antigens generated *in vivo* from bone marrow. *Science* 2000; 290: 1779–82.
55. Jackson KA, Majka SM, Wang H *et al.* Regeneration of ischemic cardiac muscle and vascular endothelium by adult stem cells. *J Clin Invest* 2001; 107: 1395–402.
56. Makino S, Fukuda K, Miyoshi S *et al.* Cardiomyocytes can be generated from marrow stromal cells *in vitro*. *J Clin Invest* 1999; 103: 697–705.
57. Mangi AA, Noiseux N, Kong D *et al.* Mesenchymal stem cells modified with Akt prevent remodeling and restore performance in infarcted hearts. *Nat Med* 2003; 9: 1195–201.
58. Bittira B, Kuang JQ, Al-Khaldi A *et al.* *In-vitro* preprogramming of marrow stromal cells for myocardial regeneration. *Ann Thorac Surg* 2002; 74: 1154–60.
59. Wakitani S, Saito T, Caplan A. Myogenic cells derived from rat bone marrow mesenchymal stem cells exposed to 5-azacytidine. *Muscle Nerve* 1995; 18: 1417–26.
60. Xu W, Zhang X, Qian H, Zhu W, Sun X, Hu J, Zhou H, Chen Y. Mesenchymal stem cells from adult human bone marrow differentiate into a cardiomyocytes phenotype *in vitro*. *Exp Biol Med* 2004; 229: 623–34.
61. Hakuno D, Fukuda K, Makino S *et al.* Bone marrow-derived regenerated cardiomyocytes (CMG cells) express functional adrenergic and muscarinic receptors. *Circulation* 2002; 105: 93–8.
62. Tomita S, Nakatani T, Fukuhara S *et al.* Bone marrow stromal cells can differentiate into cardiac lineage and contract synchronously with cardiomyocytes by direct cell-to-cell interaction *in vitro*. Presented at the 82nd Annual Meeting of the American Association for Thoracic Surgery, Washington, DC, 2002.
63. Valiunas V, Doronin S, Valiuniene L, Potapova I, Zuckerman J, Walcott B, Robinson RB, Rosen MR, Brink PR, Cohen IS. Human mesenchymal stem cells make cardiac connexins and form functional gap junctions. *J Physiol* 2004; 555: 617–26.

64. Bird SD, Doevandans PA, van Rooijen MA, Brutel de la Riviere A, Hassink RJ, Passier R, Mummery CL. The human adult cardiomyocyte phenotype. *Cardiovasc Res* 2003; 58: 423–34.
65. Tomita S, Li RK, Weisel RD *et al.* Autologous transplantation of bone marrow cells improves damaged heart function. *Circulation* 1999; 100: II-247–56.
66. Chiu RCJ. Cellular cardiomyoplasty: the biology and clinical importance of milieu dependent differentiation. In *Handbook of cardiac cell transplantation*, 2004 Kipshidzen N, Serruys PW (eds), Martin Dunitz Ltd, London UK: 15–31.
67. Wang JS, Shum-Tim D, Chedrawy E, Chiu RCJ. The coronary delivery of marrow stromal cells for myocardial regeneration: pathophysiologic and therapeutic implications. *J Thorac Cardiovasc Surg* 2001; 122: 699–705.
68. Shake JG, Gruber PJ, Baumgartner WA, Senechal G, Meyers J, Redmond JM, Pittenger MF, Martin BJ. Mesenchymal stem cell implantation in a swine myocardial infarct model: engraftment and functional effects. *Ann Thorac Surg* 2002; 73: 1919–26.
69. Davani S, Marandin A, Mersin N *et al.* Mesenchymal progenitor cells differentiate into an endothelial phenotype, enhance vascular density, and improve heart function in a rat cellular cardiomyoplasty model. *Circulation* 2003; 108(Suppl 1): II253–8.
70. Dick AJ, Guttman MA, Raman VK, Peters DC, Pessanha BS, Hill JM, Smith S, Scott G, McVeigh ER, Lederman RJ. Magnetic resonance fluoroscopy allows targeted delivery of mesenchymal stem cells to infarct borders in swine. *Circulation* 2003; 108: 2899–904.
71. Kawada H, Fujita J, Kinjo K *et al.* Nonhematopoietic mesenchymal stem cells can be mobilized and differentiate into cardiomyocytes after myocardial infarction. *Blood* 2004; 104: 3581–7.
72. Laflamme MA, Myerson D, Saffitz JE, Murry CE. Evidence for cardiomyocyte repopulation by extracardiac progenitors in transplanted human hearts. *Circ Res* 2002; 90: 634–40.
73. Quaini F, Urbanek K, Beltrami AP, Finato N, Beltrami CA, Nadal-Ginard B, Kajstura J, Leri A, Anversa P. Chimerism of the transplanted heart. *New Engl J Med* 2002; 346: 5–15.
74. Taylor DA, Hruban R, Rodriguez R *et al.* Cardiac chimerism as a mechanism for self repair; does it happen and if so to what degree? *Circulation* 2002; 106: 2–4.
75. Botta R, Gao E, Stassi G *et al.* Heart infarct in NOD-SCID mice: therapeutic vasculogenesis by transplantation of human CD34+ cells and low dose CD34+ KDR+ cells. *FASEB J* 2004; 18: 1392–4.
76. Kinnaird T, Stabile E, Burnett MS, Lee CW, Barr S, Fuchs S, Epstein SE. Marrow-derived stromal cells express genes encoding a broad spectrum of arteriogenic cytokines and promote *in vitro* and *in vivo* arteriogenesis through paracrine mechanisms. *Circ Res* 2004; 94: 678–85.
77. Lee N, Thorne T, Losordo D, Yoon Y. Repair of ischemic heart disease with novel bone marrow-derived multipotent stem cells. *Cell Cycle* 2005; 4(7): 861–4.
78. Terada N, Hamazaki T, Oka M *et al.* Bone marrow cells adopt the phenotype of other cells by spontaneous cell fusion. *Nature* 2002; 416: 542–5.
79. Wang X, Willenbring H, Akkari Y *et al.* Cell fusion is the principal source of bone-marrow derived hepatocytes. *Nature* 2003; 422: 897–901.
80. Alvarez-Dolado M, Pardal R, Garcia-Verdugo JM *et al.* Fusion of bone marrow-derived cells with Purkinje neurons, cardiomyocytes and hepatocytes. *Nature* 2003; 425: 968–73.

81. Anversa P, Nadal-Ginard B. Myocyte renewal and ventricular remodeling. *Nature* 2002; 415: 240–3.
82. Beltrami AP, Urbanek K, Kajstura J *et al.* Evidence that human cardiac myocytes divide after myocardial infarction. *N Engl J Med* 2001; 344: 1750–7.
83. Beltrami AP, Barlucchi L, Torella D *et al.* Adult cardiac stem cells are multipotent and support myocardial regeneration. *Cell* 2003; 114: 763–76.
84. Messina E, De Angelis L, Frati G *et al.* Isolation and expansion of adult cardiac stem cells from human and murine heart. *Circ Res* 2004; 95: 911–21.
85. Martin C, Meeson A, Robertson S, Hawke T, Richardson J, Bates S, Goetsch S, Gallardo T, Garry D. Persistent expression of the ATP-binding cassette transporter, Abcg2, identifies cardiac SP cells in the developing and adult heart. *Dev Biol* 2004; 265: 262–75.
86. Chiu RCJ. Adult stem cell therapy for heart failure. *Expert Opin. Biol. Ther.* 2003; 3(2): 215–25.
87. Schofield R. The relationship between the spleen colony-forming cell and the hematopoietic stem cell. *Blood Cells* 1978; 4: 7–25.
88. Fuchs E, Tumbar T, Guasch G. Socializing with the neighbors: stem cells and their niche. *Cell* 2004; 116: 769–78.
89. Davey R, Zandstra P. Signal processing underlying extrinsic control of stem cell fate. *Curr Opin Hematol* 2004; 11: 95–101.
90. Calvi LM, Adams GB, Weibrecht KW *et al.* Osteoblastic cells regulate the hematopoietic stem cell niche. *Nature* 2003; 425: 841–6.
91. Marie-Rattis F, Voermans C, Reya T. Wnt signaling in the stem cell niche. *Curr Opin Hematol* 2004; 11: 88–94.
92. Park I, Qian D, Kiel M, Becker M, Pihalja M, Weissman I, Morrison S, Clarke M. Bmi-1 is required for maintenance of adult self-renewing hematopoietic stem cells. *Nature* 2003; 423: 302–4.
93. Duncan A, Rattis F, DiMascio L, Cougdon K, Pazianos G, Zhao C, Yoon K, Cook M, Willert K, Gaiano N, Reya T. Integration of Notch and Wnt signaling in hematopoietic stem cell maintenance. *Nature Imm* 2005; 6(3): 314–21.
94. Hirashima M, Ogawa M, Nishikawa S *et al.* A chemically defined culture of VEGFR2+ cells derived from embryonic stem cells reveals the role of VEGFR1 in tuning the threshold for VEGF in developing endothelial cells. *Blood* 2003; 101: 2261–7.
95. Takamatsu Y, Simmons PJ, Morris HA, To LB, Levesque J-P. Osteoclast-mediated bone resorption is stimulated during short term administration of granulocyte colony-stimulating factor but is not responsible for hematopoietic progenitor stem cell mobilization. *Blood* 1998; 92: 3465.
96. Peled A, Petit I, Kollet O *et al.* Dependence of human stem cell engraftment and repopulation of NOD/SCID mice on CXCR4. *Science* 1999; 283: 845.
97. Peled A, Kellet O, Ponomaryov T *et al.* The chemokine SDF-1 activates the integrins LFA-1, VLA-4 and VLA-5 on immature human CD34+ cells: role in transendothelial/stromal migration and engraftment on NOD/SCID mice. *Blood* 2000; 95: 3289.
98. Sun M, Opavsky A, Stewart D, Rabinovitch M, Dawood F, Wen W, Liu P. Temporal response and localization of integrins $\beta 1$ and $\beta 3$ in the heart after myocardial infarction. *Circulation* 2003; 107: 1046–52.
99. Fujita J, Suzuki Y, Ando K *et al.* G-CSF improves postinfarction heart failure by mobilizing bone marrow stem cells but GM-CSF increases the mortality by deteriorating heart function in mice. *Circulation* 2002; 106(Suppl.): II-15.

100. Miki T, Sakamoto J, Nakano A *et al.* Mobilization of bone marrow cells by G-CSF/M-CSF improves ventricular function in heart undergoing postinfarct remodeling. *Circulation* 2002; 106(Suppl.): II-15.
101. Fibbe WE, Pruijt JF, van Kooyk Y, Figdor CG, Opdenakker G, Willemze R. The role of metalloproteinases and adhesion molecules in IL-8 induced stem cell mobilization. *Semin Hematol* 2000; 37: 19.
102. Velders G, Fibbe W. Involvement of proteases in cytokine-induced hematopoietic stem cell mobilization. *Ann NY Acad Sci* 2005; 1044: 60–9.
103. Sweeney EA, Priestley GV, Nakamoto B, Collins RG, Beaudet AL, Papayannopoulou T. Mobilization of stem,/progenitor cells by sulfated polysaccharides does not require selectin presence. *Proc Natl Acad Sci USA* 2000; 97: 6544.
104. Yamaguchi J, Kusano K, Masuo O, Kawamoto A, Silver M, Murasawa S, Marce M, Masuda H, Losordo D, Isner J, Asahara T. SDF-1 effects on *ex vivo* expanded EPC recruitment for ischemic neovascularization. *Circulation* 2003; 107: 1322–8.
105. Papayannopoulou T, Nakamoto B. Peripheralization of hematopoietic progenitors in primates treated with anti-VLA4 integrin. *Proc Natl Acad Sci USA* 1993; 90: 9374.
106. Levesque JP, Takamatsu Y, Nilsson SK, Haylock DN, Simmons PJ. Vascular cell adhesion molecule-1 (CD106) is cleaved by neutrophil proteases in the bone marrow following hematopoetic progenitor cell mobilization by G-CSF. *Blood* 2001; 98: 1289.
107. Petit I, Szyper-Kravitz M, Nagler A *et al.* G-CSF induces stem cell mobilization by decreasing bone marrow SDF-1 and upregulating CXCR4. *Nature Immunol* 2002; 3: 687.
108. Heissig B, Hattori K, Dias S *et al.* *In vivo* inhibition of metalloproteinases blocks chemo-cytokine induced endothelial and hematopoietic stem cell mobilization. *Blood* 2000; 96: 5–40a.
109. Valenzuela-Fernandez A, Planchenault T, Baleux F *et al.* Leukocyte elastase negatively regulates stromal cell-derived factor-1 SDF/CXCR4 binding and functions by amino-terminal processing of SDF-1 and CXCR4. *J Biol Chem* 2002; 277: 15677.
110. Heissig B, Hattori K, Dias S *et al.* Matrix recruitment of stem and progenitor cells from the bone marrow niche requires MMP-9 mediated release of kit-ligand. *Cell* 2002; 109: 625.
111. Flores I, Cayuela M, Blasco M. Effects of telomerase and telomere length on epidermal stem cell behavior. *Science* 2005; 309: 1253–6.
112. Nilsson S, Simmons P. Transplantable stem cells: home to specific niches. *Curr Opin Hematol* 2004; 11: 102–6.
113. Frenette PS, Subbarao S, Mazo IB *et al.* Endothelial selectins and vascular cell adhesion molecule-1 promote hematopoietic progenitor homing to bone marrow. *Proc Natl Acad Sci USA* 1998; 95: 14423–8.
114. Katayama Y, Hidalgo A, Furie BC *et al.* PSGL-1 participates in E-selectin-mediated progenitor homing to bone marrow: evidence for cooperation between E-selectin ligands and alpha-4 integrin. *Blood* 2003; 102: 2060–7.
115. Papayannopoulou T, Craddock C, Nakamoto B *et al.* The VLA-4/VCAM-1 adhesion pathway defines contrasting mechanisms of lodgment of transplanted murine hematopoietic progenitors between bone marrow and spleen. *Proc Natl Acad Sci USA* 1995; 92: 9647–51.
116. Peled A, Petit I, Kollet O *et al.* Dependence on human stem cell engraftment and repopulation of NOD/SCID mice on CXCR4. *Science* 1999; 283: 845–8.

117. Kollet O, Spiegel A, Peled A, Petit I, Byk T, Hershkoviz R, Guetta E, Barkai G, Nagler A, Lapidot T. Rapid and efficient homing of human CD34+CD38-CXCR4+ stem and progenitor cells to the bone marrow and spleen of NOD/SCID and NOD/SCID/B2m mice. *Blood* 2001; 97: 3283–91.
118. Solanilla A, Grosset C, Duchez P *et al.* Flt3-ligand induces adhesion of hematopoietic progenitor cells via a very late antigen VLA-4 and VLA-5 dependent mechanism. *Br J Hematol* 2003; 120: 782–6.
119. Driessen RL, Johnston HM, Nilsson SK. Membrane-bound SCF is a key regulator in the intial lodgment of stem cells within the endosteal marrow region. *Exp Hematol* 2003; 31: 1284–91.
120. Mohle R, Boehmler AM, Denzlinger C *et al.* Nonpeptide mediators in the hematopoietic microenvironment. *Ann NY Acad Sci* 2003; 996: 61–6.
121. Netelenbos T, van den Born J, Kessler FL *et al.* Proteoglycans on bone marrow endothelial cells bind and present SDF-1 toward hematopoietic progenitor cells. *Leukemia* 2003; 17: 175–84.
122. Nilsson SK, Haylock DN, Johnston HM *et al.* Hyaluronan is synthesized by primitive hematopoietic cells, participates in their lodgment at the endosteum following transplantation, and is involved in the regulation of their proliferation and differentiation *in vitro*. *Blood* 2003; 101: 856–62.
123. Gazitt Y. Homing and mobilization of HSC and hematopoietic cancer cells are mirror image processes, utilizing similar signaling pathways and occurring concurrently: circulating cancer cells constitute an ideal target for concurrent treatment with chemotherapy and anti-lineage specific antibodies. *Leukemia* 2004; 18: 1–10.
124. Ahmed F, Ings SJ, Pizzey AR *et al.* Impaired bone marrow homing of cytokine-activated CD34+ cells in NOD/SCID model. *Blood* 2004; 103: 2079–87.
125. Liu B, Buckley SM, Lewis ID *et al.* Homing defect of cultured human hematopoietic cells in the NOD/SCID mouse is mediated by Fas/CD95. *Exp Hematol* 2003; 31: 824–32.
126. Zheng Y, Watanabe N, Nagamura-Inoue T *et al. Ex-vivo* manipulation of umbilical cord-blood derived hematopoietic stem/progenitor cells with recombinant human SCF can upregulate levels of homing-essential molecules to increase their transmigratory potential. *Exp Hematol* 2003; 31: 1237–46.
127. Gazzitt Y, Liu Q. Plasma levels of SDF-1 and expression of SDF-1 receptor on CD34+ cells in mobilized peripheral blood of non Hodgkin's lymphoma patients. *Stem Cells* 2001; 19: 37.
128. Pittenger MF, MacKay AM. Multipotential human mesenchymal stem cells. *Graft* 2000; 3(6): 288–93.
129. Reya T, Duncan AW, Ailles L *et al.* A role of Wnt signaling in self-renewal of hematopoietic stem cells. *Nature* 2003; 423: 409–14.
130. Zhang J, Niu C, Ye L *et al.* Identification of the hematopoietic stem cell niche and control of the niche size. *Nature* 2003; 425: 836–41.
131. Abbott J, Huang Y, Liu D, Hickey R, Krause D, Giordano F. SDF-1 plays a critical role in stem cell recruitment to the heart after myocardial infarction but is not sufficient to induce homing in the absence of injury. *Circulation* 2004; 110: 3300–5.
132. Saito T, Kuang JQ, Bittira B, Al-Khaldi A, Chiu RCJ. Xenotransplant cardiac chimera: immune tolerance of adult stem cells. *Ann Thorac Surg* 2002; 74: 19–24.
133. Wang L, Li Y, Chen J, Gautam SC, Zhang Z, Lu M, Chopp M. Ischemic cerebral tissue and MCP-1 enhance rat bone marrow stromal cell migration in interface culture. *Exp Hematol* 2002; 30: 831–6.

134. Devine SM, Bartholomew AM, Mahmud N *et al.* Mesenchymal stem cells are capable of homing to the bone marrow of non-human primates following systemic infusion. *Exp Hematol* 2001; 29: 244–55.
135. Sorger JM, Despres D, Hill J, Schimel D, Martin BJ, McVeigh ER. MRI tracking of intravenous magnetically labeled mesenchymal stem cells following myocardial infarctions in rats. *Circulation* 2002; 106: 2–16.
136. Rombouts WJ, Ploemacher RE. Primary murine MSC show highly efficient homing to the bone marrow but lose homing ability following culture. *Leukemia* 2003; 17:160–70.
137. Askari AT, Unzek S, Popovic ZB, Goldman CK, Forudi F, Kiedrowski M, Rovner A, Ellis SG, Thomas JD, Dicorleto PE, Topol EJ, Penn MS. Effect of SDF1 on stem cell homing and tissue regeneration in ischemic cardiomyopathy. *Lancet* 2003; 362: 697.
138. Takahashi T, Kalka C, Masuda H *et al.* Ischemia and cytokine-induced mobilization of bone marrow-derived endothelial progenitor cells for neovascularization. *Nat Med* 1999; 5: 434–8.
139. Harada M, Qin Y, Takano H, Minamino T, Zou Y, Toko H, Ohtsuka M, Matsuura K, Sano M, Nishi J, Iwanaga K, Akazawa H, Kunieda T, Zhu W, Hasehawa H, Kunisada K, Nagai T, Nakaya H, Yamauchi K, Komuro I. G-CSF prevents cardiac remodeling after myocardial infarction by activating the Jak-Stat pathway in cardiomyocytes. *Nat Med* 2005; 11(3): 305–11.
140. Bhatia R, Hare J. Mesenchymal stem cells: future source for reparative medicine. *CHF* 2005; 11: 87–91.
141. Liesveld JL, Rosell K, Panoskaltsis N, Belanger T, Harbol A, Abboud CN. Response of human CD34+ cells to CXC, CC and CX3C chemokines: implications for cell migration and activation. *J Hematother Stem Cell Res* 2001; 10: 643–5.
142. Eghbali-Webb M. The role of cardiac fibroblast in cardiac regeneration. In *Handbook of cardiac cell transplantation*, 2004 Kipshidzen N, Serruys PW (eds), Martin Dunitz Ltd, London UK: 73–90.
143. Heissig B, Hattori K, Dias S *et al.* Recruitment of stem and progenitor cells from the bone marrow niche requires MMP-9 mediated release of kit-ligand. *Cell* 2002; 109: 625–37.
144. Heissig B, Werb Z, Rafii S, Hattori K. Role of ckit/kit ligand signaling in regulating vasculogenesis. *Thromb Haemost* 2003; 90: 570–6.
145. MacDonald D, Burstein B, Shum-Tim D *et al.* Feasibility of intravenous adult bone marrow stromal cell administration in cellular cardiomyoplasty. Paper presented at the International Society of Stem Cell Research, 10–13 June 2004: 2nd Annual Meeting; Boston, MA.
146. Geng Y. Molecular mechanisms for cardiovascular stem cell apoptosis and growth in the hearts with atherosclerotic coronary disease and ischemic heart failure. *Ann NY Acad Sci* 2003; 1010: 687–97.
147. Griese DP, Grumbeck B, Kunz-Schughardt L *et al.* Genetic engineering of postnatal hematopoietic stem cells for gene and cell-based therapy of ischemic heart disease in rats. *Circulation* 2002; 106(Suppl.): II-15.
148. Yau TM, Fung K, Fujii T *et al.* Enhanced myocardial angiogenesis by gene transfer using transplanted cells. *Circulation* 2001; 104: 1218–22.
149. Chuah MK, Brems H, Vanslembrouck V, Collen D, Vandendriessche T. Bone marrow stromal cells as targets for gene therapy of Hemophila A. *Hum Gene Ther* 1998; 9: 353.
150. Alessandri G, Emanueli C, Madeddu P. Genetically engineered stem cell therapy for tissue regeneration. *Ann NY Acad Sci* 2004; 1015: 271–84.

151. Schwartz EJ, Alexander GM, Prockop DJ *et al.* Multipotential marrow stromal cells transduced to produce L-DOPA: engraftment in a rat model of Parkinson's disease. *Hum Gene Ther* 1999; 10: 2539–49.
152. Assmus B, Schachinger V, Teupe C *et al.* Transplantation of progenitor cells and regeneration enhancement in acute myocardial infarction (TOPCARE-AMI). *Circulation* 2002; 106: 3009–17.
153. Strauer BE, Brehm M, Zeus T *et al.* Repair of infarcted myocardium by autologous intracoronary mononuclear bone marrow cell transplantation in humans. *Circulation* 2002; 106: 1913–18.
154. Perin EC, Dohmann HF, Borojevic R *et al.* Transendocardial autologous bone marrow cell transplantation for severe chronic ischemic heart failure. *Circulation* 2003; 107: 2294–302.
155. Wollert KC, Meyer GP, Lotz J *et al.* Intracoronary autologous bone marrow cell transfer after myocardial infarction: the BOOST randomized controlled trial. *Lancet* 2004; 364: 141–8.
156. Kuethe F, Richartz BM, Sayer HG, Kasper C, Werner GS, Hoffken K, Figulla HR. Lack of regeneration of myocardium by autologous intracoronary mononuclear bone marrow cell transplantation in humans with large anterior myocardial infarctions. *Int J Cardiol* 2004; 97: 123–27.
157. Chen SL, Fang WW, Ye F, Liu YH, Qian J, Shan SJ, Zhang JJ, Chunhua RZ, Liao LM, Lin S, Sun JP. Effect on left ventricular function of intracoronary transplantation of autologous bone marrow mesenchymal stem cells in patients with acute myocardial infarction. *Am J Cardiol* 2004; 94: 92–95.
158. Lazarus H, Koc O. Culture-expanded human marrow-derived MSCs in clinical hematopoietic stem cell transplantation. *Graft* 2000; 3(6): 329–33.
159. Peschle C, Condorelli G. Stem cells for cardiomyocyte regeneration: state of the art. *Ann NY Acad Sci* 2005; 1047: 376–85.
160. Reffelmann T, Kloner R. Cellular cardiomyoplasty – cardiomyocytes, skeletal myoblasts, or stem cells for regenerating myocardium and treatment of heart failure. *Cardiovasc Res* 2003; 58: 358–68.
161. Le Blanc K, Tammik C, Rosendahl K, Zetterberg E, Ringden O. HLA expression and immunologic properties of differentiated and undifferentiated mesenchymal stem cells. *Exp Hematol* 2003; 31: 890–6.
162. Liechty KW, Mackenzie TC, Shaaban AF *et al.* Human mesenchymal stem cells engraft and demonstrate site-specific differentiation after *in utero* transplantation in sheep. *Nat Med* 2000; 11: 1282–6.
163. MacDonald D, Saito T, Shum-Tim D, Bernier PL, Chiu RCJ. Persistence of marrow stromal cells implanted into acutely infarcted myocardium: observations in a xenotransplant model. *J Thorac Cardiovasc Surg* 2005; 130(4): 1114–21.
164. Klyushnenkova E, Mosca JD, McIntosh KR. Human mesenchymal stem cells suppress allogeneic T cell responses in vitro: implications for allogeneic transplantation. *Blood* 1998; 92: 642a.
165. Chou S, Kuo T, Liu M, Lee O. *In utero* transplantation of human bone-marrow derived multipotent mesenchymal stem cells in mice. *J Ortho Res* 2006, 24(3): 301–12.
166. Matzinger P. The Danger Model: a renewed sense of self. *Science* 2002; 296(5566): 301–5.
167. Meisel R, Zibert A, Laryea M *et al.* Human bone marrow stromal cells inhibit allogeneic T-cell responses by indoleamine 2,3-deoxygenase-mediated tryptophan degradation. *Blood* 2004; 103: 4465–72.

168. Aggarwal S, Pittenger M. Human mesenchymal stem cells modulate allogeneic immune cell responses. *Blood* 2005; 105: 1815–22.
169. Jiang Y, Jahagirdar BN, Reinhardt RL *et al.* Pluripotency of mesenchymal stem cells derived from adult marrow. *Nature* 2002; 418(6893): 41–9.
170. Yoon Y, Wecker A, Heyd L, Park J, Tkebuchava T, Kusano K, Hanley A, Scadova H, Qui G, Cha D, Johnston K, Aikawa R, Asahara T, Losordo D. Clonally expanded novel multipotent stem cells from human bone amrrow regenerate myocardium after myocardial infarction. *J Clin Invest* 2005; 115(2): 326–36.
171. Noort W, Kruisselbrink A, de Paus R *et al.* Co-transplantation of mesenchymal stem cells (MSC) and UCB CD34+ cells results in enhanced hematopoietic engraftment in NOD/SCID mice without homing of MSC to the bone marrow. *Exp Hematol* 2002; 30: 870–8.
172. Wobus AM, Wallakukat G, Heschler J. Pluripotent mouse embryonic stem cells are capable to differentiate into cardiomyocytes expressing chronotropic responses to adrenergic and cholinergic agents and Ca^{2+} channel blockers. *Differentiation* 1991; 48: 173–82.
173. Tang YL, Zhao Q, Qin X *et al.* Paracrine actions enhance the effect of autologous mesenchymal stem cell transplantation on vascular regeneration in rat model of myocardial infarction. *Ann Thorac Surg* 2005; 80: 229–35.

7
Musculoskeletal tissue engineering with skeletal muscle-derived stem cells

K A CORSI and J HUARD, University of Pittsburgh, USA

7.1 Introduction

Despite numerous medical breakthroughs and the emergence of new technologies that allow faster translation of scientific discoveries from bench-top to bedside, there still remains progress to be made in alleviating the chronic pain and physical disability associated with musculoskeletal disorders and injuries.

Currently there are no cures for genetic disorders such as muscular dystrophies or osteogenesis imperfecta (OI), which affect the musculoskeletal system. Muscular dystrophies are inherited disorders characterized by defects in the dystrophin gene. The resultant lack of dystrophin, a structural protein, ultimately leads to muscle weakness followed by respiratory and cardiac failure. In the case of OI, there is a point mutation in the type I collagen gene, a necessary component of the bone matrix. The lack of this gene results in individuals having brittle bones that are prone to fractures. Researchers are investigating several approaches to restore the defective or missing gene, but the search for novel alternatives is ongoing.[1]

Large bone defects or non-union fractures that do not heal on their own are another problem confronting clinicians. The clinical management of such defects is complicated and often requires multiple procedures with unpredictable results. The same can be said of diseases or injuries affecting articular cartilage, as its avascular nature makes it unable to repair itself. The field of orthopaedics is therefore being faced with the challenge of rapidly identifying treatment options that would repair diseased or damaged tissue and return it to its original state both structurally and functionally.

One treatment option that has received particular attention as a promising future therapy is cell engineering and cell therapy for skeletal muscle, bone and articular cartilage regeneration. The first section of this review focuses on the important characteristics that must be taken into account when identifying an optimal cell source and the various methods available to engineer the isolated cells for therapeutic applications. The second section outlines the current progress made in our laboratory with an early progenitor cell source isolated from postnatal skeletal muscle termed muscle-derived stem cells (MDSCs).

7.2 Identification of the optimal cell source

Selection of the appropriate cell source for cell therapy depends largely on the application in which it will be used and the function it is expected to perform. One option is to select a specialized or tissue-specific progenitor cell that can replenish the cellular component that was damaged through injury. For example, chondrocytes would seem to be a reasonable cell source for articular cartilage repair because they are directly involved in articular cartilage formation. The transplantation of chondrocytes obtained from a non-load-bearing area into an injured area has been in clinical practice since 1987, and is referred to as autologous chondrocyte implantation.[2,3] Good to excellent clinical results have been reported, although some evidence of fibrocartilage has been observed 2 years post-operatively.[4] Other limitations also exist with this seemingly successful therapy. The isolation of the chondrocytes by arthroscopy remains an invasive procedure and, because it is difficult to expand chondrocytes from older patients, this therapy is mainly available to younger individuals.[5] Also, the development of hypertrophic tissue often necessitates a second surgery to remove the excessive tissue.[6] Because of these limitations, the alternative to the use of differentiated or tissue-committed cells is the use of stem cells, uncommitted cells that have the ability to differentiate to various lineages.

Numerous progenitor cell sources have been discovered and are currently being investigated. Postnatal stem cells have been isolated from several tissues and display varying capacities to undergo differentiation.[7] Among them, adult bone marrow-derived mesenchymal stem cells (MSCs) have received particular attention.[8–10] MSCs are capable of differentiating *in vitro* and *in vivo* along multiple lineages that include bone; cartilage; cardiac and skeletal muscle; and neural, tendon, adipose, and connective tissues. In addition to their multipotency, MSCs are relatively easy to isolate from the bone marrow, although the procedure is associated with some risk and morbidity. Stem cells harvested from adipose tissue are another interesting and readily available source.[11,12] Current studies on adipose-derived stem cells have tested their differentiation to adipocytes, osteoblasts, myoblasts, and chondroblasts.[11–14] There is also recent evidence suggesting that these cells may also be able to undergo neural differentiation.[15] Human embryonic stem cells (hESCs) are also being investigated for their pluripotency.[16] However, ethical issues surrounding their isolation and use have increased the need to look at reservoirs of stem cells derived from postnatal tissues.

Our group has investigated the biology of stem and progenitor cells contained within the postnatal skeletal muscle, a plentiful source of tissue within our bodies that is an accessible area from which to obtain biopsies. Using a modified preplate technique, we have fractioned various populations of mononucleated cells from digested skeletal muscle on the basis of their differing abilities to adhere to collagen-coated flasks. By employing this technique on mouse skeletal muscle, our group has reported the existence of a late-adhering population that was found to be positive for the stem cell markers CD34, Bcl-2 and stem cell antigen-1.[17] In addition, these cells demonstrate a capacity for long-term proliferation, self-renewal and immune-privileged behavior.[17,18] Furthermore, they can undergo multilineage differentiation toward skeletal muscle; bone; cartilage; and neural, endothelial and hematopoietic tissues.[17–24] Because of these stem cell characteristics, we consider the cells to be MDSCs.

7.3 Ability for cell expansion *ex vivo*

Self-renewal is a defining characteristic of stem cells that allows them to continuously replenish the tissues in which they reside and regenerate damaged tissue when necessary. MSCs can be isolated from bone marrow aspirates obtained from adults at a concentration of 1 in 100 000–500 000 cells.[9] Likewise, MDSCs can be found at a ratio of 1 in 100 000 cells.[17] Stem cells can overcome this limited supply because of their ability to self-renew. They are also a desirable cell source for cell therapies because of their capacity to be significantly expanded without loss of progenitor characteristics. Some groups have investigated the expansion and self-renewal capabilities of stem cells. Reports have shown that hESCs remained pluripotent when expanded to 250 population doublings (PDs) and retained a normal karyotype.[25] Human hematopoietic stem cells from cord blood have also been expanded up to 2×10^6-fold over 6 months with little evidence of differentiation.[26] The long-term expansion of MDSCs up to a PD of 300 revealed that these cells maintain their ability for high skeletal muscle regeneration until PD 200, upon which it drops dramatically.[27] Understanding the mechanisms by which stem cells self-renew and the effects of long-term expansion on stem cells will be essential for clinical applications.

7.4 Cell engineering for therapeutic applications

In addition to the appropriate cell type, growth factors are often necessary for regeneration to take place. Growth factors provide the essential signals for proliferation, migration, matrix synthesis, and cell differentiation. Administration of growth factors to the cells can be accomplished directly through *in vivo* injections or by *ex vivo* stimulation (Fig. 7.1(a) and (b)). Alternatively, gene therapy can be used to genetically engineer the cells to express the growth factors (Fig. 7.1(c) and (d)). This can also be done *in vivo* through direct injection of a vector carrying the gene of interest, or by genetically modifying the cells *ex vivo* to express the growth factor prior to their transplantation into the patient.

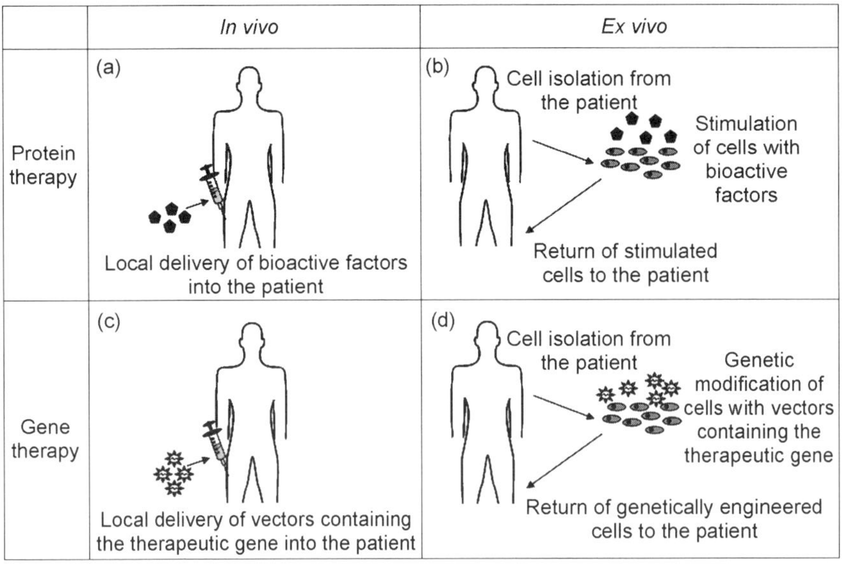

7.1 The two common approaches to engineer cells for therapeutic applications. In protein therapy, cells are stimulated (a) *in vivo* by the direct administration of bioactive factors, or (b) *ex vivo* by first isolating the cells from the patient, stimulating them and returning the stimulated cells to the patient. In gene therapy, cells can be (c) genetically modified *in vivo* by local delivery of the vectors carrying the therapeutic gene, or (d) genetically modified *ex vivo* and then returned to the host.

7.4.1 Protein therapy

The local delivery of exogenous proteins or growth factors has aided in the healing of skeletal muscle, bone, and articular cartilage. Skeletal muscle injuries have shown improved healing after direct injection of antifibrotic agents such as decorin, gamma interferon, relaxin, suramin, insulin-like growth factor-1 (IGF-1), basic fibroblast growth factor (bFGF), or nerve growth factor (NGF).[28–35] Bone fracture healing has occurred in human patients after local bone morphogenetic protein (BMP) administration to fractures of the tibia, femur, humerus, or fibula.[36–40] Similarly, osteochondral defects in animal models have displayed improved healing after direct administration of growth factors such as BMP2, IGF-1, bFGF, FGF-2, hepatocyte growth factor, and growth hormone.[41–46] The downside of delivering the protein in its active form through direct injection is that it results in a high concentration of protein at the site of injury. Owing to resorption by the surrounding environment and the protein's short half-life, maintaining a constant concentration of protein at the site of injury to promote healing remains difficult. Overcoming this challenge by increasing the initial dose could lead to cytotoxicity, and performing repeated injections would not be cost effective. An alternative to injecting solely a solution of protein is to seed it within a carrier matrix that allows it to be released in a controlled regimen.[47–50] Controlled release of growth factors is currently the focus of extensive research and may enable prolonged administration of the protein to facilitate healing.

As an alternative to direct injection of growth factors, cells can be stimulated *ex vivo* before they are implanted into a patient. For example, exogenous stimulation of osteoblasts, adipose-derived stromal cells, and bone marrow-derived stromal cells, which were seeded onto a scaffold, with rhBMP2 and retinoic acid for 4 weeks before implantation *in vivo* led to accelerated bone formation when compared with an unstimulated cell-seeded scaffold.[51] However, at longer time points *in vivo*, there was evidence of bone resorption, suggesting that this *ex vivo* stimulation remains to be optimized. Preconditioning the cells with growth factors *in vitro* can also enhance *in vivo* development of human engineered articular cartilage in an ectopic nude mouse model.[52] Preculturing adult human articular chondrocytes in a scaffold with medium containing transforming growth factor (TGF)-beta1 and insulin for 2 weeks before implantation increased *in vivo* development of articular cartilage tissue. Moreover, the primed chondrocytes generated repaired tissues that displayed higher equilibrium moduli than seen with chondrocytes cultured without the growth factors. Hence, protein therapy, whether administered *in vivo* or *ex vivo*, leads to numerous beneficial outcomes for the healing of skeletal muscle, bone, and articular cartilage. The short half-life, and the cost associated with producing large quantities of growth factors can be limiting factors associated with this type of therapy. To circumvent these limitations, gene therapy is also being investigated as a possible therapeutic option.

7.4.2 Gene therapy

Gene therapy involves the transfer of genetic material into a cell to induce it to produce a protein for therapeutic purposes. Various vehicles, both viral and non-viral, can be used to deliver the transgene into the cells. Viral vectors are created from viruses that have been genetically altered to remove all viral functions (including replication) but retain the ability to infect cells. Some examples of viral vectors are adenoviruses, adeno-associated viruses (AAV), herpes simplex viruses, lentiviruses, and retroviruses. Non-viral vectors include plasmid DNA, a simple circular form of DNA containing a gene of interest. Plasmid DNA can be delivered in its native form, although its high susceptibility to degradation has motivated research into possible methods to improve the transfection efficiency. As such, plasmid DNA can be inserted into the cell by electroporation or by complexing the DNA with cationic polymers, which protect it from degradation.[53] As with protein therapy, the introduction of therapeutic genes into cells can be accomplished by the direct injection of the vector into the body or by harvesting the cells, manipulating them *ex vivo*, and subsequently reimplanting the genetically engineered cells.

In vivo gene therapy for muscular dystrophies has been tested in animal models with various viral vectors.[54–58] The use of non-viral techniques has also been investigated.[59,60] For example, electroporation enabled the introduction of the full-length dystrophin cDNA into the skeletal muscle of adult *mdx* mice, which model Duchenne muscular dystrophy and lack the dystrophin protein.[59] Also, the complexation of DNA with polymers can facilitate the delivery of dystrophin to skeletal muscles.[60] *In vivo* gene therapy for bone repair is also an area of ongoing research. Injection of adenovirus carrying the human *BMP2* or *BMP6* gene has led to bone formation in a rabbit model.[61,62] Likewise, intra-articular administration of AAV carrying the *bFGF* gene has aided in the repair of articular cartilage defects.[63] In all applications, the safety of gene transfer is of utmost importance for effective therapy. For this reason, *in vivo* gene therapy, especially with viral vectors, raises some concerns because direct injection of the vector into the body makes it difficult to target a specific subset of cells. On the other hand, *ex vivo* gene therapy allows the cells that will be genetically engineered to be chosen and thoroughly tested before being returned to the patient.

Research into the possible use of *ex vivo* gene therapy for the treatment of muscular dystrophy has recently been carried out with the use of lentiviral vectors and myogenic progenitor cells.[64] The lentiviral vector was able to deliver the dystrophin gene to both dividing and non-dividing cells, which is important for skeletal muscle therapy because the targeted muscle fibers are terminally differentiated and do not divide. In the same study, myoblasts obtained from an *mdx* mouse were also genetically modified *ex vivo* with a lentivirus and then re-injected into the muscle. Although many newly regenerated myofibers were evident, there was some evidence of infiltrating mononuclear cells, suggesting the possibility of an immune reaction.

Ex vivo gene therapy has also been studied to promote bone repair. In a recent study, Zhang *et al.* tried to improve the osteogenic capacity of allografts by seeding BMP2-producing bone marrow stromal cells onto them.[65] This resulted in the bridging of a segmental femoral bone graft model in mice. A similar study was completed in a larger animal model, a dyaphyseal bone defect in the goat.[66] The animals received scaffolds seeded with autologous bone marrow-derived MSCs that were either transduced to express a marker gene, or genetically modified to secrete hBMP2. The cells expressing hBMP2 led to partial or complete healing of the defect, whereas non-transduced cells did not induce healing.

Adipose-derived stem cells and primary human fibroblasts can also be genetically modified with an adenovirus carrying the recombinant human *BMP7* gene.[67,68] Both cell types led to newly formed bone after implantation into the subcutaneous pocket of rats or mice. Likewise, the use of *ex vivo* gene therapy has shown promise for the regeneration of articular cartilage in animal studies. In a recent study, MSCs were isolated from the periosteum, the bone marrow, and the fat of adult rats and then genetically modified with an adenovirus to express BMP2.[69] The *BMP2*-transduced cells and their unmodified controls were implanted into an articular cartilage lesion made in the patellar groove of the rat femur. The results from this study indicate that the periosteum and bone marrow-derived cells that were genetically engineered to express BMP2 generated repair tissue that had properties similar to those of articular cartilage, whereas the *BMP2*-transduced fat stromal cells formed mostly fibrous tissue. This study is a clear indicator that not all cells will perform identically *in vitro* or *in vivo*. For this reason, it is important to investigate and compare different cell types and growth factors and identify the optimal combination.

7.5 Current progress with MDSCs

MDSCs were originally identified by researchers trying to find an optimal cell source to be used for cell replacement therapy in Duchenne muscular dystrophy. In the early stages of study, our group focused on characterizing the ability of these cells to regenerate dystrophin-expressing muscle fibers in *mdx* mice. We observed that transplanted MDSCs displayed a high degree of skeletal muscle regeneration that was superior to that of transplanted satellite cells, a skeletal muscle-committed progenitor cell.[17] In these experiments, MDSCs displayed long-term proliferation *in vivo*, a strong capacity for self-renewal, multipotent differentiation, and immune-privileged behavior.[17,18]

Ex vivo gene therapy with MDSCs has been used to deliver dystrophin to *mdx* mice.[70] MDSCs were isolated from *mdx* mice, genetically engineered with a retrovirus carrying a functional human mini-dystrophin gene and then injected into the skeletal muscle of *mdx* mice. Dystrophin-positive muscle fibers were present up to 24 weeks postinjection, suggesting that *ex vivo* gene transfer may aid in the delivery of dystrophin to skeletal muscle. To improve the transplantation efficiency of MDSCs isolated from normal mice and injected into dystrophic skeletal muscles, researchers recently used NGF to test the *ex vivo* protein stimulation and *ex vivo* gene therapy of MDSCs.[71] MDSCs were either stimulated with NGF *in vitro* before injection or were retrovirally transduced to express NGF. In both cases, cells stimulated with NGF exhibited higher engraftment in skeletal muscle than did unstimulated controls. Interestingly, the *ex vivo* stimulation of MDSCs with NGF led to a higher regeneration of muscle fibers than elicited by *NGF*-transduced MDSCs, suggesting that, in some cases, simply preconditioning the cells with growth factors may be sufficient to induce a beneficial effect.

In numerous other studies, MDSCs genetically engineered to express BMP2 or BMP4 have led to bone formation and healed critical-sized calvarial and long bone defects.[21–24,72,73] *In vivo* tracking of the cells has demonstrated that more than 95% of the implanted MDSCs expressing BMP2 were found in the newly formed bone.[21] Some also appeared to co-localize with osteocalcin, suggesting that the cells differentiated into osteogenic cells. These studies indicate that MDSCs genetically engineered to express BMP excel in two crucial activities for bone formation. First, they can function as osteoprogenitor cells. Second, they can act as gene delivery vehicles by secreting the growth factors necessary to recruit surrounding cells that can participate in new bone formation.

It is hypothesized that the recruitment of progenitor cells to further promote bone regeneration can also be achieved by increasing the blood supply to the site of injury. VEGF is a potent angiogenic factor. Members of our laboratory investigated the synergistic effect of VEGF on the promotion of bone formation by BMP-expressing MDSCs.[22,73] In this study, MDSCs transduced to express BMP4 and VEGF were implanted into the muscle pockets or calvarial defects in mice. We observed that BMP4 and VEGF work synergistically together to produce a greater amount of bone formation than observed after implantation of only BMP4-expressing MDSCs. Interestingly, the dosage of BMP4- to VEGF-expressing cells played a large role in the amount and quality of bone formation. A ratio of one-fifth VEGF to BMP4 resulted in complete bone union, whereas the presence of five times more VEGF than BMP4 led to only small patches of mineralized bone. A more recent study investigated the effect of BMP2 and VEGF on the ability of MDSCs to form bone *in vivo*.[73] Similar results to the previous study with BMP4 and VEGF were obtained, namely, VEGF in combination with BMP2 accelerated and enhanced the bone formation elicited by BMP2-expressing cells alone. Increased bone formation was also evident at specific ratios of VEGF- to BMP2-producing MDSCs. Five times more VEGF- than BMP2-expressing cells caused a decrease in bone formation. Interestingly, when comparing the two studies, VEGF had a greater effect on BMP4-induced bone formation than on BMP2-induced bone formation.[22,73]

Although our studies have shown that MDSCs can promote bone formation through delivery of BMP4 and by direct participation in the newly formed bone, in some particular cases, the goal is to reduce ossification. One clinical problem faced by orthopaedic surgeons is that of heterotopic ossification of muscles, tendons, and ligaments. In an attempt to inhibit such bone formation, we investigated the ability of Noggin, a BMP antagonist, to inhibit heterotopic ossification.[74] To accomplish this, MDSCs were transduced to express Noggin and were co-implanted with MDSCs transduced to secrete BMP4 into the hind limbs of mice. Groups with different ratios of Noggin to BMP4 were evaluated for heterotopic ossification. The results showed that Noggin inhibited the amount of bone formation in a dose-dependent manner.

Interestingly, controlling bone growth with Noggin is attractive for bone tissue engineering applications. In the previous studies involving MDSCs transduced to express BMP2 or BMP4 and implanted into critical-sized calvarial defects, there was often evidence of bone overgrowth. In an attempt to regulate bone formation, an inducible tet-on retroviral vector was designed to confer high *BMP4* gene expression by transduced MDSCs only upon exposure to doxycycline.[72] Addition of doxycycline to the drinking water of the mice implanted with MDSCs expressing inducible BMP4 would lead to *BMP4* gene expression and bone formation, while an absence of doxycycline in the drinking water should block *BMP4* gene expression and bone formation. It was, however, noticed that there were very small isolated bone nodules at the defect sites even in the absence of doxycycline. Therefore, to control for the overgrowth of bone seen in the earlier studies and the leakage of the self-inactivating tet-on retroviral vector, MDSCs expressing inducible BMP4 were co-implanted into critical-sized calvarial defects with MDSCs expressing Noggin.[75] At specific dosages, this combinatorial therapy led to regulation of bone formation. It prevented the overgrowth of bone previously seen in skull defects, it blocked the basal level of bone formation seen with the inducible BMP4 vector, and, even more interestingly, it led to the regeneration of bone that was histologically similar to normal bone. In fact, the bone regenerated in the BMP4 with Noggin group resembled the original bone more closely than did the bone regenerated with BMP4 only. Therapies including BMP4 and Noggin would closely mimic the physiological environment of the injury site, as both have been shown to be present at the fracture repair site during healing.[76] The interactions between BMP4 and VEGF, and between BMP4 and Noggin and how these agents can improve and regulate bone formation warrant further investigation into cell engineering and its ability to lead to new approaches to enhance and regulate tissue repair, regeneration or replacement.

Genetic engineering of MDSCs with *BMP4* has also been used for articular cartilage repair.[20] We have shown that MDSCs expressing BMP4 acquired a chondrocytic phenotype *in vitro*, in both monolayer and micromass pellet culture, whereas control MDSCs did not undergo chondrogenic differentiation. After implantation of BMP4-expressing MDSCs into full-thickness articular cartilage defects, donor cells could be seen within the defect and co-localized with collagen type II, a marker of chondrogenic differentiation. In addition, the repaired tissue in the defect was well integrated with the adjacent normal articular cartilage for up to 24 weeks. The group that received control MDSCs did not display such a high degree of regeneration, suggesting that BMP4 is beneficial for cartilage repair. Thus, cell engineering of MDSCs to secrete BMP4 and other agents appears to be important for optimal regeneration of injured or diseased bone and articular cartilage tissue.

7.6 Future trends

To develop efficient cell therapies, it will be necessary to continue identifying cell sources that are easy to isolate and expand. It will also be important to determine if progenitor cells can be found to the same degree in young patients versus older patients. Another area that must be investigated is the possibility of sexual dimorphism. It is already known that females are more affected than males by musculoskeletal problems such as osteoarthritis, osteoporosis, spinal disorders, and fractures.[77] This is probably due to differences at the molecular and cellular level between females and males. For this reason, the recommendations brought forth by the American Academy of Orthopaedic Surgeons and the National Institutes of Health included a proposition to 'Promote research on sex differences at the molecular, cellular, and tissue levels with specific emphasis on musculoskeletal diseases and conditions'.[77] The need to study sex-related differences associated with stem cells and tissue engineering was also mentioned. By investigating the effects of expansion, donor age, and sex on cellular behavior and understanding the underlying mechanisms, cells can be optimized to further increase their therapeutic potential.

As presented in this chapter, the optimization of stem cells can occur by growth factor-guided cellular differentiation. This can be done by using bioactive factors or genetically engineering the cells to produce the bioactive factors. However, both approaches are faced with important challenges. In the case of protein therapy, the large-scale production and cost effectiveness must be optimized. In the case of gene therapy, scientists must continue their efforts to develop safer and more efficient vectors. More research into the area of controlled release is also required to enable the controlled release of proteins or vectors at physiological or therapeutic doses, at the proper location, and at the optimal time to promote healing and regeneration.

In conclusion, due to an aging population and an increasing life expectancy, the occurrence of musculoskeletal disorders and injuries is anticipated to rise. However, cell therapy will also continue to evolve and could be the therapy of choice for musculoskeletal tissue engineering. In particular, cells isolated from the skeletal muscle, a readily available source, possess potential for muscle, bone and articular cartilage regeneration and warrant further investigation.

7.7 Acknowledgements

The authors wish to thank T. R. Payne for his critical reading of the manuscript and R. Sauder for his outstanding editorial assistance with the manuscript.

7.8 References

1. Millington-Ward, S., McMahon, H.P. and Farrar, G.J. 2005. Emerging therapeutic approaches for osteogenesis imperfecta. *Trends Mol Med* 11: 299–305.
2. Brittberg, M., Lindahl, A., Nilsson, A., Ohlsson, C., Isaksson, O. and Peterson, L. 1994. Treatment of deep cartilage defects in the knee with autologous chondrocyte transplantation. *N Engl J Med* 331: 889–895.
3. Peterson, L., Minas, T., Brittberg, M. and Lindahl, A. 2003. Treatment of osteochondritis dissecans of the knee with autologous chondrocyte transplantation: results at two to ten years. *J Bone Joint Surg Am* 85-A Suppl 2: 17–24.
4. Clar, C., Cummins, E., McIntyre, L., Thomas, S., Lamb, J., Bain, L., Jobanputra, P.

and Waugh, N. 2005. Clinical and cost-effectiveness of autologous chondrocyte implantation for cartilage defects in knee joints: systematic review and economic evaluation. *Health Technol Assess* 9: 1–98.

5. Giannoni, P., Pagano, A., Maggi, E., Arbico, R., Randazzo, N., Grandizio, M., Cancedda, R. and Dozin, B. 2005. Autologous chondrocyte implantation (ACI) for aged patients: development of the proper cell expansion conditions for possible therapeutic applications. *Osteoarthritis Cartilage* 13: 589–600.
6. Bartlett, W., Skinner, J.A., Gooding, C.R., Carrington, R.W., Flanagan, A.M., Briggs, T.W. and Bentley, G. 2005. Autologous chondrocyte implantation versus matrix-induced autologous chondrocyte implantation for osteochondral defects of the knee: a prospective, randomised study. *J Bone Joint Surg Br* 87: 640–645.
7. Young, H.E., Duplaa, C., Katz, R., Thompson, T., Hawkins, K.C., Boev, A.N., Henson, N.L., Heaton, M., Sood, R., Ashley, D., *et al.* 2005. Adult-derived stem cells and their potential for use in tissue repair and molecular medicine. *J Cell Mol Med* 9: 753–769.
8. Arinzeh, T.L. 2005. Mesenchymal stem cells for bone repair: preclinical studies and potential orthopedic applications. *Foot Ankle Clin* 10: 651–665, viii.
9. Caplan, A.I. 2005. Review. Mesenchymal stem cells: cell-based reconstructive therapy in orthopedics. *Tissue Eng* 11: 1198–1211.
10. Reiser, J., Zhang, X.Y., Hemenway, C.S., Mondal, D., Pradhan, L. and La Russa, V.F. 2005. Potential of mesenchymal stem cells in gene therapy approaches for inherited and acquired diseases. *Expert Opin Biol Ther* 5: 1571–1584.
11. Rodriguez, A.M., Elabd, C., Amri, E.Z., Ailhaud, G. and Dani, C. 2005. The human adipose tissue is a source of multipotent stem cells. *Biochimie* 87: 125–128.
12. Strem, B.M., Hicok, K.C., Zhu, M., Wulur, I., Alfonso, Z., Schreiber, R.E., Fraser, J.K. and Hedrick, M.H. 2005. Multipotential differentiation of adipose tissue-derived stem cells. *Keio J Med* 54: 132–141.
13. Rodriguez, A.M., Pisani, D., Dechesne, C.A., Turc-Carel, C., Kurzenne, J.Y., Wdziekonski, B., Villageois, A., Bagnis, C., Breittmayer, J.P., Groux, H., *et al.* 2005. Transplantation of a multipotent cell population from human adipose tissue induces dystrophin expression in the immunocompetent mdx mouse. *J Exp Med* 201: 1397–1405.
14. Xu, Y., Malladi, P., Wagner, D.R. and Longaker, M.T. 2005. Adipose-derived mesenchymal cells as a potential cell source for skeletal regeneration. *Curr Opin Mol Ther* 7: 300–305.
15. Kokai, L.E., Rubin, J.P. and Marra, K.G. 2005. The potential of adipose-derived adult stem cells as a source of neuronal progenitor cells. *Plast Reconstr Surg* 116: 1453–1460.
16. Przyborski, S.A. 2005. Differentiation of human embryonic stem cells after transplantation in immune-deficient mice. *Stem Cells* 23: 1242–1250.
17. Qu-Petersen, Z., Deasy, B., Jankowski, R., Ikezawa, M., Cummins, J., Pruchnic, R., Mytinger, J., Cao, B., Gates, C., Wernig, A., *et al.* 2002. Identification of a novel population of muscle stem cells in mice: potential for muscle regeneration. *J Cell Biol* 157: 851–864.
18. Jankowski, R.J., Deasy, B.M., Cao, B., Gates, C. and Huard, J. 2002. The role of CD34 expression and cellular fusion in the regeneration capacity of myogenic progenitor cells. *J Cell Sci* 115: 4361–4374.
19. Cao, B., Zheng, B., Jankowski, R.J., Kimura, S., Ikezawa, M., Deasy, B., Cummins, J., Epperly, M., Qu-Petersen, Z. and Huard, J. 2003. Muscle stem cells differentiate into haematopoietic lineages but retain myogenic potential. *Nat Cell Biol* 5: 640–646.

20. Kuroda, R., Usas, A., Kubo, S., Corsi, K., Peng, H., Rose, T., Cummins, J., Fu, F.H. and Huard, J. 2006. Cartilage repair using bone morphogenetic protein 4 and muscle-derived stem cells. *Arthritis Rheum* 54: 433–442.
21. Lee, J.Y., Qu-Petersen, Z., Cao, B., Kimura, S., Jankowski, R., Cummins, J., Usas, A., Gates, C., Robbins, P., Wernig, A., *et al.* 2000. Clonal isolation of muscle-derived cells capable of enhancing muscle regeneration and bone healing. *J Cell Biol* 150: 1085–1100.
22. Peng, H., Wright, V., Usas, A., Gearhart, B., Shen, H.C., Cummins, J. and Huard, J. 2002. Synergistic enhancement of bone formation and healing by stem cell-expressed VEGF and bone morphogenetic protein-4. *J Clin Invest* 110: 751–759.
23. Shen, H.C., Peng, H., Usas, A., Gearhart, B., Cummins, J., Fu, F.H. and Huard, J. 2004. *Ex vivo* gene therapy-induced endochondral bone formation: comparison of muscle-derived stem cells and different subpopulations of primary muscle-derived cells. *Bone* 34: 982–992.
24. Wright, V., Peng, H., Usas, A., Young, B., Gearhart, B., Cummins, J. and Huard, J. 2002. BMP4-expressing muscle-derived stem cells differentiate into osteogenic lineage and improve bone healing in immunocompetent mice. *Mol Ther* 6: 169–178.
25. Amit, M., Carpenter, M.K., Inokuma, M.S., Chiu, C.P., Harris, C.P., Waknitz, M.A., Itskovitz-Eldor, J. and Thomson, J.A. 2000. Clonally derived human embryonic stem cell lines maintain pluripotency and proliferative potential for prolonged periods of culture. *Dev Biol* 227: 271–278.
26. Piacibello, W., Sanavio, F., Garetto, L., Severino, A., Bergandi, D., Ferrario, J., Fagioli, F., Berger, M. and Aglietta, M. 1997. Extensive amplification and self-renewal of human primitive hematopoietic stem cells from cord blood. *Blood* 89: 2644–2653.
27. Deasy, B.M., Gharaibeh, B.M., Pollett, J.B., Jones, M.M., Lucas, M.A., Kanda, Y. and Huard, J. 2005. Long-term self-renewal of postnatal muscle-derived stem cells. *Mol Biol Cell* 16: 3323–3333.
28. Sato, K., Li, Y., Foster, W., Fukushima, K., Badlani, N., Adachi, N., Usas, A., Fu, F.H. and Huard, J. 2003. Improvement of muscle healing through enhancement of muscle regeneration and prevention of fibrosis. *Muscle Nerve* 28: 365–372.
29. Negishi, S., Li, Y., Usas, A., Fu, F.H. and Huard, J. 2005. The effect of relaxin treatment on skeletal muscle injuries. *Am J Sports Med* 33: 1816–1824.
30. Menetrey, J., Kasemkijwattana, C., Day, C.S., Bosch, P., Vogt, M., Fu, F.H., Moreland, M.S. and Huard, J. 2000. Growth factors improve muscle healing *in vivo*. *J Bone Joint Surg Br* 82: 131–137.
31. Li, Y., Negishi, S., Sakamoto, M., Usas, A. and Huard, J. 2005. The use of relaxin improves healing in injured muscle. *Ann NY Acad Sci* 1041: 395–397.
32. Fukushima, K., Badlani, N., Usas, A., Riano, F., Fu, F. and Huard, J. 2001. The use of an antifibrosis agent to improve muscle recovery after laceration. *Am J Sports Med* 29: 394–402.
33. Foster, W., Li, Y., Usas, A., Somogyi, G. and Huard, J. 2003. Gamma interferon as an antifibrosis agent in skeletal muscle. *J Orthop Res* 21: 798–804.
34. Chan, Y.S., Li, Y., Foster, W., Horaguchi, T., Somogyi, G., Fu, F.H. and Huard, J. 2003. Antifibrotic effects of suramin in injured skeletal muscle after laceration. *J Appl Physiol* 95: 771–780.
35. Chan, Y.S., Li, Y., Foster, W., Fu, F.H. and Huard, J. 2005. The use of suramin, an antifibrotic agent, to improve muscle recovery after strain injury. *Am J Sports Med* 33: 43–51.
36. Friedlaender, G.E. 2004. Osteogenic protein-1 in treatment of tibial nonunions:

current status. *Surg Technol Int* 13: 249–252.
37. Friedlaender, G.E., Perry, C.R., Cole, J.D., Cook, S.D., Cierny, G., Muschler, G.F., Zych, G.A., Calhoun, J.H., LaForte, A.J. and Yin, S. 2001. Osteogenic protein-1 (bone morphogenetic protein-7) in the treatment of tibial nonunions. *J Bone Joint Surg Am* 83-A Suppl 1: S151–158.
38. Geesink, R.G., Hoefnagels, N.H. and Bulstra, S.K. 1999. Osteogenic activity of OP-1 bone morphogenetic protein (BMP-7) in a human fibular defect. *J Bone Joint Surg Br* 81: 710–718.
39. Johnson, E.E., Urist, M.R. and Finerman, G.A. 1990. Distal metaphyseal tibial nonunion. Deformity and bone loss treated by open reduction, internal fixation, and human bone morphogenetic protein (hBMP). *Clin Orthop Relat Res*: 234–240.
40. Johnson, E.E., Urist, M.R. and Finerman, G.A. 1992. Resistant nonunions and partial or complete segmental defects of long bones. Treatment with implants of a composite of human bone morphogenetic protein (BMP) and autolyzed, antigen-extracted, allogeneic (AAA) bone. *Clin Orthop Relat Res*: 229–237.
41. Cuevas, P., Burgos, J. and Baird, A. 1988. Basic fibroblast growth factor (FGF) promotes cartilage repair *in vivo*. *Biochem Biophys Res Commun* 156: 611–618.
42. Nixon, A.J., Fortier, L.A., Williams, J. and Mohammed, H. 1999. Enhanced repair of extensive articular defects by insulin-like growth factor-I-laden fibrin composites. *J Orthop Res* 17: 475–487.
43. Sellers, R.S., Peluso, D. and Morris, E.A. 1997. The effect of recombinant human bone morphogenetic protein-2 (rhBMP-2) on the healing of full-thickness defects of articular cartilage. *J Bone Joint Surg Am* 79: 1452–1463.
44. Sellers, R.S., Zhang, R., Glasson, S.S., Kim, H.D., Peluso, D., D'Augusta, D.A., Beckwith, K. and Morris, E.A. 2000. Repair of articular cartilage defects one year after treatment with recombinant human bone morphogenetic protein-2 (rhBMP-2). *J Bone Joint Surg Am* 82: 151–160.
45. Wakitani, S., Imoto, K., Kimura, T., Ochi, T., Matsumoto, K. and Nakamura, T. 1997. Hepatocyte growth factor facilitates cartilage repair. Full thickness articular cartilage defect studied in rabbit knees. *Acta Orthop Scand* 68: 474–480.
46. Yamamoto, T., Wakitani, S., Imoto, K., Hattori, T., Nakaya, H., Saito, M. and Yonenobu, K. 2004. Fibroblast growth factor-2 promotes the repair of partial thickness defects of articular cartilage in immature rabbits but not in mature rabbits. *Osteoarthritis Cartilage* 12: 636–641.
47. Desire, L., Mysiakine, E., Bonnafous, D., Couvreur, P., Sagodira, S., Breton, P. and Fattal, E. 2006. Sustained delivery of growth factors from methylidene malonate 2.1.2-based polymers. *Biomaterials* 27: 2609–2620.
48. Isogai, N., Morotomi, T., Hayakawa, S., Munakata, H., Tabata, Y., Ikada, Y. and Kamiishi, H. 2005. Combined chondrocyte-copolymer implantation with slow release of basic fibroblast growth factor for tissue engineering an auricular cartilage construct. *J Biomed Mater Res A* 74: 408–418.
49. Matsusaki, M. and Akashi, M. 2005. Novel functional biodegradable polymer IV: pH-sensitive controlled release of fibroblast growth factor-2 from a poly(gamma-glutamic acid)-sulfonate matrix for tissue engineering. *Biomacromolecules* 6: 3351–3356.
50. Robinson, S.N. and Talmadge, J.E. 2002. Sustained release of growth factors. *In Vivo* 16: 535–540.
51. Cowan, C.M., Aalami, O.O., Shi, Y.Y., Chou, Y.F., Mari, C., Thomas, R., Quarto, N., Nacamuli, R.P., Contag, C.H., Wu, B., *et al.* 2005. Bone morphogenetic protein 2 and retinoic acid accelerate *in vivo* bone formation, osteoclast recruitment, and bone turnover. *Tissue Eng* 11: 645–658.

52. Moretti, M., Wendt, D., Dickinson, S.C., Sims, T.J., Hollander, A.P., Kelly, D.J., Prendergast, P.J., Heberer, M. and Martin, I. 2005. Effects of *in vitro* preculture on *in vivo* development of human engineered cartilage in an ectopic model. *Tissue Eng* 11: 1421–1428.
53. Dean, D.A. 2005. Nonviral gene transfer to skeletal, smooth, and cardiac muscle in living animals. *Am J Physiol Cell Physiol* 289: C233–245.
54. Goncalves, M.A., van Nierop, G.P., Tijssen, M.R., Lefesvre, P., Knaan-Shanzer, S., van der Velde, I., van Bekkum, D.W., Valerio, D. and de Vries, A.A. 2005. Transfer of the full-length dystrophin-coding sequence into muscle cells by a dual high-capacity hybrid viral vector with site-specific integration ability. *J Virol* 79: 3146–3162.
55. Kobinger, G.P., Louboutin, J.P., Barton, E.R., Sweeney, H.L. and Wilson, J.M. 2003. Correction of the dystrophic phenotype by *in vivo* targeting of muscle progenitor cells. *Hum Gene Ther* 14: 1441–1449.
56. Lai, Y., Yue, Y., Liu, M., Ghosh, A., Engelhardt, J.F., Chamberlain, J.S. and Duan, D. 2005. Efficient *in vivo* gene expression by trans-splicing adeno-associated viral vectors. *Nat Biotechnol* 23: 1435–1439.
57. Menezes, K.M., Mok, H.S. and Barry, M.A. 2006. Increased transduction of skeletal muscle cells by fibroblast growth factor-modified adenoviral vectors. *Hum Gene Ther* 17: 1–7.
58. Liu, M., Yue, Y., Harper, S.Q., Grange, R.W., Chamberlain, J.S. and Duan, D. 2005. Adeno-associated virus-mediated microdystrophin expression protects young mdx muscle from contraction-induced injury. *Mol Ther* 11: 245–256.
59. Murakami, T., Nishi, T., Kimura, E., Goto, T., Maeda, Y., Ushio, Y., Uchino, M. and Sunada, Y. 2003. Full-length dystrophin cDNA transfer into skeletal muscle of adult mdx mice by electroporation. *Muscle Nerve* 27: 237–241.
60. Pitard, B., Bello-Roufai, M., Lambert, O., Richard, P., Desigaux, L., Fernandes, S., Lanctin, C., Pollard, H., Zeghal, M., Rescan, P.Y., *et al.* 2004. Negatively charged self-assembling DNA/poloxamine nanospheres for *in vivo* gene transfer. *Nucleic Acids Res* 32: e159.
61. Baltzer, A.W., Lattermann, C., Whalen, J.D., Wooley, P., Weiss, K., Grimm, M., Ghivizzani, S.C., Robbins, P.D. and Evans, C.H. 2000. Genetic enhancement of fracture repair: healing of an experimental segmental defect by adenoviral transfer of the BMP-2 gene. *Gene Ther* 7: 734–739.
62. Bertone, A.L., Pittman, D.D., Bouxsein, M.L., Li, J., Clancy, B. and Seeherman, H.J. 2004. Adenoviral-mediated transfer of human *BMP-6* gene accelerates healing in a rabbit ulnar osteotomy model. *J Orthop Res* 22: 1261–1270.
63. Hiraide, A., Yokoo, N., Xin, K.Q., Okuda, K., Mizukami, H., Ozawa, K. and Saito, T. 2005. Repair of articular cartilage defect by intraarticular administration of basic fibroblast growth factor gene, using adeno-associated virus vector. *Hum Gene Ther* 16: 1413–1421.
64. Li, S., Kimura, E., Fall, B.M., Reyes, M., Angello, J.C., Welikson, R., Hauschka, S.D. and Chamberlain, J.S. 2005. Stable transduction of myogenic cells with lentiviral vectors expressing a minidystrophin. *Gene Ther* 12: 1099–1108.
65. Zhang, X., Xie, C., Lin, A.S., Ito, H., Awad, H., Lieberman, J.R., Rubery, P.T., Schwarz, E.M., O'Keefe R, J. and Guldberg, R.E. 2005. Periosteal progenitor cell fate in segmental cortical bone graft transplantations: implications for functional tissue engineering. *J Bone Miner Res* 20: 2124–2137.
66. Dai, K.R., Xu, X.L., Tang, T.T., Zhu, Z.A., Yu, C.F., Lou, J.R. and Zhang, X.L. 2005. Repairing of goat tibial bone defects with BMP-2 gene-modified tissue-engineered bone. *Calcif Tissue Int* 77: 55–61.

67. Lin, C.Y., Schek, R.M., Mistry, A.S., Shi, X., Mikos, A.G., Krebsbach, P.H. and Hollister, S.J. 2005. Functional bone engineering using *ex vivo* gene therapy and topology-optimized, biodegradable polymer composite scaffolds. *Tissue Eng* 11: 1589–1598.
68. Yang, M., Ma, Q.J., Dang, G.T., Ma, K., Chen, P. and Zhou, C.Y. 2005. *In vitro* and *in vivo* induction of bone formation based on *ex vivo* gene therapy using rat adipose-derived adult stem cells expressing BMP-7. *Cytotherapy* 7: 273–281.
69. Park, J., Gelse, K., Frank, S., von der Mark, K., Aigner, T. and Schneider, H. 2006. Transgene-activated mesenchymal cells for articular cartilage repair: a comparison of primary bone marrow-, perichondrium/periosteum- and fat-derived cells. *J Gene Med* 8: 112–125.
70. Ikezawa, M., Cao, B., Qu, Z., Peng, H., Xiao, X., Pruchnic, R., Kimura, S., Miike, T. and Huard, J. 2003. Dystrophin delivery in dystrophin-deficient DMDmdx skeletal muscle by isogenic muscle-derived stem cell transplantation. *Hum Gene Ther* 14: 1535–1546.
71. Lavasani, M., Lu, A., Peng, H., Cummins, J. and Huard, J. 2006. Nerve growth factor improves the muscle regeneration capacity of muscle stem cells in dystrophic muscle. *Hum Gene Ther* 17: 180–192.
72. Peng, H., Usas, A., Gearhart, B., Young, B., Olshanski, A. and Huard, J. 2004. Development of a self-inactivating tet-on retroviral vector expressing bone morphogenetic protein 4 to achieve regulated bone formation. *Mol Ther* 9: 885–894.
73. Peng, H., Usas, A., Olshanski, A., Ho, A.M., Gearhart, B., Cooper, G.M. and Huard, J. 2005. VEGF improves, whereas sFlt1 inhibits, BMP2-induced bone formation and bone healing through modulation of angiogenesis. *J Bone Miner Res* 20: 2017–2027.
74. Hannallah, D., Peng, H., Young, B., Usas, A., Gearhart, B. and Huard, J. 2004. Retroviral delivery of Noggin inhibits the formation of heterotopic ossification induced by BMP-4, demineralized bone matrix, and trauma in an animal model. *J Bone Joint Surg Am* 86-A: 80–91.
75. Peng, H., Usas, A., Hannallah, D., Olshanski, A., Cooper, G.M. and Huard, J. 2005. Noggin improves bone healing elicited by muscle stem cells expressing inducible BMP4. *Mol Ther* 12: 239–246.
76. Yoshimura, Y., Nomura, S., Kawasaki, S., Tsutsumimoto, T., Shimizu, T. and Takaoka, K. 2001. Colocalization of noggin and bone morphogenetic protein-4 during fracture healing. *J Bone Miner Res* 16: 876–884.
77. Tosi, L.L., Boyan, B.D. and Boskey, A.L. 2005. Does sex matter in musculoskeletal health? The influence of sex and gender on musculoskeletal health. *J Bone Joint Surg Am* 87: 1631–1647.

Part III

Artificial cells for cell therapy

8

Artificial cells for oral delivery of live bacterial cells for therapy

S PRAKASH, J BHATHENA and A M URBANSKA,
McGill University, Canada

8.1 Introduction

Live bacteria and other microorganisms have been commonly ingested in food by way of fermentation since early times. The concept of intake of live bacterial cells was established near the beginning of the twentieth century by Metchnikoff,[1] who hypothesized that the ingestion of fermented milk products containing lactic acid-producing bacteria had a beneficial impact on health and human longevity. The gastrointestinal microflora is known to play a vital role in metabolic processes.[2,3] Specific functions of intestinal bacteria include protection against potential colonization of incoming microbes, fermentation of exogenous and endogenous carbon and energy sources, metabolism of faecal residue and immune education.[4] It is now well recognized that the metabolic capability of the intestinal microflora can be enhanced by the oral delivery of probiotics. Oral administration is the most convenient route of delivery for probiotics and viable microorganisms are commonly added to foods, especially dairy products.[5] However, to be a successful probiotic, bacteria generally must be ingested in live form, be able to colonize in the gut to maintain sufficient viable microorganisms that survive the host's digestive process and reproduce to the extent that they can be identified in the stool, and be able to attach to and adhere to the gut epithelium.

While the ingestion of live bacterial cells has proven benefits, the optimism associated with their use is counterbalanced by the fact that many so-called 'probiotic' products are unreliable in content[6–10] and unproven clinically.[11] Some have been found to play a role in disease remediation; however, mainstream medicine and science remain divided about the validity of health claims made about them. Much remains to be done to gain the acceptance of the broader medical community. Therapeutic cultures may be aimed at a variety of ailments, but there is ample evidence that oral ingestion of live microbes alleviates or prevents various disorders (Tables 8.1 and 8.2).[12,13]

Table 8.1 Potential therapies based on the oral delivery of live bacterial cells

Targeted therapy	Bacterial cells	Proposed mode of action	References
Diarrhea	*L. rhamnosus* *L. casei* *L. reuteri* *L. GG* *B. lactis* *B. bifidum* *B. lactis* Bb12	Treatment and prevention of rotavirus and acute diarrhea in children and adults; reduction of antibiotic-associated diarrhea in children and adults; prevention of traveler's diarrhea; promotion of serum and intestinal immune responses to rotavirus thus establishing immunity against rotavirus infections.	19–24, 26, 27, 29, 30, 34
Colorectal cancer	*Lactobacillus* *B. breve* *B. longum*	Enhancement of the host's immune response; binding and degrading potential carcinogens; alterations in the intestinal microflora incriminated in producing recognized carcinogens; production of anticarcinogenic or antimutagenic compounds in the colon; alteration of the metabolic activities of the resident microflora; alteration of physicochemical conditions; effects on general physiology.	78, 79
Bladder cancer	*L. casei* (Shirota)	Reduced recurrence of superficial bladder cancer. Increased percentage of T-helper cells and NK cells	177
Inflammatory bowel disease	*L. lactis* *L. GG* *L. salivarius* UCC118	Adjunct nutritional therapy for Crohn's disease. Increase in gut IgA immune response promoting the gut immunological barrier.	13, 63, 178–180
H. pylori colonization/ gastric ulcer	*L. acidophilus*	Down-regulation of *H. pylori* infection by inhibition of intestinal cell adhesion and invasion.	25, 48–51
Steatorrhea of lipids (malabsorption of lipids)	*B. plantarii*	Bacteria express lipolytic activity with substantial enzyme stability in human gastric juice leading to the increased absorption of lipids in the small intestine.	181, 182
Immune modulation	*L. acidophilus* *L. johnsonii* La1 *L. casei* (LcS) *L. crispatus* *L. GG* *B. lactis* Bb12 *B. bifidum*	Innate immunity is enhanced by stimulating the activity of splenic NK cells; antigen-feeding alone primes for an immune response, co-feeding on antigen and probiotic bacteria suppresses both antibody and cellular immune responses; potential to attenuate autoimmune diseases (by jointly dosing with myelin basic protein and probiotic bacteria).	15, 17, 62, 64–66, 69, 183

Table 8.1 Continued

Targeted therapy	Bacterial cells	Proposed mode of action	References
Lower LDL-cholesterol and triglycerides	*L. plantarum* 299V *L. acidophilus* *L. bulgaricus* *L. reuteri*	Bacteria may bind or incorporate cholesterol directly into the cell membrane. Bile salt hydrolase (BSH) enzyme deconjugates intraluminal bile acids making them less likely to be reabsorbed into the enterohepatic circulation (ECH), causing *de novo* synthesis of bile acids from blood serum cholesterol.	99, 101, 103–105, 108–110, 184
Irritable bowel syndrome	*L. plantarum* 299V *L. paracasei* F19 *B. breve* *E. coli* Nissle 1917 *B. cereus* Toyos *B. subtilis* *P. freudenreichii* JS	Restoration of normal intestinal microflora; shorter colonic transit times. Increased tolerance to certain foods.	91, 93–96
Chronic kidney failure	*L. acidophilus* *Lactic acid bacteria*	Reduction in blood levels of uremic toxins produced in the intestine as bacterial putrefactive metabolites. Inhibition of bacterial proliferation; correction of the intestinal microflora.	185–187
Kidney stones	*L. acidophilus* *L. plantarum* *L. brevis* *S. thermophilus* *B. infantis* *O. faecalis*	Reduction in urinary excretion of oxalate; oxalate-degrading enzymes break down the unwanted oxalate; can be used to prevent the subsequent evolution of kidney stones.	188, 189
Allergy	*L. rhamnosus* GG *L. paracasei* 33 *E. coli* Nissle 1917 LGG; *B. lactis* Bb12; Combination of *L. rhamnosus* 19070-2 and *L. reuteri* DSM 122460	Counterbalancing the skewed T_H2 immune phenotype in the gut; reduced colonization of bacterial pathogens; boost in the systemic immune response.	54–56

Table 8.2 Clinical benefits of live bacterial cells used in human intervention studies

Clinical benefits	Bacterial culture used	References
Treatment of acute diarrhoea in children especially caused by rotavirus; reduction of the risk of antibiotic-associated symptoms in children and in adults; reduction of the risk of acute diarrhoea in children	META, LGG BB12	21, 23, 29, 190, 191
Reduction of the recurrence of *Clostridium difficile* enterocolitis	LG299v	28, 192
Relief of milk allergy/atopic dermatitis in infants; relief of allergic rhinitis; reduction in the risk of atopic diseases in infants	LGG BB12 combination of *L. rhamnosus* 19070-2 and *L. reuteri* DSM 122460 *L. paracasei* 33 LGG	54–56, 193–195
Reduction in the risk of respiratory infections	LGG	16, 17
Amelioration of the immune response	Various	196–199
Reduction in the risk of dental caries	LGG	200
Suppression of *Helicobacter pylori*	Various	44–47
Reducing the recurrence of pouchitis	VSL#3; LGG	201, 202
Relief of IBS symptoms	299v	94
Relief of rheumatoid arthritis symptoms	LGG	203

META, indicates meta-analyses and LGG, 299v and BB12 refer to *L. rhamnosus* GG, *L. plantarum* 299v and *B. lactis* BB12, respectively.

8.2 Current therapies based on the oral delivery of live bacterial cells

There is considerable clinical interest in the utility of probiotic therapy – the feeding of (live) non-pathogenic bacteria, for disease treatment or health promotion. Recent evidence illustrates numerous benefits elicited by probiotics:

- pathogen interference, exclusion and antagonism;
- immunostimulation and immunomodulation;
- anticarcinogenic and antimutagenic activities in animal models;
- alleviation of symptoms of lactose intolerance;
- vaginal/urinary tract health;
- reduction in blood pressure in hypertensive subjects;
- decreased incidence and duration of diarrhea (antibiotic-associated diarrhoea, *Clostridium difficile*, travellers and rotaviral); and
- maintenance of mucosal integrity.

The mechanism of probiotic action, however, is not fully understood. Overall, the bacteria are believed to have two effects: the first, called the 'nutritional effect', is a reduction in the number of toxin-producing bacteria due to competition from probiotics. The second, called the 'sanitary effect', involves resistance to colonization and stimulation of immune response. The mechanisms of probiotic action can be divided into direct and indirect effects. Direct effects occur at the epithelium itself. Indirect mechanisms involve reactions that do not take place directly on the epithelium, e.g. probiotics may also induce the intestine to secrete substances that interfere with the adherence process. Established uses of live bacterial cells are described briefly below.

8.2.1 Live bacterial cell therapy and modulation of the microflora

The novel idea with bacterial therapy has always been to change the composition of the normal intestinal microflora from a potentially detrimental composition towards a microflora that would be advantageous for the host. Probiotic cells that endure gastrointestinal (GI) transit are liable to cause an augmentation in fecal levels of that particular genus, especially when initial levels were low. Owing to antagonism for adhesion sites and nutrients, and possibly the production of antimicrobial substances, levels of certain less desirable genera can decrease. A concomitant increase in fecal levels of genera other than the probiotic consumed has also been observed for certain live bacterial cells, e.g. consumption of *L. rhamnosus* GG has been observed to be associated with an increase in fecal bifidobacteria[14] and consumption of *L. salivarius* UCC118 caused an increase in fecal *Enterococcus* levels.[15]

A probiotic fermented milk drink has been shown to lower the nasal mucosal colonization by potentially pathogenic bacteria (*Staphylococcus aureus*, *Streptococcus pneumoniae*, β-hemolytic streptococci and *Haemophilus influenzae*).[16] This result supports the finding that milk enriched with *L. rhamnosus* GG reduced respiratory infections with complications and decreased the need for antibiotic treatment.[17]

It is apparent that evading colonization by pathogens and reducing the risk for over growth of potential pathogenic bacteria is beneficial to the host. However, in some cases much emphasis is placed on this change in microflora composition without considering the actual health benefit.

8.2.2 Preventive therapy of diarrhoeal diseases using live bacterial cell therapy

The gut microflora and the host animal make up a complex ecosystem whose stability is important to the host's gastrointestinal health. However, the gut microflora can become unbalanced, weakening the host's defenses against pathogens.[18] The concentration of bacteria is highest in the colon where it reaches 10^{11} to 10^{12} colony-forming units (CFU) per milliliter. The intestinal microflora is important for maturation of the immune system, development of normal intestinal morphology and maintenance of an immunologically balanced inflammatory response. The microflora reinforces the barrier function of the intestinal mucosa, helping to prevent the attachment of pathogenic microorganisms and controlling the entry of allergens. In addition, some species may contribute to the body's requirements for certain vitamins including biotin, pantothenic acid and vitamin B_{12}. The composition of the microbial flora varies with age and can be disrupted by antibiotics. A number of clinical trials have tested the effectiveness of live bacteria (as probiotics) in the deterrence of acute diarrheal conditions.[19]

Two meta-analyses concluded that probiotics are useful in preventing antibiotic associated diarrhea.[20,21] Supplementation of an infant formula with *Bifidobacterium bifidum* and *Streptococcus thermophilus* reduced the incidence of diarrhea and rotavirus shedding in hospitalized infants and *L. rhamnosus* GG has been shown to be beneficial in preventing nosocomial diarrheas.[22,23] In uncontrolled studies, *Bifidobacterium lactis, L. reuteri* and *L. rhamnosus* GG have also been investigated as preventive therapy to decrease the incidence of episodes of acute diarrhea in community-based groups of children.[23,24] In a recent review article, the effect of live bacterial cell therapy in acute diarrhoea in children has been systematically analyzed.[25]

Some initial clinical studies demonstrated the effectiveness of *L. rhamnosus* GG in reducing the reversion of *C. difficile* diarrhea following treatment with antibiotics.[26,27] In small, double-blind, placebo-controlled studies, the impact of *L. plantarum* (299V) and *L. rhamnosus* strain GG has been evaluated in the prevention of recurrent episodes of *C. difficile* associated diarrhea.[28,29] Several studies have shown that selected live bacterial cells, such as *L. rhamnosus* GG, *L. reuteri*, *L. casei* Shirota and *B. lactis* Bb12, can shorten the duration of rotavirus diarrhea by approximately 1 day.[22,30–32]

One related area in which the evidence for beneficial preventive effects is inconsistent is in traveler's diarrhea.[33,34] Prophylaxis with a mixture of *L. acidophilus*, *B. bifidum*, *L. bulgaricus* and *S. thermophilus* has been shown to significantly reduce the frequency of diarrhea in travellers in Egypt.[34] Prevention of traveler's diarrhea with *L. rhamnosus* GG has been shown in two separate studies by Hilton *et al.* and Oksanen *et al.*[35,36] A mixture of *L. acidophilus* and *L. bulgaricus* was tested in tourists traveling to Mexico.[37] There were no significant differences in the incidence of diarrhea between groups of British soldiers deployed to Belize after 3 weeks when they were randomly administered *L. fermentum* strain KLD, *L. acidophilus* (LA) or placebo.[38]

A common complication in cancer patients treated with radiotherapy is acute diarrhea.[39] A probiotic preparation, VSL#3 (VSL Pharmaceuticals, Fort Lauderdale, MD, USA), has been assessed for the preventive effect in patients who had postoperative radiotherapy after surgery for cancer.[40] The probiotic product VSL#3 contained strains of *L. casei*, *L. plantarum*, *L. acidophilus*, *L. delbruekii* ssp. *bulgaricus*, *B. longum*, *B. breve*, *B. infantis* and *S. salivarius* ssp. *thermophilus*. A double-blind and placebo-controlled study determined the efficacy of *L. rhamnosus* (Antibiophilus®, Germania Pharmazeutika GmbH, Vienna, Austria) compared with a placebo product in the treatment of patients suffering from mild to moderate radiation-induced diarrhea.[41] *L. acidophilus* and *L. bulgaricus* (Lactinex) were used for the prevention of diarrhea in patients that were tube-fed for less than 5 days.[42]

An analysis by Huang *et al.* reviewed 18 well-designed studies and concluded that bacterial probiotic therapy reduced the duration of acute diarrheal illness by approximately 1 day.[43]

8.2.3 Live bacterial cells as a strategy for treatment of *H. pylori* infections

Live bacterial cells do not eradicate the pathogenic bacterium *Helicobacter pylori*, which is the main causative agent of gastritis and gastric ulcer, but suppress its growth and reduce inflammation in the stomach significantly.[44–47] Antagonistic actions of some *Lactobacillus* strains against *H. pylori in vitro* have been reported.[48] A significant reduction of the urease activity has been reported in patients treated with a supernatant of *L. johnsonii* LA1 (Nestlé, Switzerland) associated with omeprazole.[49] Two trials have reported that the ingestion of a fermented dairy product containing this strain or a heat killed *Lactobacillus acidophilus* could help to decrease the gastric colonization by *H. pylori*.[50,51] However, administration of a probiotic-containing yogurt was found to be ineffective in the eradication of *H. pylori* infection in 27 subjects.[52] A number of clinical studies on the effects of probiotics on the eradication rates of *H. pylori* have been carried out; for a summary see Sullivan and Nord.[25]

8.2.4 Uses of live bacterial cells for treatment of atopic inflammations

The use of probiotics in the prevention or treatment of allergic disease is a relatively new concept.[53] Effects of probiotic therapy were reported in randomized, double blind, placebo-controlled studies with *Lactobacillus* GG and *L. reuteri* and showed significant improvement in clinical symptoms and markers of intestinal inflammation.[54–56] Studies in infants with eczema who receive formulas supplemented with *L. rhamnosus* GG have shown benefit in decreasing both gastrointestinal symptoms and eczema.[54,57] As a means of preventing allergy, a randomized controlled study by Lodinova-Zadnikova and Sonnenborn investigated the effect of at-birth colonization with non-pathogenic *Escherichia coli* Nissle 1917.[58] Effectiveness of viable but not heat-inactivated probiotics in the management of atopic eczema and cows' milk allergy and an association between the consumption of fermented milk products and a decrease in allergic symptoms has also been reported.[59–61]

8.2.5 Immune modulations using live bacterial cells

The ability of specific strains of probiotics to enhance or amend aspects of both natural and acquired immune responses in humans is well documented. *Lactobacillus johnsonii* La1, L. salivarius, *Bifidobacterium lactis* Bb12 have been shown to enhance the phagocytic capacity of peripheral blood leukocytes.[15,62] It has also been observed that different lactic acid bacteria strains differ in their capacity to influence complement receptor expression in phagocytic cells.[63] Subjects receiving milk containing *L. rhamnosus* (HN001) or *B. lactis* (HN019) were found to exhibit significantly more phagocytically active blood leukocytes (neutrophils and monocytes).[64–66] Significant improvements in NK cell activity, and increases in the percentage of NK cells in the peripheral blood of human volunteers fed yogurt, milk or sausages containing live bacterial cells have also been reported.[63–65,67] Differences in the ability of lactic acid bacteria to influence immune function are well documented.[68] The systemic effect of probiotics on the immune system has been demonstrated in placebo-controlled studies that have identified an enhanced antibody response to vaccines.[17,69,70]

8.2.6 Live bacterial cells for cancer prevention and therapy

Several animal studies have shown that supplementation with specific strains of lactic acid bacteria could prevent the establishment, growth and metastasis of transplantable and chemically induced tumors.[71–73] Studies in human subjects have also revealed that probiotic therapy may reduce the risk of colon cancer by inhibiting transformation of pro-carcinogen to active carcinogens, binding/inactivating mutagenic compounds, producing antimutagenic compounds, suppressing the growth of pro-carcinogenic bacteria, reducing the absorption of mutagens from the intestine, and enhancing immune function.[74,75] Placebo-controlled trials by Aso and colleagues revealed that *L. casei* Shirota increases the percentage of T-helper cells and NK cells in adult colorectal cancer patients and has a protective effect on the recurrence of superficial bladder cancer.[76] Furthermore, recent studies also indicate that *L. casei* Shirota may prolong survival during specific cancer treatments.[77] Additionally, select strains of lactobacilli have been shown to significantly suppress intestinal tumors by chemical mutagens.[78,79] Findings suggest that *S. choleraesuis* in combination with cisplatin, represents a promising strategy for the treatment of primary and metastatic tumors.[80,81]

8.2.7 Live bacterial cells targeted delivery of TNF-α and other therapeutic molecules for systemic infections and for use in inflammatory bowel diseases (IBD) and other GI diseases

Live bacteria have been used as treatment options for managing inflammatory bowel diseases (IBD) such as Crohn's disease, ulcerative colitis and pouchitis.[82] *Lactococcus lactis*, genetically engineered to secrete the anti-inflammatory cytokine interleukin-10, has been shown to cure or prevent experimental enterocolitis, a model designed to mimic human IBD.[83,84] Some lactobacillus strains including *L. casei* are known to down-regulate the spontaneous release of TNF-α by inflamed tissue, and also the inflammatory response induced by inherent coliforms.[85,86] Oral administration of either VSL#3 or *L. plantarum* significantly decreased histological colitis in mice deficient in the IL-10 gene.[87,88] Non-pathogenic *E. coli* strain Nissle 1917 has an effect in maintaining remission in colonic Crohn's disease and ulcerative colitis.[63] Interestingly, *E. coli* (Nissle) has been observed to be effective in alleviating the symptoms of IBD.[13] A product (BFM, Yakult Co. Ltd, Tokyo, Japan) containing *B. breve*, *B. bifidum* and *L. acidophilus* YIT 0168 was evaluated as a dietary adjunct in the treatment of ulcerative colitis[89] showed a reduced frequency of relapses, not confirmed endoscopically. Although initial open label studies showed promising prospects, with the exception of the studies on pouchitis, published results of controlled clinical trials in patients with inflammatory bowel diseases are poor (see Tamboli *et al.* for an extensive review of published trials[90]).

Efficacy of live cell therapy for IBS

In a double blind clinical trial, administration of a *L. plantarum* strain decreased pain and flatulence in patients with IBS.[91] Likewise, a recent placebo-controlled trial concluded that the VSL#3 probiotic mixture is useful for the relief of abdominal bloating in patients with diarrhoea predominant IBS.[92] Halpern *et al.*[93] performed a crossover study using a heat-killed strain (strain LB) of *L. acidophilus* and concluded that the product demonstrated a statistically significant therapeutic benefit in 50% of the patients. A fruit drink containing *L. plantarum* (299V) has been assessed;[94] *Lactobacillus plantarum* (299V) has also been evaluated for the effect on colonic fermentation in patients with IBS according to the Rome criteria.[95] Patients in a test were assigned to receive either an active preparation containing *Lactobacillus plantarum* LP0 1 or *Bifidobacterium breve* BR0; the severity score of characteristic IBS symptoms was found to have significantly decreased in probiotic group.[96] At the present time, this therapy seems to be an interesting track as at least two random controlled trials have shown that some probiotics could influence the colonic transit time of healthy humans.[97,98]

8.2.8 Lowering of cholesterol by the utilization of live bacterial cell therapy

Mann and Spoerry were the first to report that consumption of fermented milk was associated with reduced serum cholesterol levels in the Maasai people.[99] In most studies, lactobacilli and bifidobacteria have been reported to deconjugate bile salts, producing free bile acids, which are more inhibitory to susceptible bacteria than the deconjugated form.[100] Effects of live bacterial therapy using *L. acidophilus* and *L. bulgaricus* on cholesterol levels in humans remain to be verified by further studies, as many confounding factors have contributed to the effects observed in different population groups.[101–103] A meta-analysis of controlled short-term studies has shown that consumption of fermented milk/ yoghurt containing *Enterococcus faecium* (Gaio) is effective in reducing both total and low-density lipoprotein cholesterol.[104,105] Administration of Pro Viva (Probi AB, Sweden) containing *L. plantarum* 299v lowered total and low-density lipoprotein cholesterol levels in a placebo-controlled study.[106] Subjects consuming milk fermented with *L. casei* TMC 0409 and *S. thermophilus* TMC 1543 showed significant increases in high-density lipoprotein levels and significant decreases in triglyceride levels.[107] The levels of triglycerides were also reduced significantly in subjects receiving the fermented milk. Consumption of 300 g yoghurt supplemented with *L. acidophilus* 145 and *Bifidobacterium longum* 913 was also found to increase high-density lipoprotein concentration and decrease ratio of low- to high-density lipoprotein.[108] It has also been reported that *L. acidophilus* NCFM and *L. reuteri* have an appreciable ability to assimilate cholesterol.[109,110] The precise mechanisms by which probiotics affect cholesterol levels are not fully understood, however; a range of mechanisms have been suggested.[111]

8.2.9 Use of live bacterial cell therapy for the improvement of lactose intolerance

The beneficial effects of some β-galactosidase producing probiotic bacteria such as *L. acidophilus* on lactose digestion have been shown.[112,113] The results of Lin *et al.*[114] showed the importance of strain selection with respect to alleviating the symptoms of lactose intolerance. A large number of human studies in which consumption of fresh yogurt (with live yogurt cultures) was compared with consumption of a pasteurized product (with heat-killed bacteria), demonstrated better lactose digestion and absorption in subjects who consumed yogurt with live cultures as well as reduction of gastrointestinal symptoms.[115–123] In clinical practice, replacement of milk by yogurt or fermented dairy products decreases or suppresses the symptoms of lactose intolerance.[115,124] This beneficial effect is usually more associated with products fermented with *L. delbrueckii* subsp. *bulgaricus* and *S. thermophilus*.[125]

8.2.10 Application of bacterial cell therapy in the alleviation of constipation and urogenital disorders

Probiotics have been suggested to relieve constipation.[126] However, review of the literature does not substantiate this claim. This may relate to the causes of constipation; physical inactivity, low-fiber diets, insufficient liquid intake and some drugs. The altered microflora composition is more likely to be a consequence than the cause of constipation, correcting the microflora composition may therefore not be of help. Daily oral intake of probiotic strains *Lactobacillus rhamnosus* GR-1 and *Lactobacillus fermentum* RC-14, resulted in some asymptomatic bacterial vaginosis patients reverting to a normal lactobacilli-dominated vaginal microflora.[127,128] In terms of eradicating symptomatic bacterial vaginosis, *L. acidophilus* administered for six days led to an improvement compared with no improvement in the placebo group.[129] A crossover study in which patients with a history of recurrent yeast vaginitis ($\geq$5 per year) were given 8 oz (240 g) of *L. acidophilus* supplemented yoghurt daily for six months, then switched to a yoghurt-free diet, resulted in 0.4 breakthrough infections compared with 2.5 per study term.[130]

8.3 Potential of artificial cells for live therapeutic bacterial cell oral delivery

The human digestive tract contains a variety of bacterial species, both beneficial and harmful. Balanced microflora in the intestine contributes to a healthy gut barrier, which is important for ensuring appropriate immunological and inflammatory responses to antigens.[131] This balance varies with age and can be disrupted by antibiotic treatment. However, probiotic bacteria, derived from such foods as yogurt, can help restore that balance. While live bacterial cell therapies show immense therapeutic potential, an important problem with such interventions is the fact that the viability of bacteria reaching the intestine is highly variable, so it is difficult to predict their therapeutic effect. When given orally, bacterial cells are exposed to severe GI conditions and experience low survival, requiring a large dosage to be given. And, large quantities of live bacterial cells when delivered orally can stimulate host immune response. Furthermore, when given orally, live bacterial cells can be retained in the intestine and repeated large doses could result in their replacing the normal intestinal flora. The risk of systemic infections, deleterious metabolic activities, adjuvant side effects, immunomodulation and risk of gene transfer further limit their widespread use.[132–136]

These limitations have raised concerns about the safety of the approach and prevented receiving approval from regulatory agencies and hence precluded the adoption of this form of therapy in regular clinical practice.

8.4 Principle of artificial cells for oral delivery of bacterial cells for therapy

Artificial cell microencapsulation is a concept wherein biologically active materials are encapsulated in specialized ultra-thin semipermeable polymer membranes.[137–141] Designed for a plethora of applications, artificial cells have been made to combine properties of biological systems such as nanoscale efficiency, self-organization and adaptability at relatively low cost. The polymer membrane can protect encapsulated materials from harsh external environments while at the same time allowing for the metabolism of selected solutes capable of passing into and out of the microcapsule. In this manner, the enclosed material (in this case live bacteria) can be retained inside and separated from the external environment, making microencapsulation particularly useful for biomedical and clinical applications (Fig. 8.1). Consequently, when they are

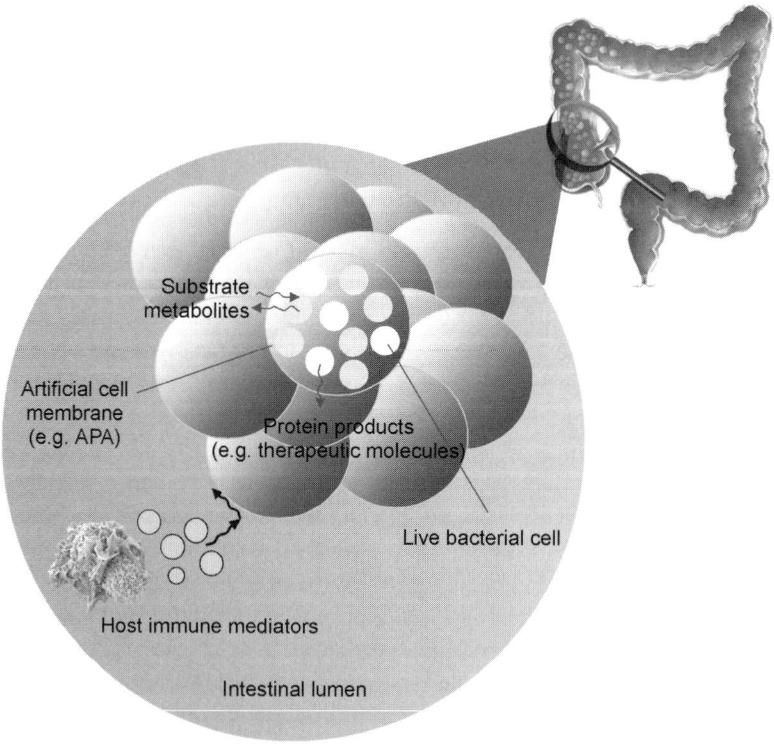

8.1 Principle for oral administration of microencapsulated live bacterial cells for therapy. Live bacterial cells can be encased in a polymeric matrix which allows diffusion of nutrients, wastes and protein products while acting as a barrier to the conduits of the immune system once the cells are ingested by a patient. Also, small molecules (including some peptides) produced by the enclosed bacterial cells can be designed to diffuse out into the body for therapeutic purposes.

administered orally, a large controlled number of viable bacteria reach the intestine and colon. The membranes of the microcapsules are permeable to smaller molecules, and thus the cells inside the microcapsules metabolize small molecules found within the gut during passage through the intestine. For these reasons, artificial cells containing bacterial cells are very good candidates for oral artificial cell therapy (Table 8.3). This concept has immediate application for microencapsulation-based oral therapy in renal failure and liver failure, physiologically responsive gene therapy and somatic gene therapy.[142]

Table 8.3 Potential therapies based on the oral delivery of microencapsulated bacterial cells

Targeted therapy	Encapsulated bacterial cells	Proposed mode of action	References
Tuberculosis	*M. bovis* BCG	Tuberculin-specific cell-mediated immune responses, system-level lymphocyte sensitization evidenced by *in vitro* tuberculin-specific IFN-γ production in splenocyte cultures, establishment of protective immunity against TB.	144, 204, 205
Kidney dialysis	*E. coli* DH5	Inserted *Klebsiella aerogenes* urease gene, in genetically engineered *E. coli DH5* cells, causes over-expression of the urease enzyme; subsequent lowering of elevated blood levels of urea and ammonia; thus *E. coli* DH5 cells normalize elevated levels of several metabolites during renal failure.	206,207
Kidney stones	*O. formigenes*	Oxalate-degrading enzymes produced by *Oxalobacter formigenes* break down unwanted oxalate, a major risk factor for renal stone formation and growth in patients with idiopathic calcium-oxalate urolithiasis; can be used to prevent subsequent evolution of kidney stones.	208, 209
Immuno-modulation	*B. bifidum* Bb-11	Significantly induces total IgA and IgM synthesis by both mesenteric lymph nodes (MLN) and Peyer's patch (PP) cells; regulates the synthesis of IgA by mucosal lymphoid cells; significantly increases the number of Ig (IgM, IgG and IgA) secreting cells.	145

Table 8.3 Continued

Targeted therapy	Encapsulated bacterial cells	Proposed mode of action	References
Elevated blood levels of cholesterol	*Genetically engineered Lactobacillus plantarum* LP80 (pCBH1) and *L. reuteri*	Overproduced BSH enzyme deconjugates intraluminal bile acids making them less likely to be reabsorbed into the ECH, causing *de novo* synthesis of bile acids in the liver from blood serum cholesterol. *L. reuteri* can be microencapsulated in combination with LP80 (pCBH1), as it has been shown to precipitate and bind bile acids, making them less bioavailable which may be important to their carcinogenic potential.	210
Preventative therapy for colon cancer	*Lactobacillus plantarum* LP80 (pCBH1) and *L. reuteri*	The BSH enzyme is overproduced by LP80 (pCBH1) cells and hydrolyzes available conjugated bile acids in the intestinal lumen. *L. reuteri*, shown to precipitate and bind bile acids, then binds the deconjugated bile acids, making them incapable of leaving the microcapsule and thus less bioavailable for exfoliation of the GI and any potential carcinogenic damage.	210
Disease of the bowel (elevated intraluminal levels of bile acids)	*Lactobacillus plantarum* LP80 (pCBH1) and *L. reuteri*	LP80 (pCBH1) and *L. reuteri* deconjugate, precipitate, and then bind conjugated bile acids within microcapsules, mitigating the problems associated with excessive electrolyte and water secretion associated clinically with diarrhea and dehydration.	210
Probiotics	*Lactobacillus Bifidobacterium*	Live microorganisms used as dietary supplements with the aim of benefiting the health of the consumer by positively influencing their intestinal microbial balance. Microencapsulated probiotic bacteria should help alleviate diarrhea, lower cholesterol, modulate immunity and prevent colon cancer.	211–215
Urogenital infections (urinary tract infection (UTI), bacterial vaginosis (BV), or yeast vaginitis (YV)	*L. rhamnosus* GR-1 *L. fermentum* RC-14 *L. acidophilus*	Restoration and maintenance of a healthy urogenital tract possibly by competition with pathogenic microbes, protect the host from recurrence of UTI; creation of an environment better able to support indigenous lactobacilli growth.	146

Table 8.3 Continued

Targeted therapy	Encapsulated bacterial cells	Proposed mode of action	References
Elevated blood levels of nitrogen and hydrogen gas	Hydrogen metabolizing (*M. smithii*) and N_2 fixing (Entero-bacteriaceae)	Live microorganisms delivered orally to a diver's large intestine during hyperbaric exposure to a gas mixture containing H_2 or N_2 metabolizes the H_2 or N_2 gas to other compounds such as methane or water for hydrogen and ammonia for nitrogen to prevent decompression sickness or reduce decompression time.	216

8.5 Design and preparation: different immobilization technologies

Three methods of preparing artificial cell microcapsules using APA membrane containing live probiotics bacterial cells have been previously described by Prakash and Jones.[142]

Table 8.4 summarizes various membrane types for preparation of artificial cells and their brief characteristics. Representative photomicrographs of artificial cells composed of alginate and poly-L-lysine that have been shown to be useful in kidney failure and managing elevated blood cholesterol are depicted in Fig. 8.2. Empty as well as loaded with live bacterial cells microcapsules preserve their shape and integrity and were able to protect live cells.

Table 8.4 Artificial cells composed of biological membrane materials

Membrane type/ technique used	Characteristics	References
Cellulose acetate phthalate (CAP) + beeswax	Used to encapsulated freeze-dried *B. pseudolongum* Able to survive GI tract	147, 148
Spray-drying CAP	Lactobacilli, bifidobacteria encapsulation Tolerance to low pH values Tolerance to simulated human bile concentration Good protection and resistance to acidic cond.	149
In butter-oil-and-whey-based medium	Bifidobacteria encapsulation Inefficient in bacteria protection against acidic conditions	150

Table 8.4 Continued

Membrane type/ technique used	Characteristics	References
κ-carrageenan–locust bean gum	Good resistance to low acid (used in lactic fermentation) Requires high concentration of potassium ions (not recommended for diet and can damage encapsulated cells)	151–154
Calcium alginate	Low acid resistance Shrinkage Decreased mechanical strength	155, 156
Gellan–xanthan	Good acid stability enhanced by Ca ions Do not provide immune protection	157, 158
Agarose Emulsification/internal gelatinization	Stable up to 12 hours *in vitro* Agarose not sufficiently immunoprotective	157, 159
Gelatin (polymer-coated gelatin)	Good *in vitro* stability Uncoated gelatin beads unstable after 15 min upon ingestion	160–163
Alginate–chitosan (AC)	Better biocompatibility in polycation component than APA Improved bio- and immuno-compatibility when coated with PEG Superior capsule strength, flexibility and biocompatibility with glutaraldehyde coating Mechanical instability in uncoated capsule	164–166
Alginate–poly-L-lysine–alginate (APA)	Established synthesis protocols, *in-vivo* effectiveness for oral bacterial delivery Mild encapsulation conditions, cell compatible core, flexible membrane permeability Strong short/mid-term mechanical stability in implantation studies Low inflammatory response if coated with PEG Mechanical instability in harsh conditions of GI, questionable permeability Inflammatory response if polycation inadequately neutralized	167–169
Alginate–cellulose sulfate–polymethylene co-guanidine–alginate (A-PMCG-A)	Flexible membrane permeability, better mechanical stability than APA	170–172

Table 8.4 Continued

Membrane type/ technique used	Characteristics	References
	Independent adjustment of capsule size, wall thickness and mechanical strength Immunogenicity, mechanical stability *in-vivo* yet to be studied	
Hydroxymethyl methacrylate–methyl methacrylate (HEMA-MMA)	Improved mechanical stability due to insolubility in aqueous conditions Easy scale-up, control over desired properties Reaction conditions potentially damaging to probiotics Diffusion properties of aqueous solutes limited	173, 174
Alginate/poly-L-lysine/pectin/ poly-L-lycine/alginate (APPPA) Alginate/poly-L-lysine/pectin/ poly-L-lycine/pectin (APPPP) Alginate/poly-L-lysine/chitosan/ poly-L-lycine/alginate (APCPA)	High GI tract stability Increased resistance to complete dissolution in aqueous solution, acidic and alkali conditions Presence of ion chelators in the membrane preparation procedures allows for control of membrane permeability	175, 176

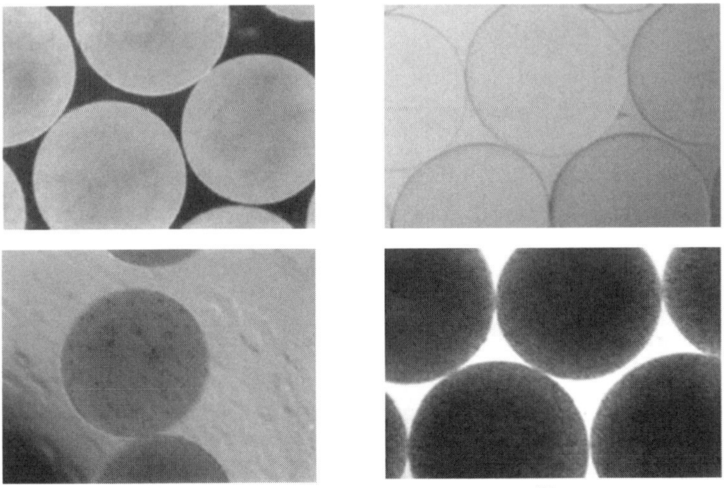

(a) (b)

8.2 Photomicrographs of freshly prepared alginate–poly-L-lysine–alginate (APA) artificial cell microcapsules, empty (top) and containing live bacteria (bottom) for therapy in (a) kidney failure (with permission from *Nature America*[143]); (b) elevated blood cholesterol (with permission from Hindawi Publishing Corporation[142]).

In a study by Aldwell *et al.*[144] *Mycobacterium bovis BCG* bacilli was lipid encapsulated and fed to mice infected with *M. tuberculosis* H37Rv. This report shows that probiotic strain provides significant pulmonary and systemic protection against aerosol challenge with *M. tuberculosis* H37Rv (Fig. 8.3a). In another study, a probiotic bacteria *Bifidobacterium bifidum* was tested to determine whether it regulates the synthesis of IgA by mucosal lymphoid cells and mucosal antibody production following peroral administration of *B. bifidum* to mice.[145] The results in Fig. 8.3b show that ingestion of *B. bifidum* significantly increased the number of Ig (IgM, IgG, and IgA) secreting cells in the

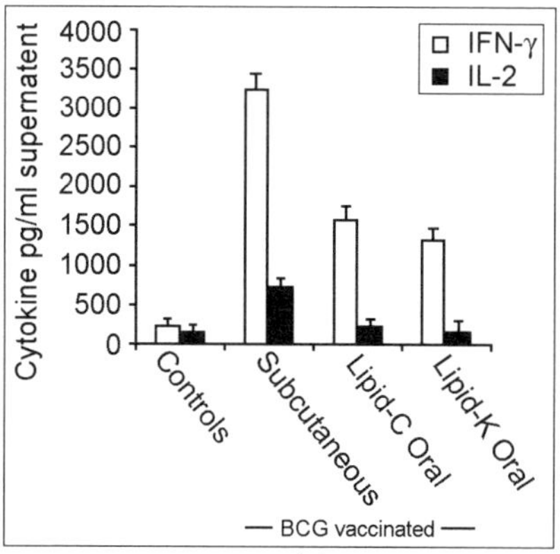

(a)

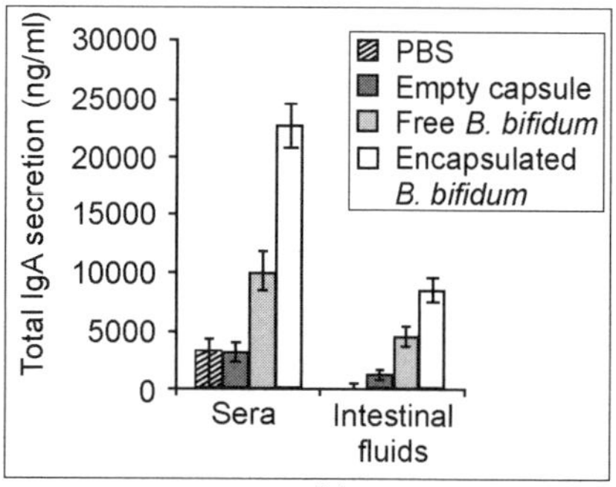

(b)

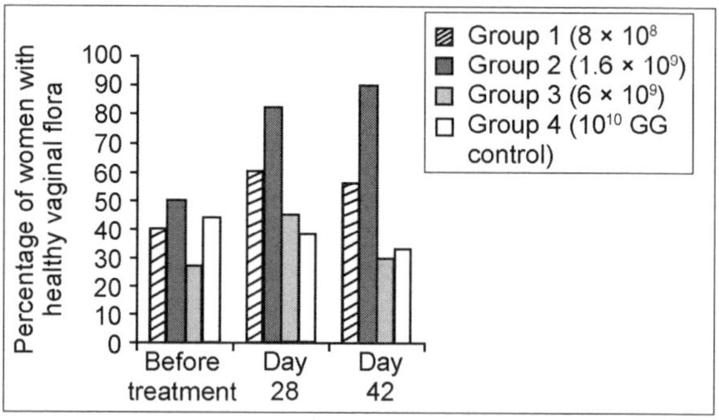

(c)

8.3 Immunomodulatory effect of (a) orally delivered lipid-encapsulated *M. bovis* BCG on cytokine production for pulmonary and systemic protection. Results indicate effective vaccination in animals receiving encapsulated *M. bovis* BCG compared to control animals at $P < 0.5$. Non-vaccinated control animals (sham injection of PBS).[144] (b) Peroral administration of encapsulated *B. bifidum* – reinforcement of the intestinal and systemic IgA production, graph designed from data.[145] (c) Potential efficacy of orally administered encapsulated *L. rhamnosus* GR-1 and *L. fermentum* RC-14 (Groups 1–3) or *L. rhamnosus* GG (control) as a non-chemotherapeutic means to restore and maintain a normal urogenital flora, graph designed from data.[146]

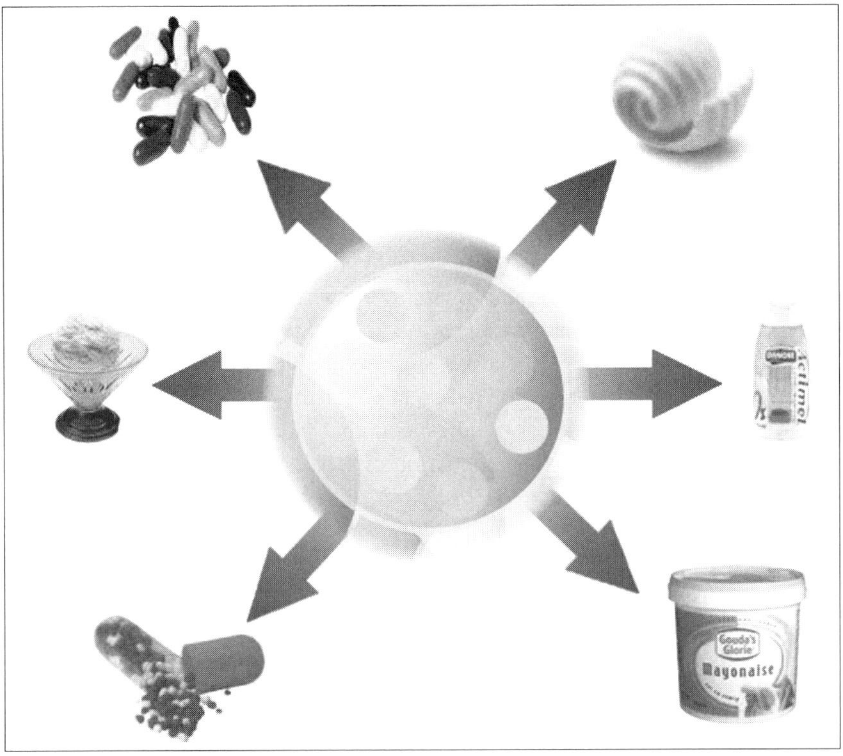

8.4 Probable delivery vehicles for microencapsulated live bacteria.

culture of both MLN and spleen cells. In addition, encapsulation of *B. bifidum* further augments the total IgA production in the culture of both MLN and spleen cells. Subsequent trials worked on three different encapsulated *Lactobacillus rhamnosus GR-1* plus *Lactobacillus fermentum* RC-14 probiotic dosage regimens or *L. rhamnosus GG* with 42 healthy women to see the effect on urogenital flora.[146] This study confirmed that the potential efficacy of orally administered lactobacilli as a non-chemotherapeutic means restores and maintains a normal urogenital flora, and showed that over 10(8) viable organisms per day is the required dose (Fig. 8.3c). There are various methods that can be exploited for delivery of artificial cells containing live bacterial cells. Figure 8.4 depicts artificial cells containing live biotherapeutics that can be delivered in a variety of platforms such as yogurts, capsules, ice-cream, etc.

8.6 Conclusions

Live bacterial cell therapy is increasingly studied and used in humans. Thus far, it has been shown to be effective in the treatment of diarrhea, inflammatory bowel disease, irritable bowel syndrome and food allergies. Other potentially preventable conditions range from necrotizing enterocolitis to urogenital infections, atopic diseases and dental caries. Various leaders in the field propose future applications in cystic fibrosis, rheumatoid arthritis and cancers. Live bacterial cell formulations are mainly used to influence the composition or activity of the intestinal microflora. However, in principle any part of the body that harbours a normal microflora (e.g. oral cavity, urogenital tract, skin, nasopharynx) can be a potential target for specific live bacterial cell therapy. Furthermore, recent advances in genetic engineering technology show that bacterial cells can be engineered to synthesize a wide array of disease-modifying substrates such as cytokines, vaccines and antibodies. These biologically active molecules can target specific disease sites, infectious agents or diseased cells. These bacterial cells can be delivered on targeted sites directly or using artificial cell microcapsule formulations. Studies in animal models also suggest that these findings could have a major impact in human medicine. Thus, alongside conceptual proofs, the development of a robust system for biological method of disease containment will allow for the examination of genetically modified bacterial cells as vectors of therapeutic protein delivery in human healthcare.

However, one cannot overemphasize the importance of carefully conducted double-blind, placebo-controlled studies to document the individual efficacy of each specific organism for each potential clinical application. The success of one species of *Lactobacillus* in a certain application does not imply that all related strains of this species will be capable of producing a comparable response. Live bacterial cell formulations should be administered carefully and cautiously, and only on the basis of strong scientific evidence. Such evidence should direct the cautious, deliberate addition of clinically proven bacterial formulations to commonly consumed food products to allow consumers to conveniently benefit from these organisms.

Determination of potential safety evaluation criteria has until now mainly been done *in vitro* or in animal studies. However, validation of such studies has not been performed, though this is clearly needed. Feeding of novel live beneficial bacterial cells to healthy adult volunteers is usually concerned with efficacy. It is often the first safety study of a new strain, but is not well recognized. These feeding trials could be used as a starting point for determining the safety of novel probiotics in humans. Even though bacterial cells such as lactobacilli and bifidobacteria are generally regarded as safe, for novel products it is necessary to perform pre-marketing determination of the safety of the strain. Genera with known potential risks can then be excluded or assessed carefully for their properties. For example, *Lactobacillus* infections are extremely rare but few infections have been associated with foodborne lactobacilli. At present, only a small percentage of physicians either know of beneficial live bacterial cells or understand their potential applicability to patient care. Thus, live bacterial cells are not yet part of the clinical arsenal for prevention and treatment of disease or maintenance of health. The establishment of accepted standards and guidelines, proposed by the US Food and Drug Administration (FDA), the Food and Agriculture Organization (FAO) of the United Nations and the World Health Organization (WHO), represents a key step in ensuring that reliable products with suitable, informative health claims become available. Based upon the evidence to date, future advances with single- and multiple-strain therapies are on the horizon for the management of a number of debilitating and even fatal conditions.

Live bacterial cells orally delivered therapeutics have the potential to offer major benefits to human health. A critical, tailored approach to safety evaluation can ensure that such benefits are accessible to consumers.

8.7 Acknowledgements

This work was supported by research grants from Canadian Institute of Health Research (CIHR), Natural Sciences and Engineering Research Council (NSERC) of Canada, and Dairy Farmers of Canada (DFC), Canada.

8.8 References

1. Metchnikoff E: *The Prolongation of Life*. (1908) G.P. Putnam's Sons.
2. Schauf CL, Moffet SB, Moffet DF: *Human Physiology: Foundations and Frontiers*. (1990), Times Mirror/Mosby College Publishing, St Louis.
3. Wrong OM, Chadwick VS, Edmonds CJE: *The Large Intestine*. (1981) MTP Press, Lancaster, UK.
4. Isolauri E, Salminen S, Ouwehand AC: Microbial–gut interactions in health and disease. Probiotics. *Best Practices and Research in Clinical Gastroenterology* (2004) **18**: 299–313.
5. Stanton C, Gardiner G, Lynch PB, Collins JK, Fitzgerald G, Ross RP: Probiotic cheese. *International Dairy Journal* (1998) **8**: 491–496.
6. Fasoli S, Marzotto M, Rizzotti L, Rossi F, Dellaglio F, Torriani S: Bacterial composition of commercial probiotic products as evaluated by PCR-DGGE analysis. *International Journal of Food Microbiology*(2003) **82**: 59–70.
7. Hamilton-Miller JMT: Deficiencies in microbiological quality and labelling of probiotic supplements. *International Journal of Food Microbiology* (2002) **72**: 175–176.
8. Hamilton-Miller JM, Shah S, Winkler JT: Public health issues arising from microbiological and labelling quality of foods and supplements containing probiotic microorganisms. *Public Health Nutrition* (1999) **2**(2): 223–229.
9. Temmerman R, Scheirlinck I, Huys G, Swings J: Culture-independent analysis of probiotic products by denaturing gradient gel electrophoresis. *Applied and Environmental Microbiology* (2003) **69**: 220–226.
10. Weese JS: Microbiologic evaluation of commercial probiotics. *Journal of the American Veterinary Medical Association* (2002) **220**: 794–797.
11. Klaenhammer TR, Kullen MJ: Selection and design of probiotics. *International Journal of Food Microbiology* (1999) **50**: 45–57.
12. Naidu AS, Bidlack WR, Clemens RA: Probiotic spectra of lactic acid bacteria (LAB). *Critical Reviews in Food Science and Nutrition* (1999) **39**: 13–126.

13. Ouwehand AC, Salminen S, Isolauri E: Probiotics: an overview of beneficial effects. *Antonie Van Leeuwenhoek International Journal of General and Molecular Microbiology* (2002) **82**: 279–289.
14. Benno Y, He F, Hosoda M *et al.*: Effects of *Lactobacillus* GG yogurt on human intestinal microecology in Japanese subjects. *Nutrition Today* (Supplement) (1996) **31**(6): 95–115.
15. Mattila-Sandholm T, Blum S, Collins JK *et al.*: Probiotics: towards demonstrating efficacy. *Trends in Food Science & Technology* (1999) **10**: 393–399.
16. Gluck U, Gebbers JO: Ingested probiotics reduce nasal colonization with pathogenic bacteria (*Staphylococcus aureus*, *Streptococcus pneumoniae*, and beta-hemolytic streptococci). *American Journal of Clinical Nutrition* (2003) **77**: 517–520.
17. Hatakka K, Savilahti E, Ponka A *et al.*: Effect of long term consumption of probiotic milk on infections in children attending day care centres: double blind, randomised trial. *British Medical Journal* (2001) **322**: 1327–1329.
18. Guillot JF: Consequences of probiotics release in the intestine of animals. (2005) http://resources.ciheam.org/om/pdf/c54/01600006.pdf
19. Marteau P, Seksik P, Jian R: Probiotics and intestinal health effects: a clinical perspective. *British Journal of Nutrition* (2002) **88**: S51–S57.
20. D'Souza AL, Rajkumar C, Cooke J, Bulpitt CJ: Probiotics in prevention of antibiotic associated diarrhoea: meta-analysis. *British Medical Journal* (2002) **324**: 1361–1364.
21. Cremonini F, Di Caro S, Nista EC *et al.*: Meta-analysis: the effect of probiotic administration on antibiotic-associated diarrhoea. *Alimentary Pharmacology & Therapeutics* (2002) **16**: 1461–1467.
22. Saavedra JM, Bauman NA, Oung I, Perman JA, Yolken RH: Feeding of *Bifidobacterium bifidum* and *Streptococcus thermophilus* to infants in hospital for prevention of diarrhea and shedding of rotavirus. *Lancet* (1994) **344**: 1046–1049.
23. Szajewska H, Mrukowicz JZ: Probiotics in the treatment and prevention of acute infectious diarrhea in infants and children: a systematic review of published randomized, double-blind, placebo-controlled trials. *Journal of Pediatric Gastroenterology and Nutrition* (2001) **33**: S17–S25.
24. Duggan C, Gannon J, Walker WA: Protective nutrients and functional foods for the gastrointestinal tract. *American Journal of Clinical Nutrition* (2002) **75**: 789–808.
25. Sullivan A, Nord CE: Probiotics and gastrointestinal diseases. *Journal of Internal Medicine* (2005) **257**: 78–92.
26. Biller JA, Katz AJ, Flores AF, Buie TM, Gorbach SL: Treatment of recurrent *Clostridium difficile* colitis with *Lactobacillus* Gg. *Journal of Pediatric Gastroenterology and Nutrition* (1995) **21**: 224–226.
27. Surawicz CM: Probiotics, antibiotic-associated diarrhoea and *Clostridium difficile* diarrhoea in humans. *Best Practice & Research in Clinical Gastroenterology* (2003) **17**: 775–783.
28. Wullt M, Hagslatt MLJ, Odenholt I: Lactobacillus plantarum 299v for the treatment of recurrent *Clostridium difficile*-associated diarrhoea: a double-blind, placebo-controlled trial. *Scandinavian Journal of Infectious Diseases* (2003) **35**: 365–367.
29. Szajewska H, Kotowska M, Mrukowicz JZ, Armanska M, Mikolajczyk W: Efficacy of *Lactobacillus* GG in prevention of nosocomial diarrhea in infants. *Journal of Pediatrics* (2001) **138**: 361–365.
30. Kaila M, Isolauri E, Soppi E, Virtanen E, Laine S, Arvilommi H: Enhancement of the circulating antibody secreting cell response in human diarrhea by a human

Lactobacillus strain. *Pediatric Research* (1992) **32**: 141–144.
31. Shornikova AV, Casas IA, Mykkanen H, Salo E, Vesikari T: Bacteriotherapy with *Lactobacillus reuteri* in rotavirus gastroenteritis. *Pediatric Infectious Disease Journal* (1997) **16**: 1103–1107.
32. Sugita T, Togawa M: Efficacy of *Lactobacillus* preparation Biolactis powder in children with rotavirus enteritis. *Japan Journal of Pediatrics* (1994) **47**: 2755–2762.
33. Clements ML, Levine MM, Black RE *et al.*: *Lactobacillus* prophylaxis for diarrhea due to entero-toxigenic *Escherichia coli*. *Antimicrobial Agents and Chemotherapy* (1981) **20**: 104–108.
34. Black FT, Anderson PL, Orskow J *et al.*: Prophylactic efficacy of lactobacilli on travellers diarrhea. In *Travel Medicine Conference on International Travel Medicine.* Steffen R: Berlin: Springer (1989) 333–335.
35. Hilton E, Kolakowski P, Singer C, SM: Efficacy of *Lactobacillus* GG as a diarrheal preventive in travelers. *Journal of Travel Medicine* (1997) **4**: 41–43.
36. Oksanen PJ, Salminen S, Saxelin M *et al.*: Prevention of travelers diarrhea by *Lactobacillus* Gg. *Annals of Medicine* (1990) **22**: 53–56.
37. De Dios Pozo-Olano J, Warram JH Jr, Gomez RG, Cavazos MG: Effect of a lactobacilli preparation on traveler's diarrhea. A randomized, double blind clinical trial. *Gastroenterology* (1978) **74** (5 Pt 1): 829–830.
38. Katelaris PH, Salam I, Farthing MJ.: Lactobacilli to prevent traveler's diarrhea? *New England Journal of Medicine* (1995) **333** (20): 1360–1361.
39. Donner CS: Pathophysiology and therapy of chronic radiation-induced injury to the colon. *Digestive Diseases* (1998) **16**: 253–261.
40. Delia P, Sansotta G, Donato V *et al.*: Prevention of radiation-induced diarrhea with the use of VSL#3, a new high-potency probiotic preparation. *American Journal of Gastroenterology* (2002) **97**: 2150–2152.
41. Urbancsek H, Kazar T, Mezes I, Neumann K: Results of a double-blind, randomized study to evaluate the efficacy and safety of Antibiophilus (R) in patients with radiation-induced diarrhoea. *European Journal of Gastroenterology & Hepatology* (2001) **13**: 391–396.
42. Heimburger DC, Sockwell DG, Geels WJ: Diarrhea with enteral feeding – prospective reappraisal of putative causes. *Nutrition* (1994) **10**: 392–396.
43. Huang JS, Bousyaros A, Davidson EJ: Efficacy of probiotic use in acute diarrhea in children: a meta-analysis. *Pediatric Research* (2002) **51**: 141A–142A.
44. Wang KY, Li SN, Liu CS *et al.*: Effects of ingesting *Lactobacillus*- and *Bifidobacterium*-containing yogurt in subjects with colonized *Helicobacter pylori*. *American Journal of Clinical Nutrition* (2004) **80**: 737–741.
45. Sakamoto I, Igarashi M, Kimura K, Takagi A, Miwa T, Koga Y: Suppressive effect of *Lactobacillus gasseri* OLL 2716 (LG21) on *Helicobacter pylori* infection in humans. *Journal of Antimicrobial Chemotherapy* (2001) **47**: 709–710
46. Linsalata M, Russo F, Berloco P *et al.*: The influence of *Lactobacillus brevis* on ornithine decarboxylase activity and polyamine profiles in *Helicobacter pylori*-infected gastric mucosa. *Helicobacter* (2004) **9**: 165–172.
47. Cruchet S, Obregon MC, Salazar G, Diaz E, Gotteland M: Effect of the ingestion of a dietary product containing *Lactobacillus johnsonii* La1 on *Helicobacter pylori* colonization in children. *Nutrition* (2003) **19**: 716–721.
48. Coconnier MH, Lievin V, Hemery E, Servin AL: Antagonistic activity against *Helicobacter* infection *in vitro* and *in vivo* by the human *Lactobacillus acidophilus* strain LB. *Applied and Environmental Microbiology* (1998) **64**: 4573–4580.
49. Michetti P, Dorta G, Wiesel PH *et al.*: Effect of whey-based culture supernatant of

Lactobacillus acidophilus (johnsonii) La1 on *Helicobacter pylori* infection in humans. *Digestion* (1999) **60**: 203–209.
50. Canducci F, Armuzzi A, Cremonini F *et al.*: A lyophilized and inactivated culture of *Lactobacillus acidophilus* increases *Helicobacter pylori* eradication rates. *Alimentary Pharmacology & Therapeutics* (2000) **14**: 1625–1629
51. Felley CP, Corthesy-Theulaz I, Rivero JLB *et al.*: Favourable effect of an acidified milk (LC-1) on *Helicobacter pylori* gastritis in man. *European Journal of Gastroenterology & Hepatology* (2001) **13**: 25–29.
52. Wendakoon CN, Thomson ABR, Ozimek L: Lack of therapeutic effect of a specially designed yogurt for the eradication of *Helicobacter pylori* infection. *Digestion* (2002) **65**: 16–20.
53. Petschow BF, Harris C, Ziegler E, Goldin B, Moran J, Vanderhoof J: Comparison of intestinal colonization and tolerance following oral administration of different levels of *lactobacillus rhamnosus* strain GG (LGG) in healthy term infants. *Journal of Pediatric Gastroenterology and Nutrition* (2003) **36**: 566.
54. Isolauri E, Arvola T, Sutas Y, Moilanen E, Salminen S: Probiotics in the management of atopic eczema. *Clinical and Experimental Allergy* (2000) **30**: 1604–1610.
55. Majamaa H, Isolauri E: Probiotics: a novel approach in the management of food allergy. *Journal of Allergy and Clinical Immunology* (1997) **99**: 179–185.
56. Rosenfeldt V, Benfeldt E, Nielsen SD *et al.*: Effect of probiotic *Lactobacillus* strains in children with atopic dermatitis. *Journal of Allergy and Clinical Immunology* (2003) **111**: 389–395.
57. Kalliomaki M, Salminen S, Arvilommi H, Kero P, Koskinen P, Isolauri E: Probiotics in primary prevention of atopic disease: a randomised placebo-controlled trial. *Lancet* (2001) **357**: 1076–1079.
58. Lodinova-Zadnikova R, Sonnenborn U: Effect of preventive administration of a nonpathogenic *Escherichia coli* strain on the colonization of the intestine with microbial pathogens in newborn infants. *Biology of Neonate* (1997) **71**: 224–232.
59. Trapp CL, Chang CC, Halpern GM, Keen CL, Gershwin ME: The influence of chronic yogurt consumption on populations of young and elderly adults. *International Journal of Immunotherapy* (1993) **9**: 53–64.
60. Kirjavainen PV, Salminen SJ, Isolauri E: Probiotic bacteria in the management of atopic disease: underscoring the importance of viability. *Journal of Pediatric Gastroenterology and Nutrition* (2003) **36**: 223–227.
61. Pessi T, Sutas Y, Hurme H, Isolauri E: Interleukin-10 generation in atopic children following oral *Lactobacillus rhamnosus* GG. *Clinical and Experimental Allergy* (2000) **30**: 1804–1808.
62. Schiffrin EJ, Rochat F, Linkamster H, Aeschlimann JM, Donnethughes A: Immunomodulation of human blood-cells following the ingestion of lactic-acid bacteria. *Journal of Dairy Science* (1995) **78**: 491–497.
63. Gill HS, Guarner F: Probiotics and human health: a clinical perspective. *Postgraduate Medical Journal* (2004) **80**: 516–526.
64. Gill HS, Cross ML, Rutherfurd KJ, Gopal PK: Dietary probiotic supplementation to enhance cellular immunity in the elderly. *British Journal of Biomedical Science* (2001) **58**: 94–96.
65. Gill HS, Rutherfurd KJ, Cross ML, Gopal PK: Enhancement of immunity in the elderly by dietary supplementation with the probiotic *Bifidobacterium lactis* HN019. *American Journal of Clinical Nutrition* (2001) **74**: 833–839.
66. Sheih YH, Chiang BL, Wang LH: Demonstration of systemic immunity enhancing

effects in healthy subjects following dietary consumption of the lactic acid bacterium *Lactobacillus rhamnosus* HN001. *Am. Coll. Nutr.* (2001) **20** (2 Suppl.) 149–156.
67. Spanhaak S, Havenaar R, Schaafsma G: The effect of consumption of milk fermented by *Lactobacillus casei* strain Shirota on the intestinal microflora and immune parameters in humans. *European Journal of Clinical Nutrition* (1998) **52**: 899–907.
68. Gill HS: Stimulation of the immune system by lactic cultures. *International Dairy Journal* (1998) **8**: 535–544.
69. Jung LK: *Lactobacillus* GC augments the immune response to typhoid vaccination: a double-blinded, placebo-controlled study. *Faseb Journal* (1999) **13**: A872.
70. Isolauri E, Joensuu J, Suomalainen H, Luomala M, Vesikari T: Improved immunogenicity of oral Dxrrv reassortant rotavirus vaccine by *Lactobacillus casei* Gg. *Vaccine* (1995) **13**: 310–312.
71. Rafter J: Lactic acid bacteria and cancer: mechanistic perspective. *British Journal of Nutrition* (2002) **88**: S89–S94.
72. Boutron MC, Faivre J, Marteau P, Couillault C, Senesse P, Quipourt V: Calcium, phosphorus, vitamin D, dairy products and colorectal carcinogenesis: a French case control study. *British Journal of Cancer* (1996) **74**: 145–151.
73. Reddy BS, Rivenson A: Inhibitory effect of *Bifidobacterium longum* on colon, mammary, and liver carcinogenesis induced by 2-amino-3-methylimidazo[4,5-F]quinoline, a food mutagen. *Cancer Research* (1993) **53**: 3914–3918.
74. Wollowski I, Rechkemmer G, Pool-Zobel BL: Protective role of probiotics and prebiotics in colon cancer. *American Journal of Clinical Nutrition* (2001) **73**: 451S–455S.
75. Vantveer P, Dekker JM, Lamers JWJ *et al.*: Consumption of fermented milk-products and breast-cancer – a case-control study in the Netherlands. *Cancer Research* (1989) **49**: 4020–4023.
76. Aso Y, Akaza H, Kotake T, Tsukamoto T, Imai K, Naito S: Preventive effect of a *Lactobacillus casei* preparation on the recurrence of superficial bladder-cancer in a double-blind trial. *European Urology* (1995) **27**: 104–109.
77. Okawa T, Niibe H, Arai T *et al.*: Effect of Lc9018 combined with radiation-therapy on carcinoma of the uterine cervix – a Phase-Iii, multicenter, randomized, controlled-study. *Cancer* (1993) **72**: 1949–1954.
78. Singh J, Rivenson A, Tomita M, Shimamura S, Ishibashi N, Reddy BS: *Bifidobacterium longum*, a lactic acid-producing intestinal bacterium inhibits colon cancer and modulates the intermediate biomarkers of colon carcinogenesis. *Carcinogenesis* (1997) **18**: 833–841.
79. McIntosh GH, Royle PJ, Playne MJ: A probiotic strain of *L-acidophilus* reduces DMH-induced large intestinal tumors in male Sprague-Dawley rats. *Nutrition and Cancer – An International Journal* (1999) **35**: 153–159.
80. Lee CH, Wu CL, Tai YS, Shiau AL: Systemic administration of attenuated *Salmonella choleraesuis* in combination with cisplatin for cancer therapy. *Molecular Therapy* (2005) **11**: 707–716.
81. Lee CH, Wu CL, Shiau AL: Systemic administration of attenuated *Salmonella choleraesuis* carrying thrombospondin-1 gene leads to tumor-specific transgene expression, delayed tumor growth and prolonged survival in the murine melanoma model. *Cancer Gene Therapy* (2005) **12**: 175–184.
82. Marteau P, Boutron-Ruault MC: Nutritional advantages of probiotics and prebiotics. *British Journal of Nutrition* (2002) **87**: S153–S157.

83. Steidler L, Hans W, Schotte L *et al.*: Treatment of murine colitis by *Lactococcus lactis* secreting interleukin-10. *Science* (2000) **289**: 1352–1355.
84. Steidler L: *In situ* delivery of cytokines by genetically engineered *Lactococcus lactis*. *Antonie Van Leeuwenhoek* (2002) **82**: 323–331.
85. Borruel N, Casellas F, Antolin M *et al.*: Increased mucosal TNF-alpha production in Crohn's disease can be modulated locally by probiotics. *Gastroenterology* (2001) **120**: A278–A279.
86. Borruel N, Casellas F, Antolin M *et al.*: Effects of nonpathogenic bacteria on cytokine secretion by human intestinal mucosa. *American Journal of Gastroenterology* (2003) **98**: 865–870.
87. Schultz M, Veltkamp C, Dieleman LA *et al.*: *Lactobacillus plantarum* 299V in the treatment and prevention of spontaneous colitis in interleukin-10-deficient mice. *Inflammatory Bowel Diseases* (2002) **8**: 71–80.
88. Madsen K: Discussion on probiotic bacteria enhance murine and human intestinal epithelial function. *Gastroenterology* (2002) **123**: 391–392.
89. Ishikawa H, Akedo I, Umesaki Y, Tanaka R, Imaoka A, Otani T: Randomized controlled trial of the effect of bifidobacteria-fermented milk on ulcerative colitis. *Journal of the American College of Nutrition* (2003) **22**: 56–63.
90. Tamboli CP, Caucheteux C, Cortot A, Colombel JF, Desreumaux P: Probiotics in inflammatory bowel disease: a critical review. *Best Practice & Research in Clinical Gastroenterology* (2003) **17**: 805–820.
91. Nobaek S, Johansson ML, Molin G, Ahrne S, Jeppsson B: Alteration of intestinal microflora is associated with reduction in abdominal bloating and pain in patients with irritable bowel syndrome. *American Journal of Gastroenterology* (2000) **95**: 1231–1238.
92. Kim HJ, Camilleri M, McKinzie S *et al.*: A randomized controlled trial of a probiotic, VSL#3, on gut transit and symptoms in diarrhoea-predominant irritable bowel syndrome. *Alimentary Pharmacology & Therapeutics* (2003) **17**: 895–904.
93. Halpern GM, Prindiville T, Blankenburg M, Hsia T, Gershwin ME: Treatment of irritable bowel syndrome with Lacteol Fort: A randomized, double-blind, cross-over trial. *American Journal of Gastroenterology* (1996) **91**: 1579–1585.
94. Niedzielin K, Kordecki H, Birkenfeld B: A controlled, double-blind, randomized study on the efficacy of *Lactobacillus plantarum* 299V in patients with irritable bowel syndrome. *European Journal of Gastroenterology & Hepatology* (2001) **13**: 1143–1147.
95. Sen S, Mullan MM, Parker TJ, Woolner JT, Tarry SA, Hunter JO: Effect of *Lactobacillus plantarum* 299v on colonic fermentation and symptoms of irritable bowel syndrome. *Digestive Diseases and Sciences* (2002) **47**: 2615–2620.
96. Saggioro A: Probiotics in the treatment of irritable bowel syndrome. *Journal of Clinical Gastroenterology* (2004) **38**: S104–S106.
97. Marteau P, Cuillerier E, Meance S *et al.*: *Bifidobacterium animalis* strain DN-173 010 shortens the colonic transit time in healthy women: a double-blind, randomized, controlled study. *Alimentary Pharmacology & Therapeutics* (2002) **16**: 587–593.
98. Bougle D, Roland N, Lebeurrier F, Arhan P: Effect of propionibacteria supplementation on fecal bifidobacteria and segmental colonic transit time in healthy human subjects. *Scandinavian Journal of Gastroenterology* (1999) **34**: 144–148.
99. Mann GV, Spoerry A: Studies of a surfactant and cholesteremia in Maasai. *American Journal of Clinical Nutrition* (1974) **27**: 464–469.
100. Klaver FAM, Vandermeer R: The assumed assimilation of cholesterol by

lactobacilli and *Bifidobacterium bifidum* is due to their bile salt-deconjugating activity. *Applied and Environmental Microbiology* (1993) **59**: 1120–1124.
101. Lin SY, Ayres JW, Winkler W, Sandine WE: *Lactobacillus* effects on cholesterol – *in vitro* and *in vivo* results. *Journal of Dairy Science* (1989) **72**: 2885–2899.
102. Schaafsma G, Meuling WJA, Van Dokkum W, Bouley C: Effects of a milk product, fermented by *Lactobacillus acidophilus* and with fructo-oligosaccharides added, on blood lipids in male volunteers. *European Journal of Clinical Nutrition* (1998) **52**: 436–440.
103. De Roos NM, Schouten G, Katan MB: Yoghurt enriched with *Lactobacillus acidophilus* does not lower blood lipids in healthy men and women with normal to borderline high serum cholesterol levels. *European Journal of Clinical Nutrition* (1999) **53**: 277–280.
104. Agerholm-Larsen L, Bell ML, Grunwald GK: The effect of probiotic milk product on plasma cholesterol: a meta-analysis of short-term intervention studies. *European Journal of Clinical Nutrition* (2000) **54** (11): 856–860.
105. Richelsen B, Kristensen K, Pedersen SB: Long-term (6 months) effect of a new fermented milk product on the level of plasma lipoproteins – a placebo-controlled and double blind study. *European Journal of Clinical Nutrition* (1996) **50**: 811–815.
106. Bukowska H, Pieczulmroz J, Chelstowski K, Naruszewicz M: Effect of *Lactobacillus plantarum* (*pro viva*) on LDL-cholesterol and fibrinogen levels in subjects with moderately elevated cholesterol. *Atherosclerosis* (1997) **134**: 325.
107. Kawase M, Hashimoto H, Hosoda M, Morita H, Hosono A: Effect of administration of fermented milk containing whey protein concentrate to rats and healthy men on serum lipids and blood pressure. *Journal of Dairy Science* (2000) **83**: 255–263.
108. Kiebling G, Schneider J, Jahreis G: Long-term consumption of fermented dairy products over 6 months increases HDL cholesterol. *European Journal of Clinical Nutrition* (2002) **56**: 843–849.
109. Anderson JW, Gilliland SE: Effect of fermented milk (yogurt) containing *Lactobacillus acidophilus* L1 on serum cholesterol in hypercholesterolemic humans. *Journal of the American College of Nutrition* (1999) **18**: 43–50.
110. Taranto MP, Medici M, Perdigon G, Holgado APR, Valdez GF: Effect of *Lactobacillus reuteri* on the prevention of hypercholesterolemia in mice. *Journal of Dairy Science* (2000) **83**: 401–403.
111. Lovegrove J, Jackson K: Coronary heart disease. In *Functional dairy products*. Mattila Sandholm T & Saarela M. Cambridge, UK: Woodhead Publishing (2003) 54–87.
112. Kim HS, Gilliland SE: *Lactobacillus acidophilus* as a dietary adjunct for milk to aid lactose digestion in humans. *Journal of Dairy Science* (1983) **66**: 959–966.
113. Lin MY, Yen CL, Chen SH: Management of lactose maldigestion by consuming milk containing lactobacilli. *Digestive Diseases and Sciences* (1998) **43**: 133–137.
114. Lin MY, Savaiano D, Harlander S: Influence of nonfermented dairy-products containing bacterial starter cultures on lactose maldigestion in humans. *Journal of Dairy Science* (1991) **74**: 87–95.
115. De Vrese M, Stegelmann A, Richter B, Fenselau S, Laue C, Schrezenmeir J: Probiotics – compensation for lactase insufficiency. *American Journal of Clinical Nutrition* (2001) **73**: 421S–429S.
116. Marteau P, Flourie B, Franchisseur C, Pochart P, Desjeux JF, Rambaud JC: Role of the microbial lactase activity from yogurt on the intestinal-absorption of lactose – an *in vivo* study in lactase-deficient subjects. *Gastroenterology* (1988) **94**: A284.

117. Savaiano DA, Abouelanouar A, Smith DE, Levitt MD: Lactose-malabsorption from yogurt, pasteurized yogurt, sweet *Acidophilus* milk, and cultured milk in lactase-deficient individuals. *American Journal of Clinical Nutrition* (1984) **40**: 1219–1223.
118. Dewit O, Pochart P, Desjeux JF: Breath hydrogen concentration and plasma-glucose, insulin and free fatty-acid levels after lactose, milk, fresh or heated yogurt ingestion by healthy-young adults with or without lactose-malabsorption. *Nutrition* (1988) **4**: 131–135.
119. Lerebours E, Ndam CN, Lavoine A, Hellot MF, Antoine JM, Colin R: Yogurt and fermented-then-pasteurized milk – effects of short-term and long-term ingestion on lactose absorption and mucosal lactase activity in lactase-deficient subjects. *American Journal of Clinical Nutrition* (1989) **49**: 823–827.
120. Varelamoreiras G, Antoine JM, Ruizroso B, Varela G: Effects of yogurt and fermented-then-pasteurized milk on lactose absorption in an institutionalized elderly group. *Journal of the American College of Nutrition* (1992) **11**: 168–171.
121. Rizkalla SW, Luo J, Kabir M, Chevalier A, Pacher N, Slama G: Chronic consumption of fresh but not heated yogurt improves breath-hydrogen status and short-chain fatty acid profiles: a controlled study in healthy men with or without lactose maldigestion. *American Journal of Clinical Nutrition* (2000) **72**: 1474–1479.
122. Labayen I, Forga L, Gonzalez A, Lenoir-Wijnkoop I, Nutr R, Martinez JA: Relationship between lactose digestion, gastrointestinal transit time and symptoms in lactose malabsorbers after dairy consumption. *Alimentary Pharmacology & Therapeutics* (2001) **15**: 543–549.
123. Pelletier X, Laure-Boussuge S, Donazzolo Y: Hydrogen excretion upon ingestion of dairy products in lactose-intolerant male subjects: importance of the live flora. *European Journal of Clinical Nutrition* (2001) **55**: 509–512.
124. Vesa TH, Marteau P, Korpela R: Lactose intolerance. *Journal of the American College of Nutrition* (2000) **19**: 165S–175S.
125. Kolars JC, Levitt MD, Aouji M, Savaiano DA: Yogurt – an autodigesting source of lactose. *New England Journal of Medicine* (1984) **310**: 1–3.
126. Goldin BR: Health benefits of probiotics. *British Journal of Nutrition* (1998) **80**: S203–S207.
127. Reid G, Bruce AW, Frase N, Heinemann C, Owen J, Henning B: Oral probiotics can resolve urogenital infectious. *Fems Immunology and Medical Microbiology* (2001) **30**: 49–52.
128. Reid G, Charbonneau D, Erb J *et al.*: Oral use of *Lactobacillus rhamnosus* GR-1 and *L. fermentum* RC-14 significantly alters vaginal flora: randomized, placebo-controlled trial in 64 healthy women. *Fems Immunology and Medical Microbiology* (2003) **35**: 131–134.
129. Hallen A, Jarstrand C, Pahlson C: Treatment of bacterial vaginosis with lactobacilli. *Sexually Transmitted Diseases* (1992) **19**: 146–148.
130. Hilton E, Isenberg HD, Alperstein P, France K, Borenstein MT: Ingestion of yogurt containing *Lactobacillus acidophilus* as prophylaxis for candidal vaginitis. *Annals of Internal Medicine* (1992) **116**: 353–357.
131. Isolauri E: Probiotics in human disease *American Journal of Clinical Nutrition* (2001) **73**: 1142S–1146S.
132. Chin J, Turner B, Barchia I, Mullbacher A: Immune response to orally consumed antigens and probiotic bacteria. *Immunological Cell Biology* (2000) **78**: 55–66.
133. Kirjavainen PV, Tuomola EM, Crittenden RG *et al.*: *In vitro* adhesion and platelet

aggregation properties of bacteremia-associated lactobacilli. *Infections and Immunity* (1999) **67**: 2653–2655.

134. Reuter G: Probiotics–possibilities and limitations of their application in food, animal feed, and in pharmaceutical preparations for men and animals. *Berliner Munchener Tierarztliche Wochenschrift* (2001) **114**: 410–419.
135. Salminen S, Von Wright A, Morelli L *et al.*: Demonstration of safety of probiotics – a review *International Journal of Food Microbiology* (1998) **44**: 93–106.
136. Wadolkowski EA, Laux DC, Cohen PS: Colonization of the streptomycin-treated mouse large intestine by a human fecal *Escherichia coli* strain: role of adhesion to mucosal receptors *Infection and Immunity* (1988) **56**: 1036–1043.
137. Chang TMS: Semipermeable microcapsules. *Science* (1964) **146**: 524–525.
138. Chang TMS: Biodegradable semipermeable microcapsules containing enzymes, hormones, vaccines, and other biologicals. *Journal of Bioengineering* (1976) **1**: 25–31.
139. Chang TMS: Bioencapsulation in biotechnology. *Biomaterials Artificial Cells and Immobilization Biotechnology* (1993) **21**: 291–297.
140. Chang TMS: Pharmaceutical and therapeutic applications of artificial cells including microencapsulation. *European Journal of Pharmaceutics and Biopharmaceutics* (1998) **45**: 3–8.
141. Chang TMS: Therapeutic applications of polymeric artificial cells. *Nature Reviews Drug Discovery* (2005) **4**: 221–235.
142. Prakash S, Jones ML: Artificial cell therapy: new strategies for the therapeutic delivery of live bacteria. *Journal of Biomedicine and Biotechnology* (2005) **2005** (1): 44–56.
143. Chang TMS, Prakash S: Therapeutic uses of microencapsulated genetically engineered cells. *Molecular Medicine Today* (1998) **4**: 221–227.
144. Aldwell FE, Brandt L, Fitzpatrick C, Orme IM: Mice fed lipid-encapsulated *Mycobacterium bovis* BCG are protected against aerosol challenge with *Mycobacterium tuberculosis*. *Infection and Immunity* (2005) **73**: 1903–1905.
145. Park JH, Um JI, Lee BJ *et al.*: Encapsulated *Bifidobacterium bifidum* potentiates intestinal IgA production. *Cellular Immunology* (2002) **219**: 22–27.
146. Reid G, Beuerman D, Heinemann C, Bruce AW: Probiotic *Lactobacillus* dose required to restore and maintain a normal vaginal flora. *Fems Immunology and Medical Microbiology* (2001) **32**: 37–41.
147. Agarwal N, Kamra DN, Chaudhary LC, Sahoo A, Pathak NN: Selection of *Saccharomyces cerevisiae* strains for use as a microbial feed additive. *Letters in Applied Microbiology* (2000) **31**: 270–273.
148. Woo CJ, Lee KY, Heo TR: Improvement of *Bifidobacterium longum* stability using cell-entrapment technique. *Journal of Microbiology and Biotechnology* (1999) 132–139.
149. Favaro-Trindade CS, Grosso CR: Microencapsulation of *L. acidophilus* (La-05) and *B. lactis* (Bb-12) and evaluation of their survival at the pH values of the stomach and in bile. *Journal of Microencapsulation* (2002) **19**: 485–494.
150. Modler HW, Villa-Garcia L: The growth of *Bifidobacterium longum* in a whey-based medium and viability of this organism in frozen yogurt with low and high levels of developed acidity. *Cultured Dairy Products Journal* (1993) **28**: 4–8.
151. Audet P, Paquin C, Lacroix C: Effect of medium and temperature of storage on viability of lactic-acid bacteria immobilized in kappa-carrageenan–locust bean gum gel beads. *Biotechnology Techniques* (1991) **5** (4): 307–312.
152. Audet P, Paquin C, Lacroix C: Immobilized growing lactic acid bacteria with

κ-carrageenan–locust bean gum gel. *Applied Microbiology and Biotechnology* (1988) **29** (1): 11–18.
153. Audet P, Paquin C, Lacroix C: Batch fermentation with a mixed culture of lactic acid bacteria immobilized separately in κ-carrageenan locust bean gum gel beads. *Applied Microbiology and Biotechnology* (1990) **32** (6): 662–668.
154. Weisz RR: Hypokalemia-induced clinical neurophysiological changes in the human neuromuscular-junction. *Clinical Research* (1980) **28**: A752.
155. Eikmeier H, Westmeier F, Rehm HJ: Morphological development of *Aspergillus niger* immobilized in Ca-alginate and κ-carrageenan. *Applied Microbiology and Biotechnology* (1984) **19** (1): 53–57.
156. Truelstrup Hansen L, Allan-Wojtas PM, Lin YL, Paulson AT: Survival of Ca-alginate microencapsulated *Bifidobacterium* spp. in milk and simulated gastrointestinal conditions. *Food Microbiology* (2002) **19**: 35–45.
157. Norton S, Lacroix C, Vuillemard J-C: Effect of pH on the morphology of *Lactobacillus helveticus* in free cell batch and immobilized-cell continuous fermentation. *Food Biotechnology* (1993) **7**: 235–251.
158. Sanderson GR: Gellan gum. In: *Food Gels*. Elsevier Applied Science (1990) 201–232.
159. Esquisabel A, Hernandez RM, Igartua M, Gascon AR, Calvo B, Pedraz JL: Preparation and stability of agarose microcapsules containing BCG. *Journal of Microencapsulation* (2002) **19**: 237–244.
160. Kailasapathy K: Microencapsulation of probiotic bacteria: technology and potential applications. *Current Issues in Intestinal Microbiology* (2002) **3**: 39–48.
161. Lian WC, Hsiao HC, Chou CC: Survival of bifidobacteria after spray-drying. *International Journal of Food Microbiology* (2002) **74**: 79–86.
162. Winzer A, Lutze V: Radiochemical studies of the influence of photographically active substances on the kinetics of the mass-transfer at silver-halide crystals. 9. activation parameters for the mass-transfer in the course of the Ostwald-ripening. *Journal of Information Recording Materials* (1994) **22** (1): 65–77.
163. Narayani R, Rao KP: Gelatin microsphere cocktails of different sizes for the controlled release of anti-cancer drugs. *International Journal of Pharmaceuticals* (1996) **143** (2): 255–258.
164. Bartkowiak A, Hunkeler D: New microcapsules based on oligoelectrolyte complexation. *Annals of the New York Academy of Science* (1999) **875**: 36–45.
165. Bartkowiak A: Optimal conditions of transplantable binary polyelectrolyte microcapsules. *Annals of the New York Academy of Science* (2001) **944**: 120–134.
166. Bartkowiak A, Hunkeler D: Alginate–oligochitosan microcapsules. II. Control of mechanical resistance and permeability of the membrane. *Chemistry of Materials* (2000) **12** (1): 206–212.
167. Awrey DE, Tse M, Hortelano G, Chang PL: Permeability of alginate microcapsules to secretory recombinant gene products. *Biotechnology and Bioengineering* (1996) **52** (4): 472–484.
168. Gugerli R, Cantana E, Heinzen C, Von Stockar U, Marison IW: Quantitative study of the production and properties of alginate/poly-L-lysine microcapsules. *Journal of Microencapsulation* (2002) **19**: 571–590.
169. Strand BL, Gaserod O, Kulseng B, Espevik T, Skjak-Baek G: Alginate–polylysine–alginate microcapsules: effect of size reduction on capsule properties. *Journal of Microencapsulation* (2002) **19**: 615–630.
170. Lacik I, Brissova M, Anilkumar AV, Powers AC, Wang T: New capsule with tailored properties for the encapsulation of living cells. *Journal of Biomedical*

Materials Research (1998) **39**: 52–60.

171. Rehor A, Canaple L, Zhang Z, Hunkeler D: The compressive deformation of multicomponent microcapsules: influence of size, membrane thickness, and compression speed. *Journal of Biomaterials Science* (Polymer Edition) (2001) **12**: 157–170.
172. Wang T, Lacik I, Brissova M *et al.*: An encapsulation system for the immunoisolation of pancreatic islets. *Natural Biotechnology* (1997) **15**: 358–362.
173. Crooks CA, Douglas JA, Broughton RL, Sefton MV: Microencapsulation of mammalian cells in a HEMA-MMA copolymer: effects on capsule morphology and permeability. *Journal of Biomedical Materials Research* (1990) **24**: 1241–1262.
174. Tse M, Uludag H, Sefton MV, Chang PL: Secretion of recombinant proteins from hydroxyethyl methacrylate–methyl methacrylate capsules. *Biotechnology and Bioengineering* (1996) **51** (3): 271–280.
175. Chen H, Ouyang W, Jones ML, Metz T, Prakash S: Preparation of new multi-layer alginate–chitosan based microcapsules for biomedical applications. In *Polymers in Medicine and Biology*, Rohnert Park, CA: American Chemical Society (2002).
176. Ouyang W, Chen H, Jones ML, Metz T, Prakash S: Novel artificial cell formulation: design, preparation and morphology studies. In *Polymers in Medicine and Biology*, Rohnert Park, CA: American Chemical Society (2002).
177. Aso Y, Akaza H, Kotake T, Tsukamoto T, Imai K, Naito S: Preventive effect of a *Lactobacillus casei* preparation on the recurrence of superficial bladder-cancer in a double-blind trial. *European Urology* (1995) **27**: 104–109.
178. Prantera C, Scribano ML, Falasco G, Andreoli A, Luzi C: Ineffectiveness of probiotics in preventing recurrence after curative resection for Crohn's disease: a randomised controlled trial with *Lactobacillus* GG. *Gut* (2002) **51**: 405–409.
179. Steidler L, Hans W, Schotte L *et al.*: Treatment of murine colitis by *Lactococcus lactis* secreting interleukin-10. *Science* (2000) **289**: 1352–1355.
180. Steidler L: *In situ* delivery of cytokines by genetically engineered *Lactococcus lactis*. *Antonie Van Leeuwenhoek International Journal of General and Molecular Microbiology* (2002) **82**: 323–331.
181. Layer P, Keller J: Lipase supplementation therapy: standards, alternatives, and perspectives. *Pancreas* (2003) **26**: 1–7.
182. Raimondo M, Dimagno EP: Lipolytic-activity of bacterial lipase survives better than that of porcine lipase in human gastric and duodenal content. *Gastroenterology* (1994) **107**: 231–235.
183. Isolauri E, Joensuu J, Suomalainen H, Luomala M, Vesikari T: Improved immunogenicity of oral DxRRV reassortant rotavirus vaccine by *Lactobacillus casei* Gg. *Vaccine* (1995) **13**: 310–312.
184. Kawase M, Hashimoto H, Hosoda M, Morita H, Hosono A: Effect of administration of fermented milk containing whey protein concentrate to rats and healthy men on serum lipids and blood pressure. *Journal of Dairy Science* (2000) **83**: 255–263.
185. Hida M, Aiba Y, Sawamura S, Suzuki N, Satoh T, Koga Y: Inhibition of the accumulation of uremic toxins in the blood and their precursors in the feces after oral administration of Lebinin®, a lactic acid bacteria preparation, to uremic patients undergoing hemodialysis. *Nephron* (1996) **74**: 349–355.
186. Simenhoff ML, Dunn SR, Zollner GP *et al.*: Biomodulation of the toxic and nutritional effects of small bowel bacterial overgrowth in end-stage kidney disease using freeze-dried *Lactobacillus acidophilus*. *Mineral and Electrolyte Metabolism* (1996) **22**: 92–96.
187. O'Loughlin JA, Bruder JM, Lysaght MJ: *In vivo* and *in vitro* degradation of urea

and uric acid by encapsulated genetically modified microorganisms. *Tissue Engineering* (2004) **10**: 1446–1455.
188. Hokama S, Honma Y, Toma C, Ogawa Y: Oxalate-degrading *Enterococcus faecalis*. *Microbiology and Immunology* (2000) **44**: 235–240.
189. Campieri C, Campieri M, Bertuzzi V *et al.*: Reduction of oxaluria after an oral course of lactic acid bacteria at high concentration. *Kidney International* (2001) **60**: 1097–1105.
190. Van Niel CW, Feudtner C, Garrison MM, Christakis DA: *Lactobacillus* therapy for acute infectious diarrhea in children: a meta-analysis. *Pediatrics* (2002) **109**: 678–684.
191. Chouraqui JP, Van Egroo LD, Fichot MC: Acidified milk formula supplemented with *Bifidobacterium lactis*: impact on infant diarrhea in residential care settings. *Journal of Pediatric Gastroenterology and Nutrition* (2004) **38**: 288–292.
192. Plummer S, Weaver MA, Harris JC, Dee P, Hunter J: *Clostridium difficile* pilot study: effects of probiotic supplementation on the incidence of *C-difficile* diarrhoea. *International Microbiology* (2004) **7**: 59–62.
193. Viljanen M, Savilahti E, Haahtela T *et al.*: Probiotics in the treatment of atopic eczema/dermatitis syndrome in infants: a double-blind placebo-controlled trial. *Allergy* (2005) **60**: 494–500.
194. Rosenfeldt V, Benfeldt E, Valerius NH, Paerregaard A, Michaelsen KF: Effect of probiotics on gastrointestinal symptoms and small intestinal permeability in children with atopic dermatitis. *Journal of Pediatrics* (2004) **145**: 612–616.
195. Wang MF, Lin HC, Wang YY, Hsu CH: Treatment of perennial allergic rhinitis with lactic acid bacteria. *Pediatric Allergy and Immunology* (2004) **15**: 152–158.
196. Yasui H, Kiyoshima J, Hori T: Reduction of influenza virus titer and protection against influenza virus infection in infant mice fed *Lactobacillus casei shirota*. *Clinical and Diagnostic Laboratory Immunology* (2004) **11**: 675–679,
197. Schultz M, Linde HJ, Lehn N *et al.*: Immunomodulatory consequences of oral administration of *Lactobacillus rhamnosus* strain GG in healthy volunteers. *Journal of Dairy Research* (2003) **70**: 165–173.
198. Marcos A, Warnberg J, Nova E *et al.*: The effect of milk fermented by yogurt cultures plus *Lactobacillus casei* DN-114001 on the immune response of subjects under academic examination stress. *European Journal of Nutrition* (2004) **43**: 381–389.
199. De Vrese M, Rautenberg P, Laue C, Koopmans M, Herremans T, Schrezenmeir J: Probiotic bacteria stimulate virus-specific neutralizing antibodies following a booster polio vaccination. *European Journal of Nutrition* (2004) **44** (7): 406–413.
200. Nase L, Hatakka K, Savilahti E *et al.*: Effect of long-term consumption of a probiotic bacterium, *Lactobacillus rhamnosus* GG, in milk on dental caries and caries risk in children. *Caries Research* (2001) **35**: 412–420.
201. Gosselink MP, Schouten WR, Van Lieshout LMC, Hop WCJ, Laman JD, Ruseler-Van Embden JGH: Delay of the first onset of pouchitis by oral intake of the probiotic strain *Lactobacillus rhamnosus* GG. *Diseases of the Colon & Rectum* (2004) **47**: 876–884.
202. Gionchetti P, Rizzello F, Helwig U *et al.*: Prophylaxis of pouchitis onset with probiotic therapy: a double-blind, placebo-controlled trial. *Gastroenterology* (2003) **124**: 1202–1209.
203. Hatakka K, Martio J, Korpela M *et al.*: Effects of probiotic therapy on the activity and activation of mild rheumatoid arthritis – a pilot study. *Scandinavian Journal of Rheumatology* (2003) **32**: 211–215.

204. Aldwell FE, Keen DL, Parlane NA, Skinner MA, De Lisle GW, Buddle BM: Oral vaccination with *Mycobacterium bovis* BCG in a lipid formulation induces resistance to pulmonary tuberculosis in brushtail possums. *Vaccine* (2003) **22**: 70–76.
205. Aldwell FE, Tucker IG, De Lisle GW, Buddle BM: Oral delivery of *Mycobacterium bovis* BCG in a lipid formulation induces resistance to pulmonary tuberculosis in mice. *Infection and Immunity* (2003) **71**: 101–108.
206. Prakash S, Chang TMS: Microencapsulated genetically engineered live *E. coli* DH5 cells administered orally to maintain normal plasma urea level in uremic rats. *Nature Medicine* (1996) **2**: 883–887.
207. Chang TMS, Prakash S: Microencapsulated genetically engineered microorganisms for clinical application. Patent 6 217 859 17 April 2001.
208. Batich C, Vaghefi F: Process for microencapsulating cells. Patent 6 242 230 5 June 2001.
209. Duncan SH, Richardson AJ, Kaul P, Holmes RP, Allison MJ, Stewart CS: Oxalobacter formigenes and its potential role in human health. *Applied Environmental Microbiology* (2002) **68**: 3841–3847.
210. Jones M, Chen HM, Ouyang W, Metz T, Prakash S: Deconjugation of bile acids with immobilized genetically engineered *Lactobacillus plantarum* 80 (pCBH1) cells and their potential for live cell therapy. *Cell Transplantation* (2003) **12**: 196–197.
211. Favaro-Trindade CS, Grosso CR: Microencapsulation of *L. acidophilus* (La-05) and *B. lactis* (Bb-12) and evaluation of their survival at the pH values of the stomach and in bile. *Journal of Microencapsulation* (2002) **19**: 485–494.
212. Kailasapathy K: Microencapsulation of probiotic bacteria: technology and potential applications. *Current Issues in Intestinal Microbiology* (2002) **3**: 39–48.
213. O'Riordan K, Andrews D, Buckle K, Conway P: Evaluation of microencapsulation of a *Bifidobacterium* strain with starch as an approach to prolonging viability during storage. *Journal of Applied Microbiology* (2001) **91**: 1059–1066.
214. Saarela M, Mogensen G, Fonden R, Matto J, Mattila-Sandholm T: Probiotic bacteria: safety, functional and technological properties. *Journal of Biotechnology* (2000) **84**: 197–215.
215. Sultana K, Godward G, Reynolds N, Arumugaswamy R, Peiris P, Kailasapathy K: Encapsulation of probiotic bacteria with alginate-starch and evaluation of survival in simulated gastrointestinal conditions and in yoghurt. *International Journal of Food Microbiology* (2000) **62**: 47–55.
216. Kayar SR, Axley MJ: Accelerated gas removal from divers' tissues utilizing gas metabolizing bacteria. Patent 5,922,317 13 July 1997.

9

Artificial cells in liver disease

V DIXIT, University of California Los Angeles, USA

9.1 Introduction

Microencapsulation of biological substances was first described in 1964.[1] Microcapsules, composed of ultra-thin, semipermeable membranes of cellular dimensions, can be prepared of various polymers and the contents can consist of enzymes, cells and other biologically active materials. Microcapsules or artificial cells are prepared in such a way as to prevent their contents from leaking out and causing immunological reactions, but still allowing those contents to interact freely in biochemical reactions. Hence, microencapsulation provides a unique and innovative technique for the transplantation of foreign material (tissue, cells, enzymes, etc.) without need for immunosuppression. The immuno-isolation features of microencapsulation were first demonstrated using microencapsulated enzymes for treating catalase and asparaginase deficiency in acatalasemic mice and suppression of asparagine-induced tumors in C3H mice.[2] In recent years, this feature has been shown to hold true for hepatocytes in xenotransplantation studies involving transplantation of rat hepatocytes into guinea pigs.[3,4] Subsequently, Dixit and colleagues had also demonstrated the feasibility and efficacy of microencapsulated hepatocyte transplantation in providing liver function in several animal models of liver disorders, as well as in bioartificial liver support devices.[5]

9.2 Preparation of microencapsulated hepatocytes

Hepatocyte microencapsulation is carried out at room temperature using aqueous buffers at physiological pH as previously described by O'Shea *et al.*[6] and subsequently modified.[7] Briefly, the freshly isolated hepatocytes are dispersed within a Matrigel™ (Collaborative Research Inc., Bedford, MA) matrix and then suspended in a mixture of 3% (viscosity 266 cps) sodium alginate (Kelco Gel® LV, Kelco, San Diego, CA). Matrigel™ very closely resembles the basement membrane found naturally in the liver,[8] and has been demonstrated to improve and prolong the function of hepatocytes in tissue culture and *in vivo* following transplantation.[9] The mixture is then extruded through a droplet-generating apparatus to form aerosol microdroplets of approximately 300–500 μm diameter. The alginate microdroplets are then reacted with 28 μM poly L-lysine (Sigma Chem. Co., St Louis, MO) for exactly 10 min to form an outer skin of polylysine on the surface of the alginate microdroplet. Thus, positively charged polylysine molecules are bound to negatively charged alginate molecules at the outer surface of the microcapsules to form a polyelectrolyte complex skin layer.[10] The polylysine-coated alginate microdroplets are then further reacted with a 0.3% solution of sodium alginate. This alginate coating helps to further cross-link the polylysine molecules, thereby increasing the stability of the microcapsule membrane. Each milliliter of microencapsulated hepatocytes contains approximately 10^7 hepatocytes. Microencapsulation has been shown to result in a less than 5% loss in hepatocyte viability.[7,10] Control (empty) microcapsules were prepared similarly and contain no hepatocytes.

9.3 Microencapsulated hepatocytes

Microencapsulation techniques for hepatocyte transplantation evolved after the successful experiments with microcarrier-attached hepatocytes described by Dimetriou *et al.*[11] Microencapsulation circumvents the need for immunosuppression by a semipermeable membrane that allows substrate and product to exchange freely but prevents the encapsulated hepatocytes from escaping the microcapsule. Microencapsulation technology is derived from Chang's pioneering work[1] and the subsequent application of this technique by Sun *et al.*[12] for encapsulating islets of Langerhans.

In contrast to the microcarrier studies where hepatocytes are cultured on the exterior surface of collagen-coated microcarriers, hepatocytes are microencapsulated within a three-dimensional collagen matrix and separated from the external environment by a semipermeable membrane. This three-dimensional orientation has been suggested as playing an important role in the maintenance of hepatocyte function through improved cell–cell interaction.[7] Thus, microencapsulation provides a unique and innovative technique for the transplantation of foreign tissue and cells that eliminates the need for immunosuppression. Indeed, microencapsulation techniques have now been successfully used for transplanting a variety of cells such as islets of Langerhans[12] for the treatment of diabetes in rats, parathyroid cells for a bioartificial parathyroid,[13] and adrenal cortex cells for neurological disorders.[14]

The feasibility of microencapsulated hepatocyte transplantation for the replacement of liver function has been demonstrated by several groups using various animal models of fulminant hepatic failure and congenital metabolic liver disease.[15–19] In controlled experiments with the well-characterized galactosamine-induced fulminant hepatic failure rat model,[20] transplantation of microencapsulated hepatocytes significantly increased the survival time two-fold when transplanted during late stages of liver failure.[21] Furthermore, using the same animal model transplanted microencapsulated hepatocytes significantly increased in the overall survival rate of rats when compared with untreated control animals, or control animals treated with empty microcapsules. It is noteworthy that the animals in the above studies received no immunosuppression. Since fulminant hepatic failure essentially requires support only during the critical phase of liver injury, these studies demonstrated the effectiveness of microencapsulated hepatocyte transplantation for short-term liver support.

In order to demonstrate the long-term effectiveness of microencapsulated hepatocytes, studies similar to those performed with microcarrier attached hepatocytes, were performed using the Gunn rat model. This animal model is congenitally deficient in the liver enzyme bilirubin-uridine diphosphate glucuronyltransferase and is unable to conjugate bilirubin resulting in lifelong, non-hemolytic, unconjugated hyperbilirubinemia.[22] A number of studies have now conclusively demonstrated that transplanted microencapsulated hepatocytes can significantly lower the level of hyperbilirubinemia in Gunn rats following a single transplantation of microencapsulated hepatocytes.[10,16] Additionally, as was the case with microcarrier attached hepatocytes, a decrease in serum bilirubin levels showed a parallel increase in conjugated bilirubin in bile.[7,16] These studies effectively demonstrated that transplanted microencapsulated hepatocytes were as competent as transplanted unencapsulated, microcarrier-attached hepatocytes in providing liver function in Gunn rats.[10,16,23] However, unlike the latter, microencapsulated hepatocytes offered the advantage of circumventing the need for immunosuppression.[16,23]

One of the main drawbacks observed in hepatocyte transplantation studies (with both microcarrier-attached hepatocytes and microencapsulated hepatocytes) is that transplanted hepatocytes have a limited (4 to 6 weeks) functional viability in the peritoneal cavity of Gunn rats.[23,24] It has been suggested that the decline in the functional viability of microencapsulated hepatocytes may be related to the specific attachment requirements of the hepatocytes.[7] Hence, by using attachment substrates such as Matrigel™, a mouse sarcoma-derived complex of proteins that closely resembles liver basement membrane proteins, a significant increase in the functional viability of transplanted microencapsulated hepatocytes was reported.[7] Other methods to improve the long-term functional viability may be achieved through co-culture techniques.[25]

If microencapsulated hepatocyte transplantation is to be considered a therapeutic option for acute and chronic liver disease, the long-term fate of the hepatocytes following transplantation must be considered. Investigators have now examined the fate of microencapsulated hepatocytes for 6 months following intraperitoneal transplantation of microencapsulated hepatocytes.[24] It was observed that microencapsulated hepatocytes are stable and function adequately for up to 6–8 weeks post-transplantation. Thereafter, the membrane surrounding the microcapsule deteriorates, causing physical breakdown of the microcapsule.[24] It has been speculated that degeneration of the microencapsulated hepatocytes (autolysis) leads to ionic and pH changes within the microcapsule microenvironment.[24] These pH changes may weaken the alginate–polylysine copolymer membrane of the microcapsule by altering the charges on the molecules that constitute the membrane. Thus, degenerative changes observed in the microcapsules beginning at 2 months post-transplantation suggest that repeated transplantation may be necessary for long-term effectiveness of microencapsulated hepatocytes.[16] In a subsequent 6 month study, repeated monthly transplantation of microencapsulated hepatocytes in Gunn rats demonstrated a prolonged and sustained decrease in serum bilirubin levels.[16] Repeated monthly transplantation of microencapsulated hepatocytes, performed before the previous month's microencapsulated hepatocyte transplant began to deteriorate, provided sustained reduction in hyperbilirubinemia. In contrast, the control rats that received empty microcapsules experienced no reduction in hyperbilirubinemia.

For microencapsulated hepatocyte transplantation to provide a viable form of artificial liver support (e.g. bridge-to-transplantation), it must be simple and convenient to use. Large quantities of healthy microencapsulated hepatocytes must be readily available for transplantation when required. In concept, a facility akin to a blood bank in which microencapsulated hepatocytes could be stored and used as needed would be ideal. Cryopreservation is a standard procedure for preserving various cell types and can effectively be used to store microencapsulated hepatocytes. It has been demonstrated that cryopreserved, microencapsulated hepatocytes functioned as well as freshly prepared, non-frozen microencapsulated hepatocytes both *in vitro* and *in vivo*.[26] Thus, transplantation of cryopreserved microencapsulated hepatocytes can provide a convenient therapeutic option for the management of fulminant hepatic failure as well as of inborn errors of metabolism without need for immunosuppression.

Despite encouraging studies with microencapsulated and microcarrier attached hepatocytes, it now appears that such techniques may be useful only for providing temporary liver support. The limited lifespan of primary hepatocyte culture requires multiple transplantations of microcarrier or microencapsulated hepatocytes, and may not be ideally suited for the clinical setting. The accumulation of microcarriers and breakdown products of microcapsules may present a safety concern not be suitable for patients with severe liver disease. Thus, attention has again focused on transplanting hepatocytes directly in the liver via the intraportal route. Recent clinical studies have also demonstrated that the previous fears of portal vein obstruction by transplanted hepatocytes are not justified, provided a slow controlled infusion of hepatocytes can be accomplished.

9.4 Clinical experience with hepatocyte transplantation

A tremendous number of controlled animal experimentations over the past three decades have enabled the implementation of the first of several of hepatocyte transplantation trials in humans. The early hepatocyte transplantation trials in humans were uncontrolled and resulted in a variable outcome. The patient population was also varied and did not facilitate good comparison among the various clinical studies that were conducted worldwide. Furthermore, the etiology of patients studied ranged from end stage liver cirrhosis to those with fulminant hepatic failure and congenital metabolic liver disease. The number of hepatocytes transplanted also differed significantly, as did the source of hepatocytes and site of transplantation. Nevertheless, the results were encouraging and have prompted other more carefully designed trials that have all, on the whole, validated the promises of the earlier experimental observations in animals.

The first clinical trials of hepatocyte transplantation were reported by Mito and Kusano, who injected isolated human hepatocytes into the spleen of 10 Japanese patients with liver cirrhosis or chronic hepatitis.[27] Although some transient improvement was observed initially, no obvious clinical improvement was observed in the patients studied. However, a year later in 1994, investigators in India conducted a study in which seven patients with fulminant hepatic failure of different etiologies were intraperitoneally transplanted with approximately 60 million hepatocytes isolated from aborted fetuses.[28] In this trial, fetal hepatocytes were pooled from fetuses ranging in gestational ages ranging from 26 to 34 weeks. Unmatched control patients did not receive hepatocyte transplantation and consisted of those patients who did not consent to the procedure. Investigators found a significant difference in the survival of patients treated with hepatocyte transplantation (48%) compared with matched controls (33%). In those patients that survived a decrease in blood ammonia and bilirubin levels was observed. No beneficial effects were seen with regard to prothrombin times. Investigators observed 100% survival in patients treated during early stages (Grade III) of hepatic encephalopathy as compared to those in late stages (Grade IV) who all died. The investigators of this study concluded that fetal hepatocyte transplantation had a beneficial effect on the patient's outcome when hepatocyte transplantation was used early in the clinical course of the disease.

In the United States approximately 19 acute liver failure patients have been treated with hepatocyte transplantation, primarily as a bridge to transplantation.[29,30] Patients were infused with approximately 10 million to 10 trillion allogeneic hepatocytes obtained from cadaveric livers. Hepatocytes were infused either into the splenic artery or portal vein. In some cases, hepatocytes were infused into the portal vein via transjugular catherization. As was expected from the nature of these experimental clinical trials and small patient population the results were largely inconclusive. Although functionally viable hepatocytes were identified in the spleen and liver, it was unclear whether there was any engraftment of the hepatocytes in the liver. No improvement in synthetic liver function was observed during the first few days following hepatocyte transplantation. However, some clinical benefit was observed in anecdotal reports with improvement in the level of encephalopathy, intracranial pressure, serum ammonia levels, prothrombin time, cerebral perfusion pressure and cardiovascular stability. The wide range in the number of hepatocytes transplanted further cloud the issue of possible benefit derived from the hepatocytes. Infusing very large numbers of hepatocytes directly into the splenic artery caused impairment of blood flow in the liver and may have resulted in the detrimental outcome in some patients who died of sepsis.[31] Overall of the 19 patients undergoing hepatocyte transplantation for acute liver failure, seven patients were successfully bridged to liver transplantation, two recovered completely and ten died due to complications resulting from hepatocyte transplantation. In those patients who survived long enough to receive whole organ transplants the average survival time was approximately 4 ± 3 days (mean $\pm$ SD). In patients who died waiting for transplant the mean survival time was approximately 4.5 ± 2.5 days.

Approximately 18 patients with decompensated chronic liver failure have been treated with hepatocyte transplantation therapy worldwide.[27,30,31] As was the modus operandi with patients with acute liver failure, hepatocytes were transplanted into the liver either via direct portal vein infusion or via the splenic artery, the latter being the more frequent route of administration. Hepatocyte transplantation via the splenic artery was preferable from a safety standpoint as it resulted in reduced obstruction of the liver blood flow. The hepatocyte infusions from cadaveric donors were well tolerated and ranged from 10 to 100 hepatocytes/patients. Although engraftment of hepatocyte was visualized in the spleen, the therapy did not affect the clinical outcome. Some anecdotal observations from a small group of patients who underwent transjugular intrahepatic portosystemic shunting (TIPS) showed resolution of their encephalopathy and anuria. Overall the procedure was safely performed with no morbidity or mortality.

A small group of patients with a diversity of inherited liver-based metabolic liver diseases have also been treated with hepatocyte transplantation therapy. The first attempt at treating an inherited metabolic liver disease with hepatocyte transplantation was made in the context of *ex vivo* gene therapy for familial hypercholesterolemia. This trial was based on preclinical experimental studies with Watanabe inheritable hyperlipidemic (WHHL) rabbits that were deficient in low-density lipoprotein (LDL) receptors.[32] Here, primary WHHL hepatocytes harvested from these animals were transduced with a recombinant retrovirus expressing human LDL-receptor gene and then transplanted back into the donor rabbit resulting in long-term reduction in serum LDL levels.[33–35] In the human trials primary hepatocytes isolated from surgically resected liver segments from the patient were similarly transduced with an LDL receptor expressing retroviral vector and then retransplanted back into the patient via the portal vein. A small trial involving four patients demonstrated a modest reduction of plasma cholesterol levels and persistence of the transplanted cells and transgene function. It was observed that the low level of transduction of primary hepatocytes by the retroviral vectors and small number of cells transplanted may have contributed to the low response to therapy. Although the procedure did not result in useful reduction in LDL levels, the trial established the feasibility of hepatocyte transplantation and the longevity of the transplanted cells.[36]

Subsequently, other investigators have transplanted allogeneic hepatocytes to correct liver-based metabolic disorders that include urea cycle disorder (ornithine transcarbamylase, OTC), alpha-1-antitrypsin deficiency, glycogen storage disease type 1a, infantile refsum disease and Crigler-Nijjar syndrome type 1.[37–41] Isolated hepatocyte transplantation appeared to result in temporary relief of hyperammonemia and protein intolerance attributable to OTC deficiency. A transient metabolic stability was achieved but was lost after 11 days, presumably because of rejection of the transplanted cells due to insufficient immunosuppression.[37] Hepatocyte transplantation was performed in an adult patient who had glycogen storage disease type 1a and severe fasting hypoglycemia. Nine months after infusion of 2 billion viable hepatocytes via an indwelling portal-vein catheter, the patient had achieved long-term improvement in blood glucose control, ate a normal diet and could fast for 7 h without experiencing hypoglycemia.

Refsum disease affects growth of the myelin sheath on nerve fibers in the brain.[40] The disease affects both males and females. Symptoms may include vision impairments, peripheral neuropathy, ataxia, impaired hearing, and bone and skin changes. Combinations of approximately 2 billion fresh and cryopreserved hepatocytes were successfully transplanted in a 4-year-old girl with infantile Refsum disease. Total bile acids and abnormal dihydroxycoprostanoic acid markedly decreased in the patient's serum, indicating resolution of cholestasis and re-population of liver cells. Pipecholic acid decreased by 40% and c26:c22 fatty acid ratio by 36% after 18 months. Donor chromosome sequences were detected on biopsy post-transplant, indicating engraftment.

Crigler-Nijjar syndrome is another metabolic liver disorder has been successfully treated with hepatocyte transplantation.[41] In this inherited disorder, bilirubin in the liver cannot be converted to bilirubin glucuronide because of a deficiency in bilirubin UDP-glucuronyltransferase. In the clinical trial, a 10-year-old patient was transplanted with approximately 7.5 billion hepatocytes (~5% of the hepatic mass) by three infusions through a percutaneously placed portal vein catheter. The patient's serum bilirubin levels which averaged 27 md/dl were dramatically reduced by 50–60% following hepatocyte transplantation. High-performance liquid chromatography (HPLC) analysis revealed that prior to hepatocyte transplantation therapy the patient's bile contained predominantly contained unconjugated bilirubin with only a trace of bilirubin monoglucuronide. Following hepatocyte transplantation therapy significant amounts of bilirubin monoglucuronide and diglucuronide could be detected in the bile of this patient. Furthermore, when compared with pre-transplantation biopsy specimens hepatic bilirubin UDP-glucuronyltransferase activity had increased nearly 14-fold to 110 pmol/h/g protein (~5.5% of normal level). Unequivocal evidence of function of transplanted human hepatocytes was found when long-term engraftment showed persistent bilirubin UDP-glucuronyltransferase activity, as measured by the analysis of duodenal bile aspirates even after 30 months following a single session of hepatocyte transplantation.[42] This patient has since undergone successful auxiliary liver transplantation.

These experimental and clinical studies have clearly demonstrated that hepatocyte transplantation is a convenient alternative to liver transplantation, especially in the treatment of rare inborn errors of metabolism. The efficacy of hepatocyte transplantation for treating acute liver failure and chronic liver disease is still unclear. In order to achieve meaningful conclusions in this regard, more research and controlled clinical studies are needed. In addition, several technical matters such as optimal transplantation sites also need to be resolved. The liver presently appears to be an ideal location and optimal ways to deliver substantial amounts hepatocytes to this location without causing hepatic obstruction must be found. The question of using adult versus fetal hepatocytes is controversial and needs to be resolved in a scientific setting. Furthermore, the increasing scientific interest in hepatic stem cell technology may provide yet another source for hepatocyte transplantation. This potential source should be able to provide a stable and uniform population of cell for liver replacement therapy. Finally, alternative technology such as liver tissue engineering is appearing on the horizon and could transform liver transplantation as it currently stands. Recent studies have demonstrated that this is a viable option that needs further investigation.[43,44] Certainly, whole organ liver transplantation will remain the gold standard for the foreseeable future. However, the growing shortage of liver organs that are available for transplantation, increasing cost of the transplantation procedure and increasing number of patients that are listed each year for liver transplantation will eventually drive the development of new and exciting therapies.

9.5 Artificial cells in bioartificial liver support systems

As extensions of previous cell transplantation studies by Dixit and colleagues with microencapsulated hepatocytes and also to develop methods that overcome the inherent physical limitations of current capillary hollow-fiber bioreactor devices, they have proposed a new hepatocyte bioreactor design for a bio-artificial liver using microencapsulated hepatocytes.[5] Microcapsules have a very high surface area to volume ratio and thus can provide both an enormous surface area for diffusion mass-transfer and a high capacity for hepatocyte mass. The bioartificial liver (BAL) design, therefore, involves the direct hemoperfusion of microencapsulated fetal porcine hepatocytes in an extracorporeal chamber. The cylindrical bioreactor chamber, fabricated with biocompatible polycarbonate (plastic) materials, is closed at both ends except for two external ports at either end for blood access.[5] Thus, arterial blood is perfused directly through a packed-bed column of microencapsulated hepatocytes and is then returned to the venous circulation. A peristaltic pump is used to drive the system.

Hemoperfusion through a bed of microencapsulated hepatocytes is based on the hypothesis that a spherical microcapsule membrane has the greatest surface area to volume relationship of any shape,[5,45] and, thus, maximum nutrient/product exchange across a semipermeable membrane may be achieved by using microencapsulated hepatocytes in a liver bioreactor of this configuration. This method also facilitates the use of very large amounts of hepatocytes, within a small bioreactor volume, when compared with the amount of liver tissue used in capillary hollow-fiber-based BAL devices. The fact that cryopreserved microencapsulated hepatocytes which are as metabolic active as freshly isolated hepatocytes[26] provides added incentive to this BAL design.

9.6 The artificial cell bioartificial liver – preliminary studies

The artificial cell BAL is a cylindrical device containing microencapsulated hepatocytes isolated from fetal or new born pigs. Arterial blood or plasma is perfused through this chamber in a fluidized bed configuration and returned to the venous circulation to complete the perfusion circuit. Freshly prepared or cryopreserved microencapsulated porcine hepatocytes (MPH) have been successfully utilized within such a BAL device for use during liver failure.

The BAL device was extensively evaluated using the well-characterized galactosamine-induced fulminant hepatic failure (FHF) model in rats.[10] A total of 50 FHF animals were grouped as follows: Group A, untreated controls; Group B, control BAL treated (i.e. BAL with no MPH); Group C, BAL containing MPH treated. BAL treatment was carried out for 2 h in conscious animals, 24 h after GalN injection. Survival statistics were recorded, as were plasma alanine aminotransferase (ALT) and total bilirubin levels pre- and post-BAL treatment. Experimental animals had pre-treatment ALT levels of 1345 ± 300 mg/dl and mean bilirubin levels of 3.2 ± 0.2 mg/dl. Hemodynamics were monitored continuously using a computerized physiological monitoring system. FHF animals undergoing hepatocyte BAL treatment had a significant ($p < 0.001$) improvement in survival rate compared with matched control animals. An overall, or complete survival was recorded if the animal survived at least 1 month after BAL treatment. Survival statistics are shown in Table 9.1. Most noteworthy here are the Group C FHF animals that were treated with the BAL device that

Table 9.1 Survival rates of FHF animals following artificial cell BAL treatment

Treatment group	Survival rate (%)	Statistical significance
A (Untreated)	0.0 (0/18)	A *vs.* B: $p < 1.00$
B (Control)	5.9 (1/17)	B *vs.* C: $p < 0.05$
C (BAL)	46.7 (7/15)	C *vs.* A: $p < 0.001$

contained microencapsulated porcine hepatocytes. These animals survived the experimental liver injury, resulting in a survival rate of 46.7%. In contrast, Group A animals that received no treatment all died. Also, Group B animals which were treated with the control BAL device that contained empty microcapsules (and no hepatocytes) had a survival rate of 5.9%. There was a high statistically significant difference ($p < 0.001$) when Group C was compared with either group A or B. There was no statistical difference between groups B and C.[5] No significant changes in hemodynamics were observed during bioreactor treatment and all animals tolerated bioreactor treatment without discomfort.[5]

In a subsequent follow-up study the BAL device was evaluated further in the animal model with varying severity of FHF, stratified according to serum ALT levels.[5] The experimental FHF animals ($n = 48$) in this study were grouped as follows: A, untreated controls (mean ALT 700); B, untreated controls (mean ALT 1400); C, untreated controls (mean ALT > 2500); D, BAL-treated animals (mean ALT 700); E, BAL-treated animals (mean ALT 1400); and F, BAL-treated animals (mean ALT > 2500). Duration of BAL treatment was 2 h at a blood flow rate of 1.5–2.0 ml/min. Survival statistics were recorded, as were plasma ALT and total bilirubin levels pre- and post-BAL treatment. The survival statistics are shown in Tables 9.2 and 9.3. Group F animals had a significantly ($p < 0.01$) prolonged survival time as a result of BAL treatment when compared

Table 9.2 Survival rates of FHF animals with moderate to severe FHF following artificial cell BAL treatment

Treatment group	Survival rate (%)
A	100[†]
B	0[†]
C	0[†]
D	100[†]
E	46*
F	0[†]

* $p < 0.001$ compared with matched control group B.
[†] No significant difference among matched contol groups.

Table 9.3 Survival time of FHF animals with moderate to severe FHF following artificial BAL treatment

Treatment group	Survival time (hours)
B	97.8 ± 8.5†
C	92.4 ± 6.3†
F	116.3 ± 5.3*

* $p < 0.01$ compared with groups B or C.
† No significant difference between groups B and C.

with untreated groups B and C (Table 9.2). Also, in Group F animals with very severe liver failure, BAL treatment significantly ($p < 0.01$) prolonged their overall survival time (Table 9.3). The respective control groups did not show any increase in survival time. Thus, these extensive BAL studies have demonstrated that the BAL system proposed here has better diffusion surface area and a higher capacity for hepatocytes than conventional capillary hollow-fiber based BAL devices. Also, this BAL device significantly improved the survival rate and significantly prolonged the survival time of FHF animals with very severe liver injury. BAL treatment was convenient, easy to operate and well tolerated, and did not adversely affect the animal's hemodynamics during treatment.[5] It is, therefore, suggested that this BAL may serve as a model bridge-to-transplant system for treating severe liver failure.

9.7 Future trends

Artificial liver support systems are still in their infancy. Recent biotechnological developments in the areas of cell–surface interaction and bioengineering have taken extrahepatic liver support research to an exciting level. Several groups worldwide are creating systems, with no one system yet emerging as the ideal. This is not surprising given the complexity of liver functions. It is increasingly evident that a multi-factorial approach to liver support will be necessary.[5,45] For example, detoxification and biologically based liver support systems may need to be fused for adequate liver support.

In order to compare and evaluate the efficacy of hybrid bioartificial liver support systems it is necessary to clearly define treatment endpoints. Furthermore, standardization and quantification of the number of hepatocytes, or mass of liver tissue, used in bioreactor devices is necessary. Currently researchers employ widely varying quantities of hepatocytes, or liver tissue. Controlled trials are also essential and will be most informative if well-established and reproducible animal models of liver disease are used. Finally, it is necessary to establish reliable treatment criteria for evaluating hepatocyte bioreactors that use transformed cell lines and primary hepatocyte cell cultures. To this end it is necessary to end the controversy over whether transformed cell lines exhibit liver-specific metabolic functions, such as those involving P_{450} enzyme systems, essential for the effectiveness of extracorporeal liver support devices.

Another area of future research is the use of fetal liver tissue. Fetal tissue offers significant advantages in that it may be transplanted across the histocompatibility barrier, eliminating the need for immunosuppression following transplantation. However, important ethical questions must be resolved before fetal tissue can be freely used. Also, recent tissue engineering studies suggest that by carefully selecting hepatocyte attachment substrates and polymer supports (scaffoldings) it may be possible to produce bioartificial matrices for producing liver tissue for transplantation or extracorporeal liver support systems.[44] Other future approaches for liver support in congenital metabolic liver disease may involve gene therapy techniques. In theory, a functional gene could be inserted into defective liver cells to improve a previously deficient liver function. This approach would allow autologous hepatocyte transplantation without any need for immunosuppression.

9.8 References

1. Chang, T.M.S. Semipermeable microcapsules. *Science* 1964; 146: 524–525.
2. Chang TMS, Poznansky MJ. Semipermeable microcapsules containing catalase for enzyme replacement in acatalasemic mice. *Nature* 1968; 218: 243–244.
3. Elcin YM, Dixit V, Lewin K, Gitnick G. Xenotransplantation of fetal porcine hepatocytes in rats using a tissue engineering approach. *Artif Organs* 1999; 23: 146–152.
4. Dixit V. Hepatocyte transplantation in liver disease. *J Gastro Hepatol* (Suppl) 1999; 14: 2–5.
5. Dixit V, Gitnick G. The bioartificial liver – state-of-the-art. *Euro J Surg* 1998; 582: 71–76.
6. O'Shea GM, Goosen MFA, Sun AM. Prolonged survival of islets of Langerhans encapsulated in a biocompatible membrane. *Biochem Biophys Acta* 1984; 804: 133–136.
7. Dixit V, Darvasi R, Arthur M, Lewin KJ, Gitnick G. Improved function of microencapsulated hepatocytes in a hybrid bioartificial liver support system. *Artif Organs* 1992; 16: 336–341.
8. Bissell DM, Choun MO. Role of extracellular matrix in normal liver. *Scand J Gastroenterol* 1988; 23 (Suppl 151): 1–7.
9. Enat R, Jefferson DM, Ruiz-Opazo N, Gatmaitan Z, Leinwand LA, Reid L. Hepatocyte proliferation *in vitro*: its dependence on the use of hormonally defined medium and substrata of extracellular matrix. *Proc Natl Acad Sci USA (Cell Biol)* 1984; 81: 1411–1415.
10. Dixit V, Darvasi R, Arthur M, Brezina M, Lewin K, Gitnick G. Restoration of liver function in Gunn rats without immunosuppression using transplanted microencapsulated hepatocytes. *Hepatology* 1990; 12: 1342–1349.
11. Demetriou AA, Levenson SM, Novikoff PM, Novikoff AB, Chowdhury NR, Whiting JF, Reisner A, Chowdhury JR. Survival, organization, and function of microcarrier-attached hepatocytes transplanted in rats. *Proc Natl Acad Sci USA* 1986; 83: 7475–7479.

12. Sun AM, Cai ZH, Shi ZQ, Ma FZ, O'Shea GM, Gharapetian H. Microencapsulated hepatocytes as a bioartificial liver. *Trans Am Soc Artif Intern Organs* 1986; 32: 39–41.
13. Fu XW, Sun AM. Microencapsulated parathyroid cells as a bioartificial parathyroid. *Transplantation* 1989; 47: 432.
14. Aebischer P, Tresco PA, Winn SR, Green LA, Jaeger CB. Long-term cross-species brain transplantation of a polymer encapsulated dopamine-secreting cell line. *Exp Neurol* 1991; 111: 267.
15. Demetriou AA, Reisner A, Sanchez J, Levenson SM, Moscioni AD, Chowdhury JR. Transplantation of microcarrier-attached hepatocytes into 90% partial hepatectomized rats. *Hepatology* 1988; 8: 1006–1009.
16. Dixit V, Arthur M, Gitnick G. Repeated transplantation of microencapsulated hepatocytes for sustained correction of hyperbilirubinemia in Gunn rats. *Cell Transplantation* 1992; 1: 275–279.
17. Cai ZH, Shi ZQ, Sherman M, Sun AM. Development and evaluation of a system of microencapsulation of primary rat hepatocytes. *Hepatology* 1989; 10: 855–860.
18. Bruni S, Chang TMS. Hepatocytes immobilized by microencapsulation in artificial cells: effects on hyperbilirubinemia in Gunn rats. *Biomat Artif Cells Artif Organs* 1989; 17: 403–411.
19. Miura Y, Akimoto T, Kanazawa H, Yagi K. Synthesis and secretion of protein by hepatocytes entrapped within calcium alginate. *Artif Organs* 1986; 10: 460–465.
20. Dixit V, Chang TMS. Brain edema and the blood brain barrier in galactosamine-induced fulminant hepatic failure rats: an animal model for evaluation of liver support systems. *ASAIO Trans* 1990; 36: 21–27.
21. Wong H, Chang TMS. Bio-artificial liver: implanted artificial cells micro-encapsulated living hepatocytes increases survival of liver failure rats. *Int J Artif Organs* 1986; 9: 335–336.
22. Yeary RA, Grothaus RH. The Gunn rat as an animal model in comparative medicine. *Lab Animal Sci* 1971; 21: 362–366.
23. Demetriou AA, Whiting JF, Feldman D, Levenson SM, Chowdhury NR, Moscioni AD, Kram M, Chowdhury JR. Replacement of liver function in rats by transplantation of microcarrier-attached hepatocytes. *Science* 1986; 233: 1190–1192.
24. Dixit V, Arthur M, Gitnick G. A morphological and functional evaluation of transplanted isolated encapsulated hepatocytes following long-term transplantation in Gunn rats. *Biomat Artif Cells Immob Biotech* 1993; 21: 119–133.
25. Guguen-Guillouzo C, Clement B, Baffet G, Beaumont C, Morel-Chaney E, Glaise D, Guillouzo A. Maintenance and reversibility of active albumin secretion by adult rat hepatocytes co-cultured with another liver epithelial cell type. *Exp Cell Res* 1983; 143: 47–54.
26. Dixit V, Darvasi R, Arthur M, Lewin KJ, Gitnick G. Cryopreserved micro-encapsulated hepatocytes: transplantation studies in Gunn rats. *Transplantation* 1993; 55: 616–622.
27. Mito M, Kusano M. Hepatocyte transplantation in man. *Cell Transplantation* 1993; 2: 65–74.
28. Habibullah CM, Syed IH, Qamar A, Taher-Uz Z. Human fetal hepatocyte transplantation in patients with fulminant hepatic failure. *Transplantation* 1994; 58: 951–977.
29. Strom SC, Fisher RA, Thompson MT, Sanyal AJ, Cole PE, Ham JM, Posner MP. Hepatocyte transplantation as a bridge to orthotopic liver transplantation in terminal liver failure. *Transplantation* 1997; 63: 559–569.

30. Bilir BM, Guinette D, Karrer F, Kumpe DA, Krysl J, Stephens J, McGavran L, Ostrowska A, Durham J. Hepatocyte transplantation in acute liver failure. *Liver Transpl* 2000; 6: 32–40.
31. Fisher RA, Strom SC. Human hepatocyte transplantation: biology and therapy. In: Berry M, Edwards A (eds), *The Hepatocyte Review*. London: Kluwer Academic Publishers, 2000.
32. Grossman M, Rader DJ, Muller DW, Kolansky DM, Kozarsky K, Clark BJ 3rd, Stein EA, Lupien PJ, Brewer HB Jr, Raper SE. A pilot study of *ex vivo* gene therapy for homozygous familial hypercholesterolemia. *Nature Med* 1995; 1: 1148–1154.
33. Chowdhury JR, Grossman M, Gupta S, Chowdhury NR, Baker JR Jr, Wilson JM. Long term improvement of hypercholesterolemia after *ex vivo* gene therapy in LDL-receptor deficient rabbits. *Science* 1991; 254: 1802–1805.
34. Grossman M, Raper SE, Kozarsky K, Stein EA, Engelhardt JF, Muller D, Lupien PJ, Wilson JM. Successful *ex vivo* gene therapy directed to the liver in a patient with familial hypercholesterolemia. *Nat Genet* 1994; 6: 335–341.
35. Raper SE. Hepatocyte transplantation and gene therapy. *Clin Transplant* 1995; 9(3 Pt 2): 249–254.
36. Raper SE, Grossman M, Rader DJ, Thoene JG, Clark BJ 3rd, Kolansky DM, Muller DW, Wilson JM. Safety and feasibility of liver-directed *ex vivo* gene therapy for homozygous familial hypercholesterolemia. *Ann Surg* 1995; 223: 116–126.
37. Horslen SP, McCowan TC, Goertzen TC, Warkentin PI, Cai HB, Strom SC, Fox IJ. Isolated hepatocyte transplantation in an infant with a severe urea cycle disorder. *Pediatrics* 2003; 111(6 Pt 1): 1262–1267.
38. Strom SC, Fisher RA, Rubinstein WS, Barranger JA, Towbin RB, Charron M, Mieles L, Pisarov LA, Dorko K, Thompson MT, Reyes J. Transplantation of human hepatocytes. *Transplant Proc* 1997; 29(4): 2103–2106.
39. Muraca M, Gerunda G, Neri D, Vilei MT, Granato A, Feltracco P, Meroni M, Giron G, Burlina AB. Hepatocyte transplantation as a treatment for glycogen storage disease type 1a. *Lancet* 2002; 359(9303): 317–318.
40. Sokal EM, Smets F, Bourgois A, Van Maldergem L, Buts JP, Reding R, Bernard Otte J, Evrard V, Latinne D, Vincent MF, Moser A, Soriano HE. Hepatocyte transplantation in a 4-year-old girl with peroxisomal biogenesis disease: technique, safety, and metabolic follow-up. *Transplantation* 2003; 76(4): 735–738.
41. Fox IJ, Chowdhury JR, Kaufman SS, Goertzen TC, Chowdhury NR, Warkentin PI, Dorko K, Sauter BV, Strom SC. Treatment of the Crigler-Najjar syndrome type I with hepatocyte transplantation. *N Engl J Med* 1998; 338(20): 1422–1426.
42. Lee SW, Wang X, Chowdhury NR, Roy-Chowdhury J. Hepatocyte transplantation: state of the art and strategies for overcoming existing hurdles. *Ann Hepatol* 2004; 3: 48–53.
43. Takimoto Y, Dixit V, Arthur M, Gitnick G. *De novo* liver tissue formation in rats using a novel collagen-polypropylene scaffold. *Cell Transplant* 2003; 12(4): 413–421.
44. Dixit V, Elcin YM. Liver tissue engineering: successes & limitations. *Adv Exp Med Biol* 2003; 534: 57–67.
45. Dixit V. Transplantation of isolated hepatocytes and their role in extrahepatic life support systems. *Scand J Gastroenterol* 1995; 30(208): 101–110.

10

Artificial cells as a novel approach to gene therapy

M POTTER, A LI, P CIRONE, F SHEN and P CHANG,
McMaster University, Canada

10.1 Gene therapy: a historical perspective

The development of recombinant genetic technology has made it possible for humans to engineer natural and novel proteins directly from the genetic code. The resulting proteins can now be produced en masse and can be introduced into a patient as a means of alleviating disease. Alternatively, it was conceived that instead of introducing the gene product, a protein, one might treat the disease at a genetic level, by gene therapy.

The earliest references to gene therapy were by 1958 Nobel prize winners Joshua Lederberg (Lederberg, 1966) and Edward Tatum (Tatum, 1966). Lederberg initially conceived that gene therapy would involve directly controlling the nucleotide sequences in a human chromosome, coupled with the recognition, selection and integration of desired genes (Lederberg, 1963). He postulated that one could transfer DNA from one cell line into the chromosome of another cell and that technology could in turn be used to repair genetic-metabolic disease (Lederberg, 1966).

Tatum later proposed that the introduction of new genes into defective cells or particular organs would be done by an *ex vivo* mechanism. According to Tatum, gene therapy could be accomplished by transduction of genetic material from healthy donor cells and into the cultured cells of a patient. The resulting cells with the desired change could then be selected, grown in culture and re-implanted into the patient (Tatum, 1966).

Soon after, in 1968, French Anderson postulated that for gene therapy, the desired gene from a donor chromosome would have to be isolated, amplified and then incorporated into the genome of the defective cell (Anderson, 1968). He proposed that gene therapy could be accomplished using a non-pathogenic virus capable of transferring genetic material to a cell. The first therapeutic human gene therapy clinical trial based on such concepts was approved in 1990 involving two children with a form of severe combined immunodeficiency resulting from adenosine deaminase deficiency (Blaese *et al.*, 1995).

Human gene therapies are defined by the Food and Drug Administration (FDA) as 'products that introduce genetic material into the body to replace faulty or missing genetic material' (FDA, 2005). Today, gene therapy encompasses many different strategies and the material transferred into patient cells may be genes, gene segments or oligonucleotides. The genetic material may be transferred directly into cells within a patient (*in vivo* gene therapy), or cells may be removed from the patient and the genetic material inserted into them in culture, prior to transplanting the modified cells back into the patient (*ex vivo* gene therapy). It is also important to note that current gene therapy is exclusively *somatic* gene therapy, the introduction of genes into somatic cells of an affected individual. The prospect of human germline gene therapy, essentially creating a transgenic human, raises a number of ethical concerns (Scully *et al.*, 2004), and is currently not sanctioned.

A major motivation for gene therapy research has been the need to develop novel treatments for diseases for which there is no effective conventional treatment. Although over 1000 gene therapy protocols have been tested in clinical trials (*Journal of Gene Medicine*, July 2005), and one commercial product has now been released (Peng, 2005), none has become the standard treatment for a given disease. Hence, alternative approaches such as immuno-isolation using artificial cells (Chang, 1964) are being developed. In this approach, non-genetically modified autologous cells are encapsulated in immuno-protective microcapsules (artificial cells) and implanted to deliver therapeutic recombinant proteins *in vivo* (Chang *et al.*, 1999). The advantage is that it avoids any patient-specific genetic alteration and provides a standard reagent suitable for all patients requiring the same therapeutic protein for treatment.

10.2 Experience with gene therapy

The current gene therapy clinical trials can be broadly classified into either *in vivo* or *ex vivo* gene therapy. *In vivo* gene therapy whereby the genetic material is transferred directly into the patient may be the only option when the target cells cannot be extracted and cultured from the patient (e.g. neuronal cells). In such cases, genetic material is introduced into a patient by a virally derived vector or by non-viral techniques such as DNA complexes in a liposome medium. As there is no way to select and amplify the cells that have incorporated the foreign genetic material, the success of this approach is dependent on the general efficacy of the gene transfer and expression (Breyer *et al.*, 2001, Greco *et al.*, 2002).

Ex vivo gene transfer, as originally conceived by Tatum (1966) involved the transfer of cloned genes into patient-derived cells grown in culture and selection for genetically modified cells. When these cells are implanted back to the patient, they represent a syngeneic graft and hence are well tolerated by the host immune system. This approach clearly would only be applicable to tissues that can be removed from the body, altered genetically and returned to the patient where they will engraft and survive for a long period of time. Blood or bone marrow cells are often used for this form of gene therapy because they are easy to collect and return to the patient, in addition to containing stem cells that can be propagated (Van Damme *et al.*, 2002, Wadhwa *et al.*, 2002).

10.2.1 Vectors and delivery systems for gene therapy

Gene delivery into various cell types has led to advances in several different viral and non-viral vectors. A good measure of the overall success of research using the various vectors is to examine which ones have made the transition from preclinical studies to clinical trials. The *Journal of Gene Medicine* database of gene therapy clinical trials is the most comprehensive listing of such trials. The data presented in this chapter was compiled from the January 2006 update of this website (*Journal of Gene Medicine*, 2005), and includes 43 trials (3.8% of 1145 total trials) with unknown vector type. The viral vectors that are most commonly employed in clinical trials are shown in Table 10.1.

Retroviruses continue to be one of the most common gene delivery vehicles used, though the number of trials as a percentage of the total has fallen from 34.1% in 2002 to 24.1% currently. These viruses are capable of stable expression due to insertion of the viral DNA into the host genome. However, recent events have raised concerns about insertional mutagenesis. In a clinical trial for X-linked severe combined immunodeficiency using a retroviral vector

Table 10.1 Viral vectors used in clinical trials: the number of human clinical trials using viral vectors registered in 2006 as a percentage of all gene therapy clinical trials (www.wiley.co.uk/genmed/clinical/)

Vector	Number of trials	% of total trials
Retrovirus	276	24.1
Adenovirus	288	25.2
Pox virus	60	5.3
Vaccinia virus	51	4.5
Herpes simplex virus	38	3.3
Adeno-associated virus	38	3.3
Pox virus + vaccinia virus	21	1.8
Flavivirus	5	0.4
Lentivirus	5	0.4
Adenovirus + retrovirus	3	0.3
Measles virus	2	0.2
Newcastle disease virus	1	0.1
Poliovirus	1	0.1
Semliki forest virus	1	0.1
Simian virus 40	1	0.1
Total viral	**791**	**69.2**

(Cavazzana-Calvo *et al.*, 2000, Aiuti *et al.*, 2002), at least 2 of the 11 children enrolled have developed leukemia secondary to retroviral insertion (Hacein-Bey-Abina *et al.*, 2003).

Equally popular is the adenovirus, because of the ease with which it can be generated in high titers without wild-type background. Adenovirus also has a very broad tropism in humans and mice (reviewed in Green and Seymour, 2002), facilitating preclinical experimentation on a variety of cell types. Conversely, it has also been possible to alter the tropism of adenovirus by altering the surface fiber proteins (Green and Seymour, 2002). Since the regulation of adenoviral gene expression and the function of adenoviral genes are well understood, it has been possible to substitute tissue-specific promoters to allow cancer cell-specific gene expression (Wesseling *et al.*, 2001, DeWeese *et al.*, 2001). The disadvantages for using adenoviruses for gene therapy are well known, including high immunogenicity and the lack of stable integration (Green and Seymour, 2002). Also, the use of adenoviral vectors has been under scrutiny since the death of a research participant enrolled in a gene therapy trial for ornithine transcarbamylase deficiency (Carmen, 2001, Raper *et al.*, 2003). The death was largely attributed to an adenovirus-induced systemic immune response resulting in coma and respiratory failure.

All other viral vectors together make up about 20% of the total clinical trials. Each vector has its advantages and disadvantages that make it suitable for a particular type of gene therapy target, but one vector that has received a lot of attention is the adeno-associated virus (AAV). Clinical trials using this vector have more than doubled from 15 in 2002 to 38 in 2006. These non-pathogenic viruses have many of the advantages of adenovirus but elicit less inflammatory response. AAV serotype 2 vectors have been shown to deliver long-term gene expression lasting more than 3 years in hemophilic canines (Herzog *et al.*, 1999, Snyder *et al.*, 1999). Although capable of persisting for long periods of time, such novel vectors still need long-term assessments for potential risks (Sands *et al.*, 2001). Hence, although viral vectors generally transduce genetic material efficiently, alternative non-viral methods for gene therapy should be considered to avoid unnecessary biological risks.

Non-viral delivery systems use a variety of techniques to mediate DNA transfer (see Table 10.2). The method gaining the most popularity is simply naked/plasmid DNA, whose use in clinical trials has increased from just 70 trials in 2002 to 193 in 2006. The main problem with non-viral delivery is that there is relatively low transduction efficiency in general and poor tissue targeting if administered systemically. An approach to improve these aspects is to complex the negatively charged DNA with cationic lipids to form a liposome (Felgner *et al.*, 1995). These complexes can then enter cells either by endocytosis or by fusion with the negatively charged cell membrane. This enhances the transduction efficiency, but does not address targeting problems. To improve this, modifications are in progress to coat the outside of these vesicles to target

Table 10.2 Non-viral vectors used in clinical trials: the number of human clinical trials using non-viral vectors registered in 2006 as a percentage of all gene therapy clinical trials (www.wiley.co.uk/genmed/clinical/)

Vector	Number of trials	% of total trials
Naked/plasmid DNA	192	16.8
Lipofection	95	8.3
RNA transfer	14	1.2
Gene gun	5	0.4
Salmonella typhimurium	2	0.2
Listeria monocytogenes	1	0.1
Saccharomyces cerevisiae	2	0.2
Total non-viral	**311**	**27.2**

specific tissues. An additional novel non-viral approach involves the use of bacteria or yeast to deliver gene products. This has been used, for example, in a phase I clinical trial where *Salmonella* expressing cytidine deaminase served as a prodrug activating system for patients with advanced cancer (Cunningham and Nemunaitis, 2001).

10.2.2 Immuno-isolation in gene therapy

Based on the FDA definition, implantation of genetically modified non-autologous cells in artificial cells is not strictly a gene therapy protocol, but rather an alternative cell-based genetic approach (see review, Chang *et al.*, 1999). The merit of this approach is that it allows for an 'off-the-shelf' standard cell stock suitable for the treatment of the same disease in different patients. The need for patient-specific genetic engineering is removed and the cost of treatment should be reduced. Because standard laboratory cell lines are easy to transfect, DNA-mediated transfection methods such as calcium-phosphate precipitation are adequate for gene transfer (Chang, 1994), thus avoiding the hazards of biological viral vectors commonly used in gene transfer. The clinical efficacy of this new form of therapy was initially proven by correcting the growth-retardation of mutant dwarf mice with microencapsulated cells secreting mouse growth hormone (al-Hendy *et al.*, 1995). This and similar immuno-isolation strategies have been applied in preclinical studies to a number of other disorders, and a few have advanced to human clinical trials for treatment of pancreatic cancer (Lohr *et al.*, 1999, 2001), amyotrophic lateral sclerosis (Aebischer *et al.*, 1996) and pain control (Bucher *et al.*, 1996) (see Table 10.3).

10.3 Cellular aspects of artificial cell gene therapy

A critical component of immuno-isolation in gene therapy is the cells used. There are only a limited number of diseases that can be treated with available

Table 10.3 Artificial cell applications in gene therapy – animal and human trials

Target; strategy	Immunoisolation device; implantation site	Cell type; product	Animal model/clinical trial	References
Acute myeloid leukemia; anti-angiogenesis	Alginate–poly-L-lysine; near tumor	Mouse mammary gland epithelial; SNRP-1	SCID mouse	Schuch *et al.* (2002)
Acute myeloid leukemia; anti-angiogenesis	Alginate–poly-L-lysine; s.c. near tumor	Porcine aortic endothelial cells; endostatin	SCID	Schuch *et al.* (2005)
Alzheimer's disease; neurotrophic factors	Polymer capsules; lateral ventricle	Baby hamster kidney cells; hNGF	Rhesus monkey	Emerich *et al.* (1994)
Amyotrophic lateral sclerosis; neurotrophic factors	Polyethersulfone hollow fibers	Baby hamster kidney cells; cNTF	Clinical (6 patients)	Aebisher *et al.* (1996)
Analgesia (pain relief)	Semi-permeable capsules; intrathecal	Bovine chromaffin cells; multiple analgesics	Clinical (7 patients)	Buchser *et al.* (1996)
Anemia	Polyethersulfone hollow membrane; s.c.	Mouse C2C12 (myoblasts); mEpo	Anemic mouse (transgenic)	Rinsch *et al.* (2002)
Breast cancer; antiangiogenesis	APA; i.p.	Mouse C2C12; angiostatin	C57BL/6 mouse (melanoma)	Cirone *et al.* (2003, 2005)
Breast cancer; immuotherapy	APA; i.p.	Mouse C2C12; sFV-IL-2	C57BL/6 mouse (melanoma)	Cirone *et al.* (2002)
Beta-thalassemia (chronic anemia)	Polyethersulfone hollow fibers; s.c.	Mouse C2C12; mEpo	Hbb thal 1 mouse	Dalle *et al.* (1999)
Colon & ovarian carcinoma; anti-tumor	Alginate–poly-L-lysine; s.c. near tumour	Human 293 (kidney cells); iNOS	Nude mouse	Xu *et al.* (2002)
Colon cancer; immunotherapy	Barium–alginate microcapsules; s.c.	Mouse fibroblasts NIH 3T3; mIL-12	BALB/C mouse	Zheng *et al.* (2003)
Dwarfism; hormone replacement	APA microcapsules; i.p.	Mouse C2C12; hGH	Snell Dwarf mouse	al Hendy *et al.* (1995, 1996)
Hemophilia A; factor replacement	APA microcapsules; i.p.	Mouse C2C12, canine MDCK; hFVIII	C57BL/6 and SCID mouse	Garcia-Martin *et al.* (2002)
Hemophilia B; factor replacement	APA microcapsules; i.p.	Mouse C2C12; hFIX	C57BL/6 mouse	Hortelano *et al.* (1996)
Hemophilia B; factor replacement	APA microcapsules; i.p.	Mouse C2C12; hFIX	Hemophilia B mouse	Van Raamsdonk *et al.* (2002)
Hemophilia B; factor replacement	APA microcapsules; i.p.	Mouse G8 myoblasts; hFIX	Hemophilia B mouse	Wen *et al.* (2006)
Hyperpituarism	Barium-poly-L-lysine aliginate; i.p.	Canine epithelial cells (MDCK); hGH	Dog	Peirone *et al.* (1998)
Mailgnant glioma; anti-angiogenesis	Alginate microcapsules; brain	Human 293; endostatin	C6 rat (glioma)	Bjerkvig *et al.* (2003)
Malignant glioma (BT4C cells); anti-angiogenesis	Alginate capsules; brain (cortex)	Rat 3T3 fibroblasts, human 293; BT4C	BD-IX rat	Read *et al.* (1999)
Malignant glioma (BT4C cells); anti-angiogenesis	Alginate capsules; brain (intracerebral)	Human 293; endostatin	Rat (glioma)	Read *et al.* (2001)
Malignant glioma (human xenograft); anti-angiogenesis	APA microcapsules; s.c. near tumour	Baby hamster kidney cells; endostatin	Nude mouse (glioma)	Joki *et al.* (2001)
Mucopolysaccharidosis type I; enzyme replacement	APA; intraventricular	Canine MDCK; alpha-iduronidase	MPS I dog	Barsoum *et al.* (2003)
Mucopolysaccharidosis type II; enzyme replacement	APA; i.p.	Mouse myoblasts; iduronate-2-sulfatase	MPS II mouse	Friso *et al.* (2005)
Mucopolysaccharidosis type VII; enzyme replacement	APA microcapsules; intraventricular	Mouse fibroblasts 2A50; beta-glucuronidase	MPSVII mouse	Ross *et al.* (2000)
Neurodegenerative disorders; neurotrophic factor	PAN/PVC hollow fiber; intraventricular	Rat fibroblasts; nerve growth factor	Lesioned Sprague-Dawley rat	Hoffman *et al.* (1993)
Pancreatic carcinoma; targeted chemotherapy	Cellulose sulfate microcapsules; intra-artery	Human 293; CYP2B1	Pig	Lohr *et al.* (2003)
Pancreatic carcinoma; targeted chemotherapy	Cellulose sulfate microcapsules; intra-artery	Human 293; CYP2B1	Clinical (14 patients)	Lohr *et al.* (2001)
Pancreatic carcinoma; targeted chemotherapy	Cellulose sulfate microcapsules; intra-tumor	Human 293; CYP2B1	Nude mouse	Lohr *et al.* (1998, 1999)

Note: i.p., intraperitoneal; s.c., subcutaneous.

cell types that have the desired secretory or metabolic capacity. The suitable cell types are further limited by their growth characteristics: some cell types can proliferate beyond the capacity of an artificial cell to provide adequate exchange of nutrients and wastes. However, genetic engineering has provided the first step around this hurdle. A variety of laboratory cultured cell types can be exploited for this purpose to deliver a desired gene product instead of depending on the patient's own cells. Candidate cell types that can be used as such cell stocks can be categorized as allogeneic or xenogeneic.

Allogeneic: cell lines are established from different individuals of the same species, and selected for their almost infinite growth potential. Examples would include C2C12 myoblasts for mice and 293 cells for humans. These cell lines are amenable to genetic modification, even without the use of viral vectors, e.g. electroporation, lipofection, calcium phosphate precipitation, etc. Such genetically modified cell lines secreting a desired therapeutic product can be implanted with immuno-protection to different individuals within the same species requiring the same product for treatment. These cells can be considered as 'universal' for that species and this approach is defined more appropriately as 'non-autologous' somatic gene therapy as the patient's own cells are not genetically altered (Chang, 1996).

Xenogeneic: cell lines are established from species other than that targeted for treatment. In addition to increasing the risks of cross-species endemic viral and parasitic infections, this approach increases the risk of an immune response even with immuno-isolation. A study in which encapsulated mouse cells were implanted into rats showed that a massive inflammatory response to the implanted capsules can develop as early as 14-days post-implantation (Chang, 1995). Hence for practical and biological reasons, the allogeneic 'universal' cells are preferable in current research and in their potential clinical application to humans.

10.3.1 Cell types

Cells considered for immuno-isolation should possess several properties such as robust proliferative potential necessary for gene transfection *in vitro*; capability to express and secrete the transgene product; and the ability to maintain stable transgene expression after encapsulation. Since the development of micro-encapsulation, many cell lines have been identified that possess these requirements and have successfully been used to deliver therapeutic proteins in different animal models. In general, immortalized cell lines survive better than primary cells and are more feasible for long-term applications, whereas 'reservoirs' of primary cells are hard to maintain because of their inability to proliferate indefinitely following genetic modification. Several types of proliferative cell lines derived from fibroblasts (Chang *et al.*, 1994), myoblasts (al-Hendy *et al.*, 1995), hybridoma cells (Orive *et al.*, 2001), PC12 (Aebischer *et al.*, 1994), BHK (Joki *et al.*, 2001), 293 cells (Lohr *et al.*, 1999, 2002), CHO

(Dawson *et al.*, 1987) have been used effectively in animal models and preclinical trials in various types of microcapsules.

Fibroblasts

Cultured fibroblasts are the most popular choice for microencapsulation because of their wide availability, robust proliferation and good survival within the microcapsule environment. Recombinant fibroblasts have been encapsulated to deliver adenosine deaminase for the treatment of severe combined immune deficiency (Hughes *et al.*, 1994), growth hormone for hypopituitarism (Chang *et al.*, 1993b), factor IX for hemophilia B (Liu *et al.*, 1993) and β-glucuronidase for Sly disease (MPS VII) (Ross *et al.*, 2000a). These encapsulated fibroblasts could proliferate and remained viable. The longest time that encapsulated fibroblasts have been cultured *in vitro* was about 18 weeks but by then, the viability dropped to only 5–6% (Peirone *et al.*, 1998a). Thus, continued proliferation of fibroblasts may lead to over-crowding within the limited microcapsule space, resulting in insufficient exchange of nutrients and metabolic wastes throughout the capsule, and leading to eventual cell death.

Myoblasts

Cultured myoblasts possess several advantages over fibroblasts when used as recombinant gene product delivery vehicles in immuno-isolation devices. They can proliferate continuously, hence are amenable to genetic modification, but they can also differentiate terminally. The capability of myoblasts to withdraw from the cell cycle through differentiation into myotubules circumvents the problem of continued proliferation and overcrowding of cells within the microcapsules. This permits maintenance of their long-term viability *in vivo* and allows for sustained expression and secretion of the therapeutic gene products, thus providing more long-term efficacy for disease treatment. For example, a mouse myoblast cell line, C2C12, engineered to express mouse growth hormone (mGH), and encapsulated in alginate microcapsules were implanted into Snell dwarf mice deficient in growth hormone. By day 35 post-implantation, the body weight and linear growth increased more than twofold compared with controls (al-Hendy *et al.*, 1995). The encapsulated myoblasts retrieved at this time remained viable, and continued to secrete mGH for another 143 days when cultured *in vitro*. Similarly, encapsulated factor IX-secreting myoblasts functioned well for up to 213 days post-implantation (Hortelano *et al.*, 1996), and the recovered encapsulated myoblasts were well preserved. The presence of multinucleated myotubes in capsules when erythropoietin was delivered to treat beta-thalassemia in a mouse model is an indication that myoblast differentiation has taken place (Dalle *et al.*, 1999).

Hybridoma cells

Hybridoma cells resulting from the fusion of B lymphocytes and myeloma cells produce specific monoclonal antibodies. As more monoclonal antibodies are being used in clinical applications, e.g. Muromonab-CD3 (OK3) to prevent acute rejection of organs (Shapiro *et al.*, 2003), Rituximab^{TM} to treat B-cell lymphoma, etc. (Coiffier *et al.*, 2002, Raderer *et al.*, 2003), a logical extension would be to encapsulate hybridoma cells to deliver monoclonal antibodies *in vivo*. Hence, anti-VE (vascular endothelium)-cadherin secreting 1B5 hybridoma cells were entrapped into alginate–agarose capsules (Orive *et al.*, 2001). The antibody was detectable in the medium during the 9 days of culture. The secreted antibody was able to inhibit microtube formation in an *in vitro* angiogenesis Matrigel assay. However, encapsulated hybridoma cells generated large aggregates in some beads, leading to breakdown of the microcapsules and release of the cells at day 15 (Orive *et al.*, 2003). Hence, uncontrolled proliferation is a serious problem that would limit the use of hybridoma cells in microencapsulation.

Epithelial cells

In recent years, there has been an increase in using epithelial cells from different species for microencapsulation. Recombinant proteins such as nerve growth factor have been produced from encapsulated BHK (baby hamster kidney) cells which were able to maintain high secretion for 6 months and provided complete protection of axotomized septal cholinergic neurons (Winn *et al.*, 1994). Another epithelial cell line MDCK (canine kidney epithelial cells) has also been used successfully to express luciferase after encapsulation. These cells maintained >95% viability and retained the ability to secrete luciferase into the medium 2 months *in vitro* after retrieval from a mouse peritoneal cavity (Li, Hou, Shen and Potter, unpublished observation). A human fetal kidney epithelial cell line 293 has also been encapsulated to express endostatin to treat gliomas, to secrete inducible nitric oxide synthase iNOS to suppress tumor growth *in vivo* (Xu *et al.*, 2002) and to deliver a pro-drug CYP2B1, an ifosfamide converting enzyme, for pancreatic carcinoma (Lohr *et al.*, 1999). Epithelial cell differentiation is regulated by growth and interaction between cells and the surrounding extracellular matrix. Breast epithelial cells differentiated into tubules in response to physical stimuli such as culturing in a floating three-dimension collagen matrix. In contrast, when the cells were cultured in the same collagen matrix attached to a culture dish, tubule formation did not occur. The different response is mediated by rho-kinase (ROCK)-generated contractility (Wozniak *et al.*, 2003). However, the differentiation of encapsulated epithelial cells has not been studied as well and, so far, there is no evidence to prove whether epithelial cells possess the ability to differentiate post-encapsulation or not.

Hence, in theory, myoblasts and epithelial cells are cell types with differentiation potential and may be better suited for encapsulation compared with hybridoma cells and fibroblasts.

10.3.2 Regulation of gene expression

Secretory signal

A requirement for this form of non-autologous somatic gene therapy is that the recombinant gene product for therapy must be a secretory protein. Since not all proteins follow a secretory pathway, they must be engineered to acquire this property. Signal sequences are short stretches of amino acids able to direct nascent polypeptides to dock with the signal recognition particle, and traverse the endoplasmic reticulum into the secretory pathway. Hence, fusing this signal to the amino terminus of a cytostolic protein (Simon *et al.*, 1987) should be possible to redirect the protein into a secretory pathway necessary for this form of gene therapy.

This strategy has been applied successfully to create secretable forms of adenosine deaminase (Hughes *et al.*, 1994), firefly luciferase (Li, Hou, Shen and Potter, unpublished observation), as well as endostatin (Joki *et al.*, 2001), all of which are normally cytostolic. Various signal sequences such as that from the prokaryotic β-lactamase (Hughes *et al.*, 1994) or eukaryotic human melanotransferrin (Li, Hou, Shen and Potter, unpublished observation) or growth hormone (Joki *et al.*, 2001) appeared equally effective. More important, the secreted fusion products were able to retain their biological activity and biochemical properties, even after traversing the microcapsule membranes.

Post-translational modification

In addition to the requirement of having a secretory signal, the therapeutic products may require specific post-translational modification such as phosphorylation, ubiquitination, glycosylation or carboxylation to manifest their biological activity. In case of proteins such as clotting factor IX which is normally expressed only in hepatic tissues, extensive post-translational modification and specifically γ-carboxylation by hepatic γ-carboxylase is required for its clotting activity. Even though factor IX is naturally secreted, expressing its cDNA from engineered cell lines that are non-hepatic in origin may not be clinically useful because of the lack of biological activation. However, appropriately processed and biologically active recombinant factor IX has been expressed from non-hepatic cell types such as myoblasts (Yao and Kurachi, 1992), skin fibroblasts (Palmer *et al.*, 1989) and endothelial cells (Yao *et al.*, 1991). More important, biologically active factor IX expressed from mouse fibroblasts (Ltk$^-$) was freely permeable through the alginate microcapules (Liu *et al.*, 1993). The demonstration of the secretion of biologically active factor IX from non-hepatic cells opens up the possibility of treating other diseases that normally also require liver-specific proteins, e.g. α-antitrysin, apolipoproteins and other coagulation factors. However, each product must be verified biologically for efficacy as not all products may function as well as factor IX expressed from a non-hepatic source.

Promoter and enhancer

Sustained expression of therapeutic genes *in vivo* is an important consideration in any form of gene therapy. Some viral genomes such as retrovirus and lentivirus can integrate into the host's genome *in vivo* and achieve lifetime expression. However, non-viral form of plasmid transfection generally used for the genetic modification of encapsulated cells does not alter the genome of the recipient. Inactivation or loss of plasmid DNA from transfected cells could result in declining levels of expression even to background level. The human cytomegalovirus (hCMV) promoter is one of the most powerful and popular promoters that can drive high-level expression both *in vitro* and *in vivo*. However, the high expression provided by hCMV is often unstable owing to the high ratio of CpG motifs that bind cytokines resulting in inactivation of hCMV promoter and contain repressor-binding sites (Yew *et al.*, 2001). Alternative hybrid promoters such as the mouse CMV/hEF1 promoter and hCMV/Ubb promoter, which contain CpG-free motifs, have been constructed and demonstrated to be as capable of driving expression as hCMV (Yew *et al.*, 2001). Thus, the choice of appropriate promoters and enhancers is a crucial consideration for the establishment of a 'universal' cell stock for encapsulation.

10.3.3 Design of microenvironment

When the cells are encapsulated, they enter into a new microenvironment, changing from being in a monolayer to a suspension in a polymer scaffold bound by limiting membranes. Theoretically, fibroblasts would continue proliferating, eventually breaking down the capsules, whereas myoblasts could differentiate into myotubes, thus avoiding the overcrowding problem. In reality, after encapsulation, the viability of encapsulated fibroblasts decreased to 50% after 4-week culture *in vitro* but without rupturing the capsules, while the encapsulated myoblasts did not proliferate and differentiate as well as in the unencapsulated state (Chang *et al.*, 1994). Hence, the microenvironment of the microcapsules is less than optimal for these cell types. As many groups have reported (Peirone *et al.*, 1998a; Orive *et al.*, 2003; Ponce *et al.*, 2005), after sometime within the microcapsules, the cells formed clusters that inhibited the exchange of oxygen, nutrients and metabolic waste, leading to apoptosis or necrosis within (Orive *et al.*, 2003). Such deleterious effects not only limit the therapeutic efficacy of the cells, but also enhance the host's immune response against the implant, since dying cells would release antigens more readily to provoke cytokine release by the host's immune cells. Hence, strategies must be developed to optimize the microenvironment of encapsulated cells, such as encapsulating with appropriate matrix and growth factors to maintain viability as well as facilitate myoblast differentiation (Li *et al.*, 2003).

The extracellular matrix (ECM) is a complex structural surrounding found within mammalian tissues. It is composed of three major classes of biomolecules: (1) structural proteins: collagen, elastin; (2) specialized proteins: fibrillin, fibronectin, laminin, merosin; (3) proteoglycans, all synthesized and secreted by the cells resting on them. The ECM attaches to the cells by integrins that exist widely on the cell surfaces. Most normal cells cannot survive without anchoring to the ECM. In turn, the physical and chemical properties of the ECM can influence the development of the attached cells. For example, when breast epithelial cells were cultured in a three-dimensional collagen matrix, they would differentiate and form duct-like tubules (Wozniak *et al.*, 2003). Similarly, when myoblasts are encapsulated, the desired path of differentiation into myotubes is highly influenced by the type of ECM included (Li *et al.*, 2003).

Mature myoblasts are usually more strongly associated with ECM components such as type IV collagen and laminin (Kuhl *et al.*, 1986). Laminin and merosin are synthesized by myoblasts which attach equally well to both, but they play distinct roles at specific stages of myoblast differentiation. When myoblasts were cultured *in vitro*, both ECM promoted fusion of myoblasts into myotubes, and merosin also promoted stability of myotube (Vachon *et al.*, 1996). However, after encapsulation, the combination of laminin and merosin only led to ~fourfold increase in growth compared with the control, but had no effect on the expression of differentiation markers such as creatine phosphokinase (CPK) or myosin heavy chain (MHC) expression. However, when a third ECM, collagen, was included, a minimal increase in differentiation was induced (Li *et al.*, 2003).

Growth factors may play a more important role in controlling the proliferation and differentiation of myoblasts within microcapsules. In cell culture, myogenesis can be tightly controlled by exogenous peptide growth factors such as fibroblast growth factor (FGF) and transforming growth factor-β (TGF-β). To induce myoblast differentiation into myotubes, the medium can be depleted of such growth factors by being converted from 10% fetal bovine serum to 2% horse serum (Blau *et al.*, 1993). The effects of other growth factors, such as the insulin-like growth factor family (IGF-I and II) on myoblast differentiation depend on the concentration and duration of exposure (Yoshiko *et al.*, 1996). Myoblasts exposed to IGFs for 24–48 h would proliferate, whereas longer exposure resulted in myogenin-mediated differentiation (Ewton *et al.*, 1987, Florini *et al.*, 1991). However, after the myoblasts were encapsulated, no effect on myoblast proliferation was observed with IGF-II, regardless of the concentrations used, but differentiation into myotubes at concentration of 100 ng/ml was stimulated. This is in contrast to the inclusion of basic FGF in the presence of collagen, when both proliferation and differentiation were enhanced (Li *et al.*, 2003).

It is thus evident that a variety of matrix materials and growth factors play distinct roles in myoblast proliferation and differentiation within microcapsules and have effects that may not replicate those under regular culture conditions. Hence, by a judicious choice of ECM and growth factors, the microenvironment of the microcapsules can be optimized to promote optimal proliferation of myoblasts and differentiation into myotubes (Li *et al.*, 2003).

The interaction between the biomaterials and encapsulated cells plays an important role in the morphological and physiological behavior of the cells and is often cell-type dependent. For example, when the same plasmid was transfected into two different cell line, normal murine mammary gland (NMuMG) epithelial cells and porcine aortic endothelial (PAE) cells, the former seemed to lose expression over time while the latter retained consistent expression after fabrication into the same kind of matrix (Schuch *et al.*, 2005).

Similarly, the chemistry and three-dimensional structure of the scaffold milieu can have profound effects on the delivery system. For example, solid and liquefied alginate microcapsules have been widely used. In liquefied microcapsules, the core of the solid capsules is partially dissolved, thus providing more space for the cells to grow. The cells in the liquefied alginate tend to grow together in a clump at the periphery of the capsules, and showed better cell viability but weaker mechanic stability (Chang *et al.*, 1994). However, the center of some large aggregates tended to become necrotic because of insufficient oxygen exchange. Since alginate cannot specifically interact with the cells, Rowley and Mooney (2002) modified alginates of varying mannuronic : guluronic (M:G) ratios with Arg-Gly-Asp (RDG)-containing peptide sequence to promote cell adhesive. For the myoblasts C2C12, high RDG ligand density and high-G alginate are needed to initiate the differentiation. For hepatocytes, lower adhesion ligand density rendered the cells to differentiation phenotype whereas endothelial cells differentiated at an intermediate level of ligand.

These data demonstrated the complex interaction between cells and biomaterials. Survival of the encapsulated cells is highly dependent on the geometry and chemistry of the microenvironment, and is cell-type specific. Different materials, or the same materials with different ratios of the constituents can show significant difference in mechanical stability, resistance, cell viability and expression of therapeutic product (Ponce *et al.*, 2005).

10.4 Application to Mendelian genetic diseases

There are only a handful of the hundreds of potential single gene/Mendelian disorders that have been targeted as candidates for artificial cell gene therapy.

We will examine two of these disorders in detail to highlight the important concepts in artificial cell gene therapy, as well as the pitfalls. The disorders are: hemophilia B and mucopolysaccharidosis type VII (Sly disease) – both very rare, single gene disorders. Hemophilia B is an X-linked recessive bleeding disorder that results from defects in the gene for coagulation factor IX (FIX). Patients with hemophilia B do not produce sufficient functional FIX, leading to prolonged clotting times, chronic bleeding into the joints and the constant threat of hemorrhage. In contrast, mucopolysaccharidosis type VII is an autosomal recessive disorder caused by the deficiency of the lysosomal enzyme beta-glucuronidase. Patients with this disorder cannot break down glycosamino-glycans (mucopolysaccharides), which leads to storage of the material in the lysosomes. The systemic effects of this deficiency include severe bone malformations, enlarged organs and mental retardation.

10.4.1 Hemophilia

Disease biology and treatment technology: finding the right match

A number of factors make coagulation disorders such as hemophilia good candidates for bioartificial cell gene therapy. First, the proteins exist and operate in a very accessible part of the body, the intravascular space. Second, the genes for these proteins have all been cloned and there is a significant body of knowledge about their biochemistry. Significantly, there is no need for tight regulation at the gene expression level, because most of these proteins circulate as inactive precursors that become activated in response to other stimuli. Third, there are animal models for some of these disorders, which are crucial to obtain sufficient preclinical data to justify human trials. These factors make it *possible* to undertake bioartificial cell gene therapy, but the burden of disease also makes it *desirable* to find better treatment. Hemophilia B, for instance, affects approximately 1 in 25 000 male births (Bell *et al.*, 1995). Current treatment for hemophilia B involves the delivery of either plasma derived or recombinant FIX during acute bleeding episodes. Owing to the high cost of this form of therapy, prophylactic treatment is often not economically feasible. Consequently, arthropathy and potentially fatal internal hemorrhage remain significant risks. Furthermore, the repeated delivery of replacement clotting factors is inconvenient and costly over time. Hence, the development of gene therapy offers a more promising and potentially curative approach to the treatment of hemophilia.

The final factor that makes hemophilia B an ideal candidate for initial efforts in artificial cell gene therapy is the current use of recombinant FIX as part of standard therapy. Thus, the efficacy of the recombinant protein has already been established. All that remains to be demonstrated is the artificial cell as a method to deliver the recombinant protein.

Review of enzyme replacement therapy/proof of principle

Enzyme replacement therapy (ERT) is simply a subset of protein therapy where the protein has enzymatic activity. The general principle is to over-express a desired protein in a cell line, and then to isolate and purify the protein from the cell culture. When this approach is scaled up for a commercial purpose, large-scale cell culture using large bioreactor tanks are used. The entire process must meet stringent manufacturing and purification guidelines, which drives up the cost of the final product. An alternative approach is to construct a transgenic animal that over-expresses the protein. If the protein is secreted in the animal's milk, this provides a non-invasive method of collecting the raw material for purification.

Protein (or peptide) therapy has been used in a number of disorders for several decades. The longest-standing and probably most familiar application is the use of insulin to treat insulin-dependent diabetes. This was first tried on humans over 80 years ago by Banting and Best shortly after their discovery of insulin (Banting and Best, 1922), a discovery that would earn the 1923 Nobel prize in physiology and medicine. However, the first specific administration of a protein with enzymatic activity was in the treatment of hemophilia – made possible by the discovery of cryoprecipitate in the 1960s by Judith Pool (Pool *et al.*, 1964). Cryoprecipitate was found to be very rich in factor VIII, the blood clotting factor deficient in hemophilia A, and its administration revolutionized treatment of hemophilia in the same way the discovery of insulin revolutionized diabetes care.

When the new molecular biology tools that enabled expression of a recombinant protein were first developed, one of the very first clinical applications that were studied in the early 1980s was again insulin therapy of diabetes. The intervening 60 years of insulin therapy had seen numerous advances in the production processes and purity of animal-derived insulin. Indeed, the porcine insulin available at the time that recombinant human insulin was first produced was so highly purified that there was 'no compelling reason to change patients ... to human insulin' (Brogden and Heel, 1987). Eventually, the decreased immune response to human insulin and, probably more importantly, its lower cost led to the use of recombinant human insulin over porcine-derived insulin.

Shortly after the cloning of factor IX (Choo *et al.*, 1982, Kurachi and Davie, 1982), the recombinant protein was expressed in cell lines (Anson *et al.*, 1985; Busby *et al.*, 1985; de la Salle *et al.*, 1985). This opened the way for a commercial product, but the introduction of recombinant FIX as an available treatment for hemophilia B patients was much slower than the rollout of a recombinant product for diabetes. One of the differentiating factors between these disorders that probably influenced this delay is the prevalence of the disease, and hence the size of the market. Type 1 diabetes is 250 times more common than hemophilia B (Bell *et al.*, 1995; Eiselein *et al.*, 2004) – a factor that drives up the per-unit cost of recombinant FIX. Indeed, while recombinant insulin has displaced naturally derived insulin from the market, recombinant FIX is more expensive and less available than its human-derived counterpart. One reason that recombinant FIX is considered the treatment of choice for hemophilia B, despite the high cost, is the safety profile. The recombinant product contains no human-derived material, reducing the risk of a harmful virus being transmitted to the patient.

The socioeconomic considerations around enzyme replacement therapy greatly affect the feasibility of this treatment on a population basis. Aside from the cost involved in the production/manufacturing of a product for a limited target patient pool, the requirement for frequent intravenous infusions is a significant burden, both in cost and in patient compliance.

All commercially available recombinant enzymes to date require repeated parenteral (i.e. avoiding enteral, or gastrointestinal, routes) injections. Parenteral methods include subcutaneous or intramuscular injections and intravenous infusions. The reason for this requirement is to avoid the degradation of the recombinant proteins in the acidic environment of the stomach, or the removal of the recombinant protein by the liver after transport by the portal vein (first-pass effect). Development of products to overcome this limitation is ongoing: the first commercial products are again related to insulin therapy of diabetes (Mandal, 2005), with the focus on delivery of insulin by inhalation and absorption through the lungs. For other disorders, however, artificial cell gene therapy is an attractive option, essentially miniaturizing the bioreactor and placing it inside the patient, even though artificial cell gene therapy still has hurdles to overcome, as discussed later in this chapter and throughout this book.

Experience with gene therapy

As with any gene therapy clinical trial, extensive preclinical work is necessary before attempting human trials and multiple different gene therapy approaches have been studied for the delivery of FIX to both normal and hemophilic animals. These studies have been facilitated by the availability of murine (Lin *et al.*, 1997; Wang *et al.*, 1997; Kundu *et al.*, 1998) and canine (Evans *et al.*, 1989) models of the disease. The canine model results from a naturally occurring missense mutation while the murine models were developed by homologous recombination of specific regions of the *FIX* gene. *Ex vivo* gene therapy approaches have been successful at providing long-term (> 600 days) delivery of FIX in normal rabbits (Zhou *et al.*, 1993; Chen *et al.*, 1998), even though the levels of delivery have been sub-therapeutic and only a small percentage of the treated animals had detectable levels of FIX (Chen *et al.*, 1998). Similarly, *in vivo* transduction of canine hepatocytes using retroviral vectors has achieved long-term delivery of FIX, but again the levels of FIX produced were sub-therapeutic (Kay *et al.*, 1993, 1994). *In vivo* transduction with adenoviral vectors, on the other hand, produced supra-physiologic levels of FIX in hemophilic dogs, but the levels of delivery decreased rapidly over time due to immunological clearance of viral constituents, resulting in only short-term delivery of therapeutic FIX levels (Fang *et al.*, 1995; Kung *et al.*, 1998).

The most promising results in terms of duration and level of delivery have come from *in vivo* studies with AAV vectors (Herzog *et al.*, 1997; Monahan *et al.*, 1998). *In vivo* transduction of murine hepatocytes with AAV vectors has achieved FIX levels in mice of 2 μg/ml (40% physiologic) lasting for more than 17 months (Snyder *et al.*, 1997, 1999). While the results were not as encouraging in dogs, it was still possible to deliver 100 ng FIX/ml (2% physiologic) for more than 8 months in hemophilic dogs (Snyder *et al.*, 1999). Based on these results, the first human clinical trials were initiated, and to date, at least three gene therapy human clinical trials for hemophilia B (CN-001, US-279, US-371) and three for hemophilia A (US-247, US-284, US-372) have been reported (www.wiley.co.uk/genmed/clinical/).

The first hemophilia B trial, with Dr J.L. Hsueh as the principal investigator, used *ex vivo* fibroblast transduction by retrovirus, followed by subcutaneous injection of the fibroblasts. This method showed promising results, with an over twofold increase in FIX levels and improved blood clotting in the two patients that have been reported on to date (Lu *et al.*, 1993; Qiu *et al.*, 1996). A second hemophilia B clinical trial used an *in vivo* approach with intra-muscular injection of adeno-associated virus. Successful transduction of muscle cells was achieved, but the plasma FIX levels were too low to have a significant clinical effect in most patients (Manno *et al.*, 2003; Kay *et al.*, 2000). The third trial has only been reported in abstract form to our knowledge, using AAV to express a FIX minigene (Kay *et al.*, 2002). The vector was delivered via injection into the hepatic artery. The hemophilia A clinical trials ranged from non-viral (electroporation) transduction of dermal cells (Roth *et al.*, 2001), to viral methods using retrovirus (Powell *et al.*, 2003) or adenovirus (White, 2001). Success has generally been transient or mild but, overall, the results of these trials are encouraging. Further refinement of the technology with respect to safety and efficacy is required before gene therapy becomes routinely available.

Microencapsulation

In our laboratory, we have studied artificial cell gene therapy for hemophilia B by encapsulating human FIX-secreting mouse myoblasts in alginate–poly-L-lysine–alginate (APA) microcapsules. Low levels of human FIX have been delivered to normal mice after treatment with hollow APA microcapsules and the delivery could last longer than 200 days based on a continued anti-FIX antibody response (Hortelano *et al.*, 1996). In contrast, mice treated with unencapsulated cells showed only transient anti-FIX antibody response, and at much reduced levels (Van Raamsdonk, 1999). Hemophilic mice treated with microcapsules enclosing FIX-secreting C2C12 myoblasts showed a temporary partial correction of activated thromboplastin time (aPTT) for approximately 3 weeks. Plasma FIX levels rose to a maximum on day 2, decreased to low levels until day 24 and then were unmeasurable after that. The declining FIX levels coincided with emergence of anti-hFIX antibody by day 14, increasing greater than tenfold by day 28. The time course, and the fact that capsules removed from the animals still secreted FIX, indicated the decline in measurable plasma FIX and the loss of aPTT correction were due to neutralizing antibodies (Van Raamsdonk *et al.*, 2002).

Similar loss of efficacy associated with immune response was also observed when B domain-deleted factor VIII was delivered to the transgenic hemophilia A mice, in which case the immune response was exacerbated by the adenoviral vector used in its delivery (Sarkar *et al.*, 2000). While it was initially reported that this strain of mice did not produce antibodies to the human FIX (Kung *et al.*, 1998), most of the mice receiving FIX from microcapsules did produce a detectable antibody response. In our pilot studies, in rare cases, this development of an antibody response did not occur (unpublished data), but the reason for this anergy is unknown. When this anergy occurred, however, the correction of aPTT times was maintained for almost 2 months in the relevant animals (unpublished data). The reason for the differences in immune response to the transgene product is poorly understood but such variability has also been observed in the delivery of factor VIII (Connelly *et al.*, 1998; Sarkar *et al.*, 2000).

There are multiple strategies available to try to overcome the development of neutralizing antibodies, such as immunosuppressive drugs or oral tolerization. In our laboratory, we used monoclonal anti-CD4 antibodies to deplete CD4-positive T cells. In the treated hemophilic mice, the anti-hFIX antibody response was totally suppressed to beyond day 28, accompanied by a significant decrease in activated thromboplastin times from hemophilic levels. After 38 days of implantation, the microcapsules could be recovered almost quantitatively from the intraperitoneal cavity and continued to secrete FIX at pre-implantation levels (Van Raamsdonk *et al.*, 2002).

Delivery of coagulation factors is probably one of the most susceptible treatments to the neutralizing effect of these antibodies, thus adding another level of complexity to any replacement therapy. This is likely because the biochemistry of the coagulation cascade demands a pool of pre-formed factors to circulate in the blood. This creates an opportune environment for the immune system to generate antibodies and for those antibodies to bind to and neutralize the factors. In contrast, ERT of lysosomal disorders, as discussed later, is minimally affected by neutralizing antibodies, even though a high proportion of ERT recipients form them.

10.4.2 Lysosomal storage disorders (LSD)

Why are they good candidates for artificial cell gene therapy?

With few exceptions, LSDs result from deficiencies of hydrolytic enzymes located in lysosomes. Lysosomes are membrane-bound cytoplasmic organelles involved in intracellular protein degradation in eukaryotic cells. Lysosomal degradation is required for normal cellular protein turnover, disposal of abnormal proteins, down-regulation of surface receptors, release of endocytosed nutrients, inactivation of pathogenic organisms and antigen processing (Dell'Angelica *et al.*, 2000). The lysosomal enzyme deficiency results in lysosomal storage of unhydrolyzed compounds and is thought to lead to the organ dysfunction that clinically characterizes the various LSDs. The exact pathophysiological mechanism by which this occurs remains unclear.

The first demonstration of principle of ERT for LSDs occurred over 30 years ago (Brady *et al.*, 1974). It was hypothesized that this approach could be used in LSDs because endocytotic vesicles containing extracellular protein fuse with lysosomes for degradation of the endocytosed material. Thus, a proportion of exogenous enzymes would end up in the lysosome, co-localizing precisely to the organelle where their therapeutic effect is needed. This hypothesis has been shown to be true for the majority of the lysosomal enzyme deficiencies studied. As the cell biology regulating lysosomal trafficking was unraveled, it was found that most lysosomal enzymes are specifically targeted to the lysosome via a mannose-6-phosphate oligosaccharide moiety added to the enzyme during processing in the Golgi (Campbell and Rome, 1983, Geuze *et al.*, 1985; Lemansky *et al.*, 1987). Membrane-bound mannose-6-phosphate receptors (Sahagian *et al.*, 1981; Hoflack and Kornfeld, 1985; Lobel *et al.*, 1988) can target newly synthesized lysosomal enzymes or endocytosed enzymes to the lysosome.

Review of ERT/proof of principle

Gaucher disease and Fabry disease are the LSDs where the most clinical experience has been obtained (Mapes *et al.*, 1970, Brady *et al.*, 1974). The enzymes deficient in both disorders are involved in glycosphingolipid degradation in the lysosome. Gaucher disease, or glucocerebrosidase deficiency, for example, has had specific enzyme replacement therapy clinically available for more than 14 years (Beutler *et al.*, 1991) from a placenta-derived product, Ceredase®, and as a recombinant product for over 10 years (Cerezyme®) (Grabowski *et al.*, 1995). In the past few years, Fabry disease (alpha-galactosidase A deficiency) has seen two enzyme replacement products become available (Fabrazyme® and Replagal™). Now, mucopolysaccharidosis type I (Hurler disease, MPS I) also has ERT commercially available (Aldurazyme®) with treatments for other LSDs in various stages of preclinical and clinical development. As mentioned above, a disadvantage of ERT is that the production costs are large, and when added to the research and development costs and marketing costs, the final products are very expensive. Each of the commercial products listed above costs several hundred thousand US dollars per year per patient.

It is clear from the Gaucher experience that ERT is effective in treating the systemic effects of the disease. One of the largest reviews, a report from the Gaucher registry that looked at 1028 patients, showed improvements in hemoglobin concentration, platelet count, liver and spleen volumes and in bone pain and bone crises after 2 to 5 years of therapy (Weinreb *et al.*, 2002). Similar systemic effects were seen in Fabry disease ERT clinical trials, where plasma and microvascular endothelial deposits of storage material were cleared in patients treated with ERT (Eng *et al.*, 2001), and in MPS I (alpha-iduronidase deficiency), where hepatosplenomegaly, joint range of motion, respiratory and cardiac status, and growth improved with ERT (Kakkis *et al.*, 2001). In contrast with the success on treating systemic disease, there is no clear evidence that intravenously administered ERT treats central nervous system (CNS) disease. While there are anecdotal accounts of ERT halting the progression or even improving neurological disease in chronic neuronopathic Gaucher (Schiffmann *et al.*, 1997; Vellodi *et al.*, 2001), there are many accounts of progression of CNS disease while on ERT (Erikson *et al.*, 1997).

The previously mentioned limitations in therapeutic outcome and the inconvenience and expense of repeated intravenous infusions have led to continued searching for a more optimal therapy. One approach is gene therapy, which addresses some of the limitations of enzyme replacement therapy by delivering a gene which is then used by the body to produce an enzyme. The enzyme does not have to be purified, and, in theory, a single treatment may effect a cure. Over the lifetime of a patient, it would probably be a less expensive treatment than traditional ERT, making it more palatable to heathcare payers.

Experience with gene therapy

There are literally hundreds of preclinical studies with almost every known mode of gene transfer in animals and in culture for LSD. One of the most studied models is mucopolysaccharidosis VII (MPS VII, Sly Syndrome, 253220), which is a progressively degenerative autosomal-recessive LSD caused by deficient β-glucuronidase. Its incidence is rare, occurring in fewer than 1 in 216 000 live births (Neufeld and Muenzer 1995). Clinically, children with MPS VII have short stature, dysostosis multiplex, severe mental retardation, coarse facies, corneal clouding, hepatosplenomegaly and cardiomyopathy (Sly *et al.*, 1973). Animal models of MPS VII closely resemble the human disease with hepatomegaly and vacuolated cytoplasm in many cell types including neurons (Haskins *et al.*, 1984, 1991; Schuchman *et al.*, 1989; Birkenmeier *et al.*, 1989; Vogler *et al.*, 1990a; Sands and Birkenmeier, 1993; Sheridan *et al.* 1994; Gitzelmann *et al.*, 1994; Sands *et al.*, 1995; Ray *et al.*, 1998, 1999; Fyfe *et al.*, 1999). The MPS

VII mouse, a naturally occurring mutant, is the most extensively characterized model which follows a degenerative neurological course similar to that of human MPS VII (Kyle *et al.*, 1990; Chang *et al.*, 1993a; Bastedo *et al.*, 1994; Levy *et al.*, 1996). Almost all the current experimental protocols and gene therapy vectors have been tested in this model, making it extremely valuable for the direct comparison of different treatment strategies. Recent creations of transgenic MPSVII mice, however, have provided even more precise models with the desired genotype–phenotype correlations (Tomatsu *et al.*, 2002, 2003). Moreover, the existence of larger canine and feline models of MPS VII enable direct scale-up to a large animal model of the same disease.

Both *ex vivo* and *in vivo* methods of gene therapy have been attempted in the MPS VII mice. Re-implantation of *ex vivo* retroviral-modified, β-glucuronidase-expressing bone marrow, macrophages, or fibroblasts to the mutant mice have resulted in diminished lysosomal storage in the liver and spleen, and low levels of peripheral β-glucuronidase activity (Wolfe *et al.*, 1992; Marechal *et al.*, 1993; Moullier *et al.*, 1993a,b; Freeman *et al.*, 1999), whereas re-implanted, retroviral-modified primary myoblasts led to genetic modification of up to 60% of the recipient's myofibrils after intramuscular injection, resulting in low levels of β-glucuronidase in the liver and spleen for up to 8 months (Naffakh *et al.*, 1993, 1994, 1995, 1996). *In vivo* gene therapy using retrovirus vector resulted in low and transient enzyme activity and reduction in lysosomal storage in some peripheral organs (Lau *et al.*, 1995; Gao *et al.*, 2000a). Similar results were obtained with high dose intravenous adenovirus-mediated gene transfer for up to 16 weeks (Ohashi *et al.*, 1997) or longer (Stein *et al.*, 1999).

Another vector tried for *in vivo* gene transfer is AAV. β-Glucuronidase delivery to newborn MPS VII mice with AAV also resulted in low levels (1%) of enzyme restoration for up to 16 weeks, but was only sufficient to reduce storage in the liver, and not in the spleen (Watson *et al.*, 1998; Daly *et al.*, 1999a,b). Furthermore, dramatic decline in β-glucuronidase expression in organs such as the liver and spleen during the first 12 weeks limited its long-term therapeutic value. Lentiviral vector has also been administered intravenously to adult MPS VII mice, resulting in high levels of enzyme in peripheral tissues after 3 weeks with partial reductions in storage pathology. However, antibodies against β-glucuronidase developed in all of the treated mice (Stein *et al.*, 2001).

Although none of these studies was able to correct the neurodegenerative manifestations in adult animals because of the blood–brain barrier, similar *ex vivo* or *in vivo* gene transfer have been attempted through direct CNS injection. In the *ex vivo* studies, retroviral-modified neural progenitor cells transplanted into the CNS of neonatal mice integrated and reduced lysosomal storage in neurons and glial cells near the graft (Snyder *et al.*, 1995). When retroviral-modified autologous fibroblasts were similarly grafted into the brains of adult MPS VII mice (Taylor and Wolfe, 1997), only transient and often low levels of enzyme delivery were achieved. Similar results were obtained using adenoviral vector-transduced primary adult human astrocytes transplanted into the striatum of athymic nude mice (Serguera *et al.*, 2001).

In vivo gene therapy, using herpes (Wolfe *et al.*, 1996; Zhu *et al.*, 2000), adenovirus (Ohashi *et al.*, 1997; Ghodsi *et al.*, 1998, 1999; Stein *et al.*, 1999), and AAV (Elliger *et al.*, 1999) administered directly to the CNS also resulted in some short-term and low-level delivery, varying from extremely low to 40% of normal. However, none of these vectors was able to sustain enzyme expression and pathology correction for the long term.

Aside from mice, canine MPS VII has also been treated, using intraperitoneal transplantation of retroviral-corrected autologous fibroblasts embedded in Gore-Tex fibres and collagen. This resulted in low levels of β-glucuronidase restoration (0.1–0.3%), but since enzyme levels of 2.5% were known to almost normalize lysosomal storage, such low levels of β-glucuronidase may be therapeutically effective in human patients (Wolfe *et al.*, 2000).

For human gene therapy clinical trails, several LSDs have been targeted, including the three diseases mentioned before (Gaucher, Fabry and MPS I) that already have commercial ERT therapies available. In addition, other LSDs such as Canavan disease, neuronal ceroid lipofuscinosis and MPS II (Hunter disease) have also had gene therapy clinical trials.

While the combined incidence of all the lysosomal storage diseases is approximately 1 in 1500 live births (Neufeld and Muenzer, 1995), the most common lysosomal storage disease is Gaucher disease, with a remarkably high incidence in some ethnic groups (e.g. 1/1000 in Ashkenazi Jews). With the relatively high disease frequency and the proven success of ERT, it is not surprising that Gaucher disease clinical trials lead the way for LSD gene therapy. According to the *Journal of Gene Medicine* on-line database (www.wiley.co.uk/genmed/clinical/), there have been three gene therapy trials for Gaucher disease, all using *ex vivo* retroviral transduction of autologous CD34+ cells. Most of the other LSDs with ERT products currently available or in late stages of clinical development are also the subjects of human gene therapy clinical trials. The same database lists one trial for Fabry disease, again using *ex vivo* retroviral transduction, but instead targeting mesenchymal stem cells. MPS IH (Hurler syndrome) has two trials underway, using an *ex vivo* retroviral transduction of stem cells or fibroblasts, while another MPS disorder, MPS II (Hunter syndrome) – with ERT nearing commercial availability – has one trial using *ex vivo* retrovirus transduction of autologous lymphocytes. MPS I and MPS II patients are currently being treated in a phase I/II clinical trial of retroviral-modified autologous peripheral blood lymphocyte re-infusion. Although the therapeutic effects remain to be determined, initial results have shown that less than 2.5% of the re-introduced cells were transduced (Whitley *et al.*, 1996; Stroncek *et al.*, 1999a).

Microencapsulation

As an alternative to the above gene therapy strategies, β-glucuronidase-secreting fibroblasts enclosed in alginate microcapsules were implanted into mutant MPS VII mice. After 24 h, β-glucuronidase activity was detected in the plasma, reaching 66% of physiological levels by 2 weeks post-implantation. Significant β-glucuronidase activity was detected in the liver and spleen for the duration of the 8-week experiment. Concomitantly, the intralysosomal accumulation of undegraded glycosaminoglycans was dramatically reduced in liver and spleen tissues and urinary glycosaminoglycan content was reduced to normal levels. Secondary elevation of the lysosomal enzymes β-hexosaminidase and α-galactosidase were also reduced. However, implanted mutant MPS VII mice developed antibodies against the murine β-glucuronidase as observed in other methods of ERT. When the antibody response was transiently circumvented with a single treatment of purified anti-CD4 antibody co-administered with the microcapsules, both levels and duration of β-glucuronidase delivery were increased. This is the proof-of-principle that cell-based therapy via micro-encapsulation is a feasible alternative to traditional gene therapy (Ross *et al.*, 2000a).

Similarly, when these β-glucuronidase-producing microencapsules were directly implanted into the ventricles of the mutant mouse, β-glucuronidase was delivered throughout most of the CNS, reversing the pathology and reducing the previously elevated levels of lysosomal enzymes β-hexosaminidase and α-galactosidase, surrogate measures of efficacy. The effectiveness of this approach was further demonstrated with improvements in the mutant circadian rhythm behavioral abnormalities. The grooming and circadian disruptions characteristic of the mutants were all significantly improved, and further behavioural deterioration was arrested. Hence, this alternative cell-based gene therapy demonstrates biochemical, histological and behavioral efficacy and provides a potentially cost-effective and nonviral treatment applicable to all lysosomal storage diseases with neurological deficits (Ross *et al.*, 2000b).

10.5 Cancer gene therapy – an expansion from Mendelian disorders

Applications of artificial cells in gene therapy are well suited to treat simple disorders that require a single therapeutic product for replacement. This is particularly relevant to Mendelian disorders described above that require product replacement for a prolonged period. Cancer, however, is a more complex multifactorial disease. Its etiology is variable and it can be mediated through many possible molecular mechanisms. Thus, it is not possible to engineer a single product replacement to treat all types of cancers, or even to treat all cases of the same type of cancer.

Therefore because of the complex nature of cancer in general, there are additional issues involved in using artificial cells for its treatment. Unlike the Mendelian disorders of lysosomal storage disease or hemophilia in which systemic long-term delivery will suffice, cancer therapy requires specific killing of tumor cells while sparing the normal tissue. Furthermore, additional issues need to be considered such as: where to place the implanted artificial cells relative to the tumor site; how to protect the normal cells from the toxicity of the product delivered by the artificial cells; and even how to protect the artificial cells if concurrent standard therapies such as radiation or chemotherapy are carried out. Once the cancer is eradicated, one also needs to consider if the artificial cells should be removed, or if they should remain for prophylactic purposes.

In spite of such complexities, encapsulation technology has been applied to the treatment of cancer with some success in selected animal models. The nature of artificial cells is best suited to provide therapeutic proteins that can be delivered systemically, but with specific apoptotic effects on tumor cells, and preferrably with tumor targeting signals. Products with some or all of these attributes that have been tested in cancer gene therapy clinical trials include cytokines, antibodies and tumor antigens (Edelstein *et al.*, 2004). Their use with the encapsulation technology to treat cancer can be broadly classified as follows.

10.5.1 Enzyme expression to enhance modification of a prodrug

The earliest application that involved encapsulated cells for cancer therapy was realized in the late 1990s. These studies involved the encapsulation of cells expressing cytochrome P450 enzymes (in particular, CYP2B1) to convert a benign prodrug (ifosfamide) into a toxic metabolite for the treatment of pancreatic cancer (Dautzenberg *et al.*, 1999; Muller *et al.*, 1999; Lohr *et al.*, 1999; Kroger *et al.*, 1999). Normally, ifosfamide is metabolized by endogenous cytochrome P450 in the liver into toxic phosphoramide mustard and acrolein that in turn alkylate DNA and protein respectively. Tumors displayed poor sensitivity to the drug since the toxic product has to travel from the liver to the tumor, thereby diluting its effectiveness. Furthermore, since the alkylation is not specific to the tumor cells, there were severe metabolic derangement and serious side effects (Fujiki *et al.*, 1997; Nishihara *et al.*, 2000). By implanting encapsulated cells with cytochrome P450 activity close to the pancreatic tumor sites, the dilution effect of delivering products from the liver to the tumor was minimized. It is important to note that the artificial cells themselves did not deliver a gene product to the tumor, but acted as a secondary metabolic center to convert the prodrug into the toxic metabolite at close proximity to the tumor. The treatment was well tolerated by the animals (Kroger *et al.*, 1999) even with the addition of ifosfamide (Kroger *et al.*, 2003). Importantly, the issue as to whether the cells could potentially escape from the capsules was addressed by polymerase chain reaction (PCR) with primers specific for the expression construct. Such analysis detected no CYP2B1-specific DNA in any organ including the brain and blood (Lohr *et al.*, 1998). This early clinical study demonstrated the product safety, the feasibility of administration and the potential therapeutic effect to patients with advanced-stage pancreatic carcinomas (Kroger *et al.*, 1999). Indeed, in a similar phase I/II study, the median survival was doubled in the treatment group by comparison with historic controls, and the 1-year survival rate was three times better (Lohr *et al.*, 2001).

An advantage of using encapsulated cells, rather than bolus injection, is that a lower dose can be used, and hence produce fewer side effects (Fujiki *et al.*, 1997). When mice bearing human PaCa-44 pancreatic cancers were treated either with ifosfamide administration alone, or ofosfamide with encapsulated CYP2B1-expressing cells (Muller *et al.*, 1999), the latter group displayed increased tumor suppression, in some cases with complete tumor disappearance (Lohr *et al.*, 1998; Muller *et al.*, 1999). More recently, this strategy was combined with standard irradiation in tumor-bearing rats and showed improved suppression of tumor and increased sensitivity of the tumor to the prodrug (Ryschich *et al.*, 2005). Thus, the combination of immuno-isolation with standard cancer therapy seems to provide a synergistic effect in tumor suppression.

10.5.2 Immunotherapy

Different immune-mediating products such as cytokines, antibodies and tumor antigens have been tested in the clinic as potential anti-cancer drugs. However, they are typically more expensive to produce than small chemicals and have relatively shorter half-lives *in vivo*, with potentially severe side effects when administered in high-dose bolus forms. Some of these products have been encapsulated in permeable microspheres to improve the pharmacokinetics of their delivery. IL-1-loaded microspheres have been implanted intratumorally to treat fibrosarcomas (Mullerad *et al.*, 2003), cytokines IL-12, TNF-alpha and GM-CSF have also been loaded into microspheres to treat breast cancer (Sabel *et al.*, 2004), while IL-2 in encapsulated form has been used to treat gliomas in rat brains (Rhines *et al.*, 2003). However, the same drawback exists in that only a limited amount of drug can be implanted in this manner, eventually resulting in the depletion of the drug. Although these studies have shown improved pharmacokinetics of the respective cytokines, continuous long-term delivery remains an unachieved goal.

To overcome the problems of transient delivery and the need for repeated implantations, a long-term solution is to implant encapsulated cells that can secrete cytokines continuously. This would defray the high costs of repeated injections of purified cytokines. Indeed, in a preclinical trial of this approach with a single implantation of capsules provided delivery of IL-2 for over 2 weeks in a mouse melanoma model (Cirone *et al.*, 2002). In this case, IL-2 was genetically fused to a single chain Fv antibody targeting to a tumor surface antigen *Her2/neu*. The efficacy of this approach to deliver cytokine continuously, as measured by sustained serum cytokine levels, was comparable to established continuous infusion protocols of IL-2 (Mier *et al.*, 2001). Therefore, a sustained low dose of cytokine delivered by microencapsulated cells is a feasible alternative to standard bolus infusions.

Other cytokine-related products such as interleukin-1 receptor antagonist (IL-1Ra) has been delivered also from microencapsulated cells to treat an IL-1-alpha-secreting fibrosarcoma in an animal model (Bar *et al.*, 2004). The rationale was to block the IL-1 receptor activation by its ligand IL-1 involved in inflammation, tumor growth and metastasis. In addition to the reduction in tumor growth with this treatment, significantly fewer blood capillaries and inflammatory cells surrounding the implant were observed, resulting in improved immune tolerance of the capsules.

Taken together, these first experiments established the feasibility of using microencapsulated cells to deliver cytokines for cancer immunotherapy. They have shown that the delivery of recombinant gene products for cancer therapy via microencapsulation is a viable concept. The pleiotropic effects of cytokines, however, are complex and their use must be mitigated by careful scrutiny. Some cytokines will undoubtedly complement artificial cell technology better than others, while some can contribute undesirable effects such as a heightened inflammatory response (Cirone *et al.*, 2002).

10.5.3 Delivery of tumor antigens in a vaccine strategy

Unlike cytokines, which may have varying biological consequences, vaccines need not have any biological activity themselves but act only as targets that ultimately elicit immunity. Although vaccination is a well-proven strategy, it has been difficult to generate a vaccine against cancer. As cancer cells are derived from a patient's own cells, it can be difficult to generate a selective immune response to tumors. However, genetic instability and aberrant antigen expression have provided opportunities to vaccinate against specific tumor markers (Coggin *et al.*, 2005; Lo *et al.*, 2005).

Proof-of-concept experiments illustrating the effective delivery of vaccine antigens via microencapsulation have been done, mostly by prolonging the exposure of antigen to the subject by immuno-isolation of the vaccine in biodegradable microcapsules. Such examples include vaccines against human T-lymphotropic virus type-1 (Frangione-Beebe *et al.*, 2000, 2001), malaria antigen SPf66 (Rosas *et al.*, 2002), human β-amyloid for treatment for Alzheimer's disease (Brayden *et al.*, 2001), as well as Ole e 1 allergen derived from olive pollen as an allergy vaccine (Batanero *et al.*, 2003). This system has been shown to be versatile and safe, providing long-lasting immunity, even after a single dose (Jiang *et al.*, 2005). The advantage of this approach is that the use of microencapsulation for controlled release of vaccine antigens slowly over time may reduce the number of doses in an immunization schedule and optimize the desired immune response. This protection has also been shown to permit peptide antigens presentation to cytotoxic T lymphocytes (CTL) by mouse and human macrophages, and by human dendritic cells *in vitro* for a longer time, as compared with the use of soluble peptides (Audran *et al.*, 2003). The encapsulation of vaccine antigens therefore represents a plausible alternative or supplement to conventional adjuvants.

With successes in delivering more effective vaccines in other disease models, these benefits can translate to a cancer vaccination strategy. Microencapsulation technology has been used to deliver vaccines against tumors in various formats. These include encapsulating the purified tumor-associated antigens in biodegradable polymers to release the antigen slowly to stimulate the immune system; to encapsulate live tumor cells in immuno-protective microcapsules to allow for shedding of tumor-associated antigens; and to encapsulate cells engineered to secrete the desired antibodies. In preclinical studies, the TheraCyte™ immuno-protective devices were implanted to deliver vaccines in the form of live tumor cells (allogeneic or xenogeneic) (Geller *et al.*, 1997). T-cell driven cytotoxicity to the encapsulated cells was generated through a slow, continuous release of antigens shed from the isolated cells. Importantly, the devices allowed a safer exposure to a broad variety of tumor antigens. It is likely that such a strategy can lead to robust and lasting immunity to several antigens. Alternatively, allogeneic cells have also been genetically modified to deliver tumor antigens relevant to a given cancer (Ma *et al.*, 1998; Audran *et al.*, 2003). This approach of course offers a greater degree of safety but allows presentation of only one or a few antigens. Although these studies are preliminary, they illustrate the modular nature of this technology. It is possible that any future cancer therapeutics with this approach can involve a number of different cell types or antigens. As cancer is a multifactorial disease with great variability among individuals, the flexibility of such a modular system to be tailored to an individual disease profile can be a significant advantage.

Currently, the use of immuno-isolation of cells to deliver vaccines is in the early stages of development. In general, vaccination studies have primarily focused on the immunopathology of these diseases rather than on micro-encapsulation as alternative forms of delivery. Vaccines are already thought of as cost-effective strategies to disease treatment. Therefore mass production with multiple dosages only represents a minor hurdle. However, immuno-isolation technology can also provide passive immunity by delivering antibodies within microcapsules to allow for their stabilization and controlled release (Ma *et al.*, 2003). The delivery of antibodies against tumor-associated antigens with microencapsulated cells has greater commercial potential since administration of purified antibodies can be an inherently expensive process. This is particularly true with regard to those antibodies generated by recombinant means, as illustrated with the recent success of Herceptin^{TM} for treatment of breast cancer and Avastin^{TM} for treatment of colorectal cancer. In the case of Herceptin^{TM} in the United Kingdom, of the 35 000 women diagnosed with breast cancer each year, about 20 000 will be suitable for HER2 testing. From this group of 20 000 women, about 5000 women may benefit from Herceptin^{TM}, although only approximately 1000 lives each year would be saved – all at an annual cost of about £100 million (www.medicalnewstoday.com, 2005). Artificial cells delivering recombinant forms of this antibody may be a more cost-effective approach. There have been some studies in the encapsulation of hybridomas to deliver antibodies in an animal model that were successful in long-term systemic antibody delivery (Pelegrin *et al.*, 1998; Dautzenberg *et al.*, 1999; Haas *et al.*, 1999; Orive *et al.*, 2001). Thus, delivery of antibodies with microencapsulation could become an important tool as more antibodies become available for a growing list of potential tumor antigens.

10.5.4 Anti-angiogenic therapy

Angiogenesis, the propagation of new blood vessels from pre-existing vasculature, is required for tumor growth and metastasis (Folkman, 1971). With respect to artificial cells for cancer gene therapy, anti-angiogenic therapy is a new approach that complements this platform technology well in several ways. Firstly, endothelial cells undergoing angiogenesis are dissimilar to those that are quiescent at the molecular level. For example, proliferating endothelial cells display certain extracellular markers such as a plasma membrane bound ATP synthase (Moser *et al.*, 1999) or integrin $\alpha_v\beta_3$ (Tarui *et al.*, 2001). Although most prevalent in development, apart from wound healing and menstruation, endothelial cells in a healthy adult are typically quiescent with only approximately 1 in 10 000 cells undergoing proliferation (Hobson and Denekamp, 1984). Indeed, cancer probably represents the most prevalent source of pathological angiogenesis. This means that targeting angiogenesis for cancer therapy would probably be well tolerated. Secondly, many of the anti-angiogenic factors available are recombinant proteins that encapsulated cells could be genetically engineered to produce. Such factors include recombinant antibodies to proangiogenic factors, signaling molecules as well as peptide fragments (Cao, 1998). Currently only a few anti-angiogenic agents have been approved for cancer therapy. These include Avastin^{TM}, a VEGF-neutralizing monoclonal antibody (McCarthy, 2003); Endostar (Endostatin) recently approved in China as a benchmark treatment for all cancers – to be administered repeatedly and potentially indefinitely. As described above, delivery of antibodies from artificial cells is a viable strategy and may be more cost effective than the traditional bolus administration of the purified product, particularly for treatments involving recombinant proteins to be delivered for an indefinite period.

Angiostatin is another anti-angiogenic factor that inhibits endothelial cell proliferation. Its efficacy in tumor suppression has been proven in several tumor models, but its delivery through microencapsulation has been demonstrated only recently (Cirone *et al.*, 2003). Angiostatin was delivered systemically into tumor-bearing mice by implanting microencapsulated cells genetically modified to express angiostatin. Tumor growth in the recipient animals was suppressed by >90% by 3 weeks post-tumor induction, while survival at this date was 100%, compared with 100% mortality in the controls. The cell viability and secretion of angiostatin by the encapsulated cells were maintained throughout the course of the study, associated with targeting of angiostatin to tumor cells and apoptosis of von Willebrand factor-positive endothelial cells (Cirone *et al.*, 2003). This suggests that the implantation of microcapsules secreting angiostatin is well tolerated and effective, as well as providing continuous delivery for a long period.

Implanting capsules intraperitoneally or subcutaneously is usually sufficient to deliver agents systemically. However, for CNS malignancies such as glioma, encapsulated cells genetically modified to deliver endostatin (an endogenous inhibitor of angiogenesis, (O'Reilly *et al.*, 1997) had to be implanted into the CNS (Read *et al.*, 2001) to overcome the blood–brain barrier. This long-term delivery of endostatin was also successful in suppressing tumor growth and metastasis in the brain (Bjerkvig *et al.*, 2003).

Since angiogenesis is implicated in metastatic growth of cancer cells (Skobe *et al.*, 1997), anti-angiogenic therapies can be used potentially as a prophylactic to prevent cancer recurrence. Drug-induced suppression of metastasis of tumors would probably require a long-term, sustained administration that can be done readily by immuno-isolation technologies. Thus, the safe nature and the capacity of microencapsulated cells for long-term delivery are well suited for such prophylactic purposes.

10.5.5 Combination therapy

The modular nature of artificial cell technology allows for simultaneous delivery of multiple agents, thereby offering a level of customization. As mentioned, this may prove to be critical owing to the pleiotropic nature of cancer. This strategy can be applied to offer multiple types of artificial cells to deliver varying therapeutic products. In our studies, we have examined two approaches to deliver a combination of angiostatin (providing anti-angiogenic treatment) and interleukin-2 (providing immunotherapy) to treat a murine melanoma model. The first approach involved a co-encapsulation protocol whereby capsules were fabricated to include a mixture of two cell types, delivering either angiostatin or interleukin-2 (co-encapsulation in Cirone *et al.*, 2005). The second approach involved implanting a mixture of two types of capsules, one containing only cells expressing angiostatin, and one containing cells expressing IL-2 (combination treatment in Cirone *et al.*, 2004). While the aim of such co-administration is to augment efficacy of tumor suppression, these studies have yielded interesting insight into the biological actions of the agents delivered.

When cells engineered to secrete IL-2 or angiostatin were co-encapsulated for implantation, the suppression of tumor growth and increase in survival were unexpectedly only of intermediate efficacy compared to the mono-therapies (Cirone *et al.*, 2005). In contrast, when the animals were implanted with two types of capsules in the combination treatment (Cirone *et al.*, 2004), the extent of tumor suppression was superior to mono-therapy alone, and was able to provide a long-lasting immunity even to further tumor challenge (Cirone *et al.*, 2006). While these studies demonstrate the feasibility of using artificial cells for treating complex diseases such as cancer, they also highlight the benefit of the modular nature of the approach, and the importance of optimizing the treatment protocol.

10.6 Future trends

Despite many years of preclinical work, there are discouragingly few examples of microencapsulation advancing to clinical trials. Some of the reasons for this are outlined in a review by Orive *et al.* (2004). For optimal function, the microcapsule should be completely biocompatible for the enclosed cells and for the host, have a defined pore size, and be stable chemically and mechanically in the *in vivo* environment. Ideally, the microcapsules should be retrievable and detectable *in vivo* for safety and monitoring purposes. They should also be made from well-characterized, pure materials, so that for mass production, they can be made in a reproducible, cost-effective manner that follows good manufacturing process guidelines. Meeting all of these targets is difficult and inevitably requires trade-offs in one area when another is improved.

10.6.1 Mechanical stability

Although there are numerous examples listed above of capsules surviving for long periods of time in smaller animals such as mice, the mechanical stresses in the peritoneum of larger animals or humans require more stable capsules (Peirone *et al.*, 1998b). One approach is to modify the basic alginate microcapsule with other polymers besides poly-L-lysine or to use completely different polymers. Many polymers have been developed for microencapsulation with properties suitable for implantation (Hoffman, 2002). Suzuki *et al.* (1999) crosslinked alginate with ethylenediamine to form an alginate gel stable enough to link a 50 mm gap in a cat sciatic nerve. Yin *et al.* (2003) encapsulated hepatocytes with an inner core of modified collagen and an outer shell of terpolymer of methyl methacrylate, methacrylate and hydroxyethyl methacrylate. Chitosan, as a natural polysaccharide produced from *N*-deacetylation of chitin under alkaline conditions, was also selected to enhance the mechanical properties of alginate based capsules (Gaserod *et al.*, 1998, 1999).

Of all the microcapsules used for cell delivery, the hydrogel alginate microcapsules, originally developed by Lim and Moss (1981), have been the most widely used because of their highly biocompatible nature, ease of manipulation and relatively well defined chemical properties (Smidsrod and Skjakbraek, 1990; Thu *et al.*, 1996b; Draget *et al.*, 1997). However, the current method of alginate microcapsule fabrication depends only on ionic linkages between Ca^{2+}, alginate, and polyelectrolyte complex force between alginate and poly-L-lysine. The absence of stable covalent linkage may be the cause of their breakdown in larger animals such as dogs (Peirone *et al.*, 1998b). Furthermore, within the physiological environment after implantation, leaching of the Ca^{2+} due to its high affinity with its carrier protein albumin would further weaken the hydrogel, since the mechanical strength of alginate capsules is directly related to the density of crosslinkers such as the Ca^{2+} ions. Calcium alginate gels usually exhibit a concentration inhomogeneity where the concentrations of alginate and calcium are considerably lower in the center of the beads. The inhomogeneity was demonstrated by Thu *et al.* (1997, 2000) with X-ray fluorescence spectroscopy, by Duez *et al.* (2000) with magnetic resonance imaging (MRI), and by Araki *et al.* (2005) with scanning transmission x-ray microscopy (STXM). This might have contributed to the relative stability of alginate capsules due to the higher alginate concentration at the periphery (Thu *et al.*, 1996a). One approach to increase the mechanical stability of the alginate hydrogel is to use barium as the crosslinker. It is a stronger crosslinker than calcium as it binds to both the guluronic and mannuronic residues in alginate. Its greater stability has been demonstrated after implantation in large animals such as dogs (Peirone *et al.*, 1998b), leading to a longer period of gene product delivery from the encapsulated cells. However, the potential neural toxicity of barium could be a barrier for its clinical application.

Other methods of enhancing the stability have also been explored. Quek *et al.* (2004) selected a synthetic anionic polyelectrolyte copolymer of HEMA (2-hydroxyethyl methacrylate) with MMA (methyl methacrylate acid) sodium salt as the outer layer, while loading methylated telopeptide-poor collagen in the core. Soon-Shiong and co-workers (Soon-Shiong *et al.*, 1994) reacted alginate with acrylic anhydride to incorporate an acrylate ester into the starting alginate, and then made capsules in the normal manner using Ca^{2+}, but subsequently exposing them to UV light for photopolymerization. Wang *et al.* (2005) fortified the alginate capsule with an interpenetrating network of sodium acrylate and *N*-vinylpyrrolidone (NVP) by creating covalent crosslinkages, also via photopolymerization. Extensive chemical mapping of this new reinforced hydrogel with scanning transmission X-ray microscopy confirmed the expected chemical modification occurring at the peripheral zone, consistent with hydrogen abstraction from the alginate, resulting in covalent crosslinking (Araki *et al.*, 2005). The final analyses for the biocompatibility and functional state of these novel capsules confirmed that the chemical modification did not have any adverse effect on the properties of these capsules for their use in gene therapy applications. This was demonstrated in a mouse cancer model in which the anti-angiogenic factor, angiostatin, was delivered from genetically modified cells within microcapsules to suppress tumor progression. It was shown that the NVP-reinforced microcapsules were similarly efficacious as the classical APA microcapsules in tumor suppression (Shen *et al.*, 2005a) and maintained their mechanical stability for a longer duration after implantation. Hence, such chemical modification shows promise for clinical application.

10.6.2 Monitoring capsules post-implantation

Alginate microcapsules have been implanted into mice, dogs, monkeys and even human beings, but once the microcapsules are implanted, the outcome of the implanted microcapsules can only be studied by killing the animals or through surgery. While the ability to surgically removal microcapsules is an important safety feature not shared by most other gene therapy protocols, it is desirable to reserve this option for serious adverse events once artificial cell gene therapy is in human clinical trials. As these capsules are intended for long-term delivery, it would be important to have a way of repeatedly monitoring the microcapsules without invasive intervention. Hence, if the implants could be visualized after implantation, their functionality and biodistribution could be evaluated more effectively.

MRI visible capsules

Magnetic resonance imaging (MRI) has evolved into a major imaging tool in clinical medicine and continues to improve in image quality and speed. The use of MRI to study carbohydrate polymers such as agarose (Belton *et al.*, 1988), alginate (Potter *et al.*, 1993, 1994; Duez *et al.*, 2000; Thu *et al.*, 2000) has been shown feasible under various experimental conditions. Simpson *et al.* (2003) reported the imaging of 800–900 μm alginate beads with a powerful MRI (with a 17.6 T vertical magnet). Potter *et al.* (1993, 1994) measured a series of homogeneous gels of varying alginate concentrations via MRI and obtained an equation which relates alginate concentration to the T_2 of the water protons within the gel. When applied to the T_2 MRI image of an inhomogeneous calcium alginate gel, the relationship gives a map of alginate concentration as a function of distance along the axis of the gel. However, because of the weak contrast of the polysaccharide gel, this work focused on alginate beads around 3–4 mm in diameter, which is approximately ten times greater than the normally implanted microcapsules of ~400 μm in diameter. Hence, such implantable alginate microcapsules would not permit tracking with MRI without increasing their contrast properties.

Ferrofluids, as described by Morneau *et al.* (1999), are stable suspensions of nanoscale magnetic iron oxide particles in appropriate non-magnetic carrier liquids. When this nanoscale magnetic technology is combined with the regular

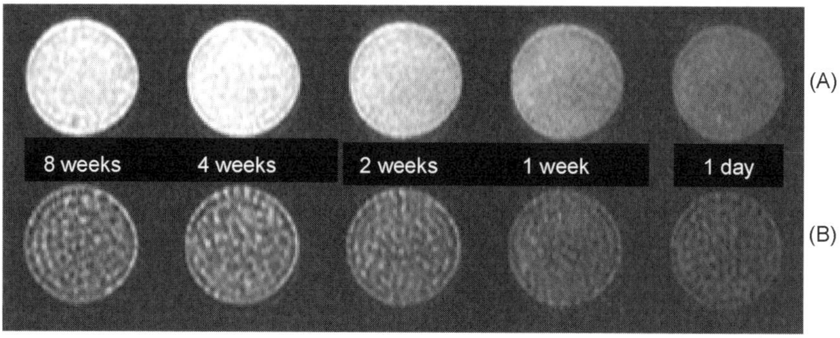

10.1 MRI I_2 images of ferrofluid alginate microcapsules: enhanced, row A; unmodified, row B. Cultured *in vitro* in culture wells of a 24-well plate. (*Source*: Shen *et al.* 2005b).

alginate-based capsule fabrication, the magnetic nano-size particles, when incorporated into the alginate as the carrier solution, enhance the contrast signals of the microcapsules so as to render them visible with MRI at high resolution (Shen *et al.*, 2003, 2005b). A method for monitoring 300–400 μm sized microcapsules was developed, under a clinically relevant setting, using a General Electric imaging machine with superconducting magnet operating at 1.5 T. This allowed the visualization of the microcapsules under both *in vitro* (Fig. 10.1) and *in vivo* (Fig. 10.2) conditions. The *in vivo* experiments thus showed that the incorporation of ferrofluid into 3 ml of the classical alginate microcapsules has provided the possibility of visualizing the implanted device with MRI without invasive surgery or impairing the functional attributes of the original hydrogel. Such development should augment the potential of using this immuno-isolation strategy for clinical applications.

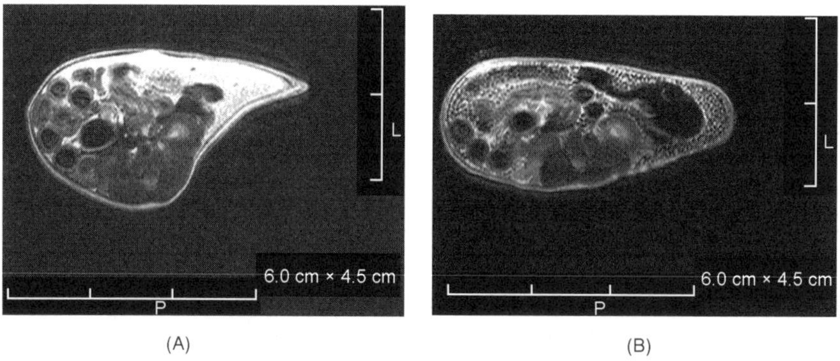

10.2 MRI of ferrofluid microcapsules *in vivo* in the abdominal cavities of mice: (A) alginate only; (B) alginate enhanced with ferrofluid. The ferrofluid microcapsules can be visualized with MRI filling the intraperitoneal space (B). Transverse section view: top, anterior; left, right of animal. (*Source*: Shen *et al.*, 2005b).

Fluorescent capsules

Confocal laser scanning microscopy (CLSM) is another powerful tool to visualize polymer gels labeled with fluorescent ligands. For APA capsules, alginate can be labeled with the fluorochrome fluoresceinamine, and poly-L-lysine can be labeled with Alexa 546 Protein Labeling Kit (Strand *et al.*, 2003). The distribution of the components in the fluorescent labeled capsules were visible with CLSM, and a 3D reconstructions of optical sections was taken throughout the capsules (Darrabie *et al.*, 2005). A more sensitive device for the detection of fluorescent signal is Cooled Charged Coupled Device (CCCD) camera. The encapsulation of genetically modified cells that express luciferase proteins rendered the alginate capsules after implantation in mice visible with a CCCD camera, as shown in Fig. 10.3 (Li, Hou, Shen and Potter, unpublished observation). The fluorescent image (center) was recorded with a CCCD camera 24 h after 3 ml of microcapsules with cells genetically modified to express luciferase were implanted in the abdominal cavity of a mouse. Hence, the *in vivo* tracking of alginate microcapsules with fluorescence markers appear to be a feasible approach, potentially offering another method of monitoring implanted device without invasive surgery.

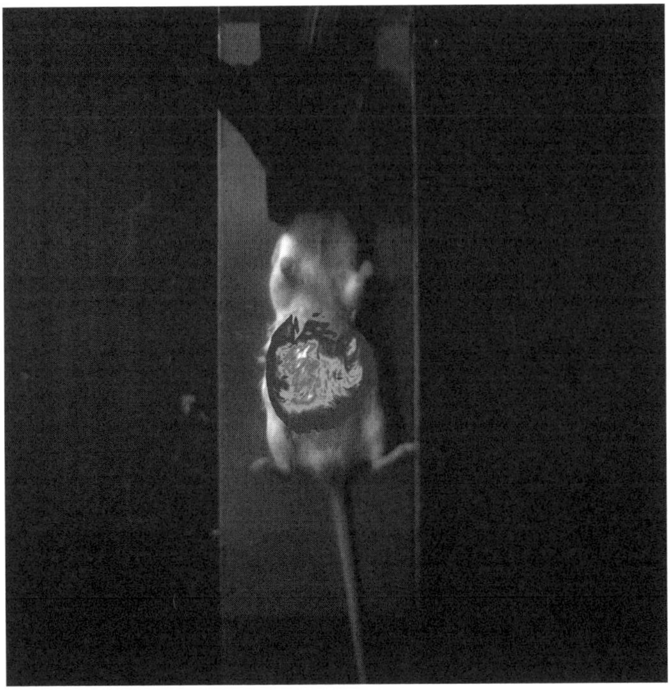

10.3 CCCD image of luciferase microcapsules *in vivo*. The fliuorescent image was combined with a white light image.

10.6.3 Molecular targeting

Delivery of the therapeutic product systemically is appropriate for the treatment of diseases such as hemophilia and many inborn errors of metabolism. However, when the product delivered has potent and even toxic side effects, e.g. cytokines, additional molecular targeting signals may be necessary, both to decrease toxicity and to improve efficacy. For example, cytokines such as IL-2, IL-12, and interferon have a clearly demonstrated anti-tumor effect but can produce severe side effects in normal cells when high levels were administrated (Siegel and Puri, 1991; Edwards *et al.*, 1992; Atkins *et al.*, 1997; Sabel and Sondak, 2003). In general, targeting complements artificial cell technology as the microcapsules are compartmentally fixed (i.e. at the site of implantation) and any secreted products would have to enter systemic circulation to approach a therapeutic target. The ability to home in to tumors would mean that potentially a higher concentration of products could reach the tumors from these capsules. Targeting to tumor-associated antigens (i.e. to *HER2*-expressing cells in breast cancer) or to the tumor vasculature are some of these potential targets.

Increasing specificity in cancer gene therapy

When the strategy of targeting via microencapsulated cells was applied with a fusion protein of IL-2 and antibody specific for the *HER2/neu* receptor, however, no enhancement of tumor-specific delivery of IL-2 was obtained (Cirone *et al.*, 2002). The fusion protein, which should have targeted the B16-F10/*neu* tumor cells, did poorly compared to other antibody-IL-2 fusion proteins targeting tumors (Becker *et al.*, 1996a,b; Lode and Reisfeld, 2000), or even other melanoma models (Xiang *et al.*, 1997; Gillies *et al.*, 1992; Niethammer *et al.*, 2001). The immunoconjugate accumulated in lymphoid tissues instead. This was possibly due to the affinity of IL-2 to its receptors that were plentiful on lymphocytes in these tissues (Foss, 2001; Yao *et al.*, 2004). Therefore, targeting by molecular fusion to antibody (or antibody fragments) would likely work best where the endogenous antigens are not widely distributed.

In spite of the above reservation, apparent targeting in a different context was claimed to have been achieved by using a common peptide motif Arg-Gly-Asp (RGD) widely distributed in the plasma membrane of many cell types. Biodegradable poly(lactic glycolic acid) (PLGA) microcapsules were developed as ultrasound contrast agents for tumors by modifying the microcapsule surface with the RGD ligand. The modified microcapsules were found adhering specifically to a breast cancer cell line MDA-MB-231 in static experiments (Lathia *et al.*, 2004). The RGD peptide sequence targets integrins expressed during angiogenesis, alphavbeta3 and alphavbeta5 as well as other integrins. These are membrane-spanning proteins in cells that play a vital role in cell attachment and many other processes. Since the integrins specific to angiogenesis are more active during cancer, this could have been the rationale of using the RGD as targeting ligands for tumor cells.

In contrast to the above targeting strategy, angiostatin, which has no known tumor-targeting ability, appeared to accumulate in tumors! When encapsulated cells secreting angiostatin were implanted into tumor-bearing mice, no angiostatin activity could be observed in serum for the first 3 weeks, whereas as similar implantation into non-tumor-bearing mice resulted in high serum level of angiostatin even from the first week (Cirone *et al.*, 2003). Therefore, fusion to angiostatin could be a favorable way to target other molecules to the tumor or even to the tumor vascular endothelium. Although several receptors for angiostatin have been identified, the exact molecular mechanism of this tumor-targeting has yet to be elucidated.

Targeting the brain

Special consideration is required when targeting recombinant molecules to the CNS. The blood–brain barrier, comprising tight junctions between capillary endothelial cells in the CNS vasculature, prevents most large molecules, including proteins, from freely moving into the CNS. While the blood–brain barrier could be physically circumvented by disruption of the junctions or direct injection into the CNS, these methods would be quite invasive. To get peripherally administered therapeutic proteins to cross the blood–brain barrier, one can take advantage of endogenous receptor-mediated transcytosis systems. These systems bypass the tight junctions between the endothelial cells by allowing trans-cellular transport via receptor-mediated endocytosis such as that for insulin, transferrin and the more recently discovered transport systems of melanotransferrin (also called P97) (Demeule *et al.*, 2002) and RAP (receptor-associated protein) (Pan *et al.*, 2004). Fusion techniques such as those described above should allow functional therapeutic proteins to be combined with ligands for these receptors to give bi-functional molecular hybrids. Such hybrids could then be secreted from engineered cells in peripherally implanted microcapsules but achieving delivery to the CNS without invasive neurosurgery.

The general approach is to use either a natural ligand to the receptor or a monoclonal antibody to the receptor as the starting point to create the fusion protein that can cross the blood–brain barrier. Because of species-specificity, the anti-human insulin receptor mouse monoclonal antibody cannot be used in mice (Pardridge *et al.*, 1995), but a 'humanized' chimeric antibody has been applied to primates and could be used in humans (Coloma *et al.*, 2000). RAP has been examined as a fusion protein with the lysosomal enzymes alpha-L-iduronidase (deficient in Hurler/Scheie syndrome) and acid alpha-glucosidase (deficient in Pompe syndrome) (Prince *et al.*, 2004), and was shown to retain its receptor binding properties.

For the transferrin receptor (TfR), the natural ligand is not effective for blood–brain barrier delivery because the receptors are almost saturated at physiological levels of transferring (Seligman, 1983). The mouse monoclonal antibody OX26 (against rat TfR) binds to a different site on the TfR from transferrin so it is not subject to competitive binding with transferrin (Jefferies *et al.*, 1984). An OX26 single chain Fv antibody-streptavidin fusion protein has been used to selectively target blood–brain barrier endothelial cells (Li *et al.*, 1999), thus demonstrating the feasibility of targeted delivery of biotinylated drugs. While the OX26 monoclonal antibody does not recognize the mouse TfR, analogous anti-mouse TfR monoclonal antibodies have been developed and tested *in vivo*. The 8D3 monoclonal antibody has been shown to selectively target the mouse brain via the TfR and could be used as a carrier protein in mouse studies (Lee *et al.*, 2000).

Melanotransferrin is a glycoprotein first discovered in malignant melanoma cells (originally called P97) (Brown *et al.*, 1981). The gene for human P97 is on chromosome 3q21 and codes for a 738 amino acid protein (Rose *et al.*, 1986) that is processed to a 694 amino acid mature protein. The mature form is membrane bound by a glycosyl phosphatidylinositol (GPI) anchor (Alemany *et al.*, 1993), but a less well-characterized soluble form of P97 has also been described (Food *et al.*, 1994) that lacks this GPI anchor. P97 has been shown to reversibly bind iron (Baker *et al.*, 1992), but the exact function of the protein is unknown. One property of the protein that has been explored is its transcytosis into the brain, both *in situ* and in an *in vitro* blood–brain barrier model. The model system used bovine brain capillary endothelial cells grown in co-culture with newborn rat astrocytes in monolayers to examine the transcytosis across the monolayer. The putative receptor for melanotransferrin is the lipoprotein receptor-related protein (LRP) (Demeule *et al.*, 2003), which is known to rapidly internalize the urokinase : plasminogen activator inhibitor : urokinase receptor complex (Conese *et al.*, 1995). Internalization of melanotransferrin by LRP has been shown to be 14-fold higher than internalization of transferrin by TfR with no intra-endothelial degradation (Demeule *et al.*, 2002), although RAP seems to have even higher transcytosis (Pan *et al.*, 2004). We have constructed a fusion protein of a marker enzyme, firefly luciferase and P97. C2C12 cells transfected with this construct were encapsulated and implanted into the peritoneal cavity of Balb/c mice. Tissue homogenates showed elevated levels of luciferase activity, including the brain (Li, Hou, Shen & Potter, unpublished observation). While more study is needed before any of these approaches can be used clinically, the preliminary reports discussed here are encouraging that the blood–brain barrier may be broached without resorting to direct implantation of artificial cells in the brain.

10.7 Conclusions

Immuno-isolation of genetically modified cells is a platform technology, potentially offering a high degree of flexibility and efficacy towards treatments of many disorders – from classical Mendelian single-gene diseases to multi-

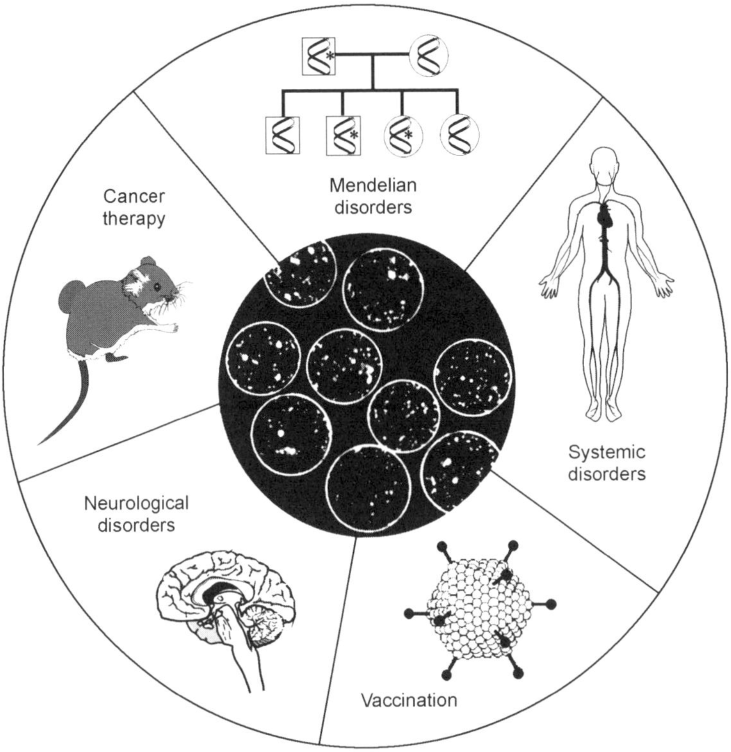

10.4 Artificial cell for gene therapy – a platform technology.

factorial cancer, for purposes of product replacement, vaccination or systemic delivery (Fig. 10.4). In the past decade, the proof-of-principle has been established for almost all of the above categories of applications. In some cases, the technology has advanced even to human clinical trials. The challenge remains in how to translate this technology into standard clinical practice. This will require concerted effort in tissue engineering, chemistry, molecular and cell biology and medicine to create an optimal combination of capsule polymer, high-expression vectors, appropriate cell types, supportive microcapsule milieu and device monitoring to meet the requirement of good manufacturing practices. This will be the challenge and opportunity for the coming decade.

10.8 References

Aebischer, P., Goddard, M., Signore, A. P. & Timpson, R. L. (1994) Functional recovery in hemiparkinsonian primates transplanted with polymer-encapsulated PC12 cells. *Exp Neurol*, 126, 151–8.

Aebischer, P., Pochon, N. A., Heyd, B., Deglon, N., Joseph, J. M., Zurn, A. D., Baetge, E. E., Hammang, J. P., Goddard, M., Lysaght, M., Kaplan, F., Kato, A. C., Schluep,

M., Hirt, L., Regli, F., Porchet, F. & de Tribolet, N. (1996) Gene therapy for amyotrophic lateral sclerosis (ALS) using a polymer encapsulated xenogenic cell line engineered to secrete hCNTF. *Hum Gene Ther*, 7, 851–60.

Aiuti, A., Slavin, S., Aker, M., Ficara, F., Deola, S., Mortellaro, A., Morecki, S., Andolfi, G., Tabucchi, A., Carlucci, F., Marinello, E., Cattaneo, F., Vai, S., Servida, P., Miniero, R., Roncarolo, M. G. & Bordignon, C. (2002) Correction of ADA-SCID by stem cell gene therapy combined with nonmyeloablative conditioning. *Science*, 296, 2410.

Alemany, R., Vila, M. R., Franci, C., Egea, G., Real, F. X. & Thomson, T. M. (1993) Glycosyl phosphatidylinositol membrane anchoring of melanotransferrin (p97): apical compartmentalization in intestinal epithelial cells. *J Cell Sci*, 104, 1155–62.

Al-Hendy, A., Hortelano, G., Tannenbaum, G. S. & Chang, P. L. (1995) Correction of the growth defect in dwarf mice with nonautologous microencapsulated myoblasts – an alternate approach to somatic gene therapy. *Hum Gene Ther*, 6, 165–75.

Al-Hendy, A., Hortelano, G., Tannenbaum, G. S. & Chang, P. L. (1996) Growth retardation – an unexpected outcome from growth hormone gene therapy in normal mice with microencapsulated myoblasts. *Hum Gene Ther*, 7, 61–70.

Anderson, W. (1968) Kennedy Foundation Symposium on Mental Retardation. *Pediatric News*, 15, 1–29.

Anson, D. S., Austen, D. E. & Brownlee, G. G. (1985) Expression of active human clotting factor IX from recombinant DNA clones in mammalian cells. *Nature*, 315, 683–5.

Araki, T., Hitchcock, A. P., Shen, F., Chang, P. L., Wang, M. & Childs, R. F. (2005) Quantitative chemical mapping of sodium acrylate- and *N*-vinylpyrrolidone-enhanced alginate microcapsules. *J Biomater Sci Polym Ed*, 16, 611–27.

Atkins, M. B., Robertson, M. J., Gordon, M., Lotze, M. T., Decoste, M., Dubois, J. S., Ritz, J., Sandler, A. B., Edington, H. D., Garzone, P. D., Mier, J. W., Canning, C. M., Battiato, L., Tahara, H. & Sherman, M. L. (1997) Phase I evaluation of intravenous recombinant human interleukin 12 in patients with advanced malignancies. *Clin Cancer Res*, 3, 409–17.

Audran, R., Peter, K., Dannull, J., Men, Y., Scandella, E., Groettrup, M., Gander, B. & Corradin, G. (2003) Encapsulation of peptides in biodegradable microspheres prolongs their MHC class-I presentation by dendritic cells and macrophages in vitro. *Vaccine*, 21, 1250.

Baker, E. N., Baker, H. M., Smith, C. A., Stebbins, M. R., Kahn, M., Hellstrom, K. E. & Hellstrom, I. (1992) Human melanotransferrin (p97) has only one functional iron-binding site. *FEBS Lett*, 298, 215–18.

Banting, F. & Best, C. (1922) The internal secretion of the pancreas. *Journal of Laboratory and Clinical Medicine*, 7, 256–71.

Bar, D., Apte, R. N., Voronov, E., Dinarello, C. A. & Cohen, S. (2004) A continuous delivery system of IL-1 receptor antagonist reduces angiogenesis and inhibits tumor development. *FASEB J*, 18, 161.

Barsoum, S. C., Milgram, W., Mackay, W., Coblentz, C., Delaney, K. H., Kwiecien, J. M., Kruth, S. A. & Chang, P. L. (2003) Delivery of recombinant gene product to canine brain with the use of microencapsulation. *J Lab Clin Med*, 142, 399–413.

Bastedo, L., Sands, M. S., Lambert, D. T. & Pisa, M. A. (1994) Behavioral consequences of bone marrow transplantation in the treatment of murine mucopolysaccharidosis type VII. *J Clin Invest*, 94(3), 1180–86.

Batanero, E., Barral, P., Villalba, M. & Rodriguez, R. (2003) Encapsulation of Ole e 1 in biodegradable microparticles induces Th1 response in mice: a potential vaccine for allergy. *J Control Release*, 92, 395.

Becker, J. C., Varki, N., Gillies, S. D., Furukawa, K. & Reisfeld, R. A. (1996a) An antibody-interleukin 2 fusion protein overcomes tumor heterogeneity by induction of a cellular immune response. *Proc. Natl. Acad. Sci. USA*, 93, 7826.

Becker, J. C., Varki, N., Gillies, S. D., Furukawa, K. & Reisfeld, R. A. (1996b) Long-lived and transferable tumor immunity in mice after targeted interleukin-2 therapy. *J. Clin. Invest*, 98, 2801.

Bell, B., Canty, D. & Audet, M. (1995) Hemophilia: an updated review. *Pediatr Rev*, 16, 290–98.

Belton, P. S., Hill, B. P. & Raimbaud, E. R. (1988) The effects of morphology and exchange on proton N.M.R. relaxation in agarose gels. *Mol Phys*, 63, 825–42.

Beutler, E., Kay, A., Saven, A., Garver, P., Thurston, D., Dawson, A. & Rosenbloom, B. (1991) Enzyme replacement therapy for Gaucher disease. *Blood*, 78, 1183–9.

Birkenmeier, E. H., Davisson, M. T., Beamer, W. G., Ganschow, R. E., Vogler, C. A., Gwynn, B., Lyford, K. A., Maltais, L. M. & Wawrzyniak, C. J. (1989) Murine mucopolysaccharidosis type VII. Characterization of a mouse with beta-glucuronidase deficiency. *J Clin Invest*, 83(4), 1258–66.

Bjerkvig, R., Read, T. A., Vajkoczy, P., Aebischer, P., Pralong, W., Platt, S., Melvik, J. E., Hagen, A. & Dornish, M. (2003) Cell therapy using encapsulated cells producing endostatin. *Acta Neurochir. Suppl*, 88, 137–41.

Blaese, R. M., Culver, K. W., Miller, A. D., Carter, C. S., Fleisher, T., Clerici, M., Shearer, G., Chang, L., Chiang, Y., Tolstoshev, P., Greenblatt, J. J., Rosenberg, S. A., Klein, H., Berger, M., Mullen, C. A., Ramsey, W. J., Muul, L., Morgan, R. A. & Anderson, W. F. (1995) T lymphocyte-directed gene therapy for ADA-SCID: initial trial results after 4 years. *Science*, 270, 475–80.

Blau, H. M., Dhawan, J. & Pavlath, G. K. (1993) Myoblasts in pattern formation and gene therapy. *Trends Genet*, 9, 269–74.

Brady, R. O., Pentchev, P. G., Gal, A. E., Hibbert, S. R. & Dekaban, A. S. (1974) Replacement therapy for inherited enzyme deficiency. Use of purified glucocerebrosidase in Gaucher's disease. *N Engl J Med*, 291, 989–93.

Brayden, D. J., Templeton, L., McClean, S., Barbour, R., Huang, J., Nguyen, M., Ahern, D., Motter, R., Johnson-Wood, K., Vasquez, N., Schenk, D. & Seubert, P. (2001) Encapsulation in biodegradable microparticles enhances serum antibody response to parenterally-delivered beta-amyloid in mice. *Vaccine*, 19, 4185–93.

Breyer, B., Jiang, W., Cheng, H., Zhou, L., Paul, R., Feng, T. & He, T. C. (2001) Adenoviral vector-mediated gene transfer for human gene therapy. *Curr Gene Ther*, 1, 149.

Brogden, R. N. & Heel, R. C. (1987) Human insulin. A review of its biological activity, pharmacokinetics and therapeutic use. *Drugs*, 34, 350–71.

Brown, J. P., Woodbury, R. G., Hart, C. E., Hellstrom, I. & Hellstrom, K. E. (1981) Quantitative analysis of melanoma-associated antigen p97 in normal and neoplastic tissues. *Proc Natl Acad Sci USA*, 78, 539–43.

Buchser, E., Goddard, M., Heyd, B., Joseph, J. M., Favre, J., De Tribolet, N., Lysaght, M. & Aebischer, P. (1996) Immunoisolated xenogenic chromaffin cell therapy for chronic pain. Initial clinical experience. *Anesthesiology*, 85, 1005–12; discussion 29A–30A.

Busby, S., Kumar, A., Joseph, M., Halfpap, L., Insley, M., Berkner, K., Kurachi, K. & Woddbury, R. (1985) Expression of active human factor IX in transfected cells. *Nature*, 316, 271–3.

Campbell, C. H. & Rome, L. H. (1983) Coated vesicles from rat liver and calf brain contain lysosomal enzymes bound to mannose 6-phosphate receptors. *J Biol Chem*, 258, 13347–52.

Cao, Y. (1998) Endogenous angiogenesis inhibitors: angiostatin, endostatin, and other proteolytic fragments. *Prog. Mol. Subcell. Biol.*, 20, 161–76.

Carmen, I. H. (2001) A death in the laboratory: the politics of the Gelsinger aftermath. *Mol Ther*, 3, 425–8.

Cavazzana-Calvo, M., Hacein-Bey, S., De Saint, B. G., Gross, F., Yvon, E., Nusbaum, P., Selz, F., Hue, C., Certain, S., Casanova, J. L., Bousso, P., Deist, F. L. & Fischer, A. (2000) Gene therapy of human severe combined immunodeficiency (SCID)-X1 disease. *Science*, 288, 669.

Chang, P. L. (1994) Calcium phosphate-mediated DNA transfection. In Wolff, J. A. (Ed.) *Gene Therapeutics*. Boston, Birkhauser Boston, Inc.

Chang, P. L. (Ed.) (1995) *Non-autologous Somatic Gene Therapy*, Boca Raton, CRC Press Inc.

Chang, P. L. (1996) Microencapsulation – an alternative approach to gene therapy. *Transfus Sci*, 17, 35–43.

Chang, P. L., Lambert, D. T. & Pisa, M. A. (1993a) Behavioural abnormalities in a murine model of a human lysosomal storage disease. *Neuroreport*, 4(5), 507–10.

Chang, P. L., Shen, N. & Westcott, A. J. (1993b) Delivery of recombinant gene products with microencapsulated cells *in vivo*. *Hum Gene Ther*, 4, 433–40.

Chang, P. L., Hortelano, G., Tse, M. & Awrey, D. E. (1994) Growth of recombinant fibroblasts in alginate microcapsules. *Biotechnol Bioeng*, 43, 925–33.

Chang, P. L., Van Raamsdonk, J. M., Hortelano, G., Barsoum, S. C., MacDonald, N. C. & Stockley, T. L. (1999) The *in vivo* delivery of heterologous proteins by microencapsulated recombinant cells. *Trends Biotechnol*, 17, 78–83.

Chang, T. M. (1964) Semipermeable microcapsules. *Science*, 146, 524–5.

Chen, L., Nelson, D. M., Zheng, Z. & Morgan, R. A. (1998) *Ex vivo* fibroblast transduction in rabbits results in long-term (>600 days) factor IX expression in a small percentage of animals. *Hum Gene Ther*, 9, 2341–51.

Choo, K. H., Gould, K. G., Rees, D. J. & Brownlee, G. G. (1982) Molecular cloning of the gene for human anti-haemophilic factor IX. *Nature*, 299, 178–80.

Cirone, P., Bourgeois, J. M., Austin, R. C. & Chang, P. L. (2002) A novel approach to tumor suppression with microencapsulated recombinant cells. *Hum Gene Ther*, 13, 1157–66.

Cirone, P., Bourgeois, J. M. & Chang, P. L. (2003) Antiangiogenic cancer therapy with microencapsulated cells. *Hum Gene Ther*, 14, 1065–77.

Cirone, P., Bourgeois, J. M., Shen, F. & Chang, P. L. (2004) Combined immunotherapy and antiangiogenic therapy of cancer with microencapsulated cells. *Hum Gene Ther*, 15, 945.

Cirone, P., Shen, F. & Chang, P. L. (2005) A multiprong approach to cancer gene therapy by coencapsulated cells. *Cancer Gene Ther*, 12, 369.

Cirone, P., Potter, M., Hirte, H. & Chang, P. (2006) Immuno-isolation in cancer gene therapy. *Curr Gene Ther*, 6, 181–91.

Coggin, J. H., Jr., Barsoum, A. L., Rohrer, J. W., Thurnher, M. & Zeis, M. (2005) Contemporary definitions of tumor specific antigens, immunogens and markers as related to the adaptive responses of the cancer-bearing host. *Anticancer Res.*, 25, 2345.

Coiffier, B., Lepage, E., Briere, J., Herbrecht, R., Tilly, H., Bouabdallah, R., Morel, P., van den Neste, E., Salles, G., Gaulard, P., Reyes, F., Lederlin, P. & Gisselbrecht, C. (2002) CHOP chemotherapy plus rituximab compared with CHOP alone in elderly

patients with diffuse large-B-cell lymphoma. *N Engl J Med*, 346, 235–42.
Coloma, M. J., Lee, H. J., Kurihara, A., Landaw, E. M., Boado, R. J., Morrison, S. L. & Pardridge, W. M. (2000) Transport across the primate blood–brain barrier of a genetically engineered chimeric monoclonal antibody to the human insulin receptor. *Pharm Res*, 17, 266–74.
Conese, M., Nykjaer, A., Petersen, C. M., Cremona, O., Pardi, R., Andreasen, P. A., Gliemann, J., Chistensen, E. I. & Blasi, F. (1995) alpha-2 Macroglobulin receptor/Ldl receptor-related protein(Lrp)-dependent internalization of the urokinase receptor. *J Cell Biol*, 131, 1609–22.
Connelly, S., Andrews, J. L., Gallo, A. M., Kayda, D. B., Qian, J., Hoyer, L., Kadan, M. J., Gorziglia, M. I., Trapnell, B. C., McClelland, A. & Kaleko, M. (1998) Sustained phenotypic correction of murine hemophilia A by *in vivo* gene therapy. *Blood*, 91, 3273–81.
Cunningham, C. & Nemunaitis, J. (2001) A phase I trial of genetically modified *Salmonella typhimurium* expressing cytosine deaminase (TAPET-CD, VNP20029) administered by intratumoral injection in combination with 5-fluorocytosine for patients with advanced or metastatic cancer. Protocol no: CL-017. Version: April 9, 2001. *Hum Gene Ther*, 12, 1594.
Dalle, B., Payen, E., Regulier, E., Deglon, N., Rouyer-Fessard, P., Beuzard, Y. & Aebischer, P. (1999) Improvement of mouse beta-thalassemia upon erythropoietin delivery by encapsulated myoblasts. *Gene Ther*, 6, 157–61.
Daly, T. M., Vogler, C., Levy, B., Haskins, M. E. & Sands, M. S. (1999a) Neonatal gene transfer leads to widespread correction of pathology in a murine model of lysosomal storage disease. *Proc Natl Acad Sci USA*, 96(5), 2296–300.
Daly, T. M., Okuyama, T., Vogler, C., Haskins, M. E., Muzyczka, N. & Sands, M. S. (1999b) Neonatal intramuscular injection with recombinant adeno-associated virus results in prolonged beta-glucuronidase expression *in situ* and correction of liver pathology in mucopolysaccharidosis type VII mice. *Hum Gene Ther*, 10(1), 85–94.
Darrabie, M. D., Kendall, W. F., Jr. & Opara, E. C. (2005) Characteristics of poly-L-ornithine-coated alginate microcapsules. *Biomaterials*, 26, 6846–52.
Dautzenberg, H., Schuldt, U., Grasnick, G., Karle, P., Muller, P., Lohr, M., Pelegrin, M., Piechaczyk, M., Rombs, K. V., Gunzburg, W. H., Salmons, B. & Saller, R. M. (1999) Development of cellulose sulfate-based polyelectrolyte complex microcapsules for medical applications. *Ann NY Acad Sci*, 875, 46.
Dawson, R. M., Broughton, R. L., Stevenson, W. T. & Sefton, M. V. (1987) Microencapsulation of CHO cells in a hydroxyethyl methacrylate-methyl methacrylate copolymer. *Biomaterials*, 8, 360–66.
De la Salle, H., Altenburger, W., Elkaim, R., Dott, K., Dieterle, A., Drillien, R., Cazenave, J. P., Tolstoshev, P. & Lecocq, J. P. (1985) Active gamma-carboxylated human factor IX expressed using recombinant DNA techniques. *Nature*, 316, 268–70.
Dell'Angelica, E. C., Mullins, C., Caplan, S. & Bonifacino, J. S. (2000) Lysosome-related organelles. *Faseb J*, 14, 1265–78.
Demeule, M., Poirier, J., Jodoin, J., Bertrand, Y., Desrosiers, R. R., Dagenais, C., Nguyen, T., Lanthier, J., Gabathuler, R., Kennard, M., Jefferies, W. A., Karkan, D., Tsai, S., Fenart, L., Cecchelli, R. & Beliveau, R. (2002) High transcytosis of melanotransferrin (P97) across the blood–brain barrier. *J Neurochem*, 83, 924–33.
Demeule, M., Bertrand, Y., Michaud-Levesque, J., Jodoin, J., Rolland, Y., Gabathuler, R. & Beliveau, R. (2003) Regulation of plasminogen activation: a role for melanotransferrin (p97) in cell migration. *Blood*, 102, 1723–31.

Deweese, T. L., Van Der, P. H., Li, S., Mikhak, B., Drew, R., Goemann, M., Hamper, U., Dejong, R., Detorie, N., Rodriguez, R., Haulk, T., Demarzo, A. M., Piantadosi, S., Yu, D. C., Chen, Y., Henderson, D. R., Carducci, M. A., Nelson, W. G. & Simons, J. W. (2001) A phase I trial of CV706, a replication-competent, PSA selective oncolytic adenovirus, for the treatment of locally recurrent prostate cancer following radiation therapy. *Cancer Res*, 61, 7464.

Draget, K. I., Skjak-Braek, G. & Smidsrod, O. (1997) Alginate based new materials. *Int J Biol Macromol*, 21, 47–55.

Duez, J.-M., Mestdagh, M., Demeure, R., Goudemant, J.-F., Hills, B. P. & Godward, J. (2000) NMR studies of calcium-induced alginate gelation. Part I – MRI tests of gelation models. *Magn Reson Chem*, 38, 324–30.

Edelstein, M. L., Abedi, M. R., Wixon, J. & Edelstein, R. M. (2004) Gene therapy clinical trials worldwide 1989–2004 – an overview. *J Gene Med*, 6, 597–602.

Edwards, M. J., Heniford, B. T., Klar, E. A., Doak, K. W. & Miller, F. N. (1992) Pentoxifylline inhibits interleukin-2-induced toxicity in C57BL/6 mice but preserves antitumor efficacy. *J Clin Invest*, 90, 637–41.

Eiselein, L., Schwartz, H. J. & Rutledge, J. C. (2004) The challenge of type 1 diabetes mellitus. *Ilar J*, 45, 231–6.

Elliger, S. S., Elliger, C. A., Aguilar, C. P., Raju, N. R. & Watson, G. L. (1999) Elimination of lysosomal storage in brains of MPS VII mice treated by intrathecal administration of an adeno-associated virus vector. *Gene Ther*, 6(6), 1175–8.

Emerich, D. F., Winn, S. R., Harper, J., Hammang, J. P., Baetge, E. E. & Kordower, J. H. (1994) Implants of polymer-encapsulated human NGF-secreting cells in the nonhuman primate: rescue and sprouting of degenerating cholinergic basal forebrain neurons. *J Comp Neurol*, 349, 148–64.

Eng, C. M., Guffon, N., Wilcox, W. R., Germain, D. P., Lee, P., Waldek, S., Caplan, L., Linthorst, G. E. & Desnick, R. J. (2001) Safety and efficacy of recombinant human alpha-galactosidase A – replacement therapy in Fabry's disease. *N Engl J Med*, 345, 9–16.

Erikson, A., Bembi, B. & Schiffmann, R. (1997) Neuronopathic forms of Gaucher's disease. *Baillieres Clin Haematol*, 10, 711–23.

Evans, J. P., Brinkhous, K. M., Brayer, G. D., Reisner, H. M. & High, K. A. (1989) Canine hemophilia B resulting from a point mutation with unusual consequences. *Proc Natl Acad Sci USA*, 86, 10095–9.

Ewton, D. Z., Falen, S. L. & Florini, J. R. (1987) The type II insulin-like growth factor (IGF) receptor has low affinity for IGF-I analogs: pleiotypic actions of IGFs on myoblasts are apparently mediated by the type I receptor. *Endocrinology*, 120, 115–23.

Fang, B., Eisensmith, R. C., Wang, H., Kay, M. A., Cross, R. E., Landen, C. N., Gordon, G., Bellinger, D. A., Read, M. S., Hu, P. C. *et al.* (1995) Gene therapy for hemophilia B: host immunosuppression prolongs the therapeutic effect of adenovirus-mediated factor IX expression. *Hum Gene Ther*, 6, 1039–44.

FDA (2005) *Cellular & Gene Therapy*. US Food and Drug Administration.

Felgner, P. L., Tsai, Y. J., Sukhu, L., Wheeler, C. J., Manthorpe, M., Marshall, J. & Cheng, S. H. (1995) Improved cationic lipid formulations for *in vivo* gene therapy. *Ann NY Acad Sci*, 772, 126–39.

Florini, J. R., Ewton, D. Z. & Roof, S. L. (1991) Insulin-like growth factor-I stimulates terminal myogenic differentiation by induction of myogenin gene expression. *Mol Endocrinol*, 5, 718–24.

Folkman, J. (1971) Tumor angiogenesis: therapeutic implications. *N Engl J Med*, 285, 1182.

Food, M. R., Rothenberger, S., Gabathuler, R., Haidl, I. D., Reid, G. & Jefferies, W. A. (1994) Transport and expression in human melanomas of a transferrin-like glycosylphosphatidylinositol-anchored protein. *J Biol Chem*, 269, 3034–40.

Foss, F. M. (2001) Interleukin-2 fusion toxin: targeted therapy for cutaneous T cell lymphoma. *Ann NY Acad Sci*, 941, 166–76.

Frangione-Beebe, M., Albrecht, B., Dakappagari, N., Rose, R. T., Brooks, C. L., Schwendeman, S. P., Lairmore, M. D. & Kaumaya, P. T. (2000) Enhanced immunogenicity of a conformational epitope of human T-lymphotropic virus type 1 using a novel chimeric peptide. *Vaccine*, 19, 1068.

Frangione-Beebe, M., Rose, R. T., Kaumaya, P. T. & Schwendeman, S. P. (2001) Microencapsulation of a synthetic peptide epitope for HTLV-1 in biodegradable poly(D,L-lactide-co-glycolide) microspheres using a novel encapsulation technique. *J Microencapsul.*, 18, 663.

Freeman, B. J., Roberts, M. S., Vogler, C. A., Nicholes, A., Hofling, A. A. & Sands, M. S. (1999) Behavior and therapeutic efficacy of beta-glucuronidase-positive mononuclear phagocytes in a murine model of mucopolysaccharidosis type VII. *Blood*, 94(6), 2142–50.

Friso, A., Tomanin, R., Alba, S., Gasparotto, N., Puicher, E. P., Fusco, M., Hortelano, G., Muenzer, J., Marin, O., Zacchelo, F. & Scarpa, M. (2005) Reduction of GAG storage in MPS II mouse model following implantation of encapsulated recombinant myoblasts. *J Gene Med*, 7, 1482–91.

Fujiki, T., Futatsuki, K., Akazawa, S., Yamamoto, K., Kanda, Y., Yamato, A., Terashi, K. & Suda, Y. (1997) [Ifosfamide chemotherapy ineffective for advanced pancreatic carcinoma]. *Gan To Kagaku Ryoho*, 24, 569.

Fyfe, J. C., Kurzhals, R. L., Lassaline, M. E., Henthorn, P. S., Alur, P. R., Wang, P., Wolfe, J. H., Giger, U., Haskins, M. E., Patterson, D. F., Sun, H., Jain, S. & Yuhki, N. (1999) Molecular basis of feline beta-glucuronidase deficiency: an animal model of mucopolysaccharidosis VII. *Genomics*, 58(2), 121–8.

Gao, C., Sands, M. S., Haskins, M. E. & Ponder, K. P. (2000a) Delivery of a retroviral vector expressing human beta-glucuronidase to the liver and spleen decreases lysosomal storage in mucopolysaccharidosis VII mice. *Mol Ther*, 2(3), 233–44.

Garcia-Martin, C., Chuah, M. K., Van Damme, A., Robinson, K. E., Vanzieleghem, B., Saint-Remy, J. M., Gallardo, D., Ofosu, F. A., van den Driessche, T. & Hortelano, G. (2002) Therapeutic levels of human factor VIII in mice implanted with encapsulated cells: potential for gene therapy of haemophilia A. *J Gene Med*, 4, 215–23.

Gaserod, O., Smidsrod, O. & Skjak-Braek, G. (1998) Microcapsules of alginate-chitosan – I. A quantitative study of the interaction between alginate and chitosan. *Biomaterials*, 19, 1815–25.

Gaserod, O., Sannes, A. & Skjak-Braek, G. (1999) Microcapsules of alginate-chitosan. II. A study of capsule stability and permeability. *Biomaterials*, 20, 773–83.

Geller, R. L., Loudovaris, T., Neuenfeldt, S., Johnson, R. C. & Brauker, J. H. (1997) Use of an immunoisolation device for cell transplantation and tumor immunotherapy. *Ann NY Acad Sci*, 831, 438.

Geuze, H. J., Slot, J. W., Strous, G. J., Hasilik, A. & von Figura, K. (1985) Possible pathways for lysosomal enzyme delivery. *J Cell Biol*, 101, 2253–62.

Ghodsi, A., Stein, C., Derksen, T., Yang, G., Anderson, R. D. & Davidson, B. L. (1998) Extensive beta-glucuronidase activity in murine central nervous system after adenovirus-mediated gene transfer to brain. *Hum Gene Ther*, 9(16), 2331–40.

Ghodsi, A., Stein, C., Derksen, T., Martins, I., Anderson, R. D. & Davidson, B.L. (1999)

Systemic hyperosmolality improves beta-glucuronidase distribution and pathology in murine MPS VII brain following intraventricular gene transfer. *Exp Neurol*, 160(1), 109–16.

Gillies, S. D., Reilly, E. B., Lo, K. M. & Reisfeld, R. A. (1992) Antibody-targeted interleukin 2 stimulates T-cell killing of autologous tumor cells. *Proc Natl Acad Sci USA*, 89, 1428.

Gitzelmann, R., Bosshard, N. U., Superti-Furga, A., Spycher, M. A., Briner, J., Wiesmann, U., Lutz, H. & Litschi, B. (1994) Feline mucopolysaccharidosis VII due to beta-glucuronidase deficiency. *Vet Pathol*, 31(4), 435–43.

Grabowski, G. A., Barton, N. W., Pastores, G., Dambrosia, J. M., Banerjee, T. K., McKee, M. A., Parker, C., Schiffmann, R., Hill, S. C. & Brady, R. O. (1995) Enzyme therapy in type 1 Gaucher disease: comparative efficacy of mannose-terminated glucocerebrosidase from natural and recombinant sources. *Ann Intern Med*, 122, 33–9.

Greco, O., Scott, S. D., Marples, B. & Dachs, G. U. (2002) Cancer gene therapy: 'delivery, delivery, delivery'. *Front Biosci.*, 7, d1516.

Green, N. K. & Seymour, L. W. (2002) Adenoviral vectors: systemic delivery and tumor targeting. *Cancer Gene Ther*, 9, 1036.

Haas, C., Herold-Mende, C., Gerhards, R. & Schirrmacher, V. (1999) An effective strategy of human tumor vaccine modification by coupling bispecific costimulatory molecules. *Cancer Gene Ther*, 6, 254.

Hacein-Bey-Abina, S., von Kalle, C., Schmidt, M., McCormack, M. P., Wulffraat, N., Leboulch, P., Lim, A., Osborne, C. S., Pawliuk, R., Morillon, E., Sorensen, R., Forster, A., Fraser, P., Cohen, J. I., de Saint Basile, G., Alexander, I., Wintergerst, U., Frebourg, T., Aurias, A., Stoppa-Lyonnet, D., Romana, S., Radford-Weiss, I., Gross, F., Valensi, F., Delabesse, E., MacIntyre, E., Sigaux, F., Soulier, J., Leiva, L. E., Wissler, M., Prinz, C., Rabbitts, T. H., le Deist, F., Fischer, A. & Cavazzana-Calvo, M. (2003) LMO2-associated clonal T cell proliferation in two patients after gene therapy for SCID-X1. *Science*, 302, 415–19.

Haskins, M. E., Wortman, J. A., Wilson, S., Wolfe, J. H. (1984) Bone marrow transplantation in the cat. *Transplantation*, 37(6), 634–6.

Haskins, M. E., Aguirre, G. D., Jezyk, P. F., Schuchman, E. H., Desnick, R. J., & Patterson, D. F. (1991) Mucopolysaccharidosis type VII (Sly syndrome). Beta-glucuronidase-deficient mucopolysaccharidosis in the dog. *Am J Pathol*, 138(6), 1553–5.

Herzog, R. W., Hagstrom, J. N., Kung, S. H., Tai, S. J., Wilson, J. M., Fisher, K. J. & High, K. A. (1997) Stable gene transfer and expression of human blood coagulation factor IX after intramuscular injection of recombinant adeno-associated virus. *Proc Natl Acad Sci USA*, 94, 5804–9.

Herzog, R. W., Yang, E. Y., Couto, L. B., Hagstrom, J. N., Elwell, D., Fields, P. A., Burton, M., Bellinger, D. A., Read, M. S., Brinkous, K. M., Podsakoff, G. M., Nichols, T. C., Kurtzman, G. J. & High, K. A. (1999) Long-term correction of canine hemophilia B by gene transfer of blood coagulation factor IX mediated by adeno-associated viral vector. *Nat Med*, 5, 56–63.

Hobson, B. & Denekamp, J. (1984) Endothelial proliferation in tumours and normal tissues: continuous labelling studies. *Br J Cancer*, 49, 405.

Hoffman, A. S. (2002) Hydrogels for biomedical applications. *Adv Drug Deliv Rev*, 54, 3–12.

Hoffman, D., Breakefield, X. O., Short, M. P. & Aebischer, P. (1993) Transplantation of a polymer-encapsulated cell line genetically engineered to release NGF. *Exp Neurol*,

122, 100–106.

Hoflack, B. & Kornfeld, S. (1985) Purification and characterization of a cation-dependent mannose 6-phosphate receptor from murine P388D1 macrophages and bovine liver. *J Biol Chem*, 260, 12008–14.

Hortelano, G., Al-Hendy, A., Ofosu, F. A. & Chang, P. L. (1996) Delivery of human factor IX in mice by encapsulated recombinant myoblasts: a novel approach towards allogeneic gene therapy of hemophilia B. *Blood*, 87, 5095–103.

Hughes, M., Vassilakos, A., Andrews, D. W., Hortelano, G., Belmont, J. W. & Chang, P. L. (1994) Delivery of a secretable adenosine deaminase through microcapsules – a novel approach to somatic gene therapy. *Hum Gene Ther*, 5, 1445–55.

Jefferies, W. A., Brandon, M. R., Hunt, S. V., Williams, A. F., Gatter, K. C. & Mason, D. Y. (1984) Transferrin receptor on endothelium of brain capillaries. *Nature*, 312, 162–3.

Jiang, W., Gupta, R. K., Deshpande, M. C. & Schwendeman, S. P. (2005) Biodegradable poly(lactic-co-glycolic acid) microparticles for injectable delivery of vaccine antigens. *Adv Drug Deliv Rev*, 57, 391.

Joki, T., Machluf, M., Atala, A., Zhu, J., Seyfried, N. T., Dunn, I. F., Abe, T., Carroll, R. S. & Black, P. M. (2001) Continuous release of endostatin from microencapsulated engineered cells for tumor therapy. *Nat Biotechnol*, 19, 35–9.

Journal of Gene Medicine (2005) 7(7), 839–987.

Kakkis, E. D., Muenzer, J., Tiller, G. E., Waber, L., Belmont, J., Passage, M., Izykowski, B., Phillips, J., Doroshow, R., Walot, I., Hoft, R. & Neufeld, E. F. (2001) Enzyme-replacement therapy in mucopolysaccharidosis I. *N Engl J Med*, 344, 182–8.

Kay, M. A., Landen, C. N., Rotherberg, S. R., Taylor, L. A., Leland, F., Wiehle, S., Fang, B., Bellinger, D., Finegold, M., Thompson, A. R. *et al.* (1994) *In vivo* hepatic gene therapy: complete albeit transient correction of factor IX deficiency in hemophilia B dogs. *Proc Natl Acad Sci USA*, 91, 2353–7.

Kay, M. A., Manno, C. S., Ragni, M. V., Larson, P. J., Couto, L. B., McClelland, A., Glader, B., Chew, A. J., Tai, S. J., Herzog, R. W., Arruda, V., Johnson, F., Scallan, C., Skarsgard, E., Flake, A. W. & High, K. A. (2000) Evidence for gene transfer and expression of factor IX in haemophilia B patients treated with an AAV vector. *Nat Genet*, 24, 257–61.

Kay, M. A., Rothenberg, S., Landen, C. N., Bellinger, D. A., Leland, F., Toman, C., Finegold, M., Thompson, A. R., Read, M. S., Brinkhous, K. M. *et al.* (1993) *In vivo* gene therapy of hemophilia B: sustained partial correction in factor IX-deficient dogs. *Science*, 262, 117–19.

Kay, M. A., High, K. A., Glader, B., Manno, C. S. & Hutchinson, S. (2002) Abstract – A phase I/II clinical trial for liver directed AAV-mediated gene transfer for severe hemophilia B. *Blood*, 100, 115a.

Kroger, J. C., Bergmeister, H., Hoffmeyer, A., Ceijna, M., Karle, P., Saller, R., Schwendenwein, I., von Rombs, K., Liebe, S., Gunzburg, W. H., Salmons, B., Hauenstein, K., Losert, U. & Lohr, M. (1999) Intraarterial instillation of microencapsulated cells in the pancreatic arteries in pig. *Ann NY Acad Sci*, 880, 374.

Kroger, J. C., Benz, S., Hoffmeyer, A., Bago, Z., Bergmeister, H., Gunzburg, W. H., Karle, P., Kloppel, G., Losert, U., Muller, P., Nizze, H., Obermaier, R., Probst, A., Renner, M., Saller, R., Salmons, B., Schwendenwein, I., von Rombs, K., Wiessner, R., Wagner, T., Hauenstein, K. & Lohr, M. (2003) Intra-arterial instillation of microencapsulated, Ifosfamide-activating cells in the pig pancreas for chemotherapeutic targeting. *Pancreatology*, 3, 55–63.

Kuhl, U., Ocalan, M., Timpl, R. & Vondermark, K. (1986) Role of laminin and fibronectin in selecting myogenic versus fibrogenic cells from skeletal-muscle cells *in vitro*. *Develop Biol*, 117, 628–35.

Kundu, R. K., Sangiorgi, F., Wu, L. Y., Kurachi, K., Anderson, W. F., Maxson, R. & Gordon, E. M. (1998) Targeted inactivation of the coagulation factor IX gene causes hemophilia B in mice. *Blood*, 92, 168–74.

Kung, S. H., Hagstrom, J. N., Cass, D., Tai, S. J., Lin, H. F., Stafford, D. W. & High, K. A. (1998) Human factor IX corrects the bleeding diathesis of mice with hemophilia B. *Blood*, 91, 784–90.

Kurachi, K. & Davie, E. W. (1982) Isolation and characterization of a cDNA coding for human factor IX. *Proc Natl Acad Sci USA*, 79, 6461–4.

Kyle, J. W., Birkenmeier, E. H., Gwynn, B., Vogler, C., Hoppe, P. C., Hoffmann, J. W., & Sly, W. S. (1990) Correction of murine mucopolysaccharidosis VII by a human beta-glucuronidase transgene. *Proc Natl Acad Sci USA*, 87(10), 3914–18.

Lathia, J. D., Leodore, L. & Wheatley, M. A. (2004) Polymeric contrast agent with targeting potential. *Ultrasonics*, 42, 763.

Lau, C., Soriano, H. E., Ledley, F. D., Finegold, M. J., Wolfe, J. H., Birkenmeier, E. H. & Henning, S. J. (1995) Retroviral gene transfer into the intestinal epithelium. *Hum Gene Ther*, 6(9), 1145–51.

Lederberg, J. (1963) Biological future of man. In Wolstenholme, G. (Ed.) *Man and His Future*. London, J. & A. Churchill Ltd.

Lederberg, J. (1966) Experimental genetics and human evolution. *The American Naturalist*, 100, 519–31.

Lee, H. J., Engelhardt, B., Lesley, J., Bickel, U. & Pardridge, W. M. (2000) Targeting rat anti-mouse transferrin receptor monoclonal antibodies through blood–brain barrier in mouse. *J Pharmacol Exp Ther*, 292, 1048–52.

Lemansky, P., Hasilik, A., von Figura, K., Helmy, S., Fishman, J., Fine, R. E., Kedersha, N. L. & Rome, L. H. (1987) Lysosomal enzyme precursors in coated vesicles derived from the exocytic and endocytic pathways. *J Cell Biol*, 104, 1743–8.

Levy, B., Galvin, N., Vogler, C., Birkenmeier, E. H. & Sly, W. S. (1996) Neuropathology of murine mucopolysaccharidosis type VII. *Acta Neuropathol (Berl)*, 92(6), 562–8.

Li, A. A., Macdonald, N. C. & Chang, P. L. (2003) Effect of growth factors and extracellular matrix materials on the proliferation and differentiation of microencapsulated myoblasts. *J Biomater Sci Polym Ed*, 14, 533–49.

Li, J. Y., Sugimura, K., Boado, R. J., Lee, H. J., Zhang, C., Duebel, S. & Pardridge, W. M. (1999) Genetically engineered brain drug delivery vectors: cloning, expression and *in vivo* application of an anti-transferrin receptor single chain antibody-streptavidin fusion gene and protein. *Protein Eng*, 12, 787–96.

Lim, F. & Moss, R. D. (1981) Microencapsulation of living cells and tissues. *J Pharm Sci*, 70, 351–4.

Lin, H. F., Maeda, N., Smithies, O., Straight, D. L. & Stafford, D. W. (1997) A coagulation factor IX-deficient mouse model for human hemophilia B. *Blood*, 90, 3962–6.

Liu, H. W., Ofosu, F. A. & Chang, P. L. (1993) Expression of human factor IX by microencapsulated recombinant fibroblasts. *Hum Gene Ther*, 4, 291–301.

Lo, H. W., Day, C. P. & Hung, M. C. (2005) Cancer-specific gene therapy. *Adv Genet*, 54, 235–55.

Lobel, P., Dahms, N. M. & Kornfeld, S. (1988) Cloning and sequence analysis of the cation-independent mannose 6-phosphate receptor. *J Biol Chem*, 263, 2563–70.

Lode, H. N. & Reisfeld, R. A. (2000) Targeted cytokines for cancer immunotherapy. *Immunol Res*, 21, 279.

Lohr, M., Muller, P., Karle, P., Stange, J., Mitzner, S., Jesnowski, R., Nizze, H., Nebe, B., Liebe, S., Salmons, B. & Gunzburg, W. H. (1998) Targeted chemotherapy by intratumour injection of encapsulated cells engineered to produce CYP2B1, an ifosfamide activating cytochrome P450. *Gene Ther*, 5, 1070.

Lohr, M., Bago, Z. T., Bergmeister, H., Ceijna, M., Freund, M., Gelbmann, W., Gunzburg, W. H., Jesnowski, R., Hain, J., Hauenstein, K., Henninger, W., Hoffmeyer, A., Karle, P., Kroger, J. C., Kundt, G., Liebe, S., Losert, U., Muller, P., Probst, A., Puschel, K., Renner, M., Renz, R., Saller, R., Salmons, B., Walter, I. *et al.* (1999b) Cell therapy using microencapsulated 293 cells transfected with a gene construct expressing CYP2B1, an ifosfamide converting enzyme, instilled intra-arterially in patients with advanced-stage pancreatic carcinoma: a phase I/II study. *J Mol Med*, 77, 393–8.

Lohr, M., Hoffmeyer, A., Kroger, J., Freund, M., Hain, J., Holle, A., Karle, P., Knofel, W. T., Liebe, S., Muller, P., Nizze, H., Renner, M., Saller, R. M., Wagner, T., Hauenstein, K., Gunzburg, W. H. & Salmons, B. (2001) Microencapsulated cell-mediated treatment of inoperable pancreatic carcinoma. *Lancet*, 357, 1591–2.

Lohr, M., Hummel, F., Faulmann, G., Ringel, J., Saller, R., Hain, J., Gunzburg, W. H. & Salmons, B. (2002) Microencapsulated, CYP2B1-transfected cells activating ifosfamide at the site of the tumor: the magic bullets of the 21st century. *Cancer Chemother Pharmacol*, 49 Suppl 1, S21–4.

Lu, D. R., Zhou, J. M., Zheng, B., Qiu, X. F., Xue, J. L., Wang, J. M., Meng, P. L., Han, F. L., Ming, B. H., Wang, X. P. *et al.* (1993) Stage I clinical trial of gene therapy for hemophilia B. *Sci China B*, 36, 1342–51.

Ma, J., Samuel, J., Kwon, G. S., Noujiam, A. A. & Madiyalakan, R. (1998) Induction of anti-idiotypic humoral and cellular immune responses by a murine monoclonal antibody recognizing the ovarian carcinoma antigen CA125 encapsulated in biodegradable microspheres. *Cancer Immunol Immunother*, 47, 13.

Ma, J., Li, Z. & Luo, D. (2003) Single chain antibody vaccination in mice against human ovarian cancer enhanced by microspheres and cytokines. *J Drug Target*, 11, 169.

Mandal, T. K. (2005) Inhaled insulin for diabetes mellitus. *Am J Health Syst Pharm*, 62, 1359–64.

Manno, C. S., Chew, A. J., Hutchison, S., Larson, P. J., Herzog, R. W., Arruda, V. R., Tai, S. J., Ragni, M. V., Thompson, A., Ozelo, M., Couto, L. B., Leonard, D. G., Johnson, F. A., McClelland, A., Scallan, C., Skarsgard, E., Flake, A. W., Kay, M. A., High, K. A. & Glader, B. (2003) AAV-mediated factor IX gene transfer to skeletal muscle in patients with severe hemophilia B. *Blood*, 101, 2963–72.

Mapes, C. A., Anderson, R. L., Sweeley, C. C., Desnick, R. J. & Krivit, W. (1970) Enzyme replacement in Fabry's disease, an inborn error of metabolism. *Science*, 169, 987–9.

Marechal, V., Naffakh, N., Danos, O. & Heard, J. M. (1993) Disappearance of lysosomal storage in spleen and liver of mucopolysaccharidosis VII mice after transplantation of genetically modified bone marrow cells. *Blood*, 82(4), 1358–65.

McCarthy, M. (2003) Antiangiogenesis drug promising for metastatic colorectal cancer. *Lancet*, 361, 1959.

Mier, J. W., Atkins, M. B., DeVita, V. T., Jr., Hellman, S. & Rosenberg, S. A. (2001) Interleukin-2. In: DeVita, V. T., Hellman, S. & Rosenberg, S. A. (Eds) *Cancer: Principles and Practice of Oncology*, 6th edition. Philadelphia, PA, Lippincott Williams & Wilkins, pp. 471–7.

Monahan, P. E., Samulski, R. J., Tazelaar, J., Xiao, X., Nichols, T. C., Bellinger, D. A., Read, M. S. & Walsh, C. E. (1998) Direct intramuscular injection with recombinant AAV vectors results in sustained expression in a dog model of hemophilia. *Gene Ther*, 5, 40–49.

Morneau, A., Pillai, V., Nigam, S., Winnik, F. M. & Ziolo, R. F. (1999) Analysis of ferrofluids by capillary electrophoresis. *Colloids Surfaces A – Physicochem Eng Aspects*, 154, 295–301.

Moser, T. L., Stack, M. S., Asplin, I., Enghild, J. J., Hojrup, P., Everitt, L., Hubchak, S., Schnaper, H. W. & Pizzo, S. V. (1999) Angiostatin binds ATP synthase on the surface of human endothelial cells. *Proc Natl Acad Sci USA*, 96, 2811.

Moullier, P., Bohl, D., Heard, J. M. & Danos, O. (1993a) Correction of lysosomal storage in the liver and spleen of MPS VII mice by implantation of genetically modified skin fibroblasts. *Nat Genet*, 4(2), 154–9.

Moullier, P., Marechal, V., Danos, O. & Heard, J. M. (1993b) Continuous systemic secretion of a lysosomal enzyme by genetically modified mouse skin fibroblasts. *Transplantation*, 56(2), 427–32.

Muller, P., Jesnowski, R., Karle, P., Renz, R., Saller, R., Stein, H., Puschel, K., von Rombs, K., Nizze, H., Liebe, S., Wagner, T., Gunzburg, W. H., Salmons, B. & Lohr, M. (1999) Injection of encapsulated cells producing an ifosfamide-activating cytochrome P450 for targeted chemotherapy to pancreatic tumors. *Ann NY Acad Sci*, 880, 337.

Mullerad, J., Cohen, S., Benharroch, D. & Apte, R. N. (2003) Local delivery of IL-1 alpha polymeric microspheres for the immunotherapy of an experimental fibrosarcoma. *Cancer Invest*, 21, 720.

Naffakh, N., Pinset, C., Montarras, D., Pastoret, C., Danos, O. & Heard, J. M. (1993) Transplantation of adult-derived myoblasts in mice following gene transfer. *Neuromuscul Disord*, 3(5–6), 413–17.

Naffakh, N., Bohl, D., Salvetti, A., Moullier, P., Danos, O. & Heard, J. M. (1994) Gene therapy for lysosomal disorders. *Nouv Rev Fr Hematol*, 36 Suppl 1, S11–16.

Naffakh, N., Henri, A., Villeval, J. L., Rouyer-Fessard, P., Moullier, P., Blumenfeld, N., Danos, O., Vainchenker, W., Heard, J. M. & Beuzard, Y. (1995) Sustained delivery of erythropoietin in mice by genetically modified skin fibroblasts. *Proc Natl Acad Sci USA*, 92(8), 3194–8.

Naffakh, N., Pinset, C., Montarras, D., Li, Z., Paulin, D., Danos, O. & Heard, J. M. (1996) Long-term secretion of therapeutic proteins from genetically modified skeletal muscles. *Hum Gene Ther*, 7(1), 11–21.

Neufeld, E.F. & Muenzer, J. (1995) The mucopolysaccharidoses. In: *The metabolic basis of inherited disease* (Sriver, C.R., Beaudet, A.L., Sly, W.S., Valle, D., Eds), pp. 2465–2494. New York: McGraw-Hill.

Niethammar, A. G., Xiang, R., Ruehlmann, J. M., Lode, H. N., Dolman, C. S., Gillies, S. D. & Reisfeld, R. A. (2001) Targeted interleukin 2 therapy enhances protective immunity induced by an autologous oral DNA vaccine against murine melanoma. *Cancer Res*, 61, 6178.

Nishihara, T., Sawada, T., Yamamoto, A., Yamashita, Y., Ho, J. J., Kim, Y. S. & Chung, K. H. (2000) Antibody-dependent cytotoxicity mediated by chimeric monoclonal antibody Nd2 and experimental immunotherapy for pancreatic cancer. *Jpn J Cancer Res*, 91, 817.

Ohashi, T., Watabe, K., Uehara, K., Sly, W. S., Vogler, C. & Eto, Y. (1997) Adenovirus-mediated gene transfer and expression of human beta-glucuronidase gene in the liver, spleen, and central nervous system in mucopolysaccharidosis type VII mice.

Proc Natl Acad Sci USA, 94(4), 1287–92.

O'Reilly, M. S., Boehm, T., Shing, Y., Fukai, N., Vasios, G., Lane, W. S., Flynn, E., Birkhead, J. R., Olsen, B. R. & Folkman, J. (1997) Endostatin: an endogenous inhibitor of angiogenesis and tumor growth. *Cell*, 88, 277.

Orive, G., Hernandez, R. M., Gascon, A. R., Igartua, M., Rojas, A. & Pedraz, J. L. (2001) Microencapsulation of an anti-VE-cadherin antibody secreting 1B5 hybridoma cells. *Biotechnol Bioeng*, 76, 285–94.

Orive, G., Hernandez, R. M., Gascon, A. R., Igartua, M. & Pedraz, J. L. (2003) Survival of different cell lines in alginate-agarose microcapsules. *Eur J Pharm Sci*, 18, 23–30.

Orive, G., Hernandez, R. M., Rodriguez Gascon, A., Calafiore, R., Chang, T. M., de Vos, P., Hortelano, G., Hunkeler, D., Lacik, I. & Pedraz, J. L. (2004) History, challenges and perspectives of cell microencapsulation. *Trends Biotechnol*, 22, 87–92.

Palmer, T. D., Thompson, A. R. & Miller, A. D. (1989) Production of human factor IX in animals by genetically modified skin fibroblasts: potential therapy for hemophilia B. *Blood*, 73, 438–45.

Pan, W., Kastin, A. J., Zankel, T. C., Van Kerkhof, P., Terasaki, T. & Bu, G. (2004) Efficient transfer of receptor-associated protein (RAP) across the blood–brain barrier. *J Cell Sci*, 117, 5071–8.

Pardridge, W. M., Kang, Y. S., Buciak, J. L. & Yang, J. (1995) Human insulin receptor monoclonal antibody undergoes high affinity binding to human brain capillaries *in vitro* and rapid transcytosis through the blood–brain barrier *in vivo* in the primate. *Pharm Res*, 12, 807–16.

Peirone, M., Ross, C. J., Hortelano, G., Brash, J. L. & Chang, P. L. (1998a) Encapsulation of various recombinant mammalian cell types in different alginate microcapsules. *J Biomed Mater Res*, 42, 587–96.

Peirone, M. A., Delaney, K., Kwiecin, J., Fletch, A. & Chang, P. L. (1998b) Delivery of recombinant gene product to canines with nonautologous microencapsulated cells. *Hum Gene Ther*, 9, 195–206.

Pelegrin, M., Marin, M., Noel, D., Del Rio, M., Saller, R., Stange, J., Mitzner, S., Gunzburg, W. H. & Piechaczyk, M. (1998) Systemic long-term delivery of antibodies in immunocompetent animals using cellulose sulphate capsules containing antibody-producing cells. *Gene Ther*, 5, 828.

Peng, Z. (2005) Current status of gendicine in China: recombinant human Ad-p53 agent for treatment of cancers. *Hum Gene Ther*, 16, 1016–27.

Ponce, S., Orive, G., Gascon, A. R., Hernandez, R. M. & Pedraz, J. L. (2005) Microcapsules prepared with different biomaterials to immobilize GDNF secreting 3T3 fibroblasts. *Int J Pharm*, 293, 1–10.

Pool, J. G., Gershgold, E. J. & Pappenhagen, A. R. (1964) High-potency antihaemophilic factor concentrate prepared from cryoglobulin precipitate. *Nature*, 203, 312.

Potter, K., Carpenter, T. A. & Hall, L. D. (1993) Mapping of the spatial variation in alginate concentration in calcium alginate gels by magnetic resonance imaging (MRI). *Carbohydr Res*, 246, 43–49.

Potter, K., Balcom, B. J., Carpenter, T. A. & Hall, L. D. (1994) The gelation of sodium alginate with calcium ions studies by magnetic resonance imaging (MRI). *Carbohydr Res*, 257, 117–26.

Powell, J. S., Ragni, M. V., White, G. C. II, Lusher, J. M., Hillman-Wiseman, C., Moon, T. E., Cole, V., Ramanathan-Girish, S., Roehl, H., Sajjadi, N., Jolly, D. J. & Hurst, D. (2003) Phase 1 trial of FVIII gene transfer for severe hemophilia A using a retroviral construct administered by peripheral intravenous infusion. *Blood*, 102, 2038–45.

Prince, W. S., McCormick, L. M., Wendt, D. J., Fitzpatrick, P. A., Schwartz, K. L., Aguilera, A. I., Koppaka, V., Christianson, T. M., Vellard, M. C., Pavloff, N., Lemontt, J. F., Qin, M., Starr, C. M., Bu, G. & Zankel, T. C. (2004) Lipoprotein receptor binding, cellular uptake, and lysosomal delivery of fusions between the receptor-associated protein (RAP) and alpha-L-iduronidase or acid alpha-glucosidase. *J Biol Chem*, 279, 35037–46.

Qiu, X., Lu, D., Zhou, J., Wang, J., Yang, J., Meng, P. & Hsueh, J. L. (1996) Implantation of autologous skin fibroblast genetically modified to secrete clotting factor IX partially corrects the hemorrhagic tendencies in two hemophilia B patients. *Chin Med J (Engl)*, 109, 832–9.

Quek, C. H., Li, J., Sun, T., Chan, M. L., Mao, H. Q., Gan, L. M., Leong, K. W. & Yu, H. (2004) Photo-crosslinkable microcapsules formed by polyelectrolyte copolymer and modified collagen for rat hepatocyte encapsulation. *Biomaterials*, 25, 3531–40.

Raderer, M., Jager, G., Brugger, S., Puspok, A., Fiebiger, W., Drach, J., Wotherspoon, A. & Chott, A. (2003) Rituximab for treatment of advanced extranodal marginal zone B cell lymphoma of the mucosa-associated lymphoid tissue lymphoma. *Oncology*, 65, 306–10.

Raper, S. E., Chirmule, N., Lee, F. S., Wivel, N. A., Bagg, A., Gao, G. P., Wilson, J. M. & Batshaw, M. L. (2003) Fatal systemic inflammatory response syndrome in a ornithine transcarbamylase deficient patient following adenoviral gene transfer. *Mol Genet Metab*, 80, 148.

Ray, J., Bouvet, A., DeSanto, C., Fyfe, J. C., Xu, D., Wolfe, J. H., Aguirre, G. D., Patterson, D. F., Haskins, M. E. & Henthorn, P. S. (1998) Cloning of the canine beta-glucuronidase cDNA, mutation identification in canine MPS VII, and retroviral vector-mediated correction of MPS VII cells. *Genomics*, 48(2), 248–53.

Ray, J., Scarpino, V., Laing, C. & Haskins, M. E. (1999) Biochemical basis of the beta-glucuronidase gene defect causing canine mucopolysaccharidosis VII. *J Hered*, 90(1), 119–23.

Read, T. A., Stensvaag, V., Vindenes, H., Ulvestad, E., Bjerkvig, R. & Thorsen, F. (1999) Cells encapsulated in alginate: a potential system for delivery of recombinant proteins to malignant brain tumours. *Int J Dev Neurosci*, 17, 653–63.

Read, T. A., Sorensen, D. R., Mahesparan, R., Enger, P. O., Timpl, R., Olsen, B. R., Hjelstuen, M. H., Haraldseth, O. & Bjerkvig, R. (2001) Local endostatin treatment of gliomas administered by microencapsulated producer cells. *Nat Biotechnol*, 19, 29.

Rhines, L. D., Sampath, P., Dimeco, F., Lawson, H. C., Tyler, B. M., Hanes, J., Olivi, A. & Brem, H. (2003) Local immunotherapy with interleukin-2 delivered from biodegradable polymer microspheres combined with interstitial chemotherapy: a novel treatment for experimental malignant glioma. *Neurosurgery*, 52, 872.

Rinsch, C., Dupraz, P., Schneider, B. L., Deglon, N., Maxwell, P. H., Ratcliffe, P. J. & Aebischer, P. (2002) Delivery of erythropoietin by encapsulated myoblasts in a genetic model of severe anemia. *Kidney Int*, 62, 1395–401.

Rosas, J. E., Pedraz, J. L., Hernandez, R. M., Gascon, A. R., Igartua, M., Guzman, F., Rodriguez, R., Cortes, J. & Patarroyo, M. E. (2002) Remarkably high antibody levels and protection against *P. falciparum* malaria in Aotus monkeys after a single immunisation of SPf66 encapsulated in PLGA microspheres. *Vaccine*, 20, 1707.

Rose, T. M., Plowman, G. D., Teplow, D. B., Dreyer, W. J., Hellstrom, K. E. & Brown, J. P. (1986) Primary structure of the human melanoma-associated antigen p97 (melanotransferrin) deduced from the mRNA sequence. *Proc Nat Acad Sci USA*, 83, 1261–5.

Ross, C. J., Bastedo, L., Maier, S. A., Sands, M. S. & Chang, P. L. (2000a) Treatment of a lysosomal storage disease, mucopolysaccharidosis VII, with microencapsulated recombinant cells. *Hum Gene Ther*, 11, 2117–27.

Ross, C. J., Ralph, M. & Chang, P. L. (2000b) Somatic gene therapy for a neurodegenerative disease using microencapsulated recombinant cells. *Exp Neurol*, 166, 276–86.

Roth, D. A., Tawa, N. E., Jr., O'Brien, J. M., Treco, D. A. & Selden, R. F. (2001) Nonviral transfer of the gene encoding coagulation factor VIII in patients with severe hemophilia A. *N Engl J Med*, 344, 1735–42.

Rowley, J. A. & Mooney, D. J. (2002) Alginate type and RGD density control myoblast phenotype. *J Biomed Mater Res*, 60, 217–23.

Ryschich, E., Jesnowski, R., Ringel, J., Harms, W., Fabian, O. V., Saller, R., Schrewe, M., Engel, A., Schmidt, J. & Lohr, M. (2005) Combined therapy of experimental pancreatic cancer with CYP2B1 producing cells: low-dose ifosfamide and local tumor irradiation. *Int J Cancer*, 113, 649.

Sabel, M. S. & Sondak, V. K. (2003) Pros and cons of adjuvant interferon in the treatment of melanoma. *Oncologist*, 8, 451–8.

Sabel, M. S., Skitzki, J., Stoolman, L., Egilmez, N. K., Mathiowitz, E., Bailey, N., Chang, W. J. & Chang, A. E. (2004) Intratumoral IL-12 and TNF-alpha-loaded microspheres lead to regression of breast cancer and systemic antitumor immunity. *Ann Surg Oncol*, 11, 147.

Sahagian, G. G., Distler, J. & Jourdian, G. W. (1981) Characterization of a membrane-associated receptor from bovine liver that binds phosphomannosyl residues of bovine testicular beta-galactosidase. *Proc Natl Acad Sci USA*, 78, 4289–93.

Sands, M. S., & Birkenmeier, E. H. (1993) A single-base-pair deletion in the beta-glucuronidase gene accounts for the phenotype of murine mucopolysaccharidosis type VII. *Proc Natl Acad Sci USA*, 90(14), 6567–71.

Sands, M. S., Erway, L. C., Vogler, C., Sly, W. S., & Birkenmeier, E. H. (1995) Syngeneic bone marrow transplantation reduces the hearing loss associated with murine mucopolysaccharidosis type VII. *Blood*, 86(5), 2033–40.

Sands, M. S., Vogler, C. A., Ohlemiller, K. K., Roberts, M. S., Grubb, J. H., Levy, B. & Sly, W. S. (2001) Biodistribution, kinetics, and efficacy of highly phosphorylated and non-phosphorylated beta-glucuronidase in the murine model of mucopolysaccharidosis VII. *J Biol Chem*, 276, 43160–5.

Sarkar, R., Gao, G. P., Chirmule, N., Tazelaar, J. & Kazazian, H. H., Jr. (2000) Partial correction of murine hemophilia A with neo-antigenic murine factor VIII. *Hum Gene Ther*, 11, 881–94.

Schiffmann, R., Heyes, M. P., Aerts, J. M., Dambrosia, J. M., Patterson, M. C., Degraba, T., Parker, C. C., Zirzow, G. C., Oliver, K., Tedeschi, G., Brady, R. O. & Barton, N. W. (1997) Prospective study of neurological responses to treatment with macrophage-targeted glucocerebrosidase in patients with type 3 Gaucher's disease. *Ann Neurol*, 42, 613–21.

Schuch, G., Machluf, M., Bartsch, G., Jr., Nomi, M., Richard, H., Atala, A. & Soker, S. (2002) *In vivo* administration of vascular endothelial growth factor (VEGF) and its antagonist, soluble neuropilin-1, predicts a role of VEGF in the progression of acute myeloid leukemia *in vivo*. *Blood*, 100, 4622–8.

Schuch, G., Oliveira-Ferrer, L., Loges, S., Laack, E., Bokemeyer, C., Hossfeld, D. K., Fiedler, W. & Ergun, S. (2005) Antiangiogenic treatment with endostatin inhibits progression of AML *in vivo*. *Leukemia*, 19, 1312–17.

Schuchman, E. H., Toroyan, T. K., Haskins, M. E. & Desnick, R. J. (1989)

Characterization of the defective beta-glucuronidase activity in canine mucopolysaccharidosis type VII. *Enzyme*, 42(3), 174–80.

Scully, J. L., Rippberger, C. & Rehmann-Sutter, C. (2004) Non-professionals' evaluations of gene therapy ethics. *Soc Sci Med*, 58, 1415.

Seligman, P. A. (1983) Structure and function of the transferrin receptor. *Prog Hematol*, 13, 131–47.

Serguera, C., Sarkis, C., Ridet, J. L., Colin, P., Moullier, P. & Mallet, J. (2001) Primary adult human astrocytes as an *ex vivo* vehicle for beta-glucuronidase delivery in the brain. *Mol Ther*, 3(6), 875–81.

Shapiro, R., Jordan, M. L., Basu, A., Scantlebury, V., Potdar, S., Tan, H. P., Gray, E. A., Randhawa, P. S., Murase, N., Zeevi, A., Demetris, A. J., Woodward, J., Marcos, A., Fung, J. J. & Starzl, T. E. (2003) Kidney transplantation under a tolerogenic regimen of recipient pretreatment and low-dose postoperative immunosuppression with subsequent weaning. *Ann Surg*, 238, 520–25; discussion 525–7.

Shen, F., Poncet-Legrand, C., Somers, S., Slade, A., Yip, C., Duft, A. M., Winnik, F. M. & Chang, P. L. (2003) Properties of a novel magnetized alginate for magnetic resonance imaging. *Biotechnol Bioeng*, 83, 282–92.

Shen, F., Li, A. A., Cornelius, R. M., Cirone, P., Childs, R. F., Brash, J. L. & Chang, P. L. (2005a) Biological properties of photocrosslinked alginate microcapsules. *J Biomed Mater Res B Appl Biomater*, 75, 425–34.

Shen, F., Li, A. A., Gong, Y. K., Somers, S., Potter, M. A., Winnik, F. M. & CHANG, P. L. (2005b) Encapsulation of recombinant cells with a novel magnetized alginate for magnetic resonance imaging. *Hum Gene Ther*, 16, 971–84.

Sheridan, O., Wortman, J., Harvey, C., Hayden, J., & Haskins, M. J. (1994) Craniofacial abnormalities in animal models of mucopolysaccharidoses I, VI, and VII. *Craniofac Genet Dev Biol*, 14(1), 7–15.

Siegel, J. P. & Puri, R. K. (1991) Interleukin-2 toxicity. *J Clin Oncol*, 9, 694–704.

Simon, K., Perara, E. & Lingappa, V. R. (1987) Translocation of globin fusion proteins across the endoplasmic reticulum membrane in *Xenopus laevis* oocytes. *J Cell Biol*, 104, 1165–72.

Simpson, N. E., Grant, S. C., Blackband, S. J. & Constantinidis, I. (2003) NMR properties of alginate microbeads. *Biomaterials*, 24, 4941–8.

Skobe, M., Rockwell, P., Goldstein, N., Vosseler, S. & Fusenig, N. E. (1997) Halting angiogenesis suppresses carcinoma cell invasion. *Nat Med*, 3, 1222.

Sly, W. S., Quinton, B. A., McAlister, W. H. & Rimoin, D. L. (1973) Beta glucuronidase deficiency: report of clinical, radiologic, and biochemical features of a new mucopolysaccharidosis. *J Pediatr*, 82, 249–57.

Smidsrod, O. & Skjakbraek, G. (1990) Alginate as immobilization matrix for cells. *Trends Biotechnol*, 8, 71–78.

Snyder, E. Y., Taylor, R. M. & Wolfe, J. H. (1995) Neural progenitor cell engraftment corrects lysosomal storage throughout the MPS VII mouse brain. *Nature*, 374(6520), 367–70.

Snyder, R. O., Miao, C. H., Patijn, G. A., Spratt, S. K., Danos, O., Nagy, D., Gown, A. M., Winther, B., Meuse, L., Cohen, L. K., Thompson, A. R. & Kay, M. A. (1997) Persistent and therapeutic concentrations of human factor IX in mice after hepatic gene transfer of recombinant AAV vectors. *Nat Genet*, 16, 270–6.

Snyder, R. O., Miao, C., Meuse, L., Tubb, J., Donahue, B. A., Lin, H. F., Stafford, D. W., Patel, S., Thompson, A. R., Nichols, T., Read, M. S., Bellinger, D. A., Brinkhous, K. M. & Kay, M. A. (1999) Correction of hemophilia B in canine and murine models using recombinant adeno-associated viral vectors. *Nat Med*, 5, 64–70.

Soon-Shiong, P., Heintz, R. E., Merideth, N., Yao, Q. X., Yao, Z., Zheng, T., Murphy, M., Moloney, M. K., Schmehl, M., Harris, M. *et al.* (1994) Insulin independence in a type 1 diabetic patient after encapsulated islet transplantation. *Lancet*, 343, 950–51.

Stein, C. S., Ghodsi, A., Derksen, T. & Davidson, B. L. (1999) Systemic and central nervous system correction of lysosomal storage in mucopolysaccharidosis type VII mice. *J Virol*, 73(4), 3424–9.

Stein, C. S., Kang, Y., Sauter, S. L., Townsend, K., Staber, P., Derksen, T. A., Martins, I., Qian, J., Davidson, B. L. & McCray, P. B. Jr. (2001) *In vivo* treatment of hemophilia A and mucopolysaccharidosis type VII using nonprimate lentiviral vectors. *Mol Ther*, 3(6), 850–6.

Strand, B. L., Morch, Y. A., Espevik, T. & Skjak-Braek, G. (2003) Visualization of alginate–poly-L-lysine–alginate microcapsules by confocal laser scanning microscopy. *Biotechnol Bioeng*, 82, 386–94.

Stroncek, D. F., Hubel, A., Shankar, R. A., Burger, S. R., Pan, D., McCullough, J. & Whitley, C. B. (1999a) Retroviral transduction and expansion of peripheral blood lymphocytes for the treatment of mucopolysaccharidosis type II, Hunter's syndrome. *Transfusion*, 39(4), 343–50.

Suzuki, Y., Tanihara, M., Ohnishi, K., Suzuki, K., Endo, K. & Nishimura, Y. (1999) Cat peripheral nerve regeneration across 50 mm gap repaired with a novel nerve guide composed of freeze-dried alginate gel. *Neurosci Lett*, 259, 75–8.

Tarui, T., Miles, L. A. & Takada, Y. (2001) Specific interaction of angiostatin with integrin alpha(v)beta(3) in endothelial cells. *J Biol Chem*, 276, 39562.

Tatum, E. L. (1966) Molecular biology, nucleic acids, and the future of medicine. *Perspect Biol Med*, 10, 19–32.

Taylor, R. M., & Wolfe, J. H. (1997) Decreased lysosomal storage in the adult MPS VII mouse brain in the vicinity of grafts of retroviral vector-corrected fibroblasts secreting high levels of beta-glucuronidase. *Nat Med*, 3(7), 771–4.

Thu, B., Bruheim, P., Espevik, T., Smidsrod, O., Soon-Shiong, P. & Skjak-Braek, G. (1996a) Alginate polycation microcapsules. I. Interaction between alginate and polycation. *Biomaterials*, 17, 1031–40.

Thu, B., Bruheim, P., Espevik, T., Smidsrod, O., Soon-Shiong, P. & Skjak-Braek, G. (1996b) Alginate polycation microcapsules. II. Some functional properties. *Biomaterials*, 17, 1069–79.

Thu, B., Skjak-Braek, G., Micali, F., Vittur, F. & Rizzo, R. (1997) The spatial distribution of calcium in alginate gel beads analysed by synchrotron-radiation induced X-ray emission (SRIXE). *Carbohydr Res*, 297, 101–5.

Thu, B., Gaserod, O., Paus, D., Mikkelsen, A., Skjak-Braek, G., Toffanin, R., Vittur, F. & Rizzo, R. (2000) Inhomogeneous alginate gel spheres: an assessment of the polymer gradients by synchrotron radiation-induced X-ray emission, magnetic resonance microimaging, and mathematical modeling. *Biopolymers*, 53, 60–71.

Tomatsu, S., Orii, K. O., Vogler, C., Grubb, J. H., Snella, E. M., Gutierrez, M. A., Dieter, T., Sukegawa, K., Orii, T., Kondo, N. & Sly, W. S. (2002) Missense models [Gustm(E536A)Sly, Gustm(E536Q)Sly, and Gustm(L175F)Sly] of murine mucopolysaccharidosis type VII produced by targeted mutagenesis. *Proc Natl Acad Sci USA*, 99, 14982–7.

Tomatsu, S., Orii, K. O., Vogler, C., Grubb, J. H., Snella, E. M., Gutierrez, M., Dieter, T., Holden, C. C., Sukegawa, K., Orii, T., Kondo, N. & Sly, W. S. (2003) Production of MPS VII mouse (Gus(tm(hE540A x mE536A)Sly)) doubly tolerant to human and mouse beta-glucuronidase. *Hum Mol Genet*, 12, 961–73.

Vachon, P. H., Loechel, F., Xu, H., Wewer, U. M. & Engvall, E. (1996) Merosin and laminin in myogenesis; specific requirement for merosin in myotube stability and survival. *J Cell Biol*, 134, 1483–97.

Van Damme, A., Vanden Driessche, T., Collen, D. & Chuah, M. K. (2002) Bone marrow stromal cells as targets for gene therapy. *Curr Gene Ther*, 2, 195.

Van Raamsdonk, J. M. (1999) The development of solid APA microcapsules for somatic gene therapeutics, McMaster University.

Van Raamsdonk, J. M., Ross, C. J., Potter, M. A., Kurachi, S., Kurachi, K., Stafford, D. W. & Chang, P. L. (2002) Treatment of hemophilia B in mice with nonautologous somatic gene therapeutics. *J Lab Clin Med*, 139, 35–42.

Vellodi, A., Bembi, B., de Villemeur, T. B., Collin-Histed, T., Erikson, A., Mengel, E., Rolfs, A. & Tylki-Szymanska, A. (2001) Management of neuronopathic Gaucher disease: a European consensus. *J Inherit Metab Dis*, 24, 319–27.

Vogler, C., Birkenmeier, E. H., Sly, W. S., Levy, B., Pegors, C., Kyle, J. W. & Beamer, W. G. (1990) A murine model of mucopolysaccharidosis VII. Gross and microscopic findings in beta-glucuronidase-deficient mice. *Am J Pathol*, 136(1), 207–17.

Wadhwa, P. D., Zielske, S. P., Roth, J. C., Ballas, C. B., Bowman, J. E. & Gerson, S. L. (2002) Cancer gene therapy: scientific basis. *Annu Rev Med*, 53, 437–52.

Wang, L., Zoppe, M., Hackeng, T. M., Griffin, J. H., Lee, K. F. & Verma, I. M. (1997) A factor IX-deficient mouse model for hemophilia B gene therapy. *Proc Natl Acad Sci USA*, 94, 11563–6.

Wang, M. S., Childs, R. F. & Chang, P. L. (2005) A novel method to enhance the stability of alginate–poly-L-lysine–alginate microcapsules. *J Biomater Sci Polym Ed*, 16, 91–113.

Watson, G. L., Sayles, J. N., Chen, C., Elliger, S. S., Elliger, C. A., Raju, N. R., Kurtzman, G. J. & Podsakoff, G. M. (1998) Treatment of lysosomal storage disease in MPS VII mice using a recombinant adeno-associated virus. *Gene Ther*, 5(12), 1642–9.

Weinreb, N. J., Charrow, J., Andersson, H. C., Kaplan, P., Kolodny, E. H., Mistry, P., Pastores, G., Rosenbloom, B. E., Scott, C. R., Wappner, R. S. & Zimran, A. (2002) Effectiveness of enzyme replacement therapy in 1028 patients with type 1 Gaucher disease after 2 to 5 years of treatment: a report from the Gaucher Registry. *Am J Med*, 113, 112–19.

Wen, J., Vargas, A. G., Ofosu, F. A. & Hortelano, G. (2006) Sustained and therapeutic levels of human factor IX in hemophilia B mice implanted with microcapsules: key role of encapsulated cells. *J Gene Med*, 8, 362–9.

Wesseling, J. G., Yamamoto, M., Adachi, Y., Bosma, P. J., Van Wijland, M., Blackwell, J. L., Li, H., Reynolds, P. N., Dmitriev, I., Vickers, S. M., Huibregtse, K. & Curiel, D. T. (2001) Midkine and cyclooxygenase-2 promoters are promising for adenoviral vector gene delivery of pancreatic carcinoma. *Cancer Gene Ther*, 8, 990.

White, G. C., II (2001) Gene therapy in hemophilia: clinical trials update. *Thromb Haemost*, 86, 172–7.

Whitley, C. B., McIvor, R. S., Aronovich, E. L., Berry, S. A., Blazar, B. R., Burger, S. R., Kersey, J. H., King, R. A., Faras, A. J., Latchaw, R. E., McCullough, J., Pan, D., Ramsay, N. K. & Stroncek, D. F. (1996) Retroviral-mediated transfer of the iduronate-2-sulfatase gene into lymphocytes for treatment of mild Hunter syndrome (mucopolysaccharidosis type II). *Hum Gene Ther*, 7(4), 537–49.

Winn, S. R., Hammang, J. P., Emerich, D. F., Lee, A., Palmiter, R. D. & Baetge, E. E. (1994) Polymer-encapsulated cells genetically modified to secrete human nerve growth factor promote the survival of axotomized septal cholinergic neurons. *Proc Natl Acad Sci USA*, 91, 2324–8.

Wolfe, J. H., Deshmane, S. L. & Fraser, N. W. (1992) Herpesvirus vector gene transfer and expression of beta-glucuronidase in the central nervous system of MPS VII mice. *Nat Genet*, 1(5), 379–84.

Wolfe, J. H., Martin, C. E., Deshmane, S. L., Reilly, J. J., Kesari, S. & Fraser, N. W. (1996) Increased susceptibility to the pathogenic effects of wild-type and recombinant herpesviruses in MPS VII mice compared to normal siblings. *J Neurovirol*, 2(6), 417–22.

Wolfe, J. H., Sands, M. S., Harel, N., Weil, M. A., Parente, M. K., Polesky, A. C., Reilly, J. J., Hasson, C., Weimelt, S. & Haskins, M. E. (2000) Gene transfer of low levels of beta-glucuronidase corrects hepatic lysosomal storage in a large animal model of mucopolysaccharidosis VII. *Mol Ther*, 2(6), 552–61.

Wozniak, M. A., Desai, R., Solski, P. A., Der, C. J. & Keely, P. J. (2003) ROCK-generated contractility regulates breast epithelial cell differentiation in response to the physical properties of a three-dimensional collagen matrix. *J Cell Biol*, 163, 583–95.

Xiang, R., Lode, H. N., Dolman, C. S., Dreier, T., Varki, N. M., Qian, X., Lo, K. M., Lan, Y., Super, M., Gillies, S. D. & Reisfeld, R. A. (1997) Elimination of established murine colon carcinoma metastases by antibody-interleukin 2 fusion protein therapy. *Cancer Res*, 57, 4948.

Xu, W., Liu, L. & Charles, I. G. (2002) Microencapsulated iNOS-expressing cells cause tumor suppression in mice. *Faseb J*, 16, 213–15.

Yao, S. N. & Kurachi, K. (1992) Expression of human factor IX in mice after injection of genetically modified myoblasts. *Proc Natl Acad Sci USA*, 89, 3357–61.

Yao, S. N., Wilson, J. M., Nabel, E. G., Kurachi, S., Hachiya, H. L. & Kurachi, K. (1991) Expression of human factor IX in rat capillary endothelial cells: toward somatic gene therapy for hemophilia B. *Proc Natl Acad Sci USA*, 88, 8101–5.

Yao, Z., Dai, W., Perry, J., Brechbiel, M. W. & Sung, C. (2004) Effect of albumin fusion on the biodistribution of interleukin-2. *Cancer Immunol Immunother*, 53, 404.

Yew, N. S., Przybylska, M., Ziegler, R. J., Liu, D. & Cheng, S. H. (2001) High and sustained transgene expression in vivo from plasmid vectors containing a hybrid ubiquitin promoter. *Mol Ther*, 4, 75–82.

Yin, C., Mien Chia, S., Hoon Quek, C., Yu, H., Zhuo, R. X., Leong, K. W. & Mao, H. Q. (2003) Microcapsules with improved mechanical stability for hepatocyte culture. *Biomaterials*, 24, 1771–80.

Yoshiko, Y., Hirao, K., Sakabe, K., Seiki, K., Takezawa, J. & Maeda, N. (1996) Autonomous control of expression of genes for insulin-like growth factors during the proliferation and differentiation of C2C12 mouse myoblasts in serum-free culture. *Life Sci*, 59, 1961–8.

Zheng, S., Xiao, Z. X., Pan, Y. L., Han, M. Y. & Dong, Q. (2003) Continuous release of interleukin 12 from microencapsulated engineered cells for colon cancer therapy. *World J Gastroenterol*, 9, 951–5.

Zhou, J. M., Qiu, X. F., Lu, D. R., Lu, J. Y. & Xue, J. L. (1993) Long-term expression of human factor IX cDNA in rabbits. *Sci China B*, 36, 1333–41.

Zhu, J., Kang, W., Wolfe, J. H. & Fraser, N. W. (2000) Significantly increased expression of beta-glucuronidase in the central nervous system of mucopolysaccharidosis type VII mice from the latency-associated transcript promoter in a nonpathogenic herpes simplex virus type 1 vector. *Mol Ther*, 2(1), 82–94.

11

Capillary devices for therapy

J W KAWIAK, L H GRANICKA, A WERYŃSKI and J M WÓJCICKI, Institute of Biocybernetics and Biomedical Engineering PAS and Medical Centre of Postgraduate Education, Poland

11.1 Hollow fiber (HF) membrane and its evaluation

11.1.1 Manufacturing of the HF membrane

Capillary membranes are typical devices among macrocapsules and they seem to be attractive for transplantation applications. Hollow fibers (HFs) ensure a repeatable composition and shape, a smooth surface, and higher mechanical properties as well as the stability that microcapsules offer.[1–5] To prepare an HF, the solvent is removed directly from the membrane during washing, after the production process (e.g. polypropylene membranes), or it diffuses directly into the coagulation bath (e.g. polysulfone, 2,5-cellulose acetate membranes). The preparation of microcapsules proceeds parallel with the cell encapsulation, and the choice of non-toxic solvent for the encapsulating cells is important. In other cases it is necessary to apply techniques that reduce the toxic impact of the solvent (e.g. pervaporation).[6–8]

The material often used for the macroencapsulation purposes is thermoplastically produced water-insoluble poly(acrylonitrile vinyl chloride) (PAN-PVC) made by the phase-inversion method elaborated by Michaels.[9] Many researchers have used the fabricated version of this polymer, Amicon XM-50, produced by Amicon, USA. Other materials, apart from PAN-PVC, often used for HF in *in vivo* applications are polyvinylidene fluoride and polyacrylonitrile-sodium methallylsulfonate.[10] The molecular weight cutoff (MWCO) values of the membranes applied in the *in vivo* uses varied from 12 to 500 kDa.[11–20] Furthermore, HFs of mixed esters of cellulose (Spectrum Laboratories Inc., USA), polysulfone (Millipore Amicon, USA) and regenerated cellulose (Spectrum Laboratories Inc., USA) of MWCO 10–13 kDa are recommended for implantation.[21]

Generally, the separated solute dimensions show a wide range of magnitude, from 10 μm to angstroms (e.g. erythrocyte (7.5 μm), polysacharides (8–20 Å), calcium ions (2.12 Å), magnesium ions (1.56 Å)). The dimensions and molecular weights of solutes separated by membranes used for immuno-isolation are presented in Table 11.1.

Table 11.1 The dimensions and the molecular weight of solutes separated by membranes used for immuno-isolation

Technique	Solute dimension	Examples of the separating solutes and their dimensions
Microfiltration	0.1–10 μm	Erythrocytes (7.5 μm) Bacteria *E. coli* (0.1–0.3 μm)
Ultrafiltration	1 nm (MW 10 000)– 100 nm (MW 1 000 000)	Albumins (about 7 nm, MW 68 000)

11.1.2 Possible regulation of the product exchange according to its MW

In biotechnological applications hydrophilization is a way of improving biocompatibility.[22,23] The membrane modification is possible after membrane formation. Hydrophilization may be obtained by grafting the functional groups to the membrane material.[24–27] Introduction of the polar phospholipid groups by grafting the 2-methacryloxyethyl phosphorylcholine groups on the surface of the dialysis cellulose capillary membranes increases their biocompatibility.[28] Sulfonation of a polystyrene surface prohibits cell adherence because of an increase in the membrane negative charge and/or hydrophilicity.[29] A sulfonated polymer, such as heparin, grafted on to the polyurethane surface through either the ozone oxidation or the photo reaction also improves biocompatibility.[30] Surface fluorination of the polyethersulfone membranes recently adopted for membrane materials in applications such as ultrafiltration and haemodialysis increases biocompatibility and membrane chemical stability.[31] Plasma polymerization allows membrane hydrophilization to be achieved by the surface treatment with low-temperature plasma.[32,33] Plasma generation is performed in a reactor filled with gas exposed to voltage. The glow discharge of the vapor or gas of organic compound (or compound mixture) produces a plasma polymer layer on the membrane surfaces exposed to it. Application of different gases produces different types of plasma modifications. Reactive gas (e.g. O_2, N_2, NH_3, $CHCl_3$) plasma application produces different functional groups on the membrane surface; non-reactive gas (e.g. Ar, He) plasma application increases the surface crosslinking degree and allows for the formation of composite systems. This method gives a low thickness layer with good mechanical strength. The membranes covered by this method are characterized by good biocompatibility.[22] Also, plasma polymerization with acrylic acid and collagen improves the poly(ϵ-caprolactan) membranes' biocompatibility.[23] The plasma technique may be used as well for immobilization of different biosolutes on the membrane such as insulin, heparin[34] and hyaluronic acid.[35]

Our Institute has elaborated surface-modified permiselective polypropylene membranes for the immuno-isolation of cells. These exhibit diffusive permeability sufficient for transporting small solutes (such as nutrients) and an MWCO prohibiting direct contact of the encapsulated cells with the host immunocompetent cells as well as with some of the complement proteins.

The polypropylene HFs commercially available for the filtration use that demonstrated satisfactory mechanical resistance were chosen for basic research on the immuno-isolation of the biological material. The polypropylene HFs (K600 Akzo-Nobel, Germany, internal diameter 0.6 mm, wall thickness 0.2 mm) were silanized[36] to improve their biocompatibility and/or to change MWCO, as compared with the original HF. The polypropylene (PP) membrane material in the form of HFs is generally used in oxygenators.[33] The application of silanization of HF avoids tissue overgrowth and improves oxygen transport.[34,35] Typical of immuno-isolation applications, the polypropylene membrane material is used in the form of diffusive chambers.[37]

Preparation technique of the silanized HF

A solution of silicone oil was introduced to the membrane structure by its exposure to the solvent in the pressure compartment (−0.1 MPa–0.1 MPa) in temperature 20 °C for 5–60 mins. The solvent may be aliphatic or aromatic hydrocarbons, alcohols, aldehydes, ketones or amines. Electron microscopy showed that the exposure time and pressure had no significant effect on silanization of the membrane surface. Initially different viscosities of the siloxanes were tested: 2200–3700 cSt and concentrations 0.1–12.2%. The best results were obtained with viscosities 2700–3000 cSt, concentration 1–2%, temperature 20 °C, time of exposure 15 min (Figs 11.1 and 11.2). The HF pore size obtained after silanization, evaluated using the 'bubble point' method,[38] was 0.47 ± 0.03 μm ($n = 10$).

Description of the technique

The conditions of porous membrane manufacture (e.g. temperature, pressure, access of air) and the remaining traces of plant oils used in the membrane manufacturing protocol are mostly responsible for existence of the epoxide or peroxide groups in the material. It was estimated that up to 0.2% of the methyl terminal groups on the membrane surface underwent oxidation. Both types of functional group are shown in Fig. 11.1.

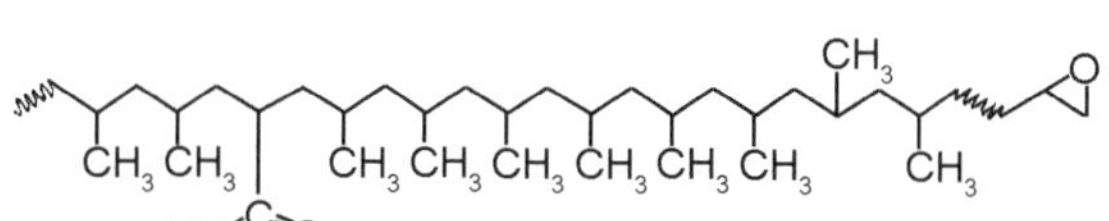

11.1 Formation of carboxylic and epoxy groups in a polypropylene chain (reprinted from *Biocybernetics and Biomedical Engineering*, 24(2): 59–67, 2004).

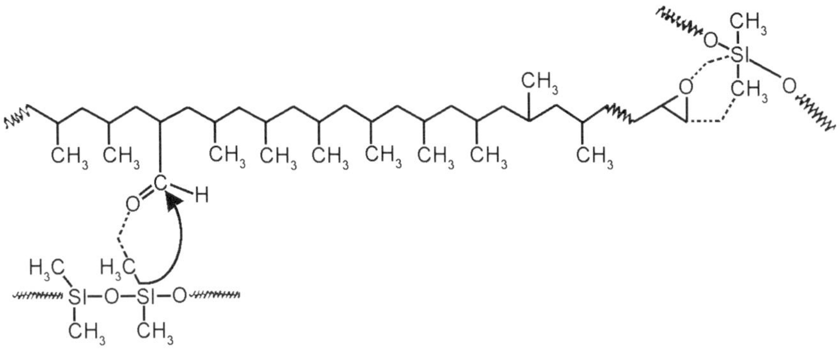

11.2 The more probable reaction of functional groups with organosilanous compounds during the contact with tissue (reprinted from *Biocybernetics and Biomedical Engineering*, 24(2): 59–67, 2004).

Part of the carboxylic groups may be reduced to carbonyl groups in contact with living cells. These groups are able to react with organosilanous compounds as well as the epoxide groups. The epoxides react with silanes, mainly with silanochloride derivatives, present in silanes in vestigial amounts. The epoxide groups present on the PP surface interact also with siloxane, but the mechanism of the interaction has not yet been explained. The carboxylic groups of the organosilanous compound (the part not reduced), being in direct contact with the blood in the biological environment, is attacked by peripheral blood plasma proteins, and react in succession with siloxanes, resulting in binding with silicone (silicone–oxygen–silicon) (Fig. 11.2).[39] This reaction is being investigated by biochemists, but it has not been finally elucidated.

The pressure application during the membrane modification by silanization delimits the diffusive transport through the membrane for large solutes without simultaneous transport change for small solutes (Fig. 11.3). The final result of the membrane silanization is desolution of the part of siloxane in the polymer and the siloxane chains joining the polymer composing the membrane. The parameters of the silanized membranes indicate that they could prohibit direct contact between the cytotoxic T lymphocytes and transplanted cells. The passage of the larger complement components necessary for its activation, e.g. the first element of the complement C1q protein (MW 410 kDa), is avoided. Cell membrane attack complex (MAC) formation (C5–C9) is prevented, which should be sufficient for immunoprotection of the encapsulated cells in the possible humoral response.

11.1.3 Assessment of the HF membrane

Biocompatibility is defined as 'the ability of a biomaterial to perform with an appropriate host response in a specific application'.[40] This definition is easy to interpret with the application of conventional artificial organs such as artificial

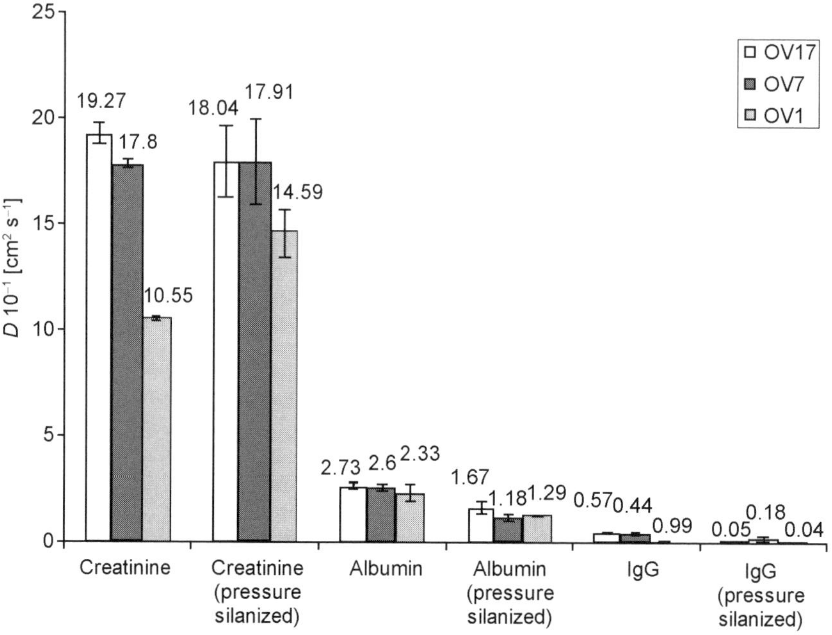

11.3 The diffuse coefficient value (D) of different modified membranes (reprinted from *Biocybernetics and Biomedical Engineering*, 24(2): 59–67, 2004).

knees or breasts, in a system consisting of the host and a biomaterial. The third element, encapsulation in a membrane biological material, makes the definition more complex. Generally, a biocompatible system of membranes is a system that does not evoke tissue overgrowth.[41–44]

Systems consisting of the host, permiselective membrane and biological material encapsulated within were analysed. In such a system the membrane biostability post-implantation must be considered as well as the membrane cytotoxity for the encapsulated biological material and ability of the biomaterial to perform with the host. The method of membrane biocompatibility evaluation was standardized. The selected three steps of the membrane evaluation, ensuring, in our opinion, membrane biocompatibility, included: (1) membrane physical and chemical stability post-implantation, (2) the biomaterial ability to perform with the host, analyzed by microscopy of cells surrounding the membrane and (3) membrane cytotoxity against biological material.

Evaluation of the physical and chemical stability of the membrane material and its ability to perform with the host

The membrane physical and chemical stability was assayed post-implantation to the animal. It was assumed that diffusive transport could be used as an indicator for physical stability, and that the Fourier transform infrared (FTIR) was representative of chemical stability. The diffusive permeability assessment is an adequate method for evaluation of membranes for cell immuno-isolation in the system HF–body fluids environment. To evaluate *in vitro* the HF diffusive permeability, the thermodynamic description of diffusive mass transport (Fick's law) and a two-compartment model were applied.[36] The HFs were filled up with the tested solute solution in saline, sealed at both ends and immersed in a beaker with continuously mixed physiological saline. The samples were taken from the beaker at various time periods to evaluate the solute concentration in the spectrophotometer at 235 and 280 nm wavelengths for creatinine and bovine IgG, respectively.

Evaluation of the spectrum of absorption for infrared irradiation (FTIR) was performed to evaluate the chemical stability of the membrane modification before and after implantation, using FTS3000 MX (BioRad Excalibur, USA) device. The 16 scans were collected with resolution of 4 cm^{-1}.

To evaluate the membrane biocompatibility in the above-mentioned terms, the HF of polypropylene K600 silanized OV1 or OV17 were implanted subcutaneously into vetbutal anesthetized mice for: 4 days or 7 days and for 4 months. The ratios of the solute diffusive coefficient in the membrane to the solute diffusive coefficient in water evaluated for K600 silanized OV1 or OV17, before implantation and after 2 months implantation into mice obtained for creatinine and IgG are presented in Fig. 11.4. Before implantation, the HF silanized with siloxane of higher viscosity (OV1) exhibited a lower permeability for the both evaluated solutes than the HF silanized with a lower viscosity siloxane (OV17). No influence of siloxanes of different polarity on HF diffusive transport was observed.

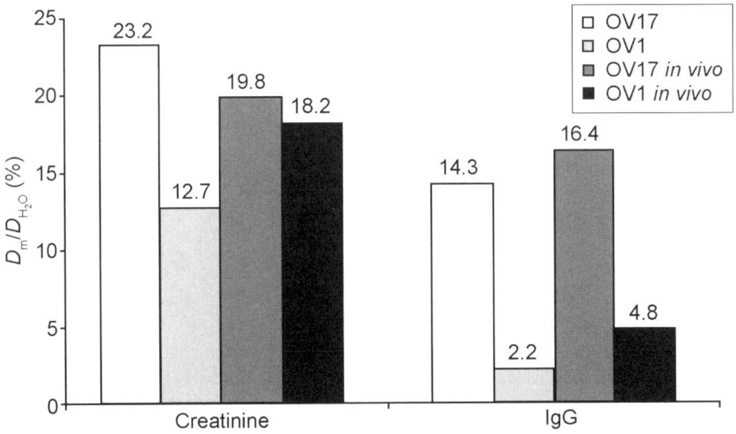

11.4 The ratio of the solute diffusive coefficient in the membrane (D_m) to the solute diffusive coefficient in water (D_{H2O}) (reprinted from *Artificial Cells, Blood Substitutes and Immobilization Biotechnology*, 32(4): 1–10, 2004).

Comparing the membrane permeability P (ml min^{-1} m^{-2}) value before and after implantation, it was found that the value of P after implantation for OV17 silanized HF decreased for creatinine, from 57.9 ± 1.4 before implantation to 49.4 ± 0.8 post-implantation. The value obtained for IgG differs in the limits of the error from 1.7 ± 0.2 before implantation to 1.9 ± 0.1 post-implantation. For OV1 silanized HF the P value increased for creatinine from 31.7 ± 0.2 before implantation to 45.4 ± 2.2 post-implantation. The value obtained for IgG differs in the limits of the error, from 0.26 ± 0.16 before implantation to 0.58 ± 0.31 after implantation.

The changes in the permeability value for creatinine before and after *in vivo* applications for OV1 and OV17 may be caused by physical changes in the membranes during the implantation period *in vivo*, like closing or stretching of pores, and by different surface deposition of proteins on the membranes modified by different siloxanes because of their different polarity. The polarity of the membranes was evaluated by contact angle of water and diiodinemethane assessement using the sessile drop method.[45] In the method the surface tension γ and its components: polar γ_p and disperse γ_d were evaluated according to the harmonic averaging. The K600 silanized OV17 HF appear to be more hydrophobic membranes than the K600 siliconized OV1 membranes, which were hydrophilic, with polarity 1.0% and 25.8%, respectively. The chemical changes in the membrane structure were excluded since no changes during the FITR evaluation were observed. The exemplary picture of FTIR spectrum for HF K600 OV1 before and 4 months after implantation is presented in Fig. 11.5.

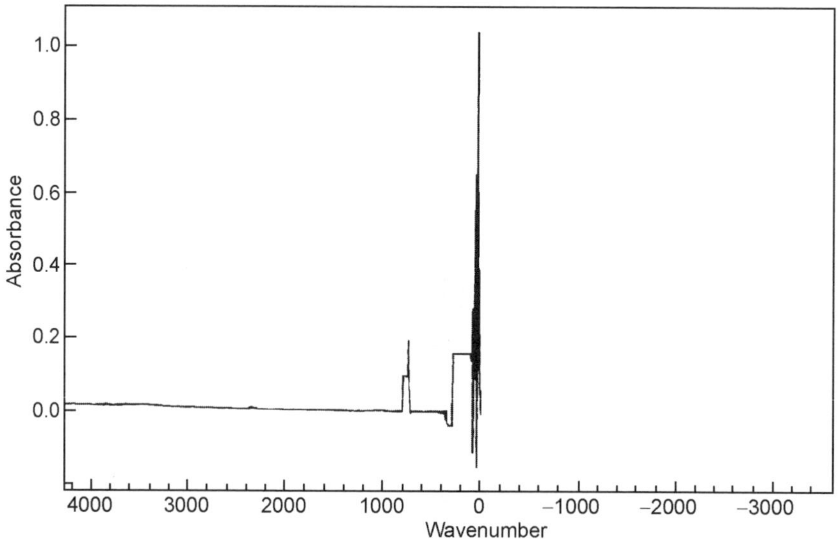

11.5 The FTIR spectrum for the difference between OV1 silanized HF before implantation and 4 months after implantation (reprinted from *Artificial Cells, Blood Substitutes and Immobilization Biotechnology*, 32(4): 1–10, 2004).

In the cut-off values for K600 silanized OV1 or OV17, no change before and after implantation was observed. There were also no differences in FTIR spectrum between K600 silanized OV1 or OV17 before and after explantation after 2, days 4 days and 4 months of implantation, it was concluded that the applied membranes had established a stable molecular structure.

After 2 or 4 months from implantation of membrane silanized OV1 or OV17 it was observed by light microscopy that the external surface of HFs was surrounded by a thin layer of multinuclear cells and by a thin layer of fibroblasts (Fig. 11.6). There was no inflammatory reaction with macrophages or lymphocyte infiltration behind the layer of fibroblasts, which separate the membrane from further host tissue.

Using electron microscopy, the morphology of cells surrounding the silanized OV1 or OV17 membrane walls was found to be comparable for selected days/months of observation. In electron microscopy pictures we may observe that the silanized material is covered with host proteins some hours after implantation. Some inflammatory cells appear near implanted membrane after a short time (4 days). However no further presence of inflammatory cells was observed at 2 or 4 months beyond the collagen layer adherent near the external membrane wall (Fig. 11.7).

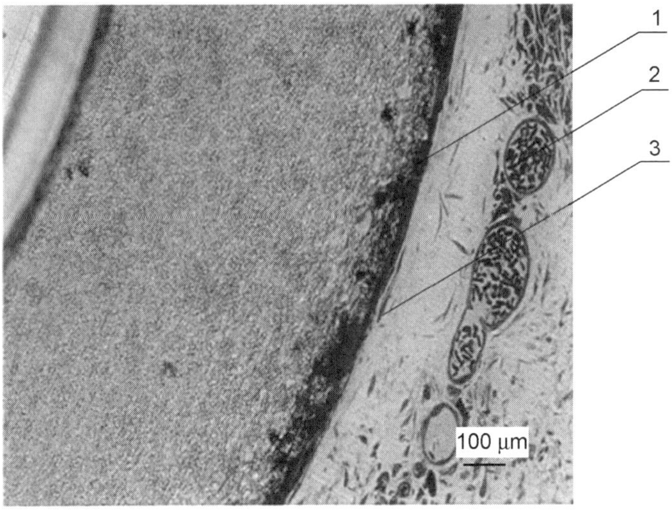

11.6 The fragment of the HF wall modified with OV1 wall after two months of subcutaneous implantation into a mouse. The external surface of HF was surrounded by a thin layer of multinuclear cells and a thin layer of fibroblasts. At some distance from the surface of HF are capillary blood vessels containing numerous erythocytes, 1 = multinuclear cells. 2 = capillary blood vessels and 3 = fibroblasts. Microscopic evaluation performed by Dr E. Jankowska (reprinted from *Artificial Cells, Blood Substitutes and Immobilization Biotechnology*, 32(4): 1–10, 2004).

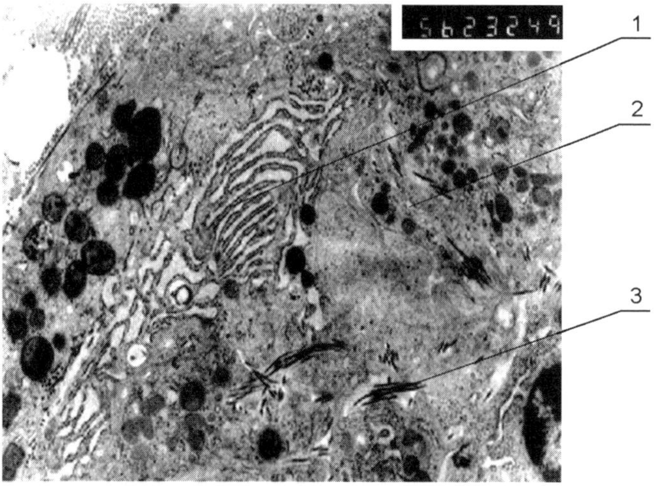

11.7 The membrane silanized OV1 external wall 2 months after implantation into mice: (1) processes of giant cell, (2) giant cell fragment, (3) collagen fibers. The electron microscopy picture (JEM 100S), magnification 2000×. Microscopic evaluation performed by Dr E. Jankowska.

Membranes undergoing silanization by different applied siloxanes do not induce *in vivo* massive tissue overgrowth and scar formation for at least 4 months after implantation. We concluded that membranes tested in the experiment are biocompatible with respect to the membrane's ability to perform with the host.

Membrane cytotoxity evaluation

The membrane cytotoxity *in vitro* for encapsulated biological material might be evaluated with different cell lines, e.g. CCL cell line recommended by Polish/European standards,[46] for human lymphocytes or for Jurkat cell line. However, before that the membrane material has to be sterilized. A very simple method was used. The HFs were stored in 70% ethanol for 30 min under UV, then washed with sterile distilled water followed by washing with sterile physiological saline, and then stored in physiological saline for at least 15 min.

Several methods of studying the membrane cytotoxity with the membranes produced in our laboratories, silanized polypropylene and polyethersulfone membranes were analyzed. The results of the evaluation of the Jurkat cell growth in 10 days culture (Fig. 11.8) in the presence of the membrane, cell viability in 72 h culture of HL-60 cells in HF lumen (Fig. 11.9) and bacteria *Escherichia coli* with green fluorescent protein (GFP) expression (Fig. 11.10) as the markers of cytotoxity are presented.

In the experiment assessing the GFP expression of the *E. coli* encapsulated in HFs during 5 days' growth, the assumed marker of cytotoxity is the change of

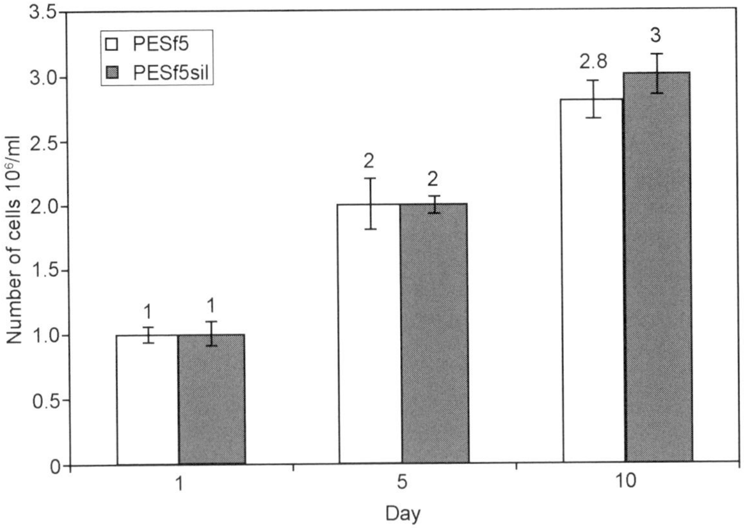

11.8 The number of Jurkat cells cultured for 10 days with polyethersulfone PESf5 original or silanized membranes ($n = 18$). The values are presented in the form: mean SD, number of experiments. The values obtained were for original or silanized membranes respectively: 1 ± 0.06 or 1 ± 0.10 on the first day, 2 ± 0.20 or 2 ± 0.06 on the fifth day or 3 ± 0.15 on the tenth day.

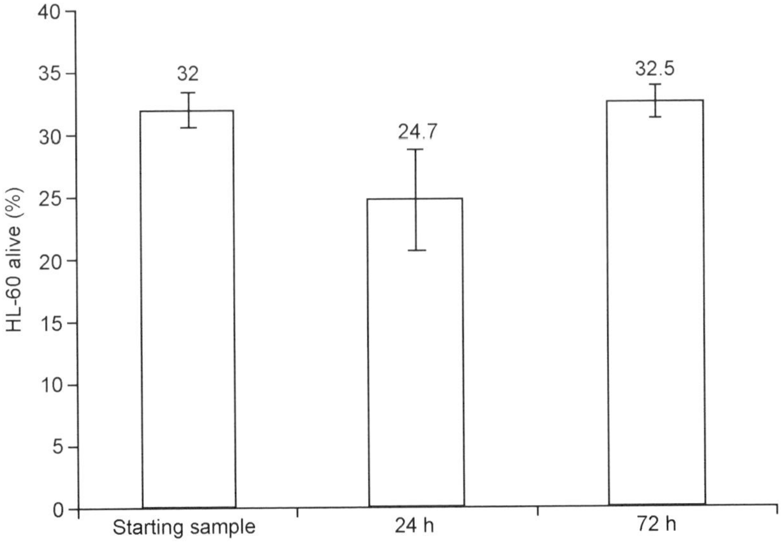

11.9 The polyethersulfone PESf1 membranes cytotoxity evaluation with HL-60 cell line ($n = 16$). The values are presented in the form: mean SD, number of experiments. The values obtained were 32.0 ± 1.41 for the starting sample, 24.7 ± 4.04 after 24 h and 32.5 ± 1.29 after 72 h.

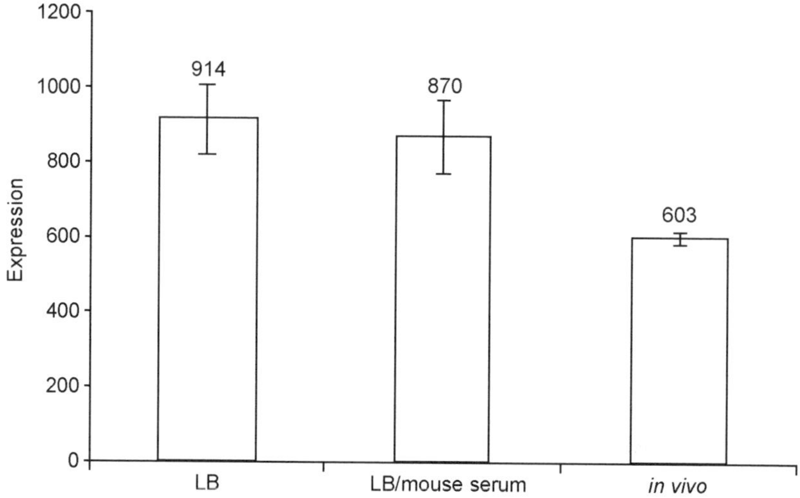

11.10 The GFP expression of E. coli encapsulated in hollow fibers during 5 days growth in different culture media: LB (culture medium mixture: Luria-Bertan (LB) and RPMI1640) ($n = 18$) or LB/mouse serum (LB supplemented with 10% mouse serum) ($n = 12$) as well as after subcutaneous implantation into mice ($n = 32$). The values are presented in the form: mean SD. The values obtained were 914 ± 93 for LB medium, 870 ± 97 for LB/mouse serum and 603 ± 17 for *in vivo* experiments.

GFP expression values after the entire experiment. The mean values of GFP expression after the experiment showed low dispersion standard deviation values (SD) for LB and LB/mouse serum medium. Also, the dispersion values of GFP expression for the entire encapsulated *E. coli* GFP 5 day implantation into the mice are comparable.

The difference between the mean GFP expression values for the *in vitro* culture (in LB or LB/mouse serum) and for implantation, where the encapsulated *E. coli* GFP after 5 days exhibited a mean expression channel value of 603 ± 17 ($n = 32$), was found. The mean expression the GFP values obtained after the culture in LB or LB/mouse serum are higher than the mean values obtained after the implantation period[47] (Fig. 11.10).

Considering the results obtained for the cytotoxity evaluation, it was observed that either of the polyethersulfone membranes, original and silanized, or silanized polypropylene membranes did not influence the cells' viability or the growth or the gene expression in *E. coli* significantly in the presence of the HF membranes. The ratio of the measured parameters for the first and the last day of observation was 0.99 for the evaluation of HL-60 cell viability; and 1.27, 1.18 and 0.94 for the evaluation of the *E. coli* expression in LB or LB/mouse serum culture or *in vivo*, respectively.

It can be concluded that the membrane cytotoxity to the biological material in the *in vitro* or *in vivo* evaluation is one of the determinants of biocompatibility.

This aspect of biocompatibility is not a constant attribute. It depends on the sort of the encapsulated biological material, the environment and the method of evaluation, so the obtained values may be different; however, the ratio of the measured parameters (except the cell growth) should be close to 1 in biocompatible systems.

11.2 Biomedical applications of hollow fiber-based devices

11.2.1 Introduction

Originally, HFs were applied as treatment for endocrine system functions deficiency as bioartificial pancreas systems. The permiselective HFs of acrylic copolymer (2–3 cm long, inner diameter (i.d.) 1.8–4.8 mm, wall thickness 69–105 μm, a nominal MWCO 50-80 kDa; W. R. Grace & Co.) were used for immunoisolation of canine, bovine or porcine pancreatic islets.[48] Pancreatic islets of Lewis rats suspended in agarose (10 000 islets/ml), encapsulated in polyacrylonitrile–sodium methallylsulfonate (AN69) HF and implanted intraperitoneally into streptozotocin-induced diabetic rats maintained glycemia below 250 mg/dl for 70 days after the implantation.[49] Rat islets immobilized in collagen gel in the presence of vascular endothelial growth factor (VEGF) to increase the cell viability, encapsulated in AN69 HF and implanted intraperitoneally into diabetic mice, maintained a stable glucose level for at least 28 days.[50] Unfortunately, such limitations as the functional cell longevity,[51] formation of fibrous capsule around the transplant delimiting the oxygen and nutrition supply,[52,53] and the presence of pre-existing natural xeno-antibodies in case of xenografts[54,55] hinder a successful application of the implants in large animals as well as in clinical trials.[53] Other studies carried out with biohybrid implants deal with the construction of a bioartificial liver,[56] the control of erythropoietin secretion by cells[57–59] and the delivery of fibroblast growth factor-2 (FGF-2) by myoblasts.[60]

Systems of genetically engineered cells encapsulated in HF can provide an efficient means for future applications involving delivery of neurotrophic factors.[61–63] Dopamine secreting cell line PC12 encapsulated in polyvinyl alcohol (PVA) foam as a matrix material in the lumen of the polyvinyl alcohol HF and implanted into rodent striatum[60,61] may serve as such a system.

At the moment, treatment with bioartificial organs using HF systems does not go beyond the animal model except preliminary clinical trials in diabetes treatment,[64] in amyotrophic lateral sclerosis (ALS) treatment[64] or in reducing chronic pain.[65,66] A clinical trial was performed in ALS patients to evaluate the safety and tolerability of the intrathecal implants of the encapsulated genetically engineered baby hamster kidney (BHK) cells releasing human ciliary neurotrophic factor (CNTF). The BHK cells, encapsulated in HF, mounted on a silicone catheter were implanted into patients for 3–4 months. The implanted cells functioned and released CNTF to the cerebrospinal fluid without immunosuppression in 9 out of 12 patients from several weeks to months.[67]

Cells encapsulated in HF after subcutaneous or intraperitoneal implantation into an animal may act as a model system for testing the activity of anticancer or anti-viral agents. Human cell lines derived from tumors of prostate, kidney, lung, central nervous system as well as melanomas, encapsulated in polyvinylidene fluoride (PVDF) HFs (i.d. 1.0 mm, MWCO 500 kDa), exhibited a significant response to some of the tested anti-cancer chemotherapeutic agents.[68,69]

Ze Qi Xu *et al.*[70] encapsulated CEM-SS cells infected with III_B strain of HIV-1 in polyvinylidene fluoride HF. The HF (i.d. 1 mm, MWCO 500 kDa) encapsulated cells were implanted subcutaneously or intraperitoneally into mice for 7 days. Calanoide A administered to mice by oral dose, was observed to produce a high anti-HIV activity in both the subcutaneous and intraperitoneal compartments. Calanoide A is a natural substance isolated from *Clausena excavata* with anti-mycobacterial and anti-retroviral activity. The viral replication was suppressed to the same level in subcutaneously and intraperitoneally implanted HF, indicating that the compound was well distributed in the host.

11.2.2 Immuno-isolation

The polypropylene silanized HFs applied for immnuno-isolation were permeable to large molecular weight solutes. However, in the opinion of many authors,[71–75] HFs should be impermeable to immunoglobulins and some complement components to avoid the immune attack. This goal was achieved by application of the membranes of MWCO lower than 100 kDa.[74–76] Nevertheless Lanza *et al.*[77] successfully used alginate, permeable for particles of MW over 600 kDa, for encapsulation of pig islets of Langerhans. Immunohistological staining of long-term islet grafts in mice (over 10 weeks old) revealed viable cells, and the external surfaces of the microreactors were free of the tissue overgrowth. In short-term experiments with implantation of different cell types encapsulated in polypropylene siliconized HFs (MWCO about 200 kDa), the cells were viable.[78–81]

Certain forms of immune attack of the host against the cells transplanted within HFs occur. T lymphocytes recognize the antigen presented by target cells in major histocompatibility complex (MHC) class I complex in direct contact. The membrane used in the experiments is the barrier to the cells, allowing to avoid direct contact between the cytotoxic T lymphocytes or macrophages and the transplanted cells. In the humoral response, the complement activation is necessary to initiate destruction of the foreign transplanted cell by specific antibodies. However, the passage of the host first element of complement C1q protein (MW 410 kDa) through the membrane may not be effective. The alternative pathway of the complement activation through C3 necessitates the presence of properdin (P) (MW 224 kDa) for stabilization of convertase C3/C5. In the absence of P, convertase C3/C5 is not protected from factors H and I, which makes MAC formation impossible and prevents cell destruction. The other antibody effector function is antibody-dependent cell-mediated cytotoxicity (ADCC). In this type of response, direct contact of the cells is indispensable, which is prevented by the membrane.

In vitro experiments enable the measurement of the permeation of individual complement proteins or complement activity[82,83] as well as the permeation or cytotoxity of locally produced inflammatory cytokines such as IL-1β, IFNγ and TNFα.[84,85] The ultimate assessment of the immune protection should come from the *in vivo* studies.

11.2.3 *In vitro* and *in vivo* HF applications

OKT3 cells encapsulation in HFs

Experiments on the function of HF-encapsulated OKT3 hybridoma cells producing anti-CD3 (anti-OKT3) antibody were performed.[78] Anti-CD3 is the mouse monoclonal antibody of IgG_{2a} class that binds to non-polymorphic polypeptide chains associated with the T-cell receptor complex on human mature T lymphocytes. The anti-CD3 prevents rejection of kidney transplants in patients.[86] The function of hybridoma OKT3 cells encapsulated in HFs and the diffuse transport of immunoglobulin through the HF membrane were tested (Fig. 11.11). It was observed that anti-CD3 antibodies produced by encapsulated OKT3 cells diffused from HF to the culture medium and were binding to T cells.

In another experiment, human peripheral blood lymphocyte activation was assessed by anti-CD3 produced by encapsulated OKT3 cells. The cells encapsulated in HF OKT3 were cultured together with human peripheral blood mononuclear cell suspension for 24 h. There was increase of the percentage number of CD69-positive T lymphocytes in the joint cultures of the HF-encapsulated OKT3 cells and lymphocytes outside the HFs as compared with the negative control (Fig. 11.12). CD69 is an early marker of T-cell activation.

Concluding, the *in vitro* experiment proved the ability of OKT3 cells encapsulated in HFs to produce the antibody. The antibody that permeates through the HF membrane binds to the T lymphocytes in a similar way to the commercial anti-CD3 antibody and effectively activates human T cells *in vitro*.

Parathyroid cells encapsulation in HFs

The survival and function of human parathyroid cells after encapsulation in HFs have been studied.[87] The parathyroid hormone (PTH) (MW 9563 Da) acts on the synthesis of 1-25 $(OH)_2$ vitamin D and on the rapid calcemia regulation *in vivo*:

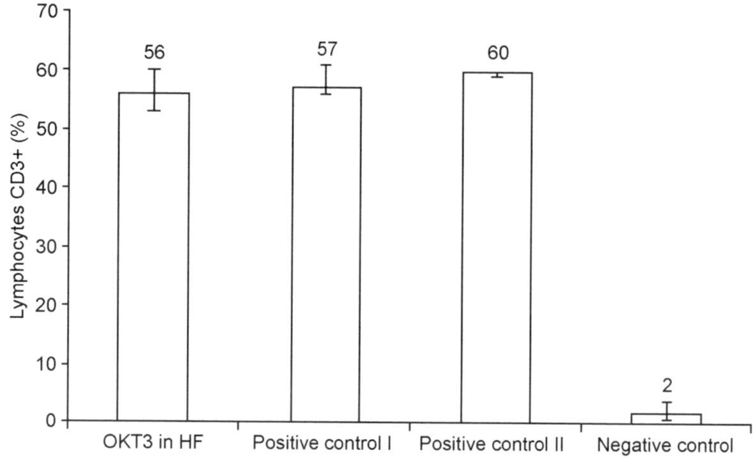

11.11 HF filled with suspension of OKT3 cells at concentration 1 × 10^6 cells/ml were cultured in the humidified atmosphere (5% CO_2, 37 °C) for 4 days. After 4 days, the culture medium was examined for the presence of anti-CD3 antibodies with human peripheral blood mononuclear cells in the immuno-cytochemical reaction. The indirect reaction with goat anti-mouse-PE antibody was applied as a second step. In positive control I, the HFs were filled up with the commercial anti-CD3 solution. In positive control II, the culture medium was supplemented with the commercial anti-CD3 solution without HF. As the negative control the empty HFs were incubated in the culture medium. The values are presented are median values (P25–P75).

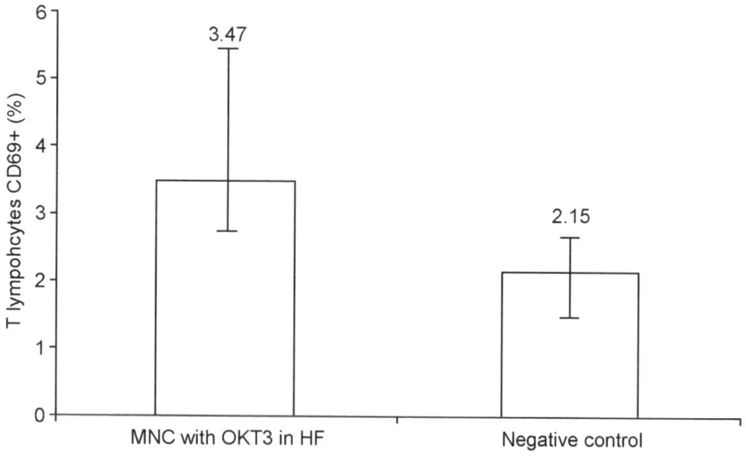

11.12 Evaluation of T lymphocyte activation during 24 h culture of human peripheral blood mononuclear cells suspension in presence of the OKT 3 cells encapsulated in HF, by assessment of the number of T lymphocytes CD69+. The values are presented as median values (P25–P75). There was statistical difference in the Mann–Whitney test, in respect of the percentage number of T lymphocytes CD69 positive as compared with the negative control (the empty HF culture) ($n = 6$).

it produces an increase in the renal tubular absorption of calcium and mobilization of the bone marrow calcium.[88,89]

A human parathyroid gland was obtained postoperatively from a patient with hyperparathyroidism, diagnosed at the Department of Pathomorphology, Children's Memorial Hospital, Warsaw. The parathyroid tissue was dispersed mechanically into pieces and resuspended in the culture medium. After the preliminary 2 week culture *in vitro*, a suspension of cells was obtained. HF were filled with parathyroid cell (PT) suspension at concentration of 1×10^6 cells/ml or 17.6×10^6 cells/ml. The encapsulated cells were cultured over 9 or 33 days in 1 ml RPMI1640 containing 10% fetal calf serum (FCS) or in Chang's medium. As the positive control, non-encapsulated PT cells were cultured. The level of intact parathormone (PTH), produced by the encapsulated cells was evaluated in the culture medium with the radioimmunoassay test (RIA).

The mean PTH production accounting for 1×10^3 PT cells encapsulated or non-encapsulated (the positive control) in HF during 9 or 33 days culture is presented in Table 11.2. The physiological level of PTH per number of PT cells, calculated for 10^3 cells (using the known level of PTH in the peripheral blood and the mass of the parathyroid gland) gives a value of less than 8 pg, which corresponds to the obtained data.[90]

Concluding, encapsulation of PT cells in HF ensures diffusion of nutrients from the culture medium to the encapsulated cells and allows for functioning of the cells for at least 33 days in the *in vitro* experiment. The PTH produced by PT cells *in vitro* diffused through the HF membrane wall, which agrees with the PTH molecular mass and the HF membrane MWCO applied in the experiment. The presence of PT cells of normal appearance and the PTH produced by encapsulated PT cells (assessed by immunocytochemistry with monoclonal rat anti-human PTH antibody) was observed in the lumen of the HF after 9 days from subcutaneous implantation into rabbits.[79]

Table 11.2 Mean PTH production calculated for 1×10^3 PT cells encapsulated or non-encapsulated (positive control) in HF during 9 or 33 days' culture

HF/time of culture	Mean PTH level, produced by 10^3 PT cells during the culture (pg/ml)
HF/9 days ($n = 18$)	3.76 ± 0.32
HF/33 days ($n = 10$)	4.31 ± 0.91
positive control/33 days ($n = 10$)	4.70 ± 1.88

The values are presented as mean values ±SD after subtraction of the background value; *n*, number of PTH assays.

Evaluation of the system producing the hemopoietic factor

Interleukin-3 (IL-3), the hematopoietic cytokine, is one of the hematopoietic bone marrow cell's growth factor.[91] This 28 kDa glycoprotein, produced primarily by the bone marrow cells and by antigen or mitogen-activated T cells, has consistent stimulatory effects on myelopoiesis.[92] The IL-3 is involved in bone marrow hemopoiesis and dendritic cell maturation in anti-viral or anti-tumor reactivity.[93–95] One of the cell lines producing IL-3 is myelomonocytic, mouse cell line WEHI-3B.[96]

Cells producing mouse IL-3, encapsulated in the permiselective membrane, *in vitro* or post *in vivo* by assessment of the IL-3 activity[80] and the phenotypic assessment of cells were evaluated.[81] The production of IL-3 by HF-encapsulated WEHI-3B cells was assessed during 2 weeks' culture and during 1 week of culture following 1 week's subcutaneous implantation into mouse. The presence of IL-3 in the culture medium was evaluated by testing the BaF3 cells viability. The growth of the mouse BaF3 cell line depends on presence of IL-3 (survival factor) in the culture medium (Fig. 11.13). The encapsulated WEHI-3B cells secreted IL-3, which diffused to the surrounding culture medium, and may be used as a source of the survival factor to BaF3 cells. The HF-encapsulated WEHI-3B cells could be implanted subcutaneously into mouse, and afterwards, after explantation of the HF cell system, the cells secreted IL-3 as well. This procedure indicates stability of the HF-encapsulated cells system, and the probable *in vivo* function of WEHI-3B cells.

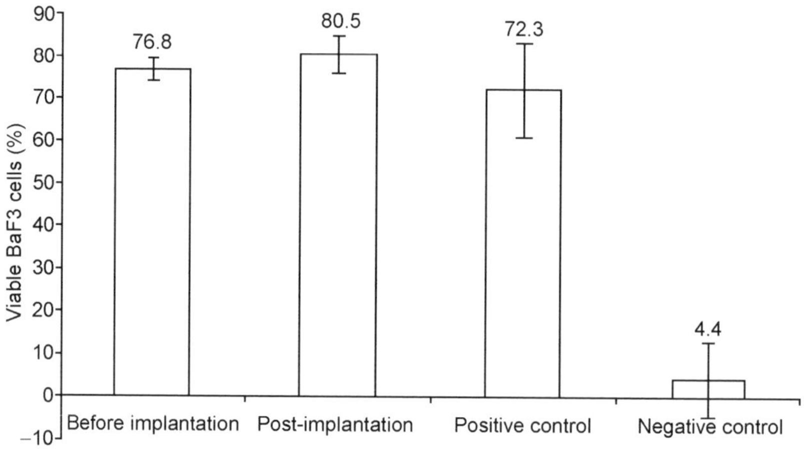

11.13 Cytometric evaluation of the BaF3 cells viability, during the culture in presence of the WEHI-3 cells encapsulated in HF. The viability of the cells depends on the presence of IL-3 in the culture medium. The BaF3 cells (1×10^6 cells/ml) were cultured over 1 day in the culture medium with addition of 15% WEHI-3B culture medium from encapsulated WEHI-3B cells culture before or after implantation or non-encapsulated WEHI-3B cells culture (the positive control) or without addition of supernatant (the negative control).

IL-3 is the growth and differentiation factor for the bone marrow cells. The culture medium of HF-encapsulated WEHI-3B cells taken from mouse were tested for ability to induce bone marrow cell differentiation. The mouse bone marrow cells at an initial concentration of 1×10^6cells/ml were cultured (5% CO_2, 37 °C) over 7 days in 1 ml culture medium in the presence of 1 week culture medium of encapsulated WEHI-3B cells taken after 2 weeks' subcutaneous implantation. The samples were tested for bone marrow cells with phenotype $CD45^+CD4^+$ and $CD45^+CD8^+$ in the immunocytochemical reaction by the flow cytometer.

The percentage number of $CD45^+CD4^+$ phenotype cells (helper T lymphocytes) did not change during 7 days of culture. Initially, there were 10.3% $CD45^+CD4^+$ cells; after 1 week of culture with encapsulated WEHI-3B cells supernatant there were $8.7 \pm 4.6\%$ $CD4^+$ cells ($n = 4$). The percentage number of $CD45^+CD8^+$ cytotoxic/suppressor T lymphocytes rose from 1.6% to $4.2 \pm 0.22\%$ ($n = 4$), after 1 week of culture with encapsulated WEHI-3B cell supernatant. The obtained values suggest the growth and differentiation of the cytotoxic T lymphocytes $CD8^+$.

Concluding, encapsulation of WEHI-3B cells in HF does not disturb their function and IL-3 production. The product of encapsulated WEHI-3B cells diffuses to the culture medium and is steadily produced after 2 weeks' implantation in the animal. It presents features characteristic of the survival and differentiation factor of IL-3.

Modification of the membrane by endothelial cells: design and development of the artificial vessel

Design and development of an artificial vessel have two main objectives: to create progress of vascular surgery by an artificial vessel endothelialization and to establish a new type of *in vitro* model, that simulates vascular biology and that can be used as a basis for diagnostics and therapy development.

The endothelial cell coating of the blood-contacting surface may reduce thrombogenicity of synthetic small diameter vascular prostheses. A combination of the synthetic implants and the endothelial cell has been studied since the late 1970s, when Herring in 1978[97] used vascular endothelial cells harvested from living animals to test artificial vessel endothelialization.

The main problem of artificial vascular endothelialization is related to the low endothelial cell attachment rate to the currently available vascular graft materials. Additionally, when the graft is exposed to the pulsatile blood flow, a high proportion of cells is washed off from the lumen. The lack of retention of cells could be partly overcome by sodding,[98] but other techniques, involving engineering the lumen to improve the endothelial cell adhesion and endothelial cell retention rate, have been developed. Precoating with EC adhesive molecules that are found in the endothelial basement membrane of blood vessels has been the most successful so far. But some authors have investigated other techniques, such as shear stress preconditioning and electrostatic charging.[99–101] Summing up, currently many research groups are working on efficient vascular endothelialization. Some of the results obtained are very promising.

The lack of good experimental models has made studies on pathophysiological mechanisms difficult or even impossible. Monitoring and assessing patient data in order to understand mechanisms of the diseases with a vascular biology are very complicated and often lead to uncertain results. Applications of the animal models are also not very successful owing to the different animal immune systems. Therefore, a cell culture model that simulates vascular biology is a necessary tool to elucidate mechanisms that can be used as a basis for diagnostics and therapy development.

Recently an endothelial cell culture model has been designed and developed, which represents the natural behaviour of human blood vessels. This enables interactions with blood cells to be studied.[102] A schematic representation of the model is presented in Fig. 11.14. The basis of this model is an original HF cartridge with polysulfone membrane. The polysulfone capillary membranes used in this bioreactor were obtained from a special installation for capillary membrane formation constructed in the Institute of Biocybernetics and Biomedical Engineering (IBBE PAS). Construction of the HF cartridge was also prepared at the IBBE PAS using polycarbonate housings. One, with an inner diameter of 9 mm and outer diameter of 11 mm and the second with 6 mm

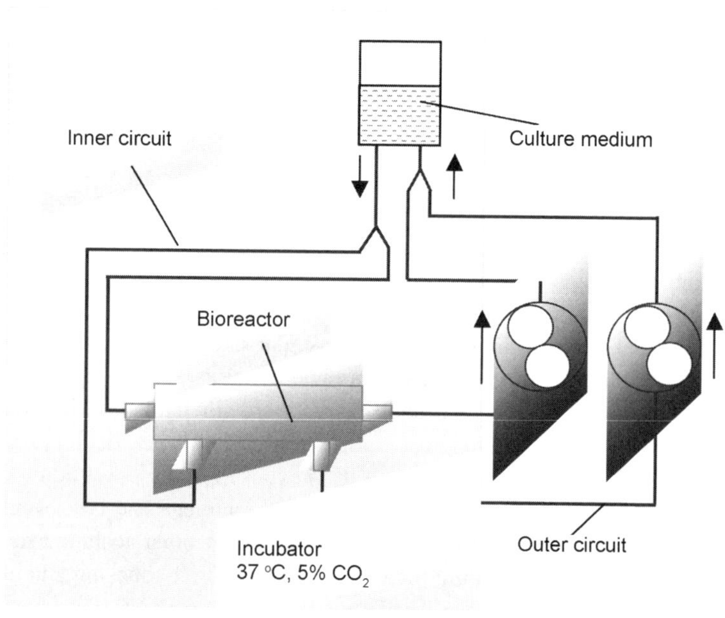

11.14 A schematic representation of the endothelial cell culture model.

Table 11.3 Specification of the capillary polysulfone filter (IBBE PAS)

Surface area (cm^2)	Number of capillaries	Inner diameter (μm)	Wall thickness (μm)	Module length (mm)
17	27	336	69	60

and 11 mm, respectively. The length of both housing types was 70 mm. The inner surface area of the tested modules was optimized from an initial value of 40 cm^2 to the final 17 cm^2. Characterization of the applied membrane is presented in Table 11.3.

The process of forming the artificial vessel model consisted of two phases: (1) endothelial cells seeding on the inner capillary surface and (2) cell culturing. The first phase was preceded by a membrane coating procedure. A number of different coating conditions were tested. Fibronectine (5 $\mu g/cm^2$) was the most adequate coating material for human umbilical vein endothelial cells (HUVEC) growth on the tested experimental polysulfone membrane. In addition, it was determined that the HUVEC cells need to be introduced to the unit by using pumping (slight pressure) due to the small diameter of the fibers. After endothelial cell introduction, the seeding phase started. This lasted 4 h and during this time the bioreactor was slowly rotated in the specially designed IBBE PAS rotating device.

The tested HF cartridge was connected to the perfusion system, as in the scheme in Fig. 11.14. The bioreactor was perfused by the cell culture medium to provide the nutrient supply. In order to simulate physiological conditions, HUVECs were grown in the bioreactor under condition of the pulsatile flow. Cell culture medium was pumped in the circuit using a peristaltic pump. For the first 24 h of culture, the flow rate was increased to set the established physiological target value of the shear stress affecting the endothelial cells. To characterize the distribution and density of the cells attached to the surface, the upper light microscopy was used in combination with hematoxylin and eosin staining and indirect fluorescence microscopy to visualize the cells on the graft surface. These methods enable us to characterize the whole graft surface within a few hours after withdrawal from the circuit.

Experimental evaluation of the developed cell culture model indicated that after both 48 and 72 h of the test, the inner surface of the membrane was almost confluently covered by endothelial cells. Continuous application of the selected value of the shear stress to the endothelial cells attached on the inner graft surface seems to enhance metabolic cellular functions, leading to an increased proliferation rate and adherence of the cells to the graft. Under these conditions, the endothelial cells have normal physiological morphology. A typical result of the obtained artificial vessel model after 72 h culturing procedure is shown in Fig. 11.15.

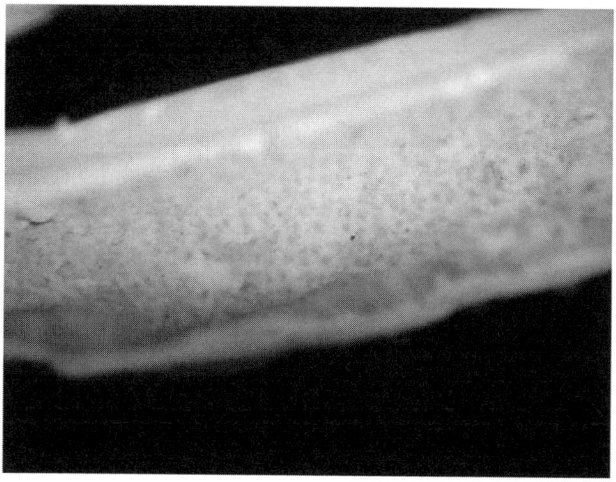

11.15 Picture (10×) of the inner capillary surface 72 h after the endothelial cell culturing.

Summing up, the results obtained from the *in vitro* experiments prove that the designed construction of the polysulfone HF bioreactor as well as the applied cell attachment and the culturing procedures are very effective. The endothelial cell model that has been developed seems to be very promising tool for studies on vascular pathophysiology in selected diseases, for example sepsis.[103]

11.3 Conclusions

The present experiments performed with HF-based systems are described to illustrate the function of the system rather than to describe its direct clinical application as a method of treatment. The value of encapsulated cell systems producing biologically active factors are as follows: possible prolonged continuous local production of regulatory factors, easy initiation of the factor application by implantation of HF with regulatory factor producing cells as well as termination of treatment by explantation of the HF, and exploitation/application of the allo- and xenogeneic cells for production of the biological factor.

Polypropylene silanized HFs, with the biological material encapsulated within, are proposed for local production of biologically active substances. An artificial vessel model could be considered as a promising tool in the studies on the pathogenesis of certain diseases.

11.4 Acknowledgements

The authors would like to express their gratitude to Ph. D. E. Jankowska from the Department of Electron Microscopy Institute of Biostructure University Medical Academy for her contribution to this work.

11.5 References

1. H. Zimmerman, M. Hillgartner, B. Manz, P. Feilen, F. Brunnenmeier, U. Leinfelder, M. Weber, H. Cramer, S. Schneider, C. Hendrich, F. Volke and U. Zimmermann, Fabrication of homogeneously cross-linked, functional alginate microcapsules validated by NMR-, CLSM-, and AFM-imaging, *Biomaterials* 24 (2003) 2083–96.
2. H. Yang, H. Iwata, H. Shimizu, T. Takagi, T. Tsuji and F. Ito, Comparative studies of *in vitro* and *in vivo* function of three different shaped bioartificial pancreas made of agarose hydrogel, *Biomaterials* 15 (1994) 113–20.
3. N. Morikawa, H. Iwata, S. Matsuda, J. I. Miyazaki and Y. Ikada, Encapsulation of mammalian cells into synthetic polymer membranes using least toxic solvents, *J Biomat Sci* 8 (1997) 575–86.
4. B.A. Zielinski, M.B. Goddard, M.J. Lysaght, in R. Lanza, R. Langer, W. Chick (Eds.), *Principles of tissue engineering*. Bioscience, Austin, Texas, USA, 1997, Chapter 22.
5. T. Kidchob, S. Kimura and Y. Imashini, Degradation and release profile of microcapsules made of poly[L-lactic acid-co-L-lysine (*Z*)], *J Contr Rel* 54 (1998) 283–92.
6. Z.Y. Zhang, Q.N. Ping and B. Xiao, Microencapsulation and characterization of tramadol-resin complexes, *J Control Rel* 66 (2000) 107–13.
7. R.M. Dawson, R.L. Broughton, W.T.K. Stevenson and M.V. Sefton, Microencapsulation of CHO cells in a hydroxyethyl methacrylate-methyl methacrylate copolymer, *Biomaterials* 8 (1987) 360–66.
8. C.A. Crooks, A.J. Douglas, R.J. Broughton and A.V. Sefton, Microencapsulation of mammalian cells in a HEMA copolymer: effects on capsule morphology and permeability, *J Biomed Mater Res* 24 (1990) 1241–62.
9. A.S. Michaels, High flow membranes. US Patent 3,615,024 (1971).
10. M. Shoichet, F. Gentile and S. Winn, The use of polymers in the treatment of neurological disorders. *Trends Polym Sci* 3 (1995) 374–80.
11. R.H. Li, S. Williams, M. White and D. Rein, Dose control with cell lines used for encapsulated cell therapy, *Tissue Eng* 5 (1999) 453–66.
12. M.G. Hollingshead, M.C. Alley, R.F. Camalier, B.J. Abbot, J.G. Mayo, L. Malspeis and MR Grever, *In vivo* cultivation of tumor cells in hollow fibers, *Life Sci* 57 (1995) 131–41.
13. R.P. Lanza, A.M. Beyer, J.E. Staruk and W.Chick, Biohybrid artificial pancreas, *Transplantation* 56 (1993) 1067–72.
14. D.W. Sharp, C.J. Swanson and B.J. Olack, Protection of encapsulated human islets implanted without immunosuppression in patients with type I or type II diabetes and in nondiabetic control subjects, *Diabetes* 43 (1994) 1167.
15. P. Aebischer, E. Buchser, J.M. Joseph, J. Favre, N. de Tribolet, M. Lysaght, S. Rudnick and M. Goddard, Transplantation in humans of encapsulated xenogenic cells without immunosuppression, *Transplantation* 58 (1994) 1275–77.
16. J. Sagen, H. Wang, P.A. Tresco and P. Aebischer, Transplants of immunologically isolated xenogenic chromaffin cells provide a long term source of neuroactive substances, *J Neurosci* 13 (1993) 2415–23.
17. Ze Qi Xu, M.G. Hollingshead, S. Borgel, C. Elder, A. Khilevich and M.T. Flavin, *In vivo* anti-HIV activity of (+)-calanolide A in the hollow fiber mouse model, *Bioorganic Medicin Chem Lett* 9 (1999) 133–8.
18. P. Prevost, S. Flori, C. Collier, E. Muscat and E. Rolland, Application of AN69

hydrogel to islet encapsulation, *Ann NY Acad Sci* 831 (1997) 344–9.

19. M.S. Schoichet, S.R. Winn, S. Athavale, J.M. Harris and F.T. Gentile, Poly(ethylene oxide)-grafted thermoplastic membranes for use as cellular hybrid bio-artificial organs in the central nervous system, *Biotechnol Bioeng* 43 (1994) 563–72.
20. M. Shoichet and D. Rein, *In vivo* biostability of a polymeric hollow fibre membrane for cell encapsulation, *Biomaterials* 17 (1996) 285–90.
21. H. Clark, T.A. Barbari, K. Stump and G. Rao, Histologic evaluation of the inflammatory responses around implanted hollow fiber membranes, *J Biomed Mater Res* 52 (2000) 183–92.
22. G.H. Hsiue, S.D. Lee and P.C. Chang, Surface modification of silicone rubber membrane by plasma induced graft copolymerization as artificial cornea, *Artif Organs* 20 (1996) 1196–207.
23. Z. Cheng and S.H. Teoh, Surface modification of ultra thin poly(epsilon-caprolactone) films using acrylic acid and collagen, *Biomaterials* 25 (2004) 1991–2001.
24. G. Poźniak, M. Bryjak and W. Trochimczuk, Sulfonated polysulfone membranes with antifouling activity, *Angew Makromol Chem* 233 (1995) 23–31.
25. M. Kabsch-Korbutowicz, G. Poźniak, W. Trochimczuk and T. Winniczki, Separation of humic substances by porous ion-exchange membranes from sulfonated polyfulfone, *Sep Sci Technol* 29 (1994) 2345–58.
26. P.C. Chang, S.D. Lee and G.H. Hsiue, Heterobifunctional membranes by plasma induced graft polymerization as an artificial organ for penetration keratoprosthesis, *J Biomed Mater Res* 39 (1998) 380–9.
27. M. Sasaki, N. Hosoya and M. Saruhashi, Vitamin E modified cellulose membrane, *Artif Organs* 24 (2000) 779–89.
28. K. Ishihaar, K. Fukumoto, H. Miyazaki and N. Nakabayashi, Improvement of hemocompatibility of a cellulose dialysis membrane with novel biomedical polymer having a phospholipid polar group, *Artif Organs* 18 (1994) 559–64.
29. Y. Wakabayashi, J. Sasaki, H. Fujita, K. Fujimoto, S. Murota and H. Kawaguchi, Effects of surface modification on materials on human neutrophil activation, *Biochem Biophys Acta* 1243 (1995) 521–8.
30. C. Nojiri, S. Kuroda, N. Saito, K.D. Park, K. Hagiwara, K. Senshu, T. Kido, T. Sugiyama, T. Kijima and Y.H. Kim, *In vitro* studies of immobilized heparin and sulfonated polyuretane using epifluorescent video microscopy, *ASAIO J* 41 (1995) M389–94.
31. J.Y. Ho, T. Matsuura and J.P. Santerre, The effect of fluorinated surface modifying macromolecules on the surface morphology of polyethersulfone membranes, *J Biomater Sci, Polymer Ed*, 11 (2000) 1085–104.
32. I. Garncarz, G. Poźniak and M. Bryjak, Modification of polysulfone membranes 1. CO_2 plasma treatment, *Eur Polym J* 35 (1999) 1419–28.
33. I. Garncarz, G. Poźniak and M. Bryjak. Modification of polysulfone membranes 1. Effect of nitrogen plasma treatment, *Eur Polym J* 36 (2000) 1563–9.
34. Y.J. Kim, I.K. Kang, M.W. Huh and S.Ch. Yoon, Surface characterization and *in vitro* blood compatibility of poly(ethylene terephthalate) immobilized with insulin and/or heparin using plasma glow discharge, *Biomaterials* 21 (2000) 121.
35. M. Mason, K.P. Vercruysse, K.R. Kirker, R. Frisch, M.D. Marecak, G.D. Prestwich and W.G. Pitt: Attachment of hyaluronic acid to polypropylene, polystyrene, and polytetrafluoroethylene, *Biomaterials* 21 (2000) 31.
36. L. Granicka, J. Kawiak, E. Głowacka and A. Weryński, Encapsulation of OKT3

cells in hollow fibers', *ASAIO J* 42 (1996) M863–6.
37. K. Takebe, T. Shimura, B. Munkhbat, M. Hagihara, H. Nakanishi and K. Tsuji, Xenogenic (pig to rat) fetal liver fragment transplantation using macrocapsules for immunoisolation, *Cell Trans* 5 (1996) S31–S33.
38. G. Reichelt, Bubble point measurements on large areas of microporous membranes, *J Membr Sci* 60 (1991) 253–9.
39. L.H. Granicka, J. Kawiak, M. Snochowski, J. M. Wójcicki, S. Sabalińska and A. Weryński, Polypropylene hollow fiber for cells isolation. methods for evaluation of diffusive transport and quality of cells encapsulation, *Artif Cells, Blood Subst Immobiliz Biotechnol* 31 (2003) 251–64.
40. D.F. Wiliams, Summary and definitions, in: *Progress in Biomedical Engineering: Definition in Biomaterials (4)*, Elsevier, Amsterdam, 1987, pp. 66–71.
41. P. Soon Shiong, M. Otterlei, G. Skjak-Brek, O. Smidsrod, R. Heintz, R.P. Lanza and T. Espevik, An immunological basis for fibrotic reaction to implanted microcapsules, *Transplant Proc* 23 (1991) 758–9.
42. B. Rihova, Biocompatibility of biomaterials: hemocompatibility, immunocompatibility and biocompatibility of solid polymeric materials and soluble targetable polymeric carriers, *Adv Drug Deliv Rev* 21 (1996) 157–76.
43. J.E. Babensee, J.M. Anderson, L.V. McIntire and A.G. Micos, Host response to tissue engineered devices, *Adv Drug Deliv Rev* 33 (1998) 111–39.
44. H. Clark, T.A. Barbari, K. Stump and G. Rao, Histologic evaluation of the inflammatory responses around implanted hollow fiber membranes, *J Biomed Mater Res* 52 (2000) 183–92.
45. S. Wu, *Polymer Interface and adhesion*, Marcel Dekker, New York, 1982.
46. Polish/European standard: Biological assessment of the medical products. The cytotoxity evaluation-methods *in vitro*. Pr PN-EN-30993-5 (Pr PN-ISO-10993-5).
47. L.H. Granicka, M. Wdowiak, A. Kosek, S. Świeżewski, D. Wasilewska, E. Jankowska, A. Weryński and J. Kawiak, Survival analysis of *Escherichia coli* encapsulated in hollow fibre membrane *in vitro* & *in vivo*. Preliminary report. *Cell Transplant* 14 (2005) 323–30.
48. R.P. Lanza, A.M. Beyer, J.E. Staruk and W. Chick, Biohybrid artificial pancreas, *Transplantation* 56 (1993) 1067–72.
49. P. Prevost, S. Flori, C. Collier, E. Muscat and E. Rolland, Application of AN69 hydrogel to islet encapsulation, *Ann NY Acad Sci* 831 (1997) 344–9.
50. S. Sigrist, A. Mechine-Neuville, K. Mandes, V. Calenda, S. Braun, G. Legeay, J.P. Bellocq, M. Pinget and L. Kessler, Influence of VEGF on the viability of encapsulated pancreatic rat islets after transplantation in diabetic mice, *Cell Transplant* 12 (2003) 627–35.
51. R.P. Lanza, R. Jackson, A. Sullivan, J. Ringeling, C. McGrath, W. Kuhtreiber and W.L. Chick, Xenotransplantation of cells using biodegradable microcapsules, *Transplantation* 67 (1999) 1105–11.
52. K.M. de Fife, M.S. Shive, K.M. Hagen, D.L. Clapper and J.M. Anderson, Effects of photochemically immobilized polymer coatings on protein adsorption, cell adhesion, and the foreign body reaction to silicone rubber, *J Biomed Mater Res* 44 (1999) 298–307.
53. J.L. Dulong and C. Legallis, Contribution of a finite element model for the geometric optimization of an implantable bioartificial pancreas, *Artif Organs* 26 (2002) 583–9.
54. A. Dorling, K. Riesbeck, A. Warrens and R. Lechler, Clinical xenotransplantation of solid organs, *Lancet* 349 (1997) 867–71.

55. U. Siebers, A. Horcher, H. Brandhorst, D. Brandhorst, B. Hering, K. Federlin, R.G. Bretzel and T. Zekorn, Analysis of the cellular reaction towards microencapsulated xenogeneic islets after intraperitoneal transplantation. *J Mol Med* 77 (1999) 215–18.
56. N. Kobayashi, T. Okitsu, S. Nakaji and N. Tanaka, Hybrid bioartificial liver: establishing a reversibly immortalized human hepatocyte line and developing a bioartificial liver for practical use, *Artif Organs* 6 (2003) 236–44.
57. C. Serguera, D. Bohl, E. Rolland, P. Prevost and J.M. Heard, Control of erythropoietin secretion by doxycyline of mifepristone in mice bearing polymer-encapsulated engineered cells, *Hum Gene Ther* 10 (1999) 375–83.
58. F. Schwenter, N. Deglon and P. Aebischer, Optimization of human erythropoietin secretion from MLV-infected human primary fibroblasts used for encapsulated cell therapy, *J Gene Med* 5 (2003) 246–57.
59. B.L. Schneider, F. Schwenter, W. F. Pralong and P. Aebischer, Prevention of the initial immuno-inflammatory response determines the long-term survival of encapsulated myoblasts genetically engineered for erythropoietin delivery, *Mol Ther* 7 (2003) 506–14.
60. C. Rinsch, P. Quinodoz, B. Pittet, N. Alizadeh, D. Baetens, D. Montandon, P. Aebischer and M.S. Pepper, Delivery of FGF-2 but not VEGF by encapsulated genetically engineered myoblasts improves survival and vascularization in a model of acute skin flap ischemia, *Gene Ther* 8 (2001) 523–33.
61. R.H. Li, M. White, S. Williams and T. Hazlett, Polyvinyl alcohol synthetic polymer foams as scaffolds for cell encapsulation, *J Biomater Sci, Polym Ed* 9 (1998) 239–58.
62. R.H. Li, S. Williams, M. White and D. Rein, Dose control with cell lines used for encapsulated cell therapy, *Tissue Eng* 5 (1999) 453–66.
63. R.H. Li, S. Williams, M. Burkstrand and E. Roos, Encapsulation matrices for neutrophic factor-secreting myoblast cells, *Tissue Eng* 6 (2000) 151–63.
64. D.W. Sharp, C.J. Swanson, B.J. Olack *et al.*, Protection of encapsulated human islets implanted without immunosuppression in patients with type I or type II diabetes and in nondiabetic control subjects, *Diabetes* 43 (1994) 1167.
65. B.A. Zielinski, M.B. Goddard, M.J. Lysaght, in R. Lanza, R. Langer, W. Chick (Eds), *Principles of Tissue Engineering*. Bioscience, Austin, Texas, USA, 1997, Chapter 22.
66. P. Aebischer, E. Buchser, J.M. Joseph, J. Favre, N. De Tribolet, M. Lysaght, S. Rudnick and M. Goddard, Transplantation in humans of encapsulated xenogenic cells without immunosuppresion, *Transplantation* 58 (1994) 1275–7.
67. A.D. Zurn, H. Henry, M. Schelup, V. Aubert, L. Winkel, B. Eilers, C. Bauchmann and P. Aebischer, Evaluation of an intrathecal immune response in amyotrophic lateral sclerosis patients implanted with encapsulated genetically engineered xenogenic cells, *Cell Transplant* 9 (2000) 471–84.
68. M.G. Hollingshead, M.C. Alley, R.F. Camalier, B.J. Abbot, J.G. Mayo, L. Malspeis and M.R. Grever, *In vivo* cultivation of tumor cells in hollow fibers, *Life Sci* 57 (1995) 131–41.
69. Q.W. Mi, D. Lantvit, E. Reyes-Lim, H.Y. Chai, W.M. Zhao, I.S. Lee, S. Peraza-Sanchez, O. Ngassapa, L.B. Kardono, S. Riswan, M.G. Hollingshead, J.G. Mayo, N.R. Farnsworth, G.A. Cordell, A.D. Kinghorn and J.M. Pezzuto, Evaluation of the potential cancer chemotherapeutic efficacy of natural product isolates employing in vivo hollow fiber tests, *J Natural Products* 65 (2002) 842–50.
70. Ze Qi Xu, M.G. Hollingshead, S. Borgel, C. Elder, A. Khilevich and M.T. Flavin, *In vivo* anti-HIV activity of (+)-calanolide A in the hollow fiber mouse model, *Bioorg Med Chem Lett* 9 (1999) 133–8.

71. S. Sakai, T. Ono, H. Ijima and K. Kawakami, Control of molecular weight cut-off for immunoisolation by multilayering glycol chitosan alginate polyion complex on alginate-based microcapsules, *J Microenc* 17 (2000) 691–9.
72. J. Brauker, L.A. Martinson, S.K. Young and R.C. Johnson, Local inflammatory response around diffusion chambers containing xenografts. Nonspecific destruction of tissues and decreased local vascularization, *Transplantation* 61 (1996) 1671–7.
73. R. Dembczynski and T. Jankowski, Determination of pore diameter and molecular weight cut-off of hydrogel-membrane liquid-core capsules for immunoisolation, *J Biomater Sci, Polym Ed* 12 (2001) 1051–8.
74. R. Robitaille, F.A. Leblond, Y. Bourgeois, N. Henley, M. Loignon and J.P. Halle, Studies on small (<350 micron) alginate-poly-L-lysine microcapsules. Determination of carbohydrate and protein permeation through microcapsules by reverse-size exclusion chromatography, *J Biomed Mater Res* 50 (2000) 420–27.
75. N. Lembert, P. Petersen, J. Wesche, P. Zschocke, A. Enderle, M. Doser, H. Planck, H.D. Becker and H.P. Ammon, *In vitro* test of a new biomaterials for development of a bioartificial pancreas, *Ann NY Acad Sci* 944 (2001) 271–6.
76. T. Orłowski, E. Godlewska, M. Mościcka and E. Sitarek, The influence of intraperitoneal transplantation of free and encapsulated Langerhans islets on the second set phenomen, *Artif Organs* 27 (2003) 1062–67.
77. R. Lanza, W. Kuhtreiber, D. Ecker, J.E. Staruk and W.L. Chick, Xenotransplantation of porcine and bovine islets without immunosuppresion using uncoated alginate microspheres, *Transplantation* 59 (1995) 1377–84.
78. L.H. Granicka, J. Kawiak, E. Głowacka and A. Weryński, Encapsulation of OKT3 cells in hollow fibers, *ASAIO J* 42 (1996) M863–6.
79. L.H. Granicka, M. Migaj, T. Zawitkowska, B. Woźniewicz, T. Tołłoczko, A. Weryński, and J. Kawiak, Evaluation of human parathyroid cells functioning encapsulated in polypropylene hollow fibers, *Cell Transplant* 8 (1999) 165.
80. L.H. Granicka, G. Hoser, A. Weryñski and J. Kawiak, Evaluation of a system producing hemopoietic factor. WEHI 3B cell line function when encapsulated in polypropylene hollow fiber, *Folia Histochem Cytobiolog* 39 (2001) 102–3.
81. L. Granicka and J. Kawiak, in A. Noworyta, A. Trusek-Hołownia (Eds), *Using Membranes to Assist in Cleaner Processes*, Wrocaw, Poland, 2001, pp. 255–8.
82. H. Iwata, N. Morikawa, T. Fuji, T. Takagi, T. Samejima and Y. Ikada, Does immunoisolation need to prevent the passage of antibodies and complements, *Transplant Proc* 27 (1995) 3224–6.
83. Y. Hagihara, Y. Saitoh, H. Iwata, T. Taki, S.I. Hirano, N. Arita and T. Hayakawa, Transplantation of xenogenic cells secreting fl-dorphin for pain treatment: analysis of the ability of components of complement to penetrate through polymer capsules, *Cell Transplant* 6 (1997) 527–30.
84. B.J. de Haan, M.M. Faas and P. de Vos, Factors influencing insulin secretion from encapsulated islets, *Cell Transplant* 12 (2003) 617–25.
85. M.V. Risbud, M.R. Bhonde, R.R. Bhonde, Effect of chitosan–polyvinyl pyrrolidone hydrogel on proliferation and cytokine expression of endothelial cells: implications in islet immunoisolation, *J Biomed Mat Res* 57 (2001) 300–5.
86. S.M. Flechner, D.A. Goldfarb, R. Fairchild, C.S. Modlin, R. Fisher, B. Mastroianni, K.J. O'Malley, D.J. Cook and A.C. Novick, A randomized prospective trial of low dose OKT3 induction therapy prevent rejection and minimize side effects in recipients of kidney transplants, *Transplantation* 69 (2000) 2374–81.
87. L.H. Granicka, M. Migaj, B. Woźniewicz, T. Zawitkowska, T. Tołłoczko, A. Weryñski and J. Kawiak, Encapsulation of parathyroid cells in hollow fibers: a

preliminary report, *Folia Histochem Cytobiolog* 38 (2000) 129–31.
88. H.T. Keutmann, M.M. Sauer, G.N. Hendy, J.L. O'Riodan and J.T. Potts, Complete amino acid sequence of human parathyroid hormone, *Biochemistry* 17 (1978) 5723–9.
89. A.M. Sun, Parathyroid, in: *Principles of Tissue Engineering*, edited by R. Lanza, R. Langer, W. Chick, Landes Company, USA, 1997.
90. F. Kokot and R. Stupnicki, *Radioimmunological and Radiocompetitive Methods in Clinical Use*. PZWL, Warsaw, 1985, pp. 194–201.
91. E.G. de Vries, M.M. van Gameren and P.H. Willemse, Recombinant human interleukin 3 in clinical oncology, *Stem Cells* 11 (1993) 72–80.
92. R.L. Cutler, D. Metcalf, N.A. Nicola and G.R. Johnson, Purification of a multipotential colony stimulating factor from pokeed mitogen-stimulated mouse spleen cell conditioned medium, *J Biol Chem* 260 (1985) 6579–87.
93. M. Martinez-Moczygemba and D.P. Huston, Biology of common beta receptor-signaling cytokines: IL-3, IL-5 and GM-CSF, *J Allergy Clin Immun* 112 (2003) 553–65.
94. W. Oster, J. Frisch and U. Nicolay, Interleukin-3. Biologic effects and clinical impact, *Cancer* 67 (1991) 2712–17.
95. H. Suzuki, N. Katayama, Y. Ikuta, K. Mukai, A. Fujieda, H. Mitani, H. Araki, H. Miyashi, N. Hoshino, H. Nishikawa, K. Nishii, N. Minami and H. Shiku, Activities of granulocyte-macrophage colony-stimulating factor and interleukin-3 on monocytes, *Am J Haematol* 75 (2004) 179–89.
96. J.C. Lee, A.J. Hapel and J.N. Ihle, Constitutive production of a unique lymphokine (IL-3) by the WEHI-3 cell line, *J Immunol* 128 (1982) 2393–8.
97. M. Herring, A. Gardner and J. Glover, A single staged technique for seeding vascular grafts with autogenous endothelium, *Surgery* 84 (1978) 498–504.
98. K.M. Ahlswede and S.K. Williams, Microvascular endothelial cell sodding of 1-mm expanded polytetrafluoroethylene vascular grafts, *Arteriosclerosis and Thrombosis* 14 (1994) 25–31.
99. L. Xiao, D. Shi and J. Chin, Role of precoating in artificial vessel endothelialization, *Traumatology* 7(5) (2004) 312–16.
100. M.D. Helmut Gulbins, M. Dauner, R. Petzold, A. Goldemund, I. Anderson, M. Doser, B. Meiser and B. Reichart, Development of an artificial vessel lined with human vascular cells, *J Thorac Cardiovasc Surg* 128 (2004) 372–7.
101. L.A. Poole-Warren, K. Schindhelm, A.R. Graham, P.R. Slowiaczek and K.R. Noble, Performance of small diameter synthetic vascular prostheses with confluent autologous endothelial cell linings, *J Biomed Mater Res* 30(2) (1996) 221–9.
102. A. Ciechanowska, D. Schwanzer-Pfeiffer, E. Rossmanith, S. Sabalinska, C. Wojciechowski, J. Hartmann, K. Hellevuo, A. Chwojnowski, P. Foltynski, D. Falkenhagen and J.M. Wojcicki, 'Artificial vessel as a basis for disease related cell culture model' Mediterranean Conference on Medical and Biological Engineering 'Health in the Information Society' Incorporating the: 2nd Health Telematic Conference, Island of Ischia, Naples, Italy, 31 July/5 August, Medicon 2004.
103. E. Rossmanith, A. Ciechanowska, S. Sabalinska, D. Schwanzer-Pfeiffer, C. Wojciechowski, J. Hartmann, K. Hellevuo, A. Chwojnowski, P. Foltynski, D. Falkenhagen and J.M. Wojcicki, Development of an endothelial cell culture model for studies on vascular pathophysiology in sepsis', *Int J Artif Organs* 27(7) (2004) 567.

Part IV

The clinical relevance of artificial cells and cell engineering

12

Artificial cells in medicine: with emphasis on blood substitutes

T M S CHANG, McGill University, Canada

12.1 Introduction

Artificial cells were first reported a number of years ago[1–5] (Fig. 12.1). Biologically active materials inside the artificial cells are prevented from coming into direct contact with external materials such as leukocytes, antibodies or tryptic enzymes. Smaller molecules can equilibrate rapidly across the ultra-thin membrane with large surface to volume relationship. A number of potential medical applications using artificial cells have been proposed or shown.[2–5] Artificial cells are studied at different scales.[6,7] They range from macro-dimensions, to micron-dimensions and to nano-dimensions (Fig. 12.2). Examples for the use of those in the nano-dimensions include red blood cell substitutes and drug and gene therapeutics. Those in the micron-dimensions include the bioencapsulation of enzymes, peptides, drugs, vaccines, biosorbents and other materials. Examples for those in the macro- and micron-dimensions include encapsulation of islets, endocrine cells, stem cells and genetic engineered cells.

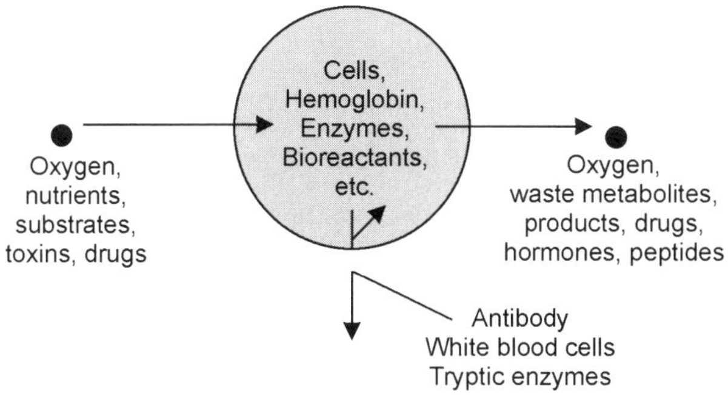

12.1 Basic principle of artificial cells. (*Source*: from Chang TMS (2004) *Artificial Cells, Blood Substitutes & Biotechnology*, 32: 1–23, with copyright permission from Marcel Dekker Publisher.)

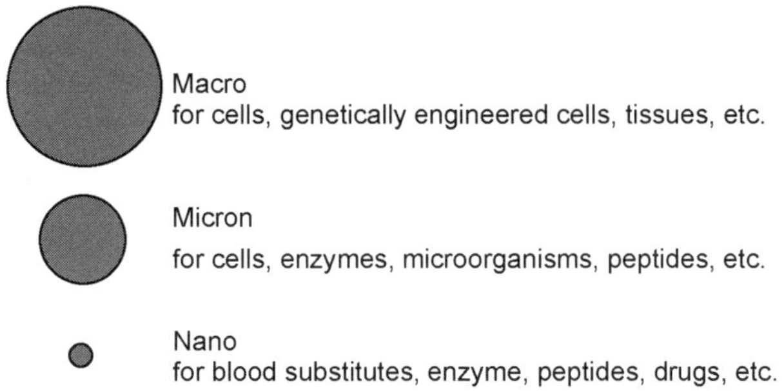

12.2 Variations in diameter for artificial cells. (*Source*: from Chang TMS (2004) *Artificial Cells, Blood Substitutes & Biotechnology*, 32: 1–23, with copyright permission from Marcel Dekker Publisher.)

12.2 Red blood cell substitutes

12.2.1 Nanobiotechnological approach: polyhemoglobin

We have extended our original approach of artificial cells containing hemoglobin and enzymes[1,2] to form nano-dimension polyhemoglobin. This is based on nanobiotechnology. Nanobiotechnology is the assembly of biological molecules into nano-dimension structures of between 1 and 100 nanometers. This is based on the use in this laboratory of bifunctional agents such as diacid[2,5] or later glutaraldehyde[8] to assemble a number of hemoglobin molecules together by crosslinking into polyhemoglobin (Fig. 12.3). With problems related to HIV

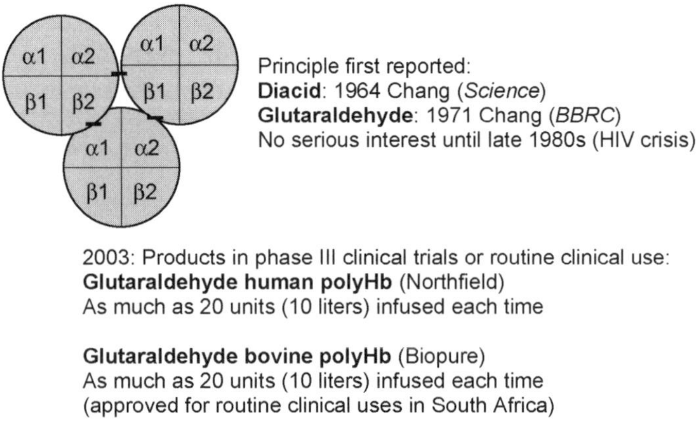

12.3 Nanobiotechnology-based red blood cell substitutes in the form of polyhemoglobin. This is formed by the intermolecular crosslinking of hemoglobin into a soluble complex. In this form, they are retained in the circulation. (*Source*: from Chang TMS (2004) *Artificial Cells, Blood Substitutes & Biotechnology*, 32: 1–23, with copyright permission from Marcel Dekker Publisher.)

in donor blood, there has been extensive development towards blood substitutes starting in the early 1990s.[8–12] At present, two of these are in the final stages of clinical trials and waiting for US Food and Drug Administration (FDA) approval. These are developed independently by two groups based on Chang's basic principle of glutaraldehyde crosslinked polyhemoglobin.[8] One is pyridoxalated glutaraldehyde human polyhemoglobin.[13,14] Gould *et al.* showed in phase III clinical trial that this can successfully replace extensive blood loss in trauma surgery by maintaining the hemoglobin level with no reported side effects.[13] They infused up to 10 liters into individual trauma surgery patients.[13] They are now carrying out further phase III clinical trials on its use in pre-hospital emergencies since no typing and cross-matching is needed and it can be used on the spot. In the United States this product has been approved for compassionate use in patients and it is waiting for regulatory decision for routine clinical uses. Another one is glutaraldehyde crosslinked bovine polyhemoglobin that has been extensively tested in phase III clinical trials.[15,16] This bovine polyhemoglobin has been approved for veterinary medicine in the USA and for routine clinical use in South Africa. The above two polyhemoglobins have been approved for compassionate uses in humans and they are waiting for regulatory approval for routine clinical uses in humans in North America. They have a number of advantages compared with donor red blood cells and they are particularly useful in surgery (Table 12.1). However, they are only oxygen carriers and do not have all the functions of red blood cells (RBC) that may be needed for certain clinical conditions, e.g. they cannot remove oxygen radicals. Furthermore their circulation times are much shorter than that of red blood cells.[12]

Table 12.1 Comparison of polyhemoglobin with donor red blood cells. Polyhemoglobin has many advantages over red blood cells and is useful for use during surgery. However, it cannot be used in a number of other clinical conditions. This is because unlike red blood cells (rbc), polyhemoglobin is only an oxygen carrier. It does not have rbc enzymes needed for many functions including the removal of oxygen radicals. Furthermore, its circulation time is much shorter than that of rbc.

	Donor red blood cells	Blood substitutes (polyhemoglobin)
Infection	Possible	Sterilized
Source	Limited	Unlimited?
Blood groups	Yes	None
Usage	Cross-match, typing	Immediately
Storage	42 days	>1 year
Functions	As red blood cells	Oxygen carrier

12.2.2 Polyhemoglobin crosslinked with RBC antioxidant enzymes

Reperfusion using oxygen carrier alone in sustained severe hemorrhagic shock or sustained ischemic organs as in stroke, myocardial infarction or organ transplantation may result in the production of oxygen radicals and tissue injury.[7–10] Chang and coworkers have used a crosslinked polyhemoglobin–superoxide dismutase–catalase (PolyHb–SOD–CAT)[17–19] (Fig. 12.4). Unlike PolyHb, PolyHb–SOD–CAT did not cause a significant increase in oxygen radicals when it was used to reperfuse ischemic rat intestine.[19] In a transient global cerebral ischemia rat model, it was found that after 60 min of ischemia, reperfusion with polyHb–SOD–CAT did not cause any disruption of blood–brain barrier or brain edema.[20] However, polyHb alone without the antioxidant enzyme caused disruption of blood–brain barrier and brain edema.[20]

12.2.3 Polyhemoglobin crosslinked with tyrosinase for melanoma

Malignant tumors have abnormal microcirculation, resulting in under-perfusion by red blood cells and therefore lower tissue oxygen tension. However, chemotherapy and radiotherapy are more effective in the presence of adequate oxygen tension during the time of therapy. Yu and Chang have recently crosslinked tyrosinase with Hb to form a soluble nanodimension polyHb–tyrosinase

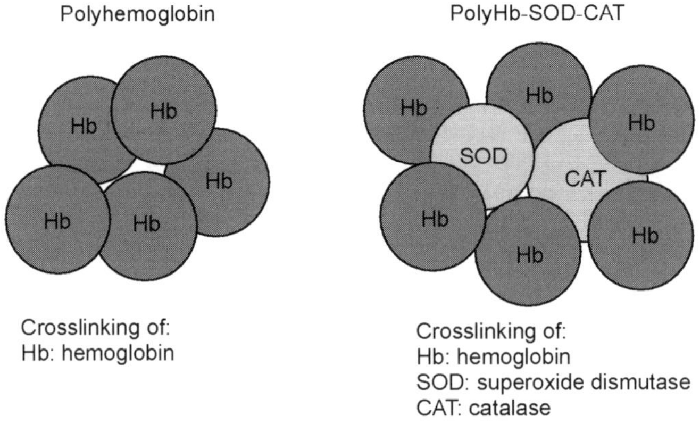

12.4 Crosslinking of hemoglobin with two RBC enzymes to form polyhemoglobin–catalase-superoxide dismutase (PolyHb-CAT-SOD). Unlike polyhemoglobin, this has RBC enzymes that can remove oxygen radicals. This is important in preventing ischemia-reperfusion injuries for ischemic conditions such as sustained severe hemorrhagic shock, stroke, myocardial infarction and organ transplantation. (*Source*: from Chang TMS (2004) *Artificial Cells, Blood Substitutes & Biotechnology*, 32: 1–23, with copyright permission from Marcel Dekker Publisher.)

complex.[21,22] This has the dual function of supplying the oxygen needed for optimal chemotherapy or radiation therapy for 24 h and also lowering the systemic levels of tyrosine. It was shown that intravenous injections of poly-hemoglobin-tyrosinase can delay the growth of the melanoma without causing adverse effects or changes in the growth of the treated animals.[22] Intravenous injections of polyhemoglobin-tyrosinase can be combined with oral administrations of microencapsulated tyrosinase to maintain a low systemic tyrosine level.[23]

12.2.4 Complete artificial red blood cells based on nanotechnology

This author's original idea of a complete artificial red blood cell[1,2] is now being developed as third generation blood substitutes.[7,12] The hemoglobin lipid vesicle is one of these approaches.[24–26] Chang and coworkers have used a different approach based on biodegradable polymer and nanotechnology resulting in an artificial red blood cell of 80–150 nm diameter.[27–29] These nano artificial red blood cells contain all the red blood cell enzymes needed for their long-term function[29] (Fig. 12.5). Recent studies show that by using a polyethylene–glycol–polylactide copolymer membrane it is possible to increase the circulation time of these nano artificial red blood cells to double that of polyhemoglobin.[29]

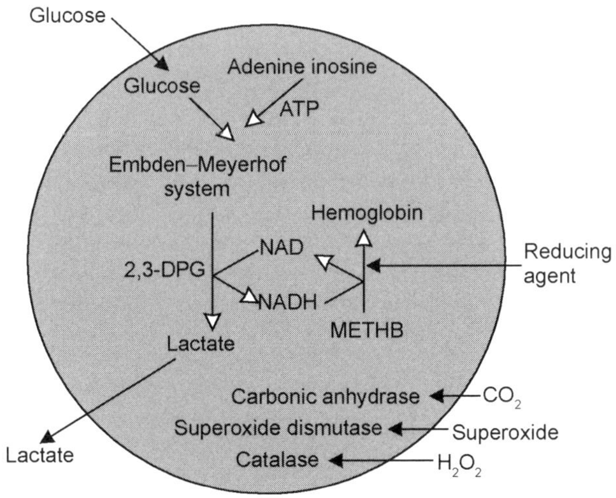

12.5 Nanodimension artificial red blood cells with polyethylene–glycol-polylactide membrane. In addition to hemoglobin, this contains the same enzymes that are normally present in red blood cells. Thus, this nano artificial red blood cell has the complete function of the red blood cells. (*Source*: from Chang TMS (2004) *Artificial Cells, Blood Substitutes & Biotechnology*, 32: 1–23, with copyright permission from Marcel Dekker Publisher.)

12.3 Delivery of enzymes, drugs and genes

12.3.1 Enzyme therapy

Enzyme inside artificial cells is protected from the extracellular environment, but substrates smaller than protein can equilibrate rapidly into the artificial cells and products can diffuse out.[2,5,6,30] Change and Poznansky earlier implanted artificial cells containing catalase into acatalesemic mice, animals with a congenitical deficiency in catalase.[31] This replaces the deficient enzymes and prevented the animals from the damaging effects of oxidants. The artificial cells protect the enclosed enzyme from immunological reactions (Fig. 12.1).[32] It was also shown that artificial cells containing asparaginase implanted into mice with lymphosarcoma delayed the onset and growth of lymphosarcoma.[33] The single problem preventing the clinical application of enzyme artificial cells is the need to repeatedly inject these enzyme artificial cells.

12.3.2 Oral administration to avoid the need for implantation

To solve the problem of repeated injections, artificial cells were given orally. As they travel through the intestine, they act as microscopic dialysers. By encapsulating enzymes and other material inside the microcapsules, they can act as a combined dialyser-bioreactor. For example, artificial cells containing urease and ammonia adsorbent were used to lower the systemic urea level.[5] It was found that microencapsulated phenylalanine ammonia lyase given orally can lower the elevated phenylalanine levels in phenylketonuria (PKU) rats.[34] This is because of the finding of an extensive recycling of amino acids between the body and the intestine.[35] This is now being developed for PKU.[36,37] In addition to PKU recent studies show that oral artificial cells containing tyrosinase is effective in lowering systemic tyrosine levels in rats for melanoma that depends on tyrosine for growth.[38] Oral microencapsulated xanthine oxidase has also been used to lower the systemic hypoxanthine levels in a patient with Lesch–Nyhan disease.[39]

12.3.3 Drug delivery systems

Artificial cell is one of the many approaches for drug and gene delivery.[6] The use of polylactide biodegradable semipermeable microcapsules containing enzymes, insulin, hormones, vaccines and other biologicals in 1976[40] is now being extended by many groups to include biodegradable nanoparticles and nanocapsules. The liposome delivery system is also being studied extensively. Other extensions of artificial cells are now being developed for nanotechnology, nanomedicine and nanobiotechnology.[41,42]

12.4 Hemoperfusion and biosorbents

The first successful use of artificial cells in routine clinical therapy was hemoperfusion.[4,5,43,44] This is because 10 ml of artificial cells have a total surface area of about 20 000 cm^2, equivalent to that of a hemodialysis machine. This, together with the ultra-thin membrane of 0.02 μm and membrane equivalent pore radius of 1.4 nm, allows for extremely fast equilibration of molecules smaller than large proteins. Thus artificial cells containing adsorbents are much more effective than standard hemodialysis in removing toxins and drugs from the blood of patients.[43,44] This is now being used extensively in many countries for the treatment of suicidal or accidental overdose of drug or poisons, especially where regular dialysis machines are not readily available.

12.5 Cell encapsulation: hepatocytes, islets, stem cells and others

12.5.1 Cell therapy

The encapsulation of biological cells was reported in 1964 based on a drop method[2] and it was proposed that 'protected from immunological process, encapsulated endocrine cells might survive and maintain an effective supply of hormone'[3,5] (Fig. 12.1). Conaught Laboratory was approached to develop this for use in islet transplantation for diabetes. Sun from Conaught and his collaborators later extended this drop-method by using milder physical cross-linking.[45] This resulted in alginate–polylysine–alginate (APA) microcapsules containing cells. They showed that after implantation, the islets inside artificial cells remain viable and continued to secrete insulin to control the glucose levels of diabetic rats.[45]

Cell encapsulation for cell therapy has been extensively developed by many groups especially using artificial cells containing endocrine tissues, hepatocytes and other cells for cell therapy.[7,45–51] Cheng and coworkers have been studying the use of implantation of encapsulated hepatocytes for liver support.[52–57] It was found that implantation increases the survival of rats in acute liver failure;[53] maintains a low bilirubin level in hyperbilirubinemic Gunn rats;[54] prevents xenograft rejection.[55] A two-step cell encapsulation method was developed to improve the APA method resulting in improved survival of implanted cells.[57]

12.5.2 Stem cells

Using this two-step method plus the use of co-encapsulation of stem cells and hepatocytes we have further increased the viability of encapsulated hepatocytes both in culture and also after implantation.[58–59] One implantation of the co-encapsulated hepatocyte-stem cells into Gunn rats lowered the systemic bilirubin levels and maintained this low level for 2 months, instead of only 1 month when not co-encapsulated with stem cells.[60] One potential problem with the use of hepatocytes is that if human hepatocytes are used, the source will be very limited. If we use hepatocytes from animals, e.g. porcine hepatocytes, the proteins secreted from the porcine hepatocytes may result in untoward immunological reactions even if the hepatocytes are immuno-isolated in the artificial cells. Using stem cells alone without hepatocytes may solve this problem. Since human bone marrow stem cells can be conveniently obtained in sufficient amounts, the following study was carried out to see if artificial cells containing only bone marrow stem cells could be used.

The use of artificial cells containing only bone marrow stem cells for rats with 90% of liver removed was studied.[61,62] It was found that one intraperitoneal injection of these artificial cells resulted in the recovery of the rats with the liver regenerating to normal weight in 2 weeks. Analysis shows that this is due to the secretion of a factor similar to hepatic growth factor that reaches the liver through the portal circulation. At the end of 2 weeks there is also trans-differentiation of bone marrow stem cells into hepatocyte-like cells positively stained with ALB, CKS, CK18, AFP and PAS. Injection of free bone marrow stem cells by themselves only show transient effect and did not increase the percentage of survival since they are not retained in the peritoneal cavity.

12.5.3 Microencapsulated genetically engineered cells

This has been studied out by many groups for potential applications in amyotrophic lateral sclerosis, dwarfism, pain treatment, IgG_1 plasmacytosis, hemophilia B, Parkinsonism and axotomized septal cholinergic neurons.[6,47,63] One group has used hollow fibers to macroencapsulated genetically engineered cells: the fibers can be inserted and then retrieved after use without being retained in the body.[64] To avoid the need for implantation, the oral use was studied of microencapsulated genetically engineered non-pathogenic *E. coli* DH5 cells containing *Klebsiella aerogenes* urease gene to lower systemic urea in renal failure rats.[65] The metabolic induction of lactobacillus similar to those for urea removal has also been investigated.[66]

12.6 Conclusions

This paper briefly summarizes some of the research from this group on artificial cells in the macro-, micro- and nano-dimensions. Bioencapsulation of biosorbent is fairly simple and therefore was in routine clinical use after only a few years of research and development. Some areas of microencapsulation of drugs for delivery are also straightforward and are in clinical application. However, in the more complicated areas, such as cell encapsulation and blood substitutes, it cannot be expected that clinical application will come after just a few years of research and development. There are many areas of applications being explored by many groups around the world.[7] The promise and potential of artificial cells also come with the need for further development towards actual applications. Much needs to be done for the more advanced uses of artificial cells to produce fruitful applications.

Artificial cells is a rapidly evolving area, and links to other groups around the world can be found on: www.artcell.mcgill.ca and www.artificialcell.info

12.7 Acknowledgements

This author is the principal investigator of the operating grant from the Canadian Institutes of Health Research; the 'Virage' Centre of Excellence in Biotechnology from the Quebec Ministry; and the MSSS-FRSQ Research Group (d'equip) award on Blood Substitutes in Transfusion Medicine from the Quebec Ministry of Health.

12.8 References

1. Chang TMS: 'Hemoglobin Corpuscles', Report of a research project for Honours Physiology, Medical Library, McGill University (1957). Also reprinted as part of '30 anniversary in Artificial Red Blood Cells Research', *J. Biomater., Artif. Cells Artif. Organs* 1988; 16: 1–9.
2. Chang TMS: Semipermeable microcapsules. *Science* 1964; 146: 524–525.
3. Chang TMS, MacIntosh FC, Mason SG: Semipermeable aqueous microcapsules: I. Preparation and properties. *Can. J. Physiol. Pharmacol.* 1966; 44: 115–128.
4. Chang TMS: Semipermeable aqueous microcapsules ['artificial cells']: with emphasis on experiments in an extracorporeal shunt system. *Trans. Am. Soc. Artif. Intern. Organs* 1966; 12: 13–19.
5. Chang TMS: Artificial Cells. Springfield: C.C. Thomas Publisher, 1972 (available for free online viewing at www.artcell.mcgill.ca).
6. Chang TMS: Therapeutic applications of polymeric artificial cells. *Nature Rev. Drug Discovery* 2005; 4: 221–235.
7. Chang TMS: *Artificial Cells: Biotechnology, Nanotechnology, Blood Substitutes, Degenerative Medicine, Bioencapsulation, Cell/Stem Cell Therapy*. World Scientific Publishing, Singapore, 2007.
8. Chang TMS: Stabilization of enzyme by microencapsulation with a concentrated protein solution or by crosslinking with glutaraldehyde. *Biochem. Biophys. Res. Com.* 1971; 44: 1531–1533.
9. Chang TMS: *Blood Substitutes: Principles, Methods, Products and Clinical Trials*. Vol. 1 Karger, Basel, 1997 (available for free online viewing at www.artcell.mcgill.ca).
10. Chang TMS. Oxygen carriers. *Curr. Opin. Investigational Drugs*, 2002; 3(8): 1187–1190.
11. R. Winslow: Current status of blood substitute research: towards a new paradigm. *J. Int. Med.* 2003; 253: 508–517.
12. Chang TMS: New generations of red blood cell substitutes. *J. Internal Med.* 2003; 253: 527–535.
13. Gould SA, Moore EE, Hoyt DB, Ness PM, Norris EJ, Carson JL, Hides GA, Freeman IHG, DeWoskin R, Moss GS: The life-sustaining capacity of human

polymerized hemoglobin when red cells might be unavailable. *J. Amer. Coll. Surgeons* 2002; 195: 445–452.
14. Gould, SA, Sehgal LR, Sehgal HL, DeWoskin R, Moss GS: The clinical development of human polymerized hemoglobin, in *Blood Substitutes: Principles, Methods, Products and Clinical Trials*. Vol. 2 (Chang TMS, ed.) Karger: Basel, 1998, pp. 12–28.
15. Sprung J, Kindscher JD, Wahr JA, Levy JH, Monk TG, Moritz MW, O'Hara PJ: The use of bovine hemoglobin glutamer-250 (Hemopure) in surgical patients: results of a multicenter, randomized, single-blinded trial. *Anesth. Analg.* 2002; 94: 799–808.
16. Pearce LB, Gawryl MS: Overview of preclinical and clinical efficacy of Biopure's HBOCs; in *Blood Substitutes: Principles, Methods, Products and Clinical Trials*. Vol. 2 (Chang TMS, ed.) Karger: Basel, 1998, pp. 82–98.
17. D'Agnillo F, Chang TMS: Polyhemoglobin-superoxide dismutase, catalase as a blood substitute with antioxidant properties. *Nature Biotechnol.* 1998; 16(7): 667–671.
18. D'Agnillo F, Chang TMS: Absence of hemoprotein-associated free radical events following oxidant challenge of crosslinked hemoglobin-superoxide dismutase-catalase. *Free Radical Biol. Med.* 1998; 24(6): 906–912.
19. Razack S, D'Agnillo F, Chang TMS: Effects of polyhemoglobin–catalase–superoxide dismutase on oxygen radicals in an ischemia-reperfusion rat intestinal model. *Artif. Cells, Blood Substit. Immobiliz. Biotechnol.* 1997; 25: 181–192.
20. Powanda D, Chang TMS: Cross-linked polyhemoglobin–superoxide dismutase–catalase supplies oxygen without causing blood brain barrier disruption or brain edema in a rat model of transient global brain ischemia-reperfusion. *Artif. Cells, Blood Substit. Immob. Biotechnol.* 2002; 30: 25–42.
21. Yu BL, Chang TMS: *In vitro* and *in vivo* enzyme studies of polyHb–tyrosinase. *J. Biotechnol. Bioeng.* 2004; 32: 311–320.
22. Yu BL, Chang TMS: *In vitro* and *in vivo* effects of polyHb–tyrosinase on murine B16F10 melanoma. *Melanoma Res.* 2004; 14: 197–202.
23. Yu BL, Chang TMS: Effects of combined oral administration and intravenous injection on maintaining decreased systemic tyrosine levels in rats. *Artif. Cells Blood Substit. Immobil. Biotechnol.* 2004; 32: 127–147.
24. Rudolph AS, Rabinovici R, Feuerstein GZ (eds): *Red Blood Cell Substitutes*. New York: Marcel Dekker, Inc., 1997.
25. Tsuchida E (ed.): *Blood Substitutes: Present and Future Perspectives*. Amsterdam: Elsevier, 1998.
26. Philips WT, Klpper RW, Awasthi VD, Rudolph AS, Cliff R, Kwasiborski VV, Goins, BA: Polyethylene glyco-modified liposome-encapsulated hemoglobin: a long circulating red cell substitute. *J. Pharm. Exp. Therapeutics* 1999; 288: 665–670.
27. Yu WP, Chang TMS: Submicron biodegradable polymer membrane hemoglobin nanocapsules as potential blood substitutes: a preliminary report. *J. Artif. Cells, Blood Substit. Immobiliz. Biotechnol.* 1994; 22: 889–894.
28. Chang TMS and Yu WP: Nanoencapsulation of hemoglobin and red blood cell enzymes based on nanotechnology and biodegradable polymer; in *Blood Substitutes: Principles, Methods, Products and Clinical Trials*. Vol. 2 (Chang TMS, ed.) Karger: Basel, 1998, pp. 216–231.
29. Chang TMS, Powanda D, Yu WP: Analysis of polyethylene–glycol–polylactide nano-dimension artificial red blood cells in maintaining systemic hemoglobin levels and prevention of methemoglobin formation. *Artif. Cells, Blood Substit. Biotechnol.* 2003; 31(3): 231–248.

30. Chang TMS: Artificial cells bioencapsulation in macro, micro, nano and molecular dimensions. *Artif. Cells, Blood Substit. Biotechnol.* 2004; 32: 1–23.
31. Chang TMS, Poznansky MJ: Semipermeable microcapsules containing catalase for enzyme replacement in acatalsaemic mice. *Nature* 1968; 218(5138): 242–245.
32. Poznansky MJ, Chang TMS: Comparison of the enzyme kinetics and immunological properties of catalase immobilized by microencapsulation and catalase in free solution for enzyme replacement. *Biochim. Biophys. Acta* 1974; 334: 103–115.
33. Chang TMS: The *in vivo* effects of semipermeable microcapsules containing L-asparaginase on 6C3HED lymphosarcoma. *Nature* 1971; 229(528): 117–118.
34. Bourget L, Chang TMS: Phenylalanine ammonia-lyase immobilized in microcapsules for the depletion of phenylalanine in plasma in phenylketonuric rat model. *Biochim. Biophys. Acta* 1986; 883: 432–438.
35. Chang TMS, Bourget L, Lister C: New theory of enterorecirculation of amino acids and its use for depleting unwanted amino acids using oral enzyme-artificial cells, as in removing phenylalanine in phenylketonuria. *Artif. Cells, Blood Substit. Immobil. Biotechnol.* 1995; 25: 1–23.
36. Sarkissian CN, Shao Z, Blain F, Peevers R, Su H, Heft R, Chang TMS, Scriver CR: A different approach to treatment of phenylketonuria: phenylalanine degradation with recombinant phenylalanine ammonia lyase. *Proc. Nat. Acad. Sci. USA* 1999; 96: 2339–2344.
37. Liu J, Jia X, Zhang J, Xiang G, Hu W, Zhou Y. Study on a novel strategy to treatment of phenylketonuria. *Artif. Cells, Blood Substit. Immobil. Biotechnol.* 2002; 30: 243–258.
38. Yu BL, Chang TMS: Effects of long term oral administration of microencapsulated tyrosinase on maintaining decreased systemic tyrosine levels in rats. *J. Pharmaceut. Sci.* 2004; 93: 831–837.
39. Palmour RM, Goodyer P, Reade T, Chang TMS: Microencapsulated xanthine oxidase as experimental therapy in Lesch–Nyhan disease. *Lancet* 1989; 2(8664): 687–688.
40. Chang TMS: Biodegradable semipermeable microcapsules containing enzymes, hormones,vaccines, and other biologicals. *J. Bioeng.* 1976; 1: 25–32.
41. Ranade VV and Hollinger MA: *Drug Delivery Systems*. New York. CRC Press Pharmacology & Toxicology Series, 2nd edition, 2003.
42. LaVan, DA, Lynn DM, Langer R: Moving smaller in drug discovery and delivery. *Nature Rev.: Drug Discovery* 2002; 1: 77–84.
43. Chang TMS: Microencapsulated adsorbent hemoperfusion for uremia, intoxication and hepatic failure. *Kidney Int.* 1975; 7: S387–S392.
44. Winchester JF: Hemoperfusion; in *Replacement of Renal Function by Dialysis*. (Maher JF, ed.). Boston: Kluwer Academic Publisher, 1988, pp. 439–592.
45. Lim F, Sun AM: Microencapsulated islets as bioartificial endocrine pancreas. *Science* 1980; 210: 908–909.
46. Chang TMS: Artificial cells with emphasis on bioencapsulation in biotechnology. *Biotechnol. Annu. Rev.* 1995; 1: 267–295.
47. Orive G, Hernandez RM, Gascon AR, Calafiore R, Chang TMS *et al.*: Cell encapsulation: promise and progress. *Nature Medicine* 2003; 9: 104–107.
48. Kulitreibez WM, Lauza PP, Cuicks WL (eds): *Cell Encapsulation Technology and Therapy*. Boston: Burkhauser, 1999.
49. Hunkeler D, Prokop A, Cherrington AD, Rajotte R, Sefton M. (eds): Bioartificial Organs A: Technology, Medicine and Material, *Ann. N.Y. Acad. Sci*, 1999, 831: 271–279.

50. Chang TMS, Prakash, S: Procedure for microencapsulation of enzymes, cells and genetically engineered microorganisms. *Molec. Biotechnol.* 2001; 17: 249–260.
51. Dionne KE, Cain BM, Li RH, Bell WJ, Doherty EJ, Rein DH, Lysaght MJ, Gentile FT: Transport characterization of membranes for immunoisolation, *Biomaterials* 1996; 17: 257–266.
52. Chang, TMS: Bioencapsulated hepatocytes for experimental liver support, *J. Hepatol.* 2001; 34: 148–149.
53. Wong H, Chang TMS: Bioartificial liver: implanted artificial cells microencapsulated living hepatocytes increases survival of liver failure rats. *Int. J. Artif. Organs* 1986; 9: 335–336.
54. Bruni S, Chang TMS: Hepatocytes immobilized by microencapsulation in artificial cells: effects on hyperbiliru-binemia in Gunn rats. *J. Biomat. Artif. Cells Artif. Organs* 1989; 17: 403–412.
55. Wong H, Chang TMS: The viability and regeneration of artificial cell microencapsulated rat hepatocyte xenograft transplants in mice. *J. Biomater. Artif. Cells Artif. Organs* 1988; 16: 731–740.
56. Wong H, Chang TMS: Microencapsulation of cells within alginate poly-L-lysine microcapsules prepared with standard single step drop technique: histologically identified membrane imperfections and the associated graft rejection. *Biomater. Artif. Cells Immobil. Biotechnol.* 1991; 19: 675–686.
57. Wong H, Chang TMS: A novel two-step procedure for immobilizing living cells in microcapsule for improving xenograft survival. *Biomater. Artif. Cells Immobil. Biotechnol.* 1991; 19: 687–698.
58. Liu Z, Chang TMS: Effects of bone marrow cells on hepatocytes: when co-cultured or co-encapsulated together. *Artif. Cells, Blood Substit. Immobil. Biotechnol.* 2000; 28 (4): 365–374.
59. Liu ZC, Chang TMS: Transplantation of co-encapsulated hepatocytes and marrow stem cells into rats. *Artif. Cells, Blood Substit. Immobil. Biotechnol.* 2002; 30: 99–112, 2002.
60. Liu ZC, Chang TMS: Coencapsulation of stem sells and hepatocytes: *in-vitro* conversion of ammonia and *in-vivo* studies on the lowering of bilirubin in Gunn rats after transplantation. *Int. J. Artif. Organs* 2003; 26(6): 491–497.
61. Liu ZC, Chang TMS: Transplantation of bioencapsulated bone marrow stem cells improves hepatic regeneration and survival of 90% hepatectomized rats: a preliminary report. *Artif. Cells, Blood Substit. Biotechnol.* 2005; 33(4): 405–410.
62. Liu, ZC, Chang TMS: Transdifferentiation of bioencapsulated bone marrow cells into hepatocyte-like cells in the 90% hepatectomized rat model. *J. Liver Transplant.* 2006; 12: 566–572.
63. Chang TMS, Prakash S: Therapeutic uses of microencapsulated genetically engineered cells. *Molec. Med. Today* 1998; 4: 221–227.
64. Aebischer P, Schluep M, Deglon N, Joseph JM, Hirt L, Heyd B, Goddard M, Hammang JP, Zurn AD, Kato AC, Regli F, Baetge EE: Intrathecal delivery of CNTF using encapsulated genetically modified xenogeneic cells in amyotrophic lateral sclerosis patients. *Nat. Med.* 1996; 2: 696–699.
65. Prakash S, Chang TMS: Microencapsulated genetically engineered live *E. coli* DH5 cells administered orally to maintain normal plasma urea level in uremic rats. *Nat. Med.* 1996; 2(8): 883–887.
66. Chow KM, Liu ZC, Prakash S, Chang TMS: Free and microencapsulated *Lactobacillus* and effects of metabolic induction on urea removal. *Artif. Cells, Blood Substit. Biotechnol.* 2003; 31(4): 425–434.

13

Bone marrow stromal cells as 'universal donor cells' for myocardial regeneration therapy

J LUO, R C-J CHIU and D SHUM-TIM,
McGill University Health Center, Canada

13.1 Current therapy for heart failure

With the advances in diagnostic techniques and pharmacological, interventional and surgical therapies, the survival rates in patients with various cardiovascular diseases have improved. Congestive heart failure (CHF) is the only cardiovascular disorder that is increasing in incidence over time in North America.[1] Moreover, patients with CHF have demonstrated to have a much poorer quality of life compared with patients with other chronic diseases, scoring poorly on various physical function, emotional well-being and overall social scales.[2,3] Current therapies such as medical treatment, mechanical assist devices and whole organ transplantation are associated with limited success, inherent complications and costly without addressing the fundamental pathophysiological problem related to the loss of functioning contractile cardiomyocytes in patients with CHF.

Regardless of the cause of CHF (ischemic heart disease, cardiomyopathy, hypertension, congenital or valvular disease), the associated decrease in left ventricular diastolic compliance and systolic ejection fraction result in impaired ventricular filling and decreased tissue perfusion, leading to progressive symptoms of fatigue, dyspnea on exertion and peripheral edema. In its advanced stage, it ultimately leads to death.[4]

Palliative therapy of CHF with medical or pharmacological agent has been recognized to have poor efficacy in reducing the symptoms and mortality of CHF patients.[5] Diuretics,[6] for preload reduction, or digoxin,[7] for positive inotropy, have been the backbone of medical therapy for many decades. The introduction of ACE-inhibitors,[8] angiotensin receptor blockers[9,10] and, more recently, calcium channel blockers,[11,12] β-blockers[13,14] or a combination of α and β-blockers[15] for regulation of cardiac afterload, remodeling or myocardial oxygen demand have promising therapeutic effects, although the clinical outcomes in the treatment of CHF remain rather poor, especially in the setting of

CHF caused by myocardial infarction which will lead to the loss of contractile cardiomyocytes. Overall, patients with CHF will live a life of low quality. Numerous ongoing clinical trials are now underway to evaluate various strategies of medical treatments. Unfortunately, more than one-fourth of the patients with end-stage heart failure are refractory to medical therapy. At present, they are potential candidates for various surgical therapies ranging from ventricular assist device (VAD) and total organ replacement.

VAD represents one of the recently established modes of therapy to prolong longevity in CHF patients.[16] It can be used as right (RVAD), left (LVAD) or biventricular assist (BiVAD) depending on the underlying pathology. However, most of the VADs, inherent with the risks of bleeding, embolism and infection, are used temporarily as a bridge to heart transplantation. Nevertheless, this costly approach does not address the fundamental pathophysiology related to the irreversible loss of contractile cardiomyocytes in patients with CHF. The survival of the patients ultimately depends on the functional reserve of the remaining cardiomyocytes or subsequent heart transplantation. It is not surprising that the best reported survival in VAD patients for the treatment of end-stage heart failure bridged to heart transplantation ranges only between 20 and 40%.[17] Although allograft heart transplantation potentially cures end-stage CHF, it requests life-long immunosuppression and has only 8 years' half-life.[18] Nevertheless donor scarcity has always been a major problem for heart transplant. There are only 2500 donors yearly available for the 40 000 end-stage CHF patients awaiting for heart donors in the United States.[18] Endeavors to increase the availability of donor hearts, such as prolonging donor organ preservation time and solid organ xenotransplantation, have had only limited success in increasing the donor pool.[19,20] Alternative therapies for patients with CHF are desperately needed.

13.2 The innate capacity for myocardial regeneration

Regeneration of terminally differentiated myocardium is limited and the adult mammalian heart is considered to be a post-mitotic organ without regenerative capacity.[21] It is assumed that shortly after birth, the heart has a relatively stable but slowly diminishing number of myocytes. Once the heart is injured, myocardial repair consists of non-contractile scar tissue replacement.

However recently, this notion has been challenged by the discovery of adult cardiac stem cells located in the heart.[22] Evidence from an observational study of post-mortem adult human hearts showed that cell division in the border zone of infarcted areas actually exists.[23] The presence of mitotic spindles, contractile rings, karyokinesis and cytokinesis were noted, suggesting cell division occurs in mature myocardium. In the infarcted heart, 4% of myocytes in the regions adjacent to the infarcts and 1% in the regions distant from the infarcts re-entered cell cycles after myocardial infarction.[23] Although this finding has shown the potential natural repairing mechanism in the heart, obviously 0.08% and 0.01% of mitotic indexes respectively in the peri-infarcted and distant region[23] were unable to improve heart function after significant cardiomyocyte loss.[22,24,25] Moreover, the origin of these mitotic myocytes is not clear. They can be derived from the resident cardiac stem cells,[22,26] from circulating stem cells[27] or from bone marrow stem cells[28] that reach the ischemic zone after infarction.[29,30]

13.3 The role of bone marrow stromal cells for myocardial regeneration

Normal bone marrow is composed of hematopoietic and non-hematopoietic cells.[31] The latter have also been termed stromal cells (BMCs) or mesenchymal stem cells (MSCs). These cells were originally thought to provide an appropriate matrix for hematopoietic cell development, but recent examination of these cell populations suggests a much broader spectrum of activity, including the generation of bone, cartilage, tendon,[32] fat,[33] muscles[34] and myocardium.[35] These findings have encouraged enormous research using MSCs for myocardial regeneration in the past few years and clinical endeavor to improve heart function in the patients with acute myocardial infarction (MI) or CHF.[36]

The years since 2000 have witnessed a plethora of new attempts to recruit endogenous and exogenous stem cells to replenish new contractile cardiomyocytes, which addresses the fundamental pathophysiological problem of CHF. Various cell types have been used for myocardial cell therapy:[37–49]

- *Embryonic stem cell*: embryonic stem cell-derived cardiomyocytes displayed structural and functional properties of early-stage cardiomyocytes that coupled electrically with host cardiomyocytes.[37,38] Its limitations include the unresolved ethical dilemma of sacrificing human embryo, allogeneic transplantation with immunological implication, and potential for teratoma formation, etc.
- *Fetal cardiomyocytes*: experimental transplantation of fetal cardiomyocytes engrafted into native heart and increased systolic function.[39,40] However, the use of fetal tissue is ethically controversial; transplantation of differentiated cells is a form of allograft transplantation that requires long-term immunosuppression. Obtaining sufficient amount of fetal cardiomyocytes to provide significant functional improvement in adult hearts is a huge logistical problem.
- *Cell-line myoblasts*: established cell-line myoblasts, such as C2C12 cells, are skeletal myoblasts genetically altered to proliferate indefinitely and differentiate into myoblasts in culture. These cells could differentiate into myofibers and express contractile proteins but no intercalated discs had been shown to exist between the implanted cells and native heart.[41] Potential tumor formation is a concern.

- *Skeletal myoblast cell*: implantation of satellite cell to regenerate new myocardium has been successfully reported by several investigators[42] but Atkins *et al.*[43] and Murry *et al.*[44] have shown them to differentiate into mature skeletal myocytes and not cardiomyocytes, when implanted into normal and injured myocardium. Despite these findings, autologous satellite cell implantation has been clinically applied in patients with improved segmental contractility and new onset of metabolic activity in the previously non-viable scar area confirmed with echocardiography and positron emission tomography, respectively.[45,46]
- *Marrow stromal cell (MSCs)*: Wakitani *et al.* first demonstrated that bone marrow cells exposed to 5-azacytidine or amphotericin B *in vitro* could differentiate into spontaneously beating myotubules.[47] Makino *et al.* further showed that a subpopulation of these cells could express various cardio-myocyte specific genes and able to fuse and contract spontaneously.[48] When 5-azacytidine pretreated MSCs were implanted into myocardial scars, they were shown to express cardiomyocytic markers and contributed to functional improvement.[49] Wang *et al.* in our group reported that MSCs could differentiate to express cardiomyocyte phenotype when injected directly into viable myocardium even without 5-azacytidine pre-treatment.[50] But the MSCs adjacent to the native cardiomyocytes appeared to differentiate more readily than those trapped within the needle tract scar at the implantation site. Orlic *et al.* had shown that implantation of MSCs resulted in formation of new myocardium that occupied 68% of the infarcted portion of the ventricle.[51] The labeled cells expressed myocyte enhancer factor-2, cardiac specific transcription factor GATA-4 and connexin-43. Recently, numerous interesting studies have shown that exogenous cells can engraft in adult myocardium and are capable of promoting meaningful repair of the LV that contributes measurable functional impact on damaged myocardium after ischemic injury in animal model.[52]

Clinical trials of MSCs implantation in patients with acute MI or CHF have confirmed its safety and potential benefits.[53–56] In most of the clinical trials, however, MSC therapy was carried out concomitantly with revascularization and therefore, its benefit could not be fully evaluated. Although most studies were small trials, enhancement of regional wall motion in the infarcted area and even increase in global LV ejection fraction (EF) were shown.[56] In a prospective randomized clinical trial, Patel *et al.* have injected autologous bone marrow derived CD34+ cells in 15 patients with NYHA class III/IV heart failure and EF<35% without coronary bypass. Early echocardiography showed significant improvement in EF in patients who received cell therapy (35%) versus control group (5%).[57]

The ideal cell sources for clinical application should be easy to obtain and expand *in vitro*, immunologically tolerated, capable of regenerating new cardiomyocytes, and readily available on the shelf.[58] With their attributes of ease of isolation, high expansion potential, genetic stability, reproducible characteristics from isolate to isolate, reproducible characteristics in widely dispersed laboratories, compatibility with tissue engineering principles and potential to enhance repair in many vital tissues, MSCs possess the unique advantages to become the preferred model for cellular therapeutic development.

13.4 Limitation of current cell therapy procedure

Despite its promising initial results from animal and clinical experience, autologous MSCs implantation has created a logistic problem clinically in patients having potential life-threatening cardiac events. Its clinical application is logistically inconvenient and expensive. In the acute setting of MI, it is limited by issues related to harvesting and culturing of sufficient quantity of autologous MSCs promptly in patients with a critical medical condition. There is also evidence that MSCs harvested from aged or cachectic patients are qualitatively and quantitatively compromised.[59] It is, therefore, ideal if a 'universal donor cell' obtained from young and healthy donors readily available for cell therapy can be used without rejection.

13.5 Unique immunological properties of MSCs

Adult stem cells are favored as the pluripotent autologous donor sources in cell therapy because autologous cells will not be rejected by the recipients. In ongoing investigations to use embryonic stem cells for tissue and organ regeneration, the need for therapeutic cloning is strongly championed by scientists in spite of the political and ethical controversies, again to avoid immunological rejection. In the rapidly advancing field of stem cell transplantation therapy for various diseases including heart failure, it is taken for granted that these donor stem cells will be subjected to immunological surveillance like any other fully differentiated somatic cells. Yet, in recent years, a number of perplexing observations to challenge this dogma have been reported. There is mounting evidence in the literature to suggest that MSCs are immune-privileged cells.[60–62] MSCs can be detected in major histocompatibility complex (MHC) mismatched recipients at extended time points without immunosuppression therapy,[63] indicating a lack of immune recognition and clearance. Liechty *et al.*[64] reported that human MSCs could engraft into various tissue organs including myocardium after intraperitoneal implantation in an immuno-competent fetal sheep. This finding was again confirmed by Airey *et al.* in a human to sheep model.[65] Toma *et al.*[66] have shown that human mesenchymal stem cells could differentiate into a cardiomyocytic phenotype in the adult murine heart. In a swine model, Poh *et al.*[67] and Makkar *et al.*[68] reported that allogeneic MSC implantation in myocardium could survive without immunosuppressive therapy.

Moreover, these cells were shown to differentiate and contribute to functional improvement of the host myocardium. Recently, our group also demonstrated that mouse MSCs transplanted intravenously into the adult rat recipients were recruited and engrafted into post-infarcted rat myocardium,[69] and contribute to functional improvement compared with the control group without MSC therapy.[70]

In view of the potential clinical implication as an alternative source for cellular and gene therapies, our lab has been investigating the immunological aspect and the fate of xenogeneic marrow stromal cells after systemic transplantation into fully immunocompetent adult recipients without immunosuppression. In our earlier study by Saito *et al.*,[69] MSCs from C57B1/6 mice were systemically given intravenously into immunocompetent Lewis rats recipients without immunosuppression. One week after xenogeneic MSCs injection, the recipient rats underwent either coronary ligation or sham operation. The lac-Z labeled cells were found to be engrafted into the bone marrow cavities without rejection. In addition, these labeled MSCs were recruited through the circulation as evidenced by the presence of lac-Z positive cells in the blood stream and subsequently engrafted into the injured myocardium at various time points. Interestingly, after homing into the injured hearts, these labeled MSCs showed evidence of phenotypic differentiation by staining positive for cardiomyocyte specific protein, and integrated in angiogenesis, and scar formation.

Encouraged by these unexpected results, we furthermore investigated the functional contribution of these xenogeneic MSCs and compared their immunological tolerance with that of terminally differentiated fibroblast from the same donors. We created a myocardial infarction model by proximal left coronary artery ligation in a series of immunocompetent adult Lewis rats and randomized them in various groups.[70] Lac-Z labeled mice MSCs were immediately injected into the peri-infarct area of the LV in one group, while Lac-Z labeled mice skin fibroblasts or plain cultured medium were injected into the peri-infarct areas of the LV in another two groups, respectively.

Recipients' heart samples harvested 4 weeks after cell injection were then stained for β-gal activity. Intense blue discoloration was seen on the infarct and peri-infarct area of the xenogeneic MSCs recipient group (Fig. 13.1). Interestingly, β-gal positive cells (black arrow) in the peri-infarcted myocardium (Fig. 13.2) were more elongated and phenotypically resemble surrounding host cardiomyocytes without any evidence of inflammatory infiltrates. Immunohistochemical staining for the cardiomyocyte-specific contractile protein troponin I-C also stained positive in the lac-Z+ cells (Fig. 13.3). In the xenogeneic fibroblast recipient group, histologic examination of left ventricular cross-sections showed extensive cellular infiltration within 1 day (Fig. 13.4), and no surviving labeled xenogeneic fibroblasts were found beyond day four after implantation.

Transthoracic echocardiography for ventricular functional studies were carried out 4 weeks after transplantation. Left ventricular ejection fraction

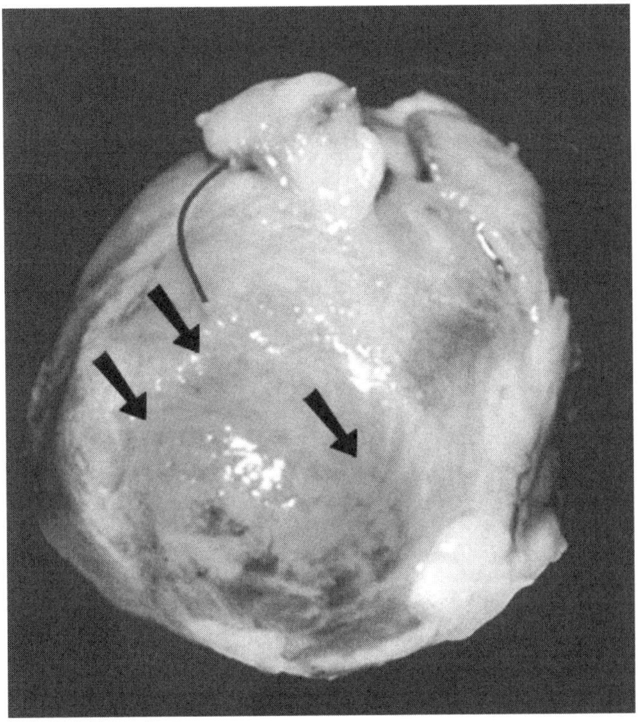

13.1 Gross heart specimen harvested at 4 weeks after xenogeneic MSCs injection were stained for β-galactosidase activity. Intense blue discoloration (arrow) was seen on infarct and peri-infarct areas of hearts even in the absence of immunosuppression.

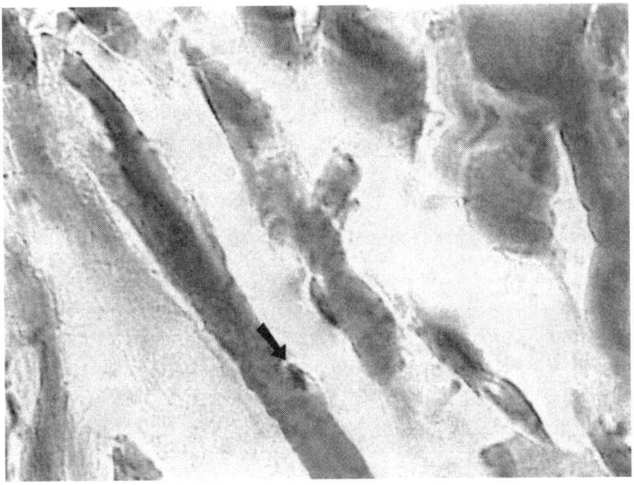

13.2 Hematoxylin and eosin staining of the myocardium showed β-galactosidase positive cells (arrow) in peri-infarcted zone that are more elongated and phenotypically resemble surrounding host cardiomyocytes.

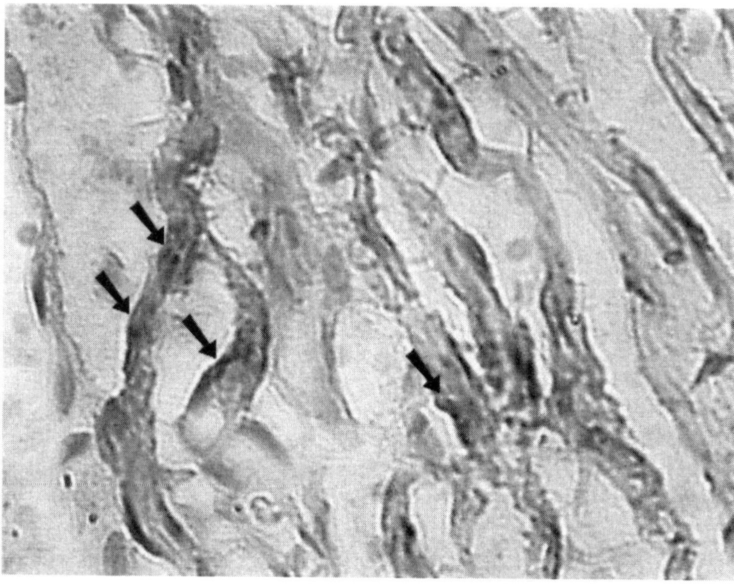

13.3 Immunohistochemical staining for cardiomyocyte-specific contractile protein troponin I-C were demonstrated in β-galactosidase positive cells (arrow).

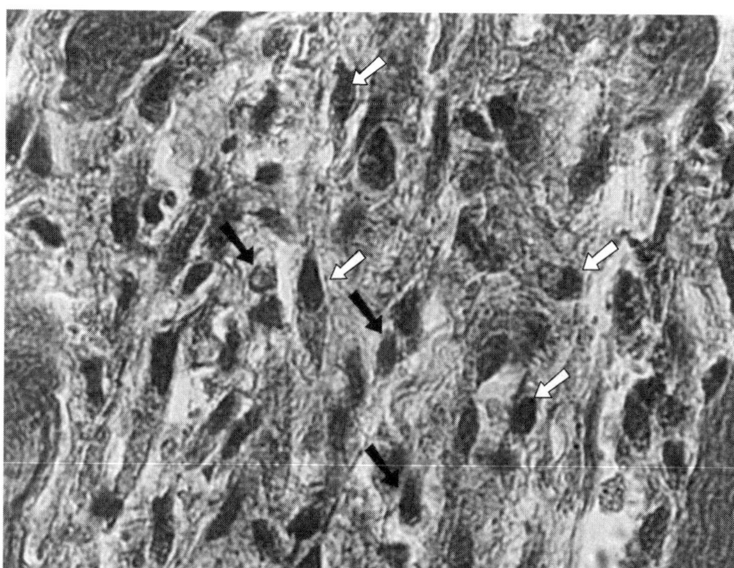

13.4 Histologic examination of LV cross-sections after injection of lac-Z labeled mouse skin fibroblasts (white arrow) showed extensive cellular infiltration (black arrow) within 1 day and no surviving labeled cells were found beyond 4 days without immunosuppression.

(EF) and fractional shortening (FS) were used as indices of left ventricular performance. Both parameters were abnormal both in the MSCs and the control groups compared with pre-operative baseline, However, at 4 weeks, both parameters were significantly better in animals that underwent MSCs implantation at the time of LAD ligation when compared with the group without cell therapy (EF: 34.0 $\pm$ 10.7% *vs.* 19.9 $\pm$ 10.0%, $p = 0.007$; FS: 24.2 $\pm$ 11.5% *vs.* 14.3 $\pm$ 5.51%, $p = 0.02$). Stroke volume was significantly better in the MSCs recipient group (0.18 $\pm$ 0.05 ml *vs.* 0.12 $\pm$ 0.05 ml, $p = 0.03$). End-diastolic volume was significantly lower in the MSC group with a strong tendency toward lower end-systolic volumes as well.

These recent studies from our laboratory further confirmed our impression that MSCs, in contrast to fibroblasts from the same donors, possessed a unique immuno-tolerance, and rejection did not occur despite the fact that immuno-suppression was not used in this rodent xenotransplantation model. In addition, these xenogeneic MSCs preserved the ability of transdifferentiation in a milieu-dependent fashion and contributed to meaningful functional improvement after ischemic injury to the hearts.

13.6 Current hypothesis of xenogeneic MSCs immunotolerance

In an attempt to understand and explain such unexpected observations, we propose a hypothesis based on recent new discoveries reported in the literature that stem cells have unique immunological properties. MSCs have very weak expression of MHC class I and II antigens. In addition, they can directly modulate immune responses to induce tolerance.[71] Le Blanc and coworkers showed in a delicate work that undifferentiated MSCs express MHC class I but not II, although the latter was present intracellularly.[72] In addition, undifferentiated and differentiated MSCs do not elicit alloreactive lymphocyte proliferative response and modulate immune responses. Aggarwal *et al.* have shown in a co-culture study of human MSCs and purified immune cells that human MSCs altered cytokine secretion profile of dendritic cells, naive and effector T cells and natural killer cells to induce a more anti-inflammatory or tolerant phenotype with increased interleukins-10 and 4, regulatory T cells, while decreased tumor necrosis factor-alpha, interferon-gamma from various immune cells.[62]

In an attempt to enhance hematopoietic engraftment and prevent graft-versus-host disease, Maitra *et al.* have demonstrated in an *in vitro* study that human MSCs did not elicit T-cell activation and suppressed T-cell activation by Tuberculin and unrelated allogeneic lymphocytes in a dose-dependent fashion.[73] Similarly, Bartholomew *et al.* have shown that MSCs could be used for their immunomodulatory properties to suppress lymphocyte proliferation, therefore, to prolong skin graft survival *in vivo*.[74] Najafian *et al.* also nicely confirmed that MSCs inhibited the response of naive and memory antigen-specific T cells to their antigenic epitopes.[77] In another co-culture study, Di Nicola *et al.* have shown that autologous or allogeneic MSCs strongly suppressed lymphocyte proliferation induced by cellular or nonspecific mitogenic stimuli.[75] The immunogenicity of MSCs was tested by Chou *et al.* using either healthy donor mononuclear cells or MSCs as stimulators in xenogeneic mixed lymphocyte reactions.[76] They showed that xenogeneic MSCs increased inhibition of allogeneic and xenogeneic lymphocyte responses with increasing cell numbers, while the control groups exhibited no inhibitory response.

Although the immunomodulary responses of MSCs are well documented, there is no explanation as to why tolerance continues even after implanted xenogeneic stem cells differentiated into their targeted tissue phenotypes. We posit that this may be explained based on recent understanding of the signaling mechanisms, known as 'immunological synapsis' between the antigen presenting cells (APCs) and T cells. There are two signaling pathways at such a synapsis. The first signal is between the MHC antigen presented by the APC and the corresponding receptor on the T-cell surface, which provides the 'recognition signal' to identify and select the T-cell subpopulation that has a specific receptor matching the MHC antigen presented. The second signaling pathway consists of a 'co-stimulant' molecule and its receptor presented at the cell surfaces of APC and T cells, respectively, such as those involving B-7 molecule and CD-28 receptor.[77] This pathway is known as the 'activation signal' and only in its co-presence will the T cells be induced to proliferate and activate to carry out immune rejection. In the presence of the 'recognition signal' but without a co-stimulant 'activation signal', the T cells will not be activated to proliferate, but rather undergo apoptosis and be eliminated, thus inducing immune tolerance.[77]

Matzinger proposed, in her 'danger model' theory, that molecules representing tissue injury or stress molecules released by cells responding to imminent danger can induce co-stimulant 'activation signal' on the APCs.[78–80] This has led to the concept that the immune system responds only to antigens perceived to be associated with a dangerous situation such as infection or cell death.[81] Danger signals, such as Freund's adjuvant, are thought to act by stimulating dendritic cells to mature so that they can present foreign antigens and stimulate T lymphocytes.[81]

Dying mammalian cells have also been found to release danger signals of unknown identity. Shi *et al.* showed that uric acid was a principal endogenous danger signal released from injured cells. Uric acid stimulated dendritic cell maturation and, when co-injected with antigen *in vivo*, significantly enhanced the generation of responses from CD8+ T cells.[82] Eliminating uric acid *in vivo* inhibited the immune response to antigens associated with injured cells.[82] Scheel *et al.*[83] suggested that protamine-condensed mRNA which was released from a damaged cell can strongly activate APC and monocytes by Toll-like receptor-dependent TLR-7 and TLR-8, leading to TNF-α and IFN-α secretion. Heil *et al.*[84] reported that double-stranded ribonucleic acid (dsRNA) and single-stranded RNA could serve as a danger signal associated with species-specific recognition and lead to stimulation of innate immune cells via TLR-7 and TLR-8. In addition, mRNA released by a damaged cell could form protamine–mRNA complexes which directly activated B cells, NK cells and granulocytes[83] and triggered an immune response.

Similarly, danger signal molecules can be released by invasive surgical procedures, such as those during heart transplantation or cell implantation. During heart transplantation, the simultaneous presence of foreign antigen 'recognition' and tissue injury 'activation' signaling pathways triggers immune rejection of the allograft. During MSC implantation, on the other hand, the immunomodulatory properties of the MSCs suppress acute rejection. While the differentiation process takes days to weeks to occur, the danger signal associated with the invasive implantation procedure will be subsided, thus inducing immunotolerance. In contrast, mice fibroblasts were terminally differentiated cells with strong MHC antigens but no immunomodulatory properties like MSCs. Therefore, an acute rejection was indeed observed early after implantation while given across different species.

While this 'Stealth Immune Tolerance'[85] hypothesis requires further investigation, it is not completely unfounded. Wekerle and associates reported that mismatched allogeneic bone marrow transplantation with co-stimulatory blockade induced macrochimerism and tolerance up to 34 weeks without cyto-reductive host treatment.[86] The immunological response of this nature is intriguing and may have tremendous clinical implication not only in creating 'universal donor cells' for organ tissue regeneration but in the field of xeno-transplantation as a whole in which minimal progress has been made for decades.

13.7 Conclusions and future trends

Further elucidation of the mechanisms and the conditions for immune tolerance of MSCs for myocardial regeneration will potentially develop 'universal donor' MSCs to be used as a source for cellular cardiomyoplasty. Such a 'universal donor cell' for cell therapy will dramatically reduce the costs of cellular cardiomyoplasty, the complications of immuno-suppression therapy and the logistic advantages of using allogeneic universal donor cells over that of autologous cells in the clinical setting of unstable and medically ill patient. The cell population that can be used as universal donor cells will also need to be characterized. The feasibility of using donor cells from a genetically normal individual to recipients with genetic defects, or using young donor MSCs for a senescent recipient whose autologous MSCs may be functionally compromised, are fascinating possibilities worthy of further exploration. Finally, understanding the basic immunological mechanisms of these observations may potentially redefine the traditional concept of self-and-non-self immunological principle.

13.8 References

1. Roger VL, Weston SA, Redfield MM, *et al.* Trends in heart failure incidence and survival in a community-based population. *JAMA* 2004; 292(3): 344–50.
2. O'Connor CM, Joynt KE. Depression: are we ignoring an important comorbidity in heart failure? *J Am Coll Cardiol* 2004; 43(9): 1550–52.
3. Hobbs FD, Kenkre JE, Roalfe AK, *et al.* Impact of heart failure and left ventricular systolic dysfunction on quality of life: a cross-sectional study comparing common chronic cardiac and medical disorders and a representative adult population. *Eur Heart J* 2002; 23(23): 1867–76.
4. Hellermann JP, Jacobsen SJ, Gersh BJ, *et al.* Heart failure after myocardial infarction: a review. *Am J Med* 2002; 113(4): 324–30.
5. Gehi AK, Pinney SP, Gass A. Recent diagnostic and therapeutic innovations in heart failure management. *Mt Sinai J Med* 2005; 72(3): 176–84.
6. Salvador DR, Rey NR, Ramos GC, Punzalan FE. Continuous infusion versus bolus injection of loop diuretics in congestive heart failure. *Cochrane Database Syst Rev* 2004(1): CD003178.
7. Wang L, Song S. Digoxin may reduce the mortality rates in patients with congestive heart failure. *Med Hypotheses* 2005; 64(1): 124–6.
8. Nieminen MS, Kupari M. The hemodynamics effects of ACE inhibitors in the treatment of congestive heart failure. *J Cardiovasc Pharmacol* 1990; 15 Suppl 2: S36–40.
9. Sica DA. Pharmacotherapy in congestive heart failure: angiotensin-converting enzyme inhibitors and angiotensin receptor blockers in congestive heart failure: do they differ in their renal effects in man? *Congest Heart Fail* 2001; 7(3): 156–61.
10. Auer JW, Berent R, Eber B. Angiotensin II type 1 receptor blockers and congestive heart failure. *Circulation* 2001; 104(15): E82.
11. Miller AB. Is there a role for calcium channel blockers in congestive heart failure? *Curr Cardiol Rep* 2001; 3(2): 114–18.
12. Francis GS. Calcium channel blockers and congestive heart failure. *Circulation* 1991; 83(1): 336–8.
13. Meng F, Yoshikawa T, Baba A, *et al.* Beta-blockers are effective in congestive heart failure patients with atrial fibrillation. *J Card Fail* 2003; 9(5): 398–403.
14. Barry WH, Gilbert EM. How do beta-blockers improve ventricular function in patients with congestive heart failure? *Circulation* 2003; 107(19): 2395–7.
15. van Zwieten PA. The rationale for the use of α-blockers in the treatment of congestive heart failure. *Am J Geriatr Cardiol* 1992; 1(3): 57–61.
16. Frazier OH, Rose EA, Oz MC, *et al.* Multicenter clinical evaluation of the HeartMate vented electric left ventricular assist system in patients awaiting heart transplantation. *J Thorac Cardiovasc Surg* 2001; 122(6): 1186–95.
17. Stevenson LW, Miller LW, Desvigne-Nickens P, *et al.* Left ventricular assist device as destination for patients undergoing intravenous inotropic therapy: a subset analysis from REMATCH (Randomized Evaluation of Mechanical Assistance in Treatment of Chronic Heart Failure). *Circulation* 2004; 110(8): 975–81.
18. Hosenpud JD, Bennett LE, Keck BM, *et al.* The Registry of the International Society for Heart and Lung Transplantation: fourteenth official report – 1997. *J Heart Lung Transplant* 1997; 16(7): 691–712.
19. Schmoeckel M, Nollert G, Shahmohammadi M, *et al.* Transgenic human decay accelerating factor makes normal pigs function as a concordant species. *J Heart Lung Transplant* 1997; 16(7): 758–64.

20. Diamond LE, McCurry KR, Martin MJ, *et al.* Characterization of transgenic pigs expressing functionally active human CD59 on cardiac endothelium. *Transplantation* 1996; 61(8): 1241–9.
21. Zak R. Development and proliferative capacity of cardiac muscle cells. *Circ Res* 1974; 35(2):suppl II: 17–26.
22. Messina E, De Angelis L, Frati G, *et al.* Isolation and expansion of adult cardiac stem cells from human and murine heart. *Circ Res* 2004; 95(9): 911–21.
23. Beltrami AP, Urbanek K, Kajstura J, *et al.* Evidence that human cardiac myocytes divide after myocardial infarction. *N Engl J Med* 2001; 344(23): 1750–57.
24. Oh H, Chi X, Bradfute SB, *et al.* Cardiac muscle plasticity in adult and embryo by heart-derived progenitor cells. *Ann NY Acad Sci* 2004; 1015: 182–9.
25. Beltrami AP, Barlucchi L, Torella D, *et al.* Adult cardiac stem cells are multipotent and support myocardial regeneration. *Cell* 2003; 114(6): 763–76.
26. Deb A, Wang S, Skelding KA, *et al.* Bone marrow-derived cardiomyocytes are present in adult human heart: a study of gender-mismatched bone marrow transplantation patients. *Circulation* 2003; 107(9): 1247–9.
27. Roufosse CA, Direkze NC, Otto WR, Wright NA. Circulating mesenchymal stem cells. *Int J Biochem Cell Biol* 2004; 36(4): 585–97.
28. Quaini F, Urbanek K, Beltrami AP, *et al.* Chimerism of the transplanted heart. *N Engl J Med* 2002; 346(1): 5–15.
29. Abbott JD, Huang Y, Liu D, *et al.* Stromal cell-derived factor-1alpha plays a critical role in stem cell recruitment to the heart after myocardial infarction but is not sufficient to induce homing in the absence of injury. *Circulation* 2004; 110(21): 3300–305.
30. Kim YH. Intramyocardial transplantation of circulating CD34+ cells: source of stem cells for myocardial regeneration. *J Korean Med Sci* 2003; 18(6): 797–803.
31. Beckman RJ, Salzman GC, Stewart CC. Classification and regression trees for bone marrow immunophenotyping. *Cytometry* 1995; 20(3): 210–17.
32. Petite H, Viateau V, Bensaid W, *et al.* Tissue-engineered bone regeneration. *Nat Biotechnol* 2000; 18(9): 959–63.
33. Krause DS. Plasticity of marrow-derived stem cells. *Gene Ther* 2002; 9(11): 754–8.
34. Ferrari G, Cusella-De Angelis G, Coletta M, *et al.* Muscle regeneration by bone marrow-derived myogenic progenitors. *Science* 1998; 279(5356): 1528–30.
35. Chiu RC. Bone-marrow stem cells as a source for cell therapy. *Heart Fail Rev* 2003; 8(3): 247–51.
36. Wollert KC, Meyer GP, Lotz J, *et al.* Intracoronary autologous bone-marrow cell transfer after myocardial infarction: the BOOST randomised controlled clinical trial. *Lancet* 2004; 364(9429): 141–8.
37. Kehat I, Kenyagin-Karsenti D, Snir M, *et al.* Human embryonic stem cells can differentiate into myocytes with structural and functional properties of cardiomyocytes. *J Clin Invest* 2001; 108(3): 407–14.
38. Kehat I, Khimovich L, Caspi O, *et al.* Electromechanical integration of cardiomyocytes derived from human embryonic stem cells. *Nat Biotechnol* 2004; 22(10): 1282–9.
39. Soonpaa MH, Koh GY, Klug MG, Field LJ. Formation of nascent intercalated disks between grafted fetal cardiomyocytes and host myocardium. *Science* 1994; 264(5155): 98–101.
40. Li RK, Jia ZQ, Weisel RD, *et al.* Cardiomyocyte transplantation improves heart function. *Ann Thorac Surg* 1996; 62(3): 654–60; discussion 660–61.
41. Koh GY, Klug MG, Soonpaa MH, Field LJ. Differentiation and long-term survival

of C2C12 myoblast grafts in heart. *J Clin Invest* 1993; 92(3): 1548–54.
42. Chiu RC, Zibaitis A, Kao RL. Cellular cardiomyoplasty: myocardial regeneration with satellite cell implantation. *Ann Thorac Surg* 1995; 60(1): 12–18.
43. Atkins BZ, Lewis CW, Kraus WE, *et al.* Intracardiac transplantation of skeletal myoblasts yields two populations of striated cells *in situ*. *Ann Thorac Surg* 1999; 67(1): 124–9.
44. Murry CE, Wiseman RW, Schwartz SM, Hauschka SD. Skeletal myoblast transplantation for repair of myocardial necrosis. *J Clin Invest* 1996; 98(11): 2512–23.
45. Menasche P, Hagege AA, Scorsin M, *et al.* Myoblast transplantation for heart failure. *Lancet* 2001; 357(9252): 279–80.
46. Menasche P, Hagege AA, Vilquin JT, *et al.* Autologous skeletal myoblast transplantation for severe postinfarction left ventricular dysfunction. *J Am Coll Cardiol* 2003; 41(7): 1078–83.
47. Wakitani S, Saito T, Caplan AI. Myogenic cells derived from rat bone marrow mesenchymal stem cells exposed to 5-azacytidine. *Muscle Nerve* 1995; 18(12): 1417–26.
48. Makino S, Fukuda K, Miyoshi S, *et al.* Cardiomyocytes can be generated from marrow stromal cells *in vitro*. *J Clin Invest* 1999; 103(5): 697–705.
49. Tomita S, Li RK, Weisel RD, *et al.* Autologous transplantation of bone marrow cells improves damaged heart function. *Circulation* 1999; 100(19 Suppl): II247–56.
50. Wang JS, Shum-Tim D, Galipeau J, *et al.* Marrow stromal cells for cellular cardiomyoplasty: feasibility and potential clinical advantages. *J Thorac Cardiovasc Surg* 2000; 120(5): 999–1005.
51. Orlic D, Kajstura J, Chimenti S, *et al.* Bone marrow cells regenerate infarcted myocardium. *Nature* 2001; 410(6829): 701–5.
52. Saito T, Kuang JQ, Lin CC, Chiu RC. Transcoronary implantation of bone marrow stromal cells ameliorates cardiac function after myocardial infarction. *J Thorac Cardiovasc Surg* 2003; 126(1): 114–23.
53. Stamm C, Westphal B, Kleine HD, *et al.* Autologous bone-marrow stem-cell transplantation for myocardial regeneration. *Lancet* 2003; 361(9351): 45–6.
54. Strauer BE, Brehm M, Zeus T, *et al.* Repair of infarcted myocardium by autologous intracoronary mononuclear bone marrow cell transplantation in humans. *Circulation* 2002; 106(15): 1913–18.
55. Assmus B, Schachinger V, Teupe C, *et al.* Transplantation of progenitor cells and regeneration enhancement in acute myocardial infarction (TOPCARE-AMI). *Circulation* 2002; 106(24): 3009–17.
56. Wollert KC, Drexler H. Clinical applications of stem cells for the heart. *Circ Res* 2005; 96(2): 151–63.
57. Patel AN PR, Brusich D, *et al.* Minimally invasive cellular therapy for congestive heart failure: a prospective randomized study (Abstract). *41st Annual Meeting of the Society of Thoracic Surgeon*s 2005: 258.
58. Reffelmann T, Kloner RA. Cellular cardiomyoplasty – cardiomyocytes, skeletal myoblasts, or stem cells for regenerating myocardium and treatment of heart failure? *Cardiovasc Res* 2003; 58(2): 358–68.
59. Zhang H, Fazel S, Tian H, *et al.* Increasing donor age adversely impacts beneficial effects of bone marrow but not smooth muscle myocardial cell therapy. *Am J Physiol Heart Circ Physiol* 2005; 289(5): H2089–2096.
60. Xiao YF, Min JY, Morgan JP. Immunosuppression and xenotransplantation of cells for cardiac repair. *Ann Thorac Surg* 2004; 77(2): 737–44.

61. Gammie JS, Pham SM. Simultaneous donor bone marrow and cardiac transplantation: can tolerance be induced with the development of chimerism? *Curr Opin Cardiol* 1999; 14(2): 126–32.
62. Aggarwal S, Pittenger MF. Human mesenchymal stem cells modulate allogeneic immune cell responses. *Blood* 2005; 105(4): 1815–22.
63. Allers C, Sierralta WD, Neubauer S, *et al.* Dynamic of distribution of human bone marrow-derived mesenchymal stem cells after transplantation into adult unconditioned mice. *Transplantation* 2004; 78(4): 503–8.
64. Liechty KW, MacKenzie TC, Shaaban AF, *et al.* Human mesenchymal stem cells engraft and demonstrate site-specific differentiation after *in utero* transplantation in sheep. *Nat Med* 2000; 6(11): 1282–6.
65. Airey JA, Almeida-Porada G, Colletti EJ, *et al.* Human mesenchymal stem cells form Purkinje fibers in fetal sheep heart. *Circulation* 2004; 109(11): 1401–7.
66. Toma C, Pittenger MF, Cahill KS, *et al.* Human mesenchymal stem cells differentiate to a cardiomyocyte phenotype in the adult murine heart. *Circulation* 2002; 105(1): 93–8.
67. Poh KK, Sperry E, Young RG, Freyman T, Barringhaus KG, Thompson CA. Repeated direct endomyocardial transplantation of allogenic mesenchymal stem cells: safety of a high dose, 'off-the-shelf', cellular cardiomyoplasty strategy. *Int J Cardiol* 2006; 2.
68. Makkar RR, Price MJ, Lill M, Frantzen M, Takizawa K, Kleisil T, *et al.* Intramyocardial injection of allogenic bone marrow-derived mesenchymal stem cells without immunosuppression preserves cardiac function in a porcine model of myocardial infarction. *J Cardiovasc Pharmacol Ther* 2005; 10(4): 225–33.
69. Saito T, Kuang JQ, Bittira B, *et al.* Xenotransplant cardiac chimera: immune tolerance of adult stem cells. *Ann Thorac Surg* 2002; 74(1): 19–24; discussion 24.
70. MacDonald DJ, Luo J, Saito T, *et al.* Persistence of marrow stromal cells implanted into acutely infarcted myocardium: observations in a xenotransplant model. *J Thorac Cardiovasc Surg* 2005; 130(4): 1114–21.
71. Zhao RC, Liao L, Han Q. Mechanisms of and perspectives on the mesenchymal stem cell in immunotherapy. *J Lab Clin Med* 2004; 143(5): 284–91.
72. Le Blanc K, Tammik C, Rosendahl K, *et al.* HLA expression and immunologic properties of differentiated and undifferentiated mesenchymal stem cells. *Exp Hematol* 2003; 31(10): 890–96.
73. Maitra B, Szekely E, Gjini K, *et al.* Human mesenchymal stem cells support unrelated donor hematopoietic stem cells and suppress T-cell activation. *Bone Marrow Transplant* 2004; 33(6): 597–604.
74. Bartholomew A, Sturgeon C, Siatskas M, *et al.* Mesenchymal stem cells suppress lymphocyte proliferation *in vitro* and prolong skin graft survival *in vivo*. *Exp Hematol* 2002; 30(1): 42–8.
75. Di Nicola M, Carlo-Stella C, Magni M, *et al.* Human bone marrow stromal cells suppress T-lymphocyte proliferation induced by cellular or nonspecific mitogenic stimuli. *Blood* 2002; 99(10): 3838–43.
76. Chou S-H, Kuo T, Liu M, Lee, OK. *In utero* transplantation of human bone marrow-derived multipotent mesenchymal stem cells in mice. *J Orth Res* 2006; 24(3): 301–12.
77. Najafian N, Khoury SJ. T cell costimulatory pathways: blockade for autoimmunity. *Expert Opin Biol Ther* 2003; 3(2): 227–36.
78. Matzinger P. The danger model: a renewed sense of self. *Science* 2002; 296(5566): 301–5.

79. Matzinger P. An innate sense of danger. *Ann NY Acad Sci* 2002; 961: 341–2.
80. Matzinger P. An innate sense of danger. *Semin Immunol* 1998; 10(5): 399–415.
81. Gallucci S, Lolkema M, Matzinger P. Natural adjuvants: endogenous activators of dendritic cells. *Nat Med* 1999; 5(11): 1249–55.
82. Shi Y, Evans JE, Rock KL. Molecular identification of a danger signal that alerts the immune system to dying cells. *Nature* 2003; 425(6957): 516–21.
83. Scheel B, Teufel R, Probst J, *et al.* Toll-like receptor-dependent activation of several human blood cell types by protamine-condensed mRNA. *Eur J Immunol* 2005; 35(5): 1557–66.
84. Heil F, Hemmi H, Hochrein H, *et al.* Species-specific recognition of single-stranded RNA via toll-like receptor 7 and 8. *Science* 2004; 303(5663): 1526–9.
85. Chiu RC. 'Stealth immune tolerance' in stem cell transplantation: potential for 'universal donors' in myocardial regenerative therapy. *J Heart Lung Transplant* 2005; 24(5): 511–16.
86. Wekerle T, Kurtz J, Ito H, *et al.* Allogeneic bone marrow transplantation with co-stimulatory blockade induces macrochimerism and tolerance without cytoreductive host treatment. *Nat Med* 2000; 6(4): 464–9.

14

Myocardial regeneration, tissue engineering and therapy

R L KAO, C E GANOTE, D G PENNINGTON and I W BROWDER, East Tennessee State University, USA

14.1 Introduction

This chapter reviews the lack of regenerative capability in adult mammalian ventricular myocardium and the development of cellular cardiomyoplasty as a possible therapy for myocardial regeneration. Section 14.2 gives a brief overview of terminal differentiation of ventricular cardiomyocytes, the possible existence of stem cells or progenitor cells in myocardium and the inability of these cells to have functional myocardial regeneration. Cellular cardiomyoplasty is a viable approach for myocardial regeneration and experimental studies using satellite cells (myoblasts) and bone marrow cells to support the clinical application of stem cells for myocardial regeneration as described. Section 14.3 summarizes the clinical outcomes using satellite cells (myoblasts) and bone marrow-derived cells for cellular cardiomyoplasty. Although the feasibility, safety and efficacy of cellular cardiomyoplasty are well established in experimental and clinical studies, the possible beneficial mechanisms lack general consensus. Section 14.4 explains that future trends are most likely to use purified stem cells or a mixture of purified stem cells to achieve myogenesis and angiogenesis without adverse effects. The development of universal donor cells, better routes and methods of cell delivery and enhancement of the retention, survival, proliferation and differentiation of stem cells will be most likely to optimize the procedure. Section 14.5 gives a brief list of web sites that cover this field, including companies that are supporting this area of research and providing supplies for the procedures.

14.2 Background

14.2.1 Proliferation and differentiation of ventricular cardiomyocytes

Ventricular muscle cells of adult mammals are terminally differentiated cells that have lost their ability to replicate to repair or reconstitute damaged myocardium.[1–7] Injury to heart consistently results in the formation of scar tissue, further attesting to the lack of regenerative capability of mammalian myocardium. Following myocardial infarction, the necrotic, apoptotic and stunned (hibernating) myocardium lead to early reduction of ventricular function. Infiltration of neutrophils and accumulation of macrophages, followed by the formation of granulation tissue and scar tissue, result in infarct expansion and ventricular remodeling that progress into ventricular dysfunction and congestive heart failure. There are 13.2 million Americans suffering coronary heart disease with 7.2 million heart attacks and 5 million heart failure patients.[8] Cellular cardiomyoplasty can be an effective treatment for patients with heart attack or heart failure.[9,10]

Observations in humans[11,12] and animals[13,14] have provided some evidence that myocyte cellular hyperplasia may occur under certain pathophysiologic conditions. However, recent studies have clearly documented that a rapid switch of cardiac myocytes from hyperplasia to hypertrophy occurred during early postnatal development.[6] Cardiomyocyte DNA synthesis in normal and injured adult mammalian hearts is extremely rare when myocyte nuclei can be reliably identified.[15,16] Even though spatially and temporally restricted endomitosis (chromosome replication leading to polyploidy) or karyokinesis (mitosis and nuclear division leading to multinucleation) can be induced in heart muscle cells,[17,18] the incomplete disassembly and presence of myofibrils at the equator region may physically impede the cytokinesis (cell division) process.[19] Adult mammalian ventricular cardiomyocytes are terminally differentiated cells that cannot effectively proliferate to regenerate the injured heart.

The possible existence of stem cells or progenitor cells in neonatal rat myocardium was suggested in 1996.[20] Recently, the presence of stem cells or progenitor cells in myocardium of adult human[21] and mature animals[22] has been reported. So far, several distinct types of stem cells or progenitor cells identified from the adult mammalian ventricular myocardium with regeneration potential have been documented by different groups. From adult rat heart the Lin^{-}c-kit^{+} cells are self-renewing, clonogenic and multipotent cells that can differentiate into myocytes, smooth muscle cells and endothelial cells to replace the acutely ischemic myocardium.[22] These cells can be delivered by intracoronary route and they home to the myocardial infarction for myocardial regeneration and functional improvement.[23] Similar cardiac stem cells have also been identified in human and dogs by the same group for possible treatment of ischemic heart failure.[24,25]

Another progenitor cell isolated from mouse heart[26] expressing Sca-1 can express cardiac specific genes *in vitro* after 5-azacytidine treatment. Intravenous delivery of these cells into mice 6 hours after myocardial infarction, engrafted cells expressing cardiac markers. Using a Cre/Lox donor/recipient pair, cardiac Sca-1^{+} cells homed to the infarct area with about half of the donor-derived cells fused with host cardiacmyocytes while the other half differentiated into heart muscle cells without fusion.[26,27] The $isl1^{+}$ progenitor cells contributed to outflow tract and some atrial and ventricular cells during normal development.[28] Cells expressing isl1 were present in the neonatal heart and to a lesser extent in the adult heart of rat, mouse and human. Purified isl1$^+$ cells proliferate in culture without differentiation, but rapidly differentiate into cardiomyocytes when co-cultured with heart muscle cells.[29] Although these cells were considered as native cardioblasts, whether they can restore an injured heart awaits future study. Isolated side population cells expressing Abcg2 from adult mouse heart also were capable of proliferation and differentiation. These cells expressed cardiac differentiation marker (α-actinin) when co-cultured with isolated heart cells.[30] Cardiac stem cells have been isolated from adult human and murine heart that grow as self-adherent clusters (cardiospheres).[31] They were capable of long-term self-renewal and differentiate into myocytes, smooth muscle cells and endothelial cells under both *in vitro* and *in vivo* conditions.[31] Why an organ well known for its lack of regenerative capacity would have multiple types of stem cells or progenitor cells as well as the origin of these cells (i.e. remained in heart through embryonic development or carried to heart by blood) requires further clarification. However, it is evident that functionally significant myocardial regeneration has not been documented in a diseased or injured heart. Adult mammalian myocardium lacks adequate endogenous regenerative capability and inducing myogenesis in an injured heart seems a viable approach to reconstitute damaged myocardium and to prevent heart failure.

14.2.2 Myocardial regeneration

The exact mechanism responsible for cardiac myocyte terminal differentiation and cell cycle arrest is currently unknown.[1–3,32] Inducing terminally differentiated cardiomyocytes to re-enter the cell cycle with controlled proliferation is not possible at the moment. Several proteins of DNA tumor viruses (SV40 large T antigen, adenovirus E1A and human papilloma virus E7) have proven to restart DNA synthesis in a broad array of cell types.[33–36] The viral proteins physically associate with the class of tumor suppressors (Rb, p107 and p130). They are known collectively as 'pocket' proteins, named after the shared domain for binding the viral gene product. These pocket proteins (Rb and p130) control cardiac cell cycle exit with overlapping functions in the heart.[37] E1A or E2F-1 also reactivated DNA synthesis in ventricular myocytes but engendered widespread apoptosis of these cells.[16,36] Although combined expression of p193 and p53 blocked E1A-induced apoptosis,[38] the proliferative response as T antigen to produce tumors and cardiac arrhythmias are undesired outcomes.[39] Transgenic mice expressing the p193 and/or the p53 dominant-interfering mutants in the heart[40] and transgenic mice with targeted expression of cyclin D2 were associated with an induction of cardiomyocyte DNA synthesis.[40,41] Mice with constitutive cyclin A2 expression in the myocardium induced cardiomyocyte mitosis and cardiac hypertrophy owing to hyperplasia.[42] In any case, molecular cardiomyoplasty cannot provide additional muscle cells by reactivating the terminally differentiated cardiomyocytes to complete the cell cycle under controlled proliferation. Genetic or molecular manipulations cannot induce adult mammalian ventricular myocyte to proliferate for myocardial regeneration.

Adult stem cells are undifferentiated cells residing in differentiated tissues capable of self-renewal and proliferation to produce differentiated cells. Adult stem cells can yield the specialized cell types of the tissue from which it originated and are capable of developing into cell types that are characteristic of other tissues (plasticity). Self-renewal and plasticity of adult stem cells have been well established in recent years.[43–46] Cell therapy has emerged as a strategy for the treatment of many human diseases. The aim of cell therapy is to replace, repair or enhance the biological function of damaged tissue or organ.[47,48] Cellular cardiomyoplasty with or without transfected specific genes has been developed to treat myocardial infarction and to augment the function of a failing heart.[44,47–54] However, directing the differentiation of the cells into cardiac lineage will be a key to the success of myocardial regeneration using autologous stem cells.

It has been more than a decade since we started myocardial regeneration by implanting autologous progenitor cells to treat myocardial infarction and heart failure in experimental animals.[55,56] That human extracardiac progenitor cells are capable of differentiating into ventricular cardiac myocytes has been shown by the observations of sex-mismatched heart transplantation samples.[57–60] These observations indicated that the mobilization, homing and differentiation of progenitor cells can occur in adult humans. Although the exact beneficial mechanisms of cellular cardiomyoplasty are under intense study of many investigators, the possibility of repairing the diseased heart with new myocardium has become a potentially valuable therapeutic procedure for patients suffering heart attack or heart failure.[44,47–54]

Fetal and adult heart muscle cells and even fetal heart tissue[61] have been transplanted into injured and normal ventricular muscle of small and large experimental animals.[62–69] The transplanted cardiomyocytes have long-term survival and form nascent intercalated discs between the transplanted cells and the host myocardium. The implanted cells survived in scar tissue and improved heart function.[62–69] Other than syngeneic hosts or inbred animals, immunosuppression therapy must be applied to the host to avoid rejection. The sensitivity of these cells to anoxia or ischemia,[70] and the ethical concerns of using fetal tissue may render this approach impractical for clinical application. The embryonic stem cells,[71,72] and their developed cardiomyocytes[73] will face the same limitations. Precursor cells or stem cells from different organs and tissues of an adult animal in addition to muscle and bone marrow-derived stem cells, such as neuronal stem cells, hematopoietic stem cells, adipose tissue stem cells, liver stem cells, blood-derived progenitor or stem cells, dental pulp stem cells, etc., all have the potential to be used for cellular cardiomyoplasty.[43,44,48–50] Cord blood and placenta-derived stem cells may also find their way into regenerative medicine.

14.2.3 Cellular cardiomyoplasty using satellite cells

Skeletal muscle satellite cells are spindle-shaped mononuclear cells located beneath the basal lamina but outside the sarcolemma of skeletal muscle.[74,75] Since their original description by Mauro,[76] muscle satellite cells have been primarily identified by their anatomical location and morphological characteristics. Satellite cells are the primary stem cells in adult skeletal muscle and are responsible for postnatal muscle growth, hypertrophy and regeneration.[74,75] Since their original identification, satellite cells have been considered as unipotent myogenic precursor cells. Recent studies demonstrating the self-renewal[77,78] and multipotential[46,79] of satellite cells supported the view that they are true stem cells. Both myofiber-associated stem cells and muscle interstitial stem cells have been identified in the skeletal muscle. Although both bone marrow-derived cells and circulation-derived cells can be localized to the same anatomic compartment as myogenic satellite cells, they display no intrinsic myogenicity.[80] Satellite cell preparations isolated by enzymatic methods not only contain other types of cells (i.e. fibroblasts) but also contain precursor cells of diverse characteristics. Whether using the mixed cell preparation or using purified satellite cells for myocardial regeneration will achieve the optimal outcome is under current investigation.

Satellite cells have been successfully isolated, labeled and implanted into injured heart with neomyocardial formation and functional improvement.[9,10,55,56] Viable muscle cells with clear labeling were found in the infarct area after cell implantation. The labeled muscle cells showed intercalated discs at cellular junctions. Significant improvement in contractile function was observed in the animals receiving successful cell transplantation.[9,10] Delivering autologous skeletal myoblasts (satellite cells) into rabbit hearts as a potential approach for myocardial repair has been confirmed by other investigators.[81] Since satellite cells were first used for myocardial regeneration in animals from our observations,[55,56] they became the first type of cells applied for clinical cellular cardiomyoplasty.

14.2.4 Cellular cardiomyoplasty using bone marrow cells

That marrow stromal cells could form bone and cartilage was reported in 1980.[82] The myogenic cells derived from bone marrow after 5-azacytidine treatment was observed in 1995.[83] The presence of mesenchymal stem cells (multipotent cells) in adult bone marrow was suggested for the potential of differentiating into different lineages of bone, cartilage, fat, tendon, muscle and marrow stroma.[84] The generation of cardiomyocytes from marrow stromal cells was also observed recently.[85,86] The marrow stromal cells were used for cellular cardiomyoplasty and improved function of damaged hearts.[87–94] The clinical application of marrow cells for myocardial regeneration had been progressed rapidly in recent years.

Adult bone marrow contains multiple cell populations of differentiated cells and undifferentiated cells (stem cells) such as hematopoietic stem cells (HSC), mesenchymal stem cells (MSC) and endothelial progenitor cells. A mixed population of bone marrow cells rather than purified stem cells is commonly used for cellular cardiomyoplasty.[87–94] The safety and ease of obtaining the cells, the observation of new muscle tissue formation, the improvement of contractile function, the enhancement of local perfusion and the prevention of remodeling and deterioration of the injured heart have been documented in studies using experimental animals.[47–54,94]

14.3 Current status

14.3.1 Clinical studies using satellite cells (myoblasts)

Clinical cellular cardiomyoplasty was first performed by Menasche's group in June 2000.[95] Since then, a number of small-scale uncontrolled clinical trials have been reported by different investigative groups (Table 14.1). Skeletal muscle satellite cells (myoblasts) have the advantages of autologous availability, can be proliferated *in vitro* to vast quantity and are highly resistant to ischemia/hypoxia environment which allows good survival and engraftment after transplantation. Early clinical applications offered highly encouraging results from others'[95,97–104] and our observations.[96] Although the feasibility, safety and encouraging outcomes have been observed; the definitive long-term efficacy requires large-scale multicenter double blind trials.

Table 14.1 Cellular cardiomyoplasty using satellite cells (myoblasts)

Coordinator	Total	Procedures	Complications
Menasche[95]	10	IMTEp (CABG)	2 D; 4 VT
Kao[96]	9	IMTEp (CABG)	None
Pagani[97]	6	IMTEp (LVAD)	2 D
Dib[98]	24	IMTEp (CABG)	1 D; 2 VT
Chachques[99]	20	IMTEp (CABG)	None
Prosper[100]	12	IMTEp (CABG)	None
Smits[101]	5	IMTEn	1 VT
Smits[102]	15	IMTEn	2 D; 1 VT
Siminiak[103]	10	IMTEp (CABG)	1 D; 2 VT
Siminiak[104]	10	IMTV	None

Unpublished and ongoing clinical trials are not listed.
CABG = coronary artery bypass graft. D = death. IM = intra-myocardium. LVAD = left ventricular assist device. TEn = trans-endocardium. TEp = Trans-epicardium. TV = transcoronary vein. VT = ventricular tachycardia (tachyarrhythmia).

14.3.2 Clinical outcomes using bone marrow cells

Despite a lack of clear understanding for the beneficial mechanisms, Phase 1 clinical studies (Table 14.2) have shown the feasibility and safety of the procedure with encouraging functional improvements.[105–107,109–114,117–121] Randomized controlled trials have only been reported recently (TOPCARE-AMI, BOOST) and the number of patients involved are relatively small.[108,115,116,122] The REPAIR-AMI[123] trial is the only randomized, double blind, placebo-controlled multicenter trial involving 204 patients at 17 medical centers. So far the better improvement in left ventricular ejection fraction, the less heart enlargement, the significantly improved blood flow reserve, the lower rate of mortality, heart attacks and hospitalization due to heart failure all indicate the efficacy of cellular cardiomyoplasty using bone marrow cells. The mechanisms regulating the differentiation of bone marrow stem cells into heart muscle cells are under investigation by several groups. The application of

Table 14.2 Cellular cardiomyoplasty using bone marrow-derived cells

Coordinator	Patient T/C	Cell type	Procedures	Complication
Hamano[105]	6/0	MNC	IMTEp (CABG)	None
Strauer[106,107]	10/10 18/18	MNC	ICA (PCI)	1 Rs
Dimmeler and Zeiher[108]	59/11	MNC (29) CPC (30)	ICA (PCI)	1 MI & D 1 MI
Stamm[109]	12/6	AC133$^+$	IMTEp (CABG)	2 VT, 1 P
Kao[110]	10/0	MNC	ICA (PCI)	None
Tse[111]	8/0	MNC	IMTEn	None
Perin and Dohmann[112]	14/9	MNC	IMTEn	2D
Siminiak[113]	2/0	CD34$^+$	ICA (PCI)	None
Fuchs[114]	10/0	BMC	IMTEn	2 CP
Wollert[115]	30/30	MNC	ICA (PCI)	None
Chen[116]	34/35	MSC	ICA (PCI)	None
Kim[117]	20/7	CPC (10) G-CSF (10)	ICA (PCI)	7 Rs
Fernandez-Aviles[118]	20/13	MNC	ICA (PCI)	2 Rs
Galinanes[119]	14/0	BMC	IMTEp (CABG)	None
Kuethe[120]	10/0	MNC	ICA (PCI)	None
Bartunek and Vanderheyden[121]	19/16	CD133$^+$	ICA (PCI)	1 Oc, 7 Rs 2 Rs
Patel[122]	10/10	CD34$^+$	IMTEp (CABG)	None
Schachinger[123]	204	MNC	ICA (PCI)	On-going

BMC = bone marrow cell. C = control. CP = chest pain. CPC = circulating progenitor cell. D = death. HF = heart failure. ICA = intra-coronary artery. IM = intra-myocardium. MI = myocardial infarction. MNC= bone marrow mononuclear cell. MSC = bone marrow mesenchymal stem cells. Oc = occlusion. P = pneumonia. PCI = percutaneous coronary intervention. Rs = restenosis. T = treated. TEn = trans-endocardium. TEp = trans-epicardium. VT = ventricular tachycardia (tachyarrhythmia).

allogenic bone marrow mesenchymal stem cells for myocardial regeneration has been approved by the FDA recently.

14.3.3 Possible beneficial mechanisms of cellular cardiomyoplasty

The technical feasibility and safety of cellular cardiomyplasty is well established, the long-term engraftment of the transplanted cells is clearly documented and the efficacy in restoration of ventricular function is unambiguously proved.[44,47–54,95–123] However, the possible beneficial mechanisms (Fig. 14.1) are lacking general consensus among investigators. Only if progenitor cells can differentiate into cardiomyocytes and have electro-mechanical coupling with other heart muscle cells, can the optimal outcomes of cellular cardiomyoplasty be expected.

14.4 Future trends

With the successful outcomes in animal studies for cellular cardiomyoplasty, clinical trials have progressed rapidly using skeletal muscle satellite cells (myoblasts) and bone marrow-derived cells. The clear benefit of cellular cardiomyoplasty has been universally observed by all investigating groups; however, the possible beneficial mechanisms have not been clearly identified. Using purified stem cells or combinations of stem cells (i.e. satellite cells for myogenesis and bone marrow cells for angiogenesis), the route and method of cell delivery, as well as possible development of universal donor cells are likely to be the future foci of research.[43,44,47–54] Stem cells delivered by needle have about 90% loss by washout or leak-out and intravascular delivery has an even lower retention rate.[124] In addition, the retained cells have close to 90% mortality after successful delivery into myocardium. The improvement of cell

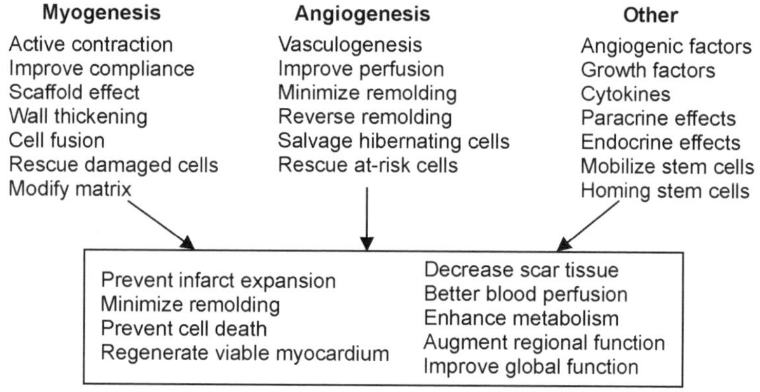

14.1 Possible beneficial mechanisms of cellular cardiomyoplasty.

retention, homing, survival, proliferation and differentiation will all enhance the outcomes of cellular cardiomyoplasty.

14.4.1 Purification of stem cell preparations

Most of the studies use enzymatic isolation of satellite cells from skeletal muscle and proliferating to significant numbers in culture without purification (contain a mixture of cell types). Whether or not using purified satellite cells or purified satellite cells mixed with other types of stem cells (i.e. bone marrow stromal cells or endothelial progenitor cells) will provide long-term optimal outcomes awaits future studies. In addition, directing the differentiation of stem cells or progenitor cells (to form cardiomyocytes, endothelial cells and smooth muscle cells instead of fibroblasts, cartilage, fat cells, etc.) may also improve cellular cardiomyoplasty. The search for optimal type or types of cells for this procedure and modifying the gene expression of the cells (angiogenic factors, growth factors, cytokines, etc.) may continue for many years.

14.4.2 Development of universal donor cells

Although beneficial outcomes have been observed using allogenic satellite cells for patients suffering from muscular dystrophy, the use of allogenic satellite cells for cellular cardiomyoplasty may not be a practical approach. Allogenic and xenogeneic[125–127] bone marrow stromal cells have been successfully applied in experimental studies for cellular cardiomyoplasty without the treatment of immunosuppressing drugs. Marrow stromal cells are mesenchymal stem cells that can differentiate into multiple cell types after transplantation to a different species without inflammatory cellular infiltration or immunorejection. The possible mechanisms for modulation of allogeneic immune cell responses by mesenchymal stem cells has been documented recently.[128] Mesenchymal stem cells have the potential to serve as universal donor cells with immediate availability of established quality, purity and quantity of stem cells for myocardial regeneration. Clinical trials for allogeneic bone marrow mesenchymal stem cells have started after receiving FDA approval.

14.4.3 Route and method of cell delivery

The poor retention (significant washout) and high degrees of cell death after cell delivery are major limitations for cellular cardiomyoplasty. The cells have been delivered into the heart muscle through epicardium (by needle and syringe), endocardium (NOGA or MyoCath catheter), or transvenous as well as through coronary artery or vein. The retention of cells is about 10% or less with significant washout and leak-out after delivery and can be much lower after intravascular delivery. Mixing the cells in a viscous solution (hydrogel or albumin) and delivering the cells through a needle with multiple side holes may improve the retention for intramyocardial injection. Improves perfusion of the cell transplanted area, transfection of the cells with anti-death strategies (Akt, HGF) and augments cell proliferation could increase survival and final numbers of the transplanted cells.

14.5 Sources of further information and advice

The National Institutes of Health (NIH) has an excellent web site (http://stemcells.nih. gov) containing much useful information. In addition the Bioheart company (www.bioheartinc.com); Diacrin (www.diacrin.com); GenVec (www.genvec.com); Genzyme (www.genzyme.com); Osiris Therapeutics, Inc. (www.osiristx.com) all have sponsored clinical trials of cellular cardiomyoplasty. Several companies also produce catheters for the procedure: Boston Scientific (www.bostonscientific.com); Johnson & Johnson (www.jnj.com); Bioheart Company (www.bioheartinc.com); BioCardai (www.biocardia.com) and Medtronic (www.medtronic.com). Cambrex (www.cambrex.com), ViaCell (www.vaicellinc.com) and Stem Cell Technologies (www.stemcell.com) are commercializing stem cells. A recent issue of *Genetic Engineering News* (Vol. 26(2), 15 January, 2006) provides a long list (>40) of suppliers for stem cell research.

14.6 References

1. Olson EN, Schneider MD. Sizing up the heart: development redux in disease. *Genes Dev* **17**: 1937–1956, 2003.
2. Pasumarthi KBS, Field LJ. Cardiomyocyte cell cycle regulation. *Circ Res* **90**: 1044–1054, 2002.
3. Anversa P, Leri A, Kajstura J, Nadal-Ginard B. Myocyte growth and cardiac repair. *J Mol Cell Cardiol* **34**: 91–105, 2002.
4. Williams RS. Cell cycle control in the terminally differentiated myocyte. A platform for myocardial repair? *Cardio Clin* **16**: 739–754, 1998.
5. Soonpaa MH, Kim KK, Pajak L, *et al.* Cardiomyocyte DNA synthesis and binucleation during murine development. *Am J Physiol* **271**: H2183–H2189, 1996.
6. Li F, Wang X, Capasso JM, *et al.* Rapid transition of cardiac myocytes from hyperplasia to hypertrophy during postnatal development. *J Mol Cell Cardiol* **28**: 1737–1746, 1996.
7. Claycomb WC. Control of cardiac muscle cell division. *Trends Cardiovasc Med* **2**: 231–236, 1992.
8. Thom T, Haase N, Rosamond W, *et al.* Heart disease and stroke statistics – 2006 update: a report from the American Heart Association Statistics Committee and Stroke Statistics Subcommittee. *Circulation* **113**: e85–e151, 2006.
9. Kao RL, Chiu RC-J, eds. *Cellular Cardiomyoplasty: Myocardial Repair with Cell Implantation*. Austin, TX: Landes, 1997.
10. Kao RL. Transplantation of satellite cells for myocardial regeneration. In:

Handbook of Cardiovascular Cell Transplantation. Kipshidae NN and Serruys PW, eds. London: Martin Dunitz, pp 91–105, 2004.
11. Kajstura J, Leri A, Finato N, *et al.* Myocyte proliferation in end-stage cardiac failure in humans. *Proc Natl Acad Sci USA* **95**: 8801–8805, 1998.
12. Beltrami AP, Urbanek K, Kajstura J, *et al.* Evidence that human cardiac myocytes divide after myocardial infarction. *N Engl J Med* **344**: 1750–1757, 2001.
13. Kajstura J, Zhang X, Reiss K, *et al.* Myocyte cellular hyperplasia and myocyte cellular hypertrophy contribute to chronic ventricular remodeling in coronary artery narrowing-induced cardiomyopathy in rats. *Circ Res* **74**: 383–400, 1994.
14. Setoguchi M, Leri A, Wang S, *et al.* Activation of cyclins and cyclin-dependent kinases, DNA synthesis and myocyte mitotic division in pacing-induced heart failure in dogs. *Lab Invest* **79**: 1545–1558, 1999.
15. Soonpaa MH, Field LJ. Assessment of cardiomyocyte DNA synthesis in normal and injured adult mouse hearts. *Am J Physiol* **272**: H220–H226, 1997.
16. Soonpaa MH, Field LJ. Survey of studies examining mammalian cardiomyocyte DNA synthesis. *Circ Res* **83**: 15–26, 1998.
17. Meckert PC, Rivello HG, Vigliano C, *et al.* Endomitosis and polyploidization of myocardial cells in the periphery of human acute myocardial infarction. *Cardiovasc Res* **67**: 116–123, 2005.
18. Yuasa S, Fukuda K, Tomita Y, *et al.* Cardiomyocytes undergo cells division following myocardial infarction is a spatially and temporally restricted event in rats. *Mol Cell Biochem* **259**: 177–181, 2004.
19. Li F, Wang X, Gerdes AM. Formation of binucleated cardiac myocytes in rat heart: II. Cytoskeletal organization. *J Mol Cell Cardiol* **29**: 1553–1565, 1997.
20. Warejcka DJ, Harvey R, Taylor BJ, Young HE, Lucas PA. A population of cells isolated from rat heart capable of differentiating into several mesodermal phenotypes. *J Surg Res* **62**: 233–242, 1996.
21. Urbanek K, Quaini F, Tasca G, *et al.* Intense myocyte formation from cadiac stem cells in human cardiac hypertrophy. *Proc Natl Acad Sci USA* **100**: 10440–10445, 2003.
22. Beltrami AP, Barlucchi L, Torella D, *et al.* Adult cardiac stem cells are multipotent and support myocardial regeneration. *Cell* **114**: 763–776, 2003.
23. Dawn B, Stein AB, Urbanek K, *et al.* Cardiac stem cells delivered intravascularly traverse the vessel barrier, regenerate infarcted myocardium and improved cardiac function. *Proc Natl Acad Sci USA* **102**: 3766–3771, 2005.
24. Urbanek K, Torella D, Sheikh F, *et al.* Myocardial regeneration by activation of myltipotent cardiac stem cells in ischemic heart failure. *Proc Natl Acad Sci USA* **102**: 8692–8697, 2005.
25. Linke A, Muller P, Nurzynska D, *et al.* Stem cells in the dog heart are self-renewing, clonogenic and multipotent and regenerate infracted myocardium, improving cardiac function. *Proc Natl Acad Sci USA* **102**: 8966–8971, 2005.
26. Oh H, Bradfute SB, Gallardo TD, *et al.* Cardiac progenitor cells from adult myocardium: homing, differentiation and fusion after infarction. *Proc Natl Acad Sci USA* **100**: 12313–12318, 2003.
27. Oh H, Chi X, Bradfute SB, *et al.* Cardiac muscle plasticity in adult and embryo by heart-derived progenitor cells. *Ann NY Acad Sci* **1015**: 182–189, 2004.
28. Cai CL, Liang X, Shi Y, *et al.* Isl1 identifies a cardiac progenitor population that proliferates prior to differentiation and contributes a majority of cells to the heart. *Dev Cell* **5**: 877–889, 2003.
29. Laugwitz KL, Moretti A, Lam J, *et al.* Postnatal isl1^{+} cardioblasts enter fully

differentiated cardiomyocyte lineages. *Nature* **433**: 647–653, 2005.
30. Martin CM, Meeson AP, Robertson SM, *et al.* Persistent expression of the ATP-binding cassette transporter, Abcg2, identifies cardiac SP cells in the developing and adult heart. *Dev Biol* **265**: 262–275, 2004.
31. Messina E, De Angelis L, Frati G, *et al.* Isolation and expansion of adult cardiac stem cells from human and murine heart. *Circ Res* **95**: 911–921, 2004.
32. MacLellan WR, Schneider MD. Genetic dissection of cardiac growth control pathways. *Annu Rev Physiol* **62**: 289–319, 2000.
33. Weinberg RA. E2F and cell proliferation a world turned upside down. *Cell* **85**: 457–459, 1996.
34. Kirshenbaum LA and Schneider MD. The cardiac cell cycle, pocket proteins and p300. *Trends Cardiovasc Med* **5**: 230–235, 1995.
35. Guo K, Walsh K. Inhibition of myogenesis by multiple cyclin-Cdk complexes. *J Biol Chem* **272**: 791–797, 1997.
36. Liu Y, Kitsis RN. Induction of DNA synthesis and apoptosis in cardiac myocytes by E1A oncoprotein. *J Cell Biol* **133**: 325–334, 1996.
37. MacLellan WR, Garcia A, Oh H, *et al.* Overlapping roles of pocket proteins in the myocardium are unmasked by germ line deletion of p130 plus heart-specific deletion of Rb. *Mol Cell Biol* **25**: 2486–2497, 2005.
38. Kishore BS, Pasumarthi SCT, Field LJ. Coexpression of mutant p53 and p193 renders embryonic stem cell-derived cardiomyocytes responsive to the growth-promoting activities of adenoviral E1A. *Circ Res* **88**: 1004–1011, 2001.
39. Field LJ. Atrial natriuretic factor-SV40 T antigen transgenes produce tumors and cardiac arrhythmias in mice. *Science* **239**: 1029–1033, 1988.
40. Nakajima H, Nakajima HO, Tsai SC, Field LJ. Expression of mutant p193 and p53 permits cardiomyocyte cell cycle reentry after myocardial infarction in transgenic mice. *Circ Res* **94**: 1606–1614, 2004.
41. Pasumarthi KBS, Nakajima H, Nakajima HO, *et al.* Targeted expression of cyclin D2 results in cardiomyocyte DNA synthesis and infarct regression in transgenic mice. *Circ Res* **96**: 110–118, 2005.
42. Chaudhry HW, Dashoush NH, Tang H, *et al.* Cyclin A2 mediates cardiomyocyte mitosis in the postmitotic myocardium. *J Biol Chem* **279**: 35858–35866, 2004.
43. Young HE, Black AC Jr. Adult stem cells. *Anat Rec* **276A**: 75–102, 2004.
44. Eisenberg LM, Eisenberg CA. Adult stem cells and their cardiac potential. *Anat Rec* **276A**: 103–112, 2004.
45. Wagers AJ, Weissman IL. Plasticity of adult stem cells. *Cell* **116**: 639–648, 2004.
46. Lakshmipathy U, Verfaillie C. Stem cell plasticity. *Blood Rev* **19**: 29–38, 2005.
47. Fukuda K. Progression in myocardial regeneration and cell transplantation. *Circ J* **69**: 1431–1446, 2005.
48. Leri A, Kajstura J, Anversa P. Cardiac stem cells and mechanisms of myocardial regeneration. *Physiol Rev* **85**:1373–1416, 2005.
49. Wollert KC, Drexler H. Clinical application of stem cells for the heart. *Circ Res* **96**: 151–163, 2005.
50. Murry CE, Field LJ, Menasche P. Cell-based cardiac repair: reflections at the 10-year point. *Circulation* **112**: 3174–3183, 2005.
51. Laflamme MA, Murry CE. Regenerating the heart. *Nat Biotechnol* **23**: 845–56, 2005.
52. Dimmeler S, Zeiher AM, Schneider MD. Unchain my heart: the scientific fundations of cardiac repair. *J Clin Invest* **115**: 572–583, 2005.

53. Dai W, Hale SL, Kloner RA. Stem cell transplantation for the treatment of myocardial infarction. *Transpl Immunol* **15**: 91–97, 2005.
54. Davani S, Deschaseaux F, Chalmers D, *et al.* Can stem cells mend a broken heart? *Cardiovasc Res* **65**: 305–316, 2005.
55. Marelli D, Desrosiers C, el-Alfy M, Kao RL, Chiu RC. Cell transplantation for myocardial repair: an experimental approach. *Cell Transplant* **1**: 383–390, 1992.
56. Chiu RC, Zibaitis A, Kao RL. Cellular cardiomyoplasty: myocardial regeneration with satellite cell implantation. *Ann Thorac Surg* **60**: 12–18, 1995.
57. Quaini F, Urbanek K, Beltrami AP, *et al.* Chimerism of the transplanted heart. *N Engl J Med* **346**: 5–15, 2002.
58. Muller P, Pfeiffer P, Koglin J, *et al.* Cardiomyocytes of noncardiac origin in myocardial biopsies of human transplanted hearts. *Circulation* **106**: 31–35, 2002.
59. Hocht-Zeisberg E, Kahnert H, Guan K, *et al.* Cellular repopulation of myocardial infarction in patients with sex-mismatched heart transplantation. *Eur Heart J* **25**: 749–758, 2004.
60. Minami E, Laflamme MA, Saffitz JE, Murry CE. Extracardiac progenitor cells repopulate most major cell types in the transplanted human heart. *Circulation* **112**: 2951–2958, 2005.
61. Leor J, Patterson M, Quinones MJ, *et al.* Transplantation of fetal myocardial tissue into the infarcted myocardium of rat. *Circulation* **94**: II332–II336, 1996.
62. Soonpaa MH, Koh GY and Klug MG. Formation of nascent intercalated disks between grafted fetal cardiomyocytes and host myocardium. *Science* **264**: 98–101, 1994.
63. Gojo S, Kitamura S, Hatano O, *et al.* Transplantation of genetically marked cardiac muscle cells. *J Thorac Cardiovasc Surg* **113**:10–18, 1997.
64. Li R-K, Jia Z-Q, Weisel RD, *et al.* Cardiomyocyte transplantation improves heart function. *Ann Thorac Surg* **62**: 654–661, 1996.
65. Li R-K, Donald AG, Mickle MD, *et al.* Natural history of fetal rat cardiomyocytes transplanted into adult rat myocardial scar tissue. *Circulation* **96**: II179–II187, 1997.
66. Li R-K, Weisel RD, Mickle DA, *et al.* Autologous porcine heart cell transplantation improved heart function after a myocardial infarction. *J Thorac Cardiovasc Surg* **119**: 62–68, 2000.
67. Koh GY, Soonpaa MH, Klug MG, *et al.* Stable fetal cardiomyocyte grafts in the hearts of dystrophic mice and dogs. *J Clin Invest* **96**: 2034–2042, 1995.
68. Matsushita T, Oyamada M, Kurata H, *et al.* Formation of cell junctions between grafted and host cardiomyocytes at the border zone of rat myocardial infarction. *Circulation* **100**: II262–II268, 1999.
69. Reinecke H, Zhang M, Bartosek T, Murry CE. Survival, integration and differentiation of caridomyocyte grafts. A study in normal and injured rat hearts. *Circulation* **100**: 193–202, 1999.
70. Zhang M, Methot D, Poppa V, *et al.* Cardiomyocyte grafting for cardiac repair: Graft cell death and anti-death strategies. *J Mol Cell Cardiol* **33**: 907–921, 2001.
71. Shamboltt MJ, Axelman J, Wang S, *et al.* Derivation of pluripotent stem cells from cultured human primordial germ cells. *Proc Natl Acad Sci USA* **95**: 13726–13731, 1998.
72. Thomson JA, Itskovitz-Eldor J, Shapiro SS, *et al.* Embryonic stem cell lines derived from human blastocysts. *Science* **282**: 1145–1147, 1998.
73. Kehat I, Khimovich L, Caspi O, *et al.* Electromechanical integration of cardio-

myocytes derived from human embryonic stem cells. *Nat Biotechnol* **22**: 1282–1289, 2004.
74. Charge SBP, Rudnicki MA. Cellular and molecular regulation of muscle regeneration. *Physiol Rev* **84**: 209–238, 2004.
75. Wozniak AC, Kong J, Bock E, Pilipowicz O, Anderson JE. Signaling satellite-cell activation in skeletal muscle: markers, models, stretch and potential alternate pathways. *Muscle Nerve* **31**: 283–300, 2005.
76. Mauro A. Satellite cell of skeletal muscle fiber. *J Biophys Biochem Cytol* **9**: 493–498, 1961.
77. Deasy BM, Gharaibeh BM, Pollett JB, Jones MM, Lucas MA, Kanda Y, Huard J. Long-term self-renewal of postnatal muscle-derived stem cells. *Mol Biol Cell* **16**: 3323–3333, 2005.
78. Collins CA, Olsen I, Zammit PS, Heslop L, Petrie A, Partridge TA, Morgan JE. Stem cell function, self-renewal and behavioral heterogeneity of cells from the adult muscle satellite cells niche. *Cell* **122**: 289–301, 2005.
79. Shefer G, Wleklinski-Lee M, Yablonka-Reuveni Z. Skeletal muscle satellite cells can spontaneously enter an alternative mesenchymal pathway. *J Cell Sci* **117**: 5393–5404, 2004.
80. Engel FB, Schebesta M, Duong MT, *et al.* p38 MAP kinase inhibition enables proliferation of adult mammalian cardiomyocytes. *Genes Dev* **19**: 1175–1187, 2005.
81. Taylor DA, Atkins BZ, Hungspreugs P, *et al.* Regenerating functional myocardium: improved performance after skeletal myoblast transplantation. *Nat Med* 4: 929–933, 1998.
82. Ashton BA, Allen TD, Howlett CR, *et al.* Formation of bone and cartilage by marrow stromal cells in diffusion chambers *in vivo*. *Clin Orthop Rel Res* **151**: 294–307, 1980.
83. Wakitani S, Saito T, Caplan AI. Myogenic cells derived from rat bone marrow mesenchymal stem cells exposed to 5-azacytidine. *Muscle Nerve* **18**: 1417–1426, 1995.
84. Pittenger MF, Mackay AM, Beck SC, *et al.* Multilineage potential of adult human mesencymal stem cells. *Science* **284**:143–146, 1999.
85. Makino S, Fukuda K, Miyoshi S, *et al.* Cardiomyocytes can be generated from marrow stromal cells *in vitro*. *J Clin Invest* **103**: 697–705, 1999.
86. Fukuda K. Development of regenerative cardiomyocytes from mesenchymal stem cells for cardiovascular tissue engineering. *Artif Organs* **25**:187–193, 2001.
87. Tomita S, Li R-K, Weisel RD, *et al.* Autologous transplantation of bone marrow cells improves damaged heart function. *Circulation* **100**: II-247–II-256, 1999.
88. Wang J-S, Shum-Tim D, Galipeau J, *et al.* Marrow stromal cells for cellular cardiomyoplasty: feasibility and potential clinical advantages. *J Thorac Cardiovasc Surg* **120**: 999–1006, 2000.
89. Wang J-S, Shum-Tim D, Chredrawy E, Chiu RC-J. The coronary delivery of marrow stromal cells for myocardial regeneration: pathophysiologic and therapeutic implications. *J Thorac Cardiovasc Surg* **122**: 699–705, 2001.
90. Orlic D, Kajstura J, Chimenti S, *et al.* Bone marrow cells regenerate infarcted myocardium. *Nature* **410**: 701–707, 2001.
91. Tomita S, Mickle DA, Weisel RD, *et al.* Improved heart function with myogenesis and angiogenesis after autologous porcine bone marrow stromal cell transplantation. *J Thorac Cardiovasc Surg* **123**: 1132–1140, 2002.
92. Kajstura J, Tota M, Whang B, *et al.* Bone marrow cells differentiate in cardiac cell

lineages after infarction independently of cell fusion. *Circ Res* **96**: 127–137, 2005.
93. Dawn B, Bolli R. Adult bone marrow-derived cells: regenerative potential, plasticity and tissue commitment. *Basic Res Cardiol* **100**: 494–503, 2005.
94. Haider HK, Ashraf M. Bone marrow stem cell transplantation for cardiac repair. *Am J Physiol* **288**: H2557–H2567, 2005.
95. Menasche P, Hagege AA, Vilquin JT, *et al.* Autologous skeletal myoblast transplantation for severe postinfarction left ventricular dysfunction. *J Am Coll Cardiol* **41**: 1078–1083, 2003.
96. Zhang F, Yiang Z, Chen Y, *et al.* Autologous satellite cells transplantation to treat coronary heart disease. *Stem Cell Cellular Therapy* **2**: 12–17, 2004.
97. Pagani FD, DerSimonian H, Zawadzka A, *et al.* Autologous skeletal myoblasts transplanted to ischemia-damaged myocardium in humans. *J Am Coll Cardiol* **41**: 879–888, 2003.
98. Dib N, Michler RE, Pagani FD, *et al.* Safety and feasibility of autologous myoblast transplantation in patients with ischemic cardiomyopathy: four-year follow-up. *Circulation* **112**: 1748–1755, 2005.
99. Chachques JC, Herreros J, Trainini J, *et al.* Autologous human serum for cell culture avoids the implantation of cardioverter-defibrillators in cellular cardiomyoplasty. *Int J Cardiol* **95**: S29–S33, 2004.
100. Herreros J, Prosper F, Perez A, *et al.* Autologous intramyocardial injection of cultured skeletal muscle-derived stem cells in patients with non-acute myocardial infarction. *Eur Heart J* **24**: 2012–2020, 2003.
101. Smits PC, van Geuns RJ, Poldermans D, *et al.* Catheter-based intramyocardial injection of autologous skeletal myoblasts as a primary treatment of ischemic heart failure: clinical experience with six-month follow-up. *Am Coll Cardiol* **42**: 2070–2072, 2003.
102. Smits PC. Myocardial repair with autologous skeletal myoblasts: a review of the clinical studies and problems. *Minerva Cardioangiol* **52**: 525–535, 2004.
103. Siminiak T, Kalawski R, Fiszer D, *et al.* Autologous skeletal myoblast transplantation for the treatment of postinfarction myocardial injury: phase 1 clinical study with 12 months of follow-up. *Am Heart J* **148**: 531–537, 2004.
104. Siminiak T, Fiszer D, Jerzykowska O, *et al.* Percutaneous trans-coronary-venous transplantation of autologous skeletal myoblasts in the treatment of post-infarction myocardial contractility impairment: the POZNAN trial. *Eur Heart J* **26**: 1188–1195, 2005.
105. Li TS, Hamano K, Hirata K, *et al.* The safety and feasibility of the local implantation of autologous bone marrow cells for ischemic heart disease. *J Card Surg* **18**(Suppl 2): S69–75, 2003.
106. Strauer BE, Brehm M, Zeus T, *et al.* Repair of infarcted myocardium by autologous intracoronary mononuclear bone marrow cell transplantation in humans. *Circulation* **106**: 1913–1918, 2002.
107. Strauer BE, Brehm M, Zeus T, *et al.* Regeneration of human infarcted heart muscle by intracoronary autologous bone marrow cell transplantation in chronic coronary artery disease: the IACT study. *J Am Coll Cardiol* **46**: 1651–1658, 2005.
108. Schachinger V, Assmus B, Britten MB, *et al.* Transplantation of progenitor cells and regeneration enhancement in acute myocardial infarction. Final one-year results of the TOPCARE-AMI trial. *J Am Coll Cardiol* **44**:1690–1699, 2004.
109. Stamm C, Kleine HD, Westphal B, *et al.* CABG and bone marrow stem cell transplantation after myocardial infarction. *Thorac Cardiovasc Surg* **52**: 152–158, 2004.

110. Zhang F, Zhou F, Yang Z, *et al.* Intracoronary delivery of autologous bone marrow stem cells repairs infarcted myocardium in humans. *Cardiovasc Reg Med* **6**: 111, 2005 (abstract).
111. Tse HF, Kwong YL, Chan JFK, *et al.* Angiogenesis in ischemic myocardium by intramyocardial autologous bone marrow mononuclear cell implantation. *Lancet* **361**: 47–49, 2003.
112. Perin EC, Dohmann HFR, Borojevic R, *et al.* Improved exercise capacity and ischemia 6 and 12 months after transendocardial injection of autologous bone marrow mononuclear cells for ischemic cardiomyopathy. *Circulation* **110**(Suppl II): II-213–II-218, 2004.
113. Siminiak T, Grygielska B, Jerzykowska O, *et al.* Autologous bone marrow stem cell transplantation in acute myocardial infarction-report of two cases. *Kardiol Pol* **59**: 502–510, 2003.
114. Fuchs S, Satler LF, Kornowski R, *et al.* Catheter-based autologous bone marrow myocardial injection in no-option patients with advanced coronary disease. A feasibility study. *J Am Coll Cardiol* **41**: 1721–1724, 2003.
115. Wollert KC, Meyer GP, Lotz J, *et al.* Intracoronary autologous bone-marrow cell transfer after myocardial infarction: the BOOST randomized controlled clinical trial. *Lancet* **364**: 141–148, 2004.
116. Chen SL, Fang WW, Ye F, *et al.* Effect on left ventricular function of intracoronary transplantation of autologous bone marrow mesenchymal stem cells in patients with acute myocardial infarction. *Am J Cardiol* **94**: 92–95, 2004.
117. Kang HJ, Kim HS, Zhang SY, *et al.* Effects of intracoronary infusion of peripheral blood stem-cells mobilized with granulocyte-colony stimulating factor on left ventricular systolic function and restenosis after coronary stenting in myocardial infarction: the MAGIC cell randomized clinical trial. *Lancet* **363**: 751–756, 2004.
118. Fernandez-Aviles F, San Roman JA, Gracia-Frade J, *et al.* Experimental and clinical regenerative capability of human bone marrow cells after myocardial infarction. *Circ Res* **95**: 742–748, 2004.
119. Galinanes M, Loubani M, Davies J, *et al.* Autotransplantation of unmanipulated bone marrow into scarred myocardium is safe and enhances cardiac function in humans. *Cell Transplant* **13**: 7–13, 2004.
120. Kuethe F, Richartz BM, Kasper C, *et al.* Autologous intracoronary mononuclear bone marrow cell transplantation in chronic ischemic cardiomyopathy in humans. *Int J Cardiol* **100**: 485–491, 2005.
121. Bartunek J, Vanderheyden M, Vandekerckhove B, *et al.* Intracoronary injection of CD133-positive enriched bone marrow progenitor cells promotes cardiac recovery after recent myocardial infarction: feasibility and safety. *Circulation* **112**(Suppl I): I-178–I-183, 2005.
122. Patel AN, Geffner L, Vina RF, *et al.* Surgical treatment for congestive heart failure with autologous adult stem cell transplantation: a prospective randomized study. *J Thorac Cardiovasc Surg* **130**: 1631–1638, 2005.
123. Schachinger V, Tonn T, Dimmeler S, Zeiher AM. Bone-marrow derived progenitor cell therapy in need of proof of concept: design of the REPAIR-AMI trial. *Nat Clin Pract Cardiovasc Med* **3**(Suppl 1): S23–S28, 2006.
124. Teng CJ, Luo J, Chiu RC, Shum-Tim D. Massive mechanical loss of microspheres with direct intramyocardial injection in the beating heart: implications for cellular cardiomyoplasty. *J Thorac Cardiovasc Surg* **132**: 628–632, 2006.

125. Chiu RCJ. Stealth immune tolerance in stem cell transplantation: potential for universal donors in myocardial regeneration therapy. *J Heart Lung Transplant* **24**: 511–516, 2005.
126. Amado L, Saliaris AP, Schuleri KH, *et al.* Cardiac repair with intramyocardial injection of allogenic mesenchymal stem cells after myocardial infarction. *Proc Natl Acad Sci USA* **102**: 11474–11479, 2005.
127. Strom TB, Field LJ, Ruediger M. Allogenic stem cell-derived repair unit therapy and the barriers to clinical development. *J Am Soc Nephrol* **15**: 1133–1139, 2004.
128. Aggarwal S, Pittenger MF. Human mesenchymal stem cells modulate allogeneic immune cell responses. *Blood* **105**: 1815–1822, 2005.

15
Kidney diseases and potentials of artificial cells

S PRAKASH, T LIM and W OUYANG,
McGill University, Canada

15.1 Introduction

In an era of growing incidence and prevalence of renal failure, it is still true that renal replacement therapies such as dialysis required to sustain lives of renal failure patients are a tremendous undertaking. Indeed, while the short-term effects related to renal replacement modalities avoid the lethal acidosis, volume overload and uremic syndromes that accompany renal failure, they do not protect the patient from the increased risks of morbidity associated with these treatment modalities. Complications such as cardiomyomorbidity resulting in cardiovascular events, a compromised immune system leading to infection episodes, or the long-term commitment to immunosuppressants in kidney transplantation each imposes substantial stress upon the renal failure patient, and as such is an appropriate target for improved renal replacement therapy. With the perpetual increase in the global renal failure population, such limitations of renal replacement therapies and donor scarcity related to kidney transplantation sparked a renewed interest in other avenues for renal replacement. This chapter aims to provide an overview of how artificial cells are gradually undertaking an important role in renal replacement therapies, with the first half of the chapter describing pathophysiologies associated with different types of renal failure, the current outlook of renal replacement protocols and their associated shortcomings, followed by the second half explaining the concept of artificial cells, its history, and new developments that have made valuable contributions to the research field.

15.2 Renal failure

Healthy kidneys perform multiple functions central to the well-being of the body, such as the removal of excess fluid, nitrogenous and toxic wastes, electrolytic balance and the maintenance of acid–base balance of the body. These functions are, however, compromised upon the onset of renal failure, which may occur either suddenly (acute) or over a period of time (chronic) due to different causes, and also dictate the type of treatment protocol administered.

15.2.1 Acute renal failure

While categorized by different etiologies, acute renal failure (ARF) is generally defined as the abrupt, uncontrolled and significant decrease of glomerular filtration rate (GFR) within a short period of time (hours to days) that results in the rapid accumulation of nitrogenous wastes in the body. In prerenal ARF, tubular and glomerular functions are healthy, but glomerular perfusion is compromised as a result of hypotension, hypovolemia caused by cardiogenic shocks, blood vessel dilatation caused by septic or anaphylactic shocks, and neurogenic shock due to spinal trauma. Another form of ARF, such as acute tubular necrosis, results from intrinsic diseases associated with the tubules of the nephron, and is typically caused by renal ischemia linked to prerenal causes, or the accumulation of drugs such as aminoglycosides, lithium, etc. Inflammation of the glomeruli resulting from infection or immunological reasons also results in glomerulonephritis which causes intrinsic renal failure. Thirdly, postrenal ARF refers to the obstruction beyond the collecting ducts, resulting in a back pressure that diminishes the pressure differential in the glomeruli and a resultant decrease in GFR. This obstruction may be caused by renal calculi or pelvic tumors. ARF episodes account not only for the sudden deterioration in chronic renal failure or end stage renal failure patients, but also for 1–25% of intensive care unit patients[1,2] who are associated with high hospital mortality.

15.2.2 Chronic renal failure

Chronic renal failure refers to the gradual and progressive loss of kidney function that happens over a prolonged period of time and may be due to systemic diseases. Intrinsic renal failure is largely associated to diseases in the renal parenchyma, typically because of blockages of the filter system at the glomeruli level. These blockages may result from hemoglobinemia, where lysis of the red blood cells releases large hemoglobin molecules that clog the glomeruli filter; myoglobinemia or the release of the oxygen carrying pigment in muscles that is released owing to the systemic breakdown of muscles; or glomerulonephritis. Post-renal obstruction due to benign prostratic hyperplasia is a common cause of chronic renal failure in the elderly population, while urethra strictures or the abnormal narrowing of the urethra may result from instrumentation trauma or urethra infection.

In the majority of cases, chronic renal failure leads to the irreversible loss of kidney functions and the ultimate progression to end stage renal disease (ESRD) where long-term renal replacement therapies are needed to sustain life. Such a trend towards ESRD is almost absolute even with therapy, owing to a recent observation in the intimate relationship between chronic renal failure, anemia and congestive heart failure.[3] Despite increased awareness about healthy lifestyles and higher standards of living in modern society, the ESRD population

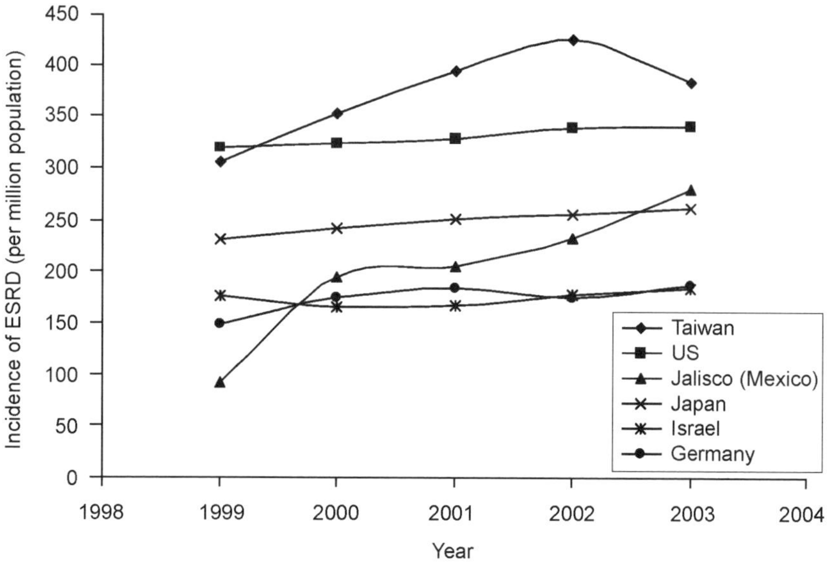

15.1 Incidence of ESRD in top six countries of highest incident rates. Incident cohorts are determined at the start of ESRD initation. (*Source*: adapted from the US Renal Data System (USRDS) 2005 Annual Data Report[4].)

is on a perpetual climb (Fig. 15.1). Causes of ESRD in the United States include, in the order of importance: diabetes, glomerulonephritis, secondary vasculitis, interstitial nephritis, hypertensive disease, cystic/congenital diseases and tumors. The pathophysiology of chronic renal failure is such as to worsen existing hypertensive symptoms.

15.3 Current renal therapy options

Current treatment modalities are largely based on the principles of renal replacement (dialysis) originally described by Schreiner and Maher in 1961.[5] Since its advent, the dialysis procedure has not only seen improvements in its technology but has also extended treatment to patients suffering from different causes of renal failure. In fact, treatment choices are more often than not dictated by the type of renal failure, due to specific procedures associated with the different modalities that may prove more efficient in one case than another.

15.3.1 Hemodialysis

Hemodialysis utilizes the hemodynamic pressure of the arteries to drive the filtration process in an extracorporeal circuit, thereby achieving high efficiency dialysis within a span of several hours. Before hemodialysis can be executed, a vascular access between the patient and the dialysis machine has to be surgically prepared. The most common is an arteriovenous (AV) fistula which abnormally connects the high pressure artery to a vein, from which the dialysis machine is connected. This is done to avoid spasms resulting from the puncture of an arterial wall, and to create high-pressure blood flow into the vein that ultimately leads to its widening for ease of repeated needle insertions and prevention of blood clots.

Patients with small veins that do not form proper fistulas will be implanted with an AV graft, which is an artificial tubing used to connect the artery to the vein. While it may not last as long as the AV fistula, if properly cared for it may last for several years.

It is pertinent for the physical connection to heal before hemodialysis may be started; this usually translates to a period of weeks to months. Therefore, the aforementioned procedures cannot be used in acute cases of renal failure. Instead, a venous catheter is temporarily connected to a large vein, where the double lumen tube allows for the flow of blood to and from the hemodialysis machine. The lower venous pressure meant that hemodialysis has to be done for a longer period of time compared with the AV connections.

15.3.2 Hemofiltration

Another form of toxin removal in the acute setting is hemofiltration. This method requires the use of a venous catheter and is typically done in intensive care settings for emergency cases such as fluid overload, a deteriorating body biochemistry or acidosis. Utilizing fluid convection principles it is less efficient than hemodialysis using a permanent fistula, and may require up to 5 hours for each cycle, three times a day. To date, there is a running debate between the choice of continuous venous hemodiafiltration or intermittent hemodialysis for better patient outcome, although a recent report highlighted the independent relationship of mortality rates to choice of technique.[6]

15.3.3 Peritoneal dialysis

Accomplished mostly by diffusion and convection forces, peritoneal dialysis (PD) is a gentler process of uremia correction than hemodialysis. The procedure involves the filling of the peritoneal cavity with dialysis solution, where a high dextrose concentration provides the necessary concentration gradient for movement of waste and excess fluid into the cavity; the dialysis solution is then completely removed after several hours of dwell time via dialysis solution drainage. This process is repeated at regular intervals and, unlike hemodialysis, is not only a daily commitment but also a continuous process. Similar to hemodialysis is the surgical implantation of a catheter in the abdomen, followed by a latency period in terms of weeks for wound healing. Two main forms of peritoneal dialysis include: continuous ambulatory peritoneal dialysis (CAPD) and continuous cyclic peritoneal dialysis (CCPD).

In CAPD, commonly known as home dialysis, the peritoneum contains an additional 2.0–2.5 liters of dialysis solution at any given time. The exchange of fluids may be done without expert supervision and gives the patient locomotive freedom and arguably a higher quality of life. Economic considerations for CAPD are another major reason for its popularity, which involves mainly the financing of fresh dialysis solutions. Contra-indications to CAPD include any infection insults to the peritoneum, which may be inflicted in previous abdominal surgery resulting in the adhesion of the peritoneum. Furthermore, because the skin barrier is permanently broken for surgical insertion of the catheter, infection is a common problem. As the worsening of such infections may ultimately lead to the loss of technique for the renal failure patient, strict aseptic practices have to be followed to maintain the integrity of the entry site, especially during fluid exchange.

Alternatively, the fluid exchange procedure may be automated by an automatic cycler in CCPD. The procedure is typically done during nocturnal hours in sleep, leaving a fresh batch of dialysis solution in the peritoneum for the start of the day. While CCPD promises an interruption-free day, it requires an 8–9 h attachment to the machine, besides the fact that additional exchanges may be necessary in some patients.

15.3.4 Kidney transplantation

While the aforementioned renal replacement therapies sustain renal failure patients' lives by performing filtration for the diseased kidneys, they do not replace the synthetic functions of the kidneys. The kidneys play an irreplaceable role in the production of essential hormones such as calcitrol for effective calcium absorption; erythropoietin essential in red blood cell production; and renin for the maintenance of blood pressure. Although additional supplements may be administered to renal failure patients receiving renal replacement therapies, it is kidney transplants that promise absolute freedom from the life-sustaining machines. Consequently, this quest for an independence of renal replacement therapies results in the waiting list for kidney transplants topping the list of total organ waiting list in the United States (Fig. 15.2). However, candidates for renal transplants are limited to a small population. Such eligibility issues as preformed antibodies from previously failed grafts, the rising age of the population at the onset of ESRD, compounded with limited donor availability, greatly limit the number of transplant recipients.[7]

The dialysis population is constantly increasing; this is especially the case for chronic or maintenance dialysis which, to a limited patient cohort, serves as a bridge to transplantation but to the majority of renal failure patients remains the only life-sustaining mechanism. As more incidence cases are uncovered with advancements in medical diagnostics, it is expected that the chronic dialysis population will reach half a million by 2010,[9] along with the complications associated with long-term dialysis.

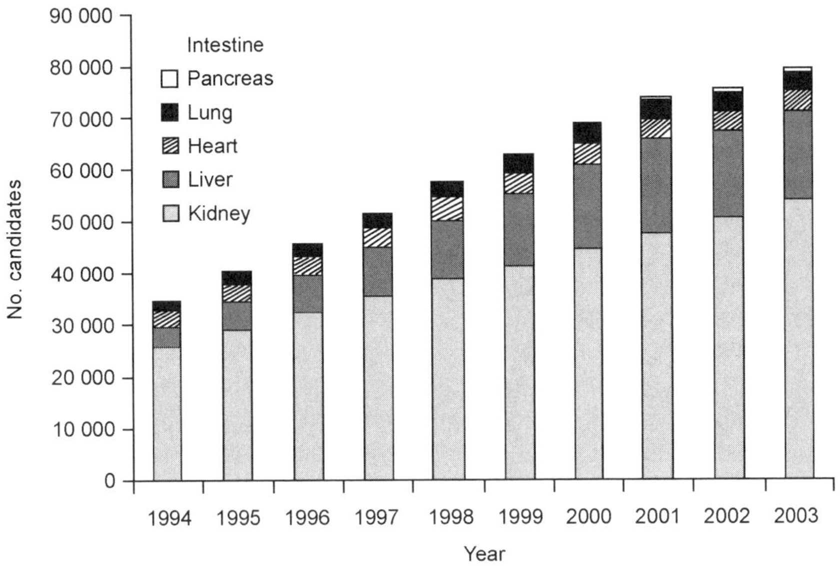

15.2 Number of waiting list candidates at the end of year, 1994–2003. (Source: 2004 Organ Procurement Transplantation Network/Scientific Registry of Transplant Recipients (OPTN/SRTR) Annual Report, Table 1.3[8].)

15.4 Limitations and complications of current treatment protocols

A major limitation of renal replacement therapies is the wide plethora of associated complications. While some of these complications are specific to a particular modality, renal failure patients undergoing renal replacement therapies are typically affected by similar side effects resulting from the uremic state. Therefore, the following sections discuss major complications and limitations of renal replacement therapies based on common morbidities in these patients, instead of issues based on the different modalities. Furthermore, complications as a result of renal replacement therapies are typically manifested in their long-term use, such as in maintenance dialysis protocols. Thus Section 15.4.1 highlighting complications of dialysis protocols refers to those associated with maintenance or chronic dialysis (used in ESRD patients). Section 15.4.2 covers issues related to kidney transplantation.

15.4.1 Limitations and complications of hemodialysis and peritoneal dialysis

Morbidity due to cardiovascular diseases

ESRD patients on hemodialysis suffer from a shorter life expectancy compared with the general population, which is largely attributed to morbidity from a

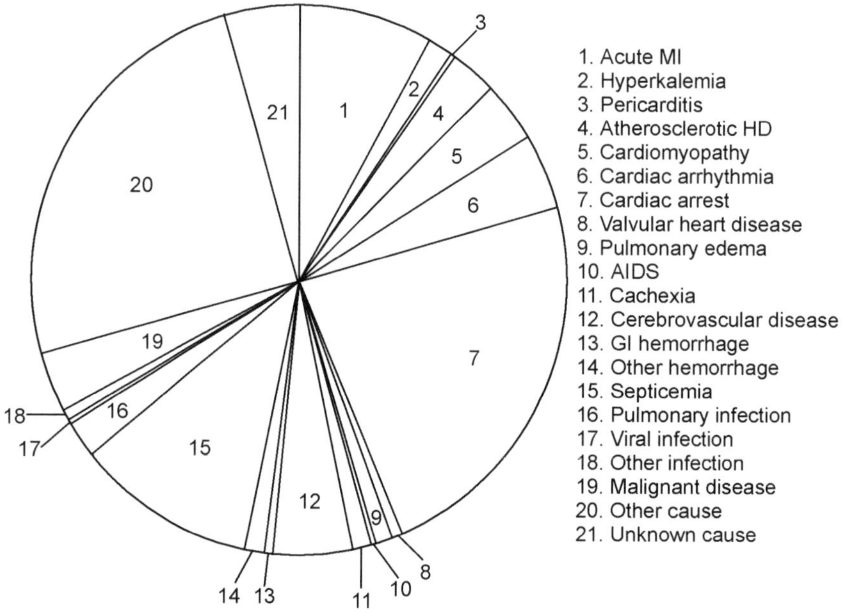

15.3 Mortality Distribution by Primary Cause of Mortality in Hemodialysis Patients (2001–2003). (*Source*: adapted from USRDS 2005 Annual Data Report[4].)

myriad of diseases as a result of chronic hemodialysis (Fig. 15.3). There is a 50% mortality rate due to cardiovascular diseases (CVD) within the first 5 years of the initiation of dialysis;[4] in a study population consisting of young and relatively illness-free patients, 30–40% of them succumbed to CVD;[10] in fact, numerous authors agree CVD is the primary cause of death in ESRD patients.[11–13] This finding was supported by many other studies which drew a unanimous link to cardiomyopathy-related mortality in dialysis patients after being stratified for age, race, gender, ethnicity, nationality and primary renal disease.[14–16] Uremic-related cardiomyopathy are manifested mostly in the form of atherosclerotic vascular diseases (reduced arterial compliance and increased intima-media thickness) and left ventricular (LV) abnormalities. These morbidities are a result of the uremic state, since LV abnormalities in dialysis patients improved after receiving transplantation.[17] Dialysis-related cardiomyopathy is a multivariate condition and may be attributed to hypertension, anemia,[11] hypoalbuminemia,[17] hyperhomocysteinemia, increased calcium-phosphate product, hypertriglyceridemia coexisting with hyper-cholesterimia, increased markers of inflammation and oxidative stress,[18] which are conditions commonly found in ESRD patients.[11,19]

Reduced immunity and its effects

The impairment of the immune system in uremia has been acknowledged since the early 1950s, as noted in early studies where allografts survived longer in subjects with uremia than non-uremic subjects.[20,21] Evidence of anergy to viruses and mycobacteria[22] led to the conclusion that most dialysis patients suffered from lymphopenia[23] and decreased absolute counts of lymphocytes.

The healthy immune system consists of an intricate network of cellular and humoral components: (1) circulating (lymphocytes, monocytes, polymorphonuclear neutrophils (PMN)) and tissue immuno-competent cells (lymphocytes, macrophages), (2) humoral substances (cytokines, chemokines, leucotrienes, prostaglandins, growth factors, complement factors, platelet activating factor, soluble receptors and receptor antagonists).[24] Upon the onset of ESRD, the uremic state alters the responses of these cells to foreign invaders, especially with regards to phagocytosis and oxidative responses. Although uremic toxicity is alleviated by dialysis, chronic dialysis reveals another set of complications, as shown by its inability to reverse the abnormalities of the immune cells and even exacerbate these deviances.[25–27] In fact, several of these defects were reversed after transplantation, indicating that metabolic activities in the renal parenchyma play a role in the proper functioning of immune mechanisms.[25] Although increased complement activation and production of pro-inflammatory substances were noted in dialysis patients,[28] the protective capacities of these immune cells were severely reduced; attenuation of autoimmune disorders, increased cancer prevalence,[29] repeated bacterial infections,[30] and depressed responses to vaccination are but several of the clinical manifestations that result in morbidity and mortality in ESRD patients. Dysfunction of the immune system contributes to about 15% of mortality in dialysis patients.[31]

In hemodialysis patients, the chemotactic process undertaken by neutrophils and monocytes is altered during an infection episode, specifically the phagocytosis capabilities and the reactive oxygen species production by PMN. A recent study found Cuprophan membranes related to the induction of apoptosis of PMN.[32] Synthetic membranes such as cellulose acetate (CA) have resulted with a poor outcome in acute renal failure.[33] In fact, aged CA membranes were found to increase inflammatory responses from neutrophils, and their significant apoptosis, compared with new CA membranes.[34]

The inadequacy of the humoral immune system in hemodialysed patients is most apparent in their lack of response to vaccinations, such as influenza, tetanus or hepatitis B.[35,36] Repeated vaccinations are required which confers a shorter-term protection against hepatitis B compared with the general population.[37,38] Recent research attributed this defect to increased levels of soluble CD40, which were normally excreted in healthy individuals[39] but not in renal failure patients.

A direct consequence of the compromised immune system is the high incident rate of infection. Dialysis patients are constantly under this threat due to the perpetual breach of the skin barrier and exposure to foreign materials at the access sites, especially since access sites require repeated manipulation. Infection-related mortality is in fact considered by many authors as the next important consequence of maintenance dialysis.[40-43] Despite the fact that such infections can be reduced by exercising precaution and stringency on the handling process, infectious diseases are still abounding.

The formation of biofilms is not a rare occurrence in silicone tubing of dialyzers, and the release of endotoxin into dialysates is a major player in inflammation. Endotoxin and bacterial contamination have been a major cause of infection related morbidity in the United States,[44–46] which explains the numerous research interventions of purification of dialysates[47–49] to produce ultrapure dialysates that have proven beneficial in the reduction of chronic inflammatory proteins such as the C-reactive protein.[50]

Despite higher standards of hygiene and improved handling techniques, peritonitis still remains one of the most frequent causes of morbidity and loss of technique in peritoneal dialysis patients.[51,52] Bacterial infections are mostly caused by the skin-dwelling bacteria *Staphylococci* because of their close proximity to the access site. More importantly, they have a remarkable ability to evolve and develop antimicrobial-resistance, which allows them to develop resistance to antibiotics such as penicillin, antistaphylococcal penicillins and vancomycin.[53] Fungal peritonitis accounts for 1–15% of infection episodes, and causes loss of technique of 40% of these patients.[54] The *Candida* species are a major culprit of most of these infections; however the recent etiologic agents include unusual and 'non-pathogenic' fungi. Since clinical manifestations of a fungal infection are not dissimilar to a bacterial infection, diagnosis is often difficult and delayed.

Effects on the joints and bone mineralization

Renal osteodystrophy in dialysis patients develops when metals, including iron, strontium, aluminum, zinc, tin and chromium, accumulate in the bone during mineralization process.[55] Aluminum overload is associated with dialysis encephalopathy and aluminum-associated bone diseases,[56] and is a major problem in developing countries, where aluminum-containing phosphate-binders are the only economically affordable choice available to renal failure patients.[57] Similarly, strontium in contaminated dialysates found their way to the mineralization front of bone formation. Not only do *in vivo* trials correlate high doses of strontium to mineralization defects,[58] there is also a synergistic effect between the accumulation of strontium and aluminum, which has proven *in vivo* to decrease bone formation.

Before the use of recombinant human erythropoietin (rHuEPO), parenteral iron and blood transfusions were common treatments for anemic patients under dialysis. The association of iron overload with bone disease[55] and atherosclerosis[59] sparked efforts for proper iron management. However serum ferritin is almost always elevated owing to inflammation, and iron overload is often misdiagnosed. In fact, autopsies showed excess iron stored in hepatic tissue in dialysis patients[60,61] as a result of inflammation-related reticuloendothelial blockage.[62] As high serum ferrritin has been determined to be a strong predicator of hospitalization in dialysis patients,[63] it is pertinent to review the upper limits of serum ferritin in iron management.[64–66]

Dialysis-related amyloidosis (DRA) – a condition where β_2-microglobulin (β2m) is deposited in bone structures and joint tissues – is an inevitable consequence of dialysis. Classical presentations include carpel tunnel syndrome (CTS), destructive arthropathy, spondyloarthropathy, and subchondral bone erosion and cysts,[67,68] which is extended to systemic and visceral distribution,[69] pleural regions[70] and in extra-osteoarticular organs such as the spleen, skin, liver, heart and gastrointestinal tract[71] with increased exposure to dialysis. According to autopsy examinations, although the prevalence of DRA is indiscriminate of the dialysis modality,[72,73] the use of highly permeable, biocompatible membranes in hemodialysis has been associated with greater clearance of β2m.[74,75] Equally important, the age of the onset of dialysis has also been identified as a major risk factor for DRA by most authors. DRA is treated symptomatically by the removal of fibrous tissue in affected areas. The delayed development is at best achieved by using high flux membranes, or an adsorbent (Lixelle) column.[67,76,77]

As diligent efforts in the dialysis technology research continues to answer questions related to morbidity of patients on maintenance dialysis, it must not be forgotten that while dialysis alleviates the uremic state, in prolonging patient life it allows the progression of other diseases not reversed by dialysis. There is no doubt dialysis does not replace the important metabolic roles of the kidney such as the production of vitamin D analogues and erythropoietin, calcium regulation, etc.

15.4.2 Limitations and complications in kidney transplantation

It is often said that solid organ transplantation is a disease exchanged for another, which is not untrue to some degree. While kidney transplantation promises a new lease of life free of renal replacement machines, it requires a lifelong commitment to immunosuppressants to avoid rejection of the transplanted graft. Rejection of the transplanted allograft is categorized into hyperacute, which is an instant response of preformed antibodies to donor human leukocyte antigens (HLA); acute, the recipient T-cell recognition of donor alloantigens that leads to the outbreak of the allo-immune response; and chronic cases, which is a more gradual scarring of the graft extracellular matrix, leading to the ultimate dysfunction of the graft. Immunosuppressive drugs are categorized into several classes with different modes of action. Earlier immunosuppressive agents such as corticosteroids and antimetabolites are less selective, by inhibiting transcription factors in multiple leukocyte lines and eliminating pre-activated memory lymphocyte cells respectively. Newer drugs such as Calcineurin inhibitors select for T cells, allowing for the preservation of other myeloid-derived cell lines by reducing the dosage of corticosteroids.[78] Other immunosuppressive agents include monoclonal antibodies, polyclonal antilymphocyte antibodies and target of Rapamycin inhibitors, each of which has different modes of action.[79]

Suppression of the immune system increases the vulnerability of the body to attacks from opportunistic infections. The incidence of infection is dependent on the degree of immunosuppression, the type of organ transplanted, and surgical or technical complications, etc.[80] Transplanted patients are also predisposed to bacterial and fungal infections, if co-infected with the Epstein–Barr virus, cytomegalovirus, HIV and hepatitis C virus.[81]

Disarming the body's immune system and tumor surveillance may be too high a price to pay in exchange for a functional graft. In the search of an answer that resolves problems related to the immune system, researchers have turned to artificial cells, where a protective semipermeable membrane serves as a means of isolating the foreign cells from the host's immune system, while allowing the exchange of nutrients and wastes. The term 'artificial cell' describes synthetic structures of cellular dimensions for the replacement or supplement of deficient cell functions. First coined by Chang several years ago,[82] artificial cells are used to compartmentalize biologically active materials either individually or in combination. These active agents are entrapped within the artificial cell matrix and are shielded from external materials such as leukocytes, antibodies or tryptic enzymes that may affect their functionality or viability. Central to its functionality as a mini bioreactor is the large surface area to volume ratio which, combined with an ultra-thin membrane of 0.02 μm thickness, allow for the rapid diffusion and equilibration of small molecules such as therapeutic products, nutrients, oxygen, hormones, peptides, small proteins and wastes. In fact, 10 ml of artificial cells each 20 μm in diameter has a total surface area of about 20 000 cm^2, equivalent to that of a hemodialysis machine. Artificial cells range in size from macro-dimension to nano-dimension,[83] and are designed to mimic specific functions of biological cells, although they are not to be mistaken for biological cell models. The permeability, composition and configuration of the artificial cell are varied using different materials. This enables excellent control and fine tuning of the artificial cell design for a wide range of applications, such as cell therapy, gene therapy, enzyme therapy, blood substitutes and drug delivery.[84] The list of therapeutic uses of artificial cells continues, although for purposes of this chapter mostly clinical applications have been described, namely implantation, oral delivery, extracorporeal devices and bioreactors,[85,86] and cell-based bioreactors for application in bioartificial assist devices.[87–89]

15.5 Artificial cells for the treatment of renal failure

15.5.1 The artificial kidney

In 1913, Abel *et al.* first devised a hemodialysis apparatus to remove toxic materials in experimental animals.[90] Extracorporeal dialysis in humans was first performed in 1943, then became a routine treatment protocol for ESRD patients in the early 1960s. In 1962, Brescia and Cimino performed the first arterio-venous anastomosis between the radial artery and the cephalic vein, which directly connected the patient to the extracorporeal circuit.[91] Through gradual developments in renal replacement techniques, the concept of an artificial kidney was progressively developed.

The use of artificial cells for the construction of an artificial kidney was first proposed in 1966.[82] Not only must the artificial kidney perform filtration functions of the hemodialysis machine, but it must do so in an efficient manner, if not more so than the conventional hemodialysis machine. This is achieved by utilizing artificial cells for this application, and has in fact, been shown that the packing of 33 ml of artificial cells each with 100 μm diameter in a shunt will amount to a total surface area greater than that of a hemodialysis machine. Since the membrane thickness is at least two orders of magnitude less than that in a hemodialysis machine, metabolites from blood flowing past these artificial cells can cross the membrane into artificial cells at least 100 times faster than a hemodialysis machine. In addition, Chang and coworkers,[2,92,93] Sparks and coworkers[94] and Gardner and Emmerlin[95] investigated the possible uses of microencapsulated urea absorbents or ammonia absorbents for the removal of urea. Microencapsulated ion exchange resins have been prepared by Chang *et al.* in 1966[92] where selection of the appropriate resins may allow for the removal of selected electrolytes. Microencapsulated activated charcoal system can be supplemented by less frequent conventional dialysis. In addition, microencapsulated activated charcoal can also be used for treatment of acute intoxication by hemoperfusion. These ideas may be incorporated with the artificial kidney, which allows for the decrease of uremic toxins in a cyclic or continuous way, control hydroelectrolytic balance and to avoid acid–base balance alterations.

15.5.2 The biological artificial kidney

Current renal replacement modalities do not perform the diverse functions of the kidney and their long-term use is associated with complications, while donor scarcity and immunosuppressant issues of kidney transplantation limit this treatment to a select population. Ideas that will resolve such issues are the utilization of actual hepatocytes in a biological artificial hepatic system capable of not only removing metabolic toxins, but also performing the necessary synthetic functions of the native kidney. This idea has become a reality with the development of artificial cells, where the entrapment and isolation of biological cells within microcapsules allows the implantation not only of allogenic cells but also of xenogenic cells without immunosuppression. Of importance is the semipermeable membrane, which despite its non-physiological nature supports normal cell physiology. Furthermore the membrane serves to protect the immobilized cells from handling damage[96] and allows for the efficient transport of metabolites and wastes, while at the same time prevents invasion by the host's immunogenic cells.

Currently, research in the area of cell encapsulation is undergoing continuous optimization stages to extend the implantation period of the microcapsules while maintaining the viability of the immobilized cells for use in renal failure applications. The use of a novel two-step encapsulation procedure marks an important milestone in implantation science. In the traditional encapsulation procedure, a small number of the immobilized cells are entrapped on the periphery of the microcapsule core extruding into the microcapsule membrane matrix. The resulting membrane structure is weakened by such structural irregularities, and may result in the exposure of these cells to the external environment eliciting an immune response. The two-step encapsulation method involves the encapsulation of first generation microcapsule cores within a second larger capsule, with about three microcapsules within each larger capsule. This method has been proven to remove the entrapment of immobilized cells in the capsule membrane matrix, thus completely isolating the immobilized cells from the host's immune system.[97] This method has thus gained popularity and recognition as a practical and efficient method for implantation of encapsulated cells. The response from the immune system was, however, reported capable of crossing the physical barrier of the membranes, which motivated the idea of generating a local immunosuppressant by the co-implantation of Sertoli cells within grafts.[98] This theory was proven in another study, where the co-encapsulation of Sertoli cells with hepatocytes showed a dramatic decrease of leukocyte proliferation in a 4 week period after peritoneal implantation.[99]

In another direction, it was shown that the co-encapsulation of hepatocytes with bone marrow stem cells prolonged hepatocyte viability in the capsules for up to 4 months, and reported higher cell viability compared with capsules containing only hepatocyte cells. It was effective in lowering ammonia levels in experimental gun rats.[100] In addition to prolonging cell viability, much work has also been focused on prolonging graft survival after transplantation, where materials such as microporous polypropylene,[101] poly-L-lysine and alginate polymers,[102] and acetonitrile methallyl sulfonate co-polymers[103] have shown some degree of efficiency.

15.5.3 Oral administration of artificial cells as a potential oral kidney substitute

Oral administration of drugs promises convenience among other advantages, and as such there has been much interest in novel methods of removing waste metabolites from the gastrointestinal tract. Such methods were not meant to replace conventional treatments methods but, rather, to act as an adjunct therapy to decrease the duration and frequency of dialysis as well as to avoid a disequilibrium syndrome caused by prolonged intradialytic periods.[104,105] Since the 1960s, there had been much interest on the use of adsorbents to remove urea and ammonia.[106–108] In particular, tests on adsorbents showed promising results for the use of oxidized starch (oxystarch), although success was thwarted by the high dosage required and a less than optimal efficacy in removing urea *in vivo* at physiological pH.[109,110] Considerable efforts were also invested in another well-established adsorbent, activated charcoal, although its capacity to bind urea is lower[111] than oxystarch as a result of its non-specificity. During the same period of the 1960s, microencapsulation technology was beginning to influence the medical field, especially in hemoperfusion research.[92,112] This technique involved the use of semipermeable microcapsules as artificial cells, which were in fact membrane-enclosed structures containing therapeutic drugs. Researchers capitalized on the idea of encapsulating adsorbents as the technique allowed for the passage of a chosen range of molecules by customizing the membrane pore size. This would improve the binding capacities of adsorbents that were otherwise undermined by the attachment of untargeted molecules. Numerous research groups have followed this idea with both adsorbents, and while oxystarch showed promise at high dosages but caused diarrhea in rats, the non-specificity of activated carbon still resulted in rapid pore saturation.

The hemoperfusion process removes endogenous toxins by passing blood over detoxificants such as activated carbon. While activated carbon is an established oral detoxificant, serious complications associated with reduced platelet levels arose when placed in direct contact with blood,[113] resulting in embolisms in the lungs, liver, spleen and kidneys. Consequently, micro-encapsulation of the detoxificant was employed using a thick membrane. While this step resolved the issue at hand, the thick membrane obstructed the efficient removal of toxins (a thin membrane may not have the platelet protection and non-thrombogenic effect).

Combinations of biomolecules and adsorbents were also used in similar ways. Biological reactants such as urease were microencapsulated together with zirconium phosphate, although its potential was undermined by the large dosage required to remove urea in human patients. Compared with the urea-removing capacities of oxystarch previously demonstrated by Giordano *et al.*,[107] Sparks *et al.*[109] and Friedman *et al.*,[108] the effect of zirconium phosphate is smaller.

From the methodological standpoint, enzyme systems are incomplete owing to the fact that an isolated enzyme catalyzes only one specific reaction, giving rise to intermediates that have to be removed by other means. In contrast, microorganisms contain the necessary enzymes that work together to maintain its internal metabolic balance.[92] To simulate a natural phenomenon such as decomposition of urea, there must be a complete set of enzymes working in cooperation not only to decompose the substrate, but to utilize for its own growth.[114] In the 1970s, Malchesky and Nose quantified the degradation of nitrogenous metabolites by selectively culturing wild-type bacteria in a series of *in vitro* batch reactor trial.[115–117] It was reported that naturally occurring strains of microorganisms were highly effective for the *in vitro* degradation of urea and other compounds found in urine, and that these bacteria could be conditioned with selected media to enhance growth and degradation efficiency.

In the late 1970s, Setala *et al.*[114,118] advanced the concept of 'microbial anti-azotemic agents' by isolating enzymes from bacteria in soils rich in organic nitrogen, to convert nitrogenous wastes including urea, creatinine and other non-protein nitrogen compounds to essential nutrients. While results looked promising, the system was unable to achieve sufficient urea removal in humans. There were other attempts to remove urea, such as the use of microencapsulated multi-enzyme systems to convert ammonia and urea into amino acids by Gu and Chang[119] in 1990.

In 1996, genetically engineered *Escherichia coli* DH5 cells containing the urea-inducible *Klebsiella aerogenes* gene was microencapsulated and tested both *in vitro* and *in vivo* by Prakash and Chang.[120,121] This system proved highly efficient compared with its ancestors, where 100 mg of bacterial cells lowered about 87.89% of plasma urea within 20 min and 99.99% within 30 min, and removed ammonia from 975.14 μM/l to about 81.15 μM/l in 30 minutes. However, the use of a genetically engineered microorganism raises safety and environmental questions and would have to undergo more rigorous testing before being recognized as a clinically approved drug. Prakash and Chang demonstrated the effectiveness of orally administered, genetically modified bacteria for uric acid degradation.[122] In the search for a safer drug, Chow *et al.*[123] successfully induce the gut bacteria *Lactobacilli delbrueckii* metabolically to remove urea from uremic plasma. About 1 g microencapsulated *L. delbrueckii* cells removed an average of 105 mg of urea nitrogen. Although the urea removal was about a tenth of that achieved by genetically engineered *E. coli* DH5 cells, there were no implications for safety.

In recent years, O'Loughlin *et al.*[124–126] characterized a single alginate microcapsule containing either two strains of genetically modified bacteria, one expressing urease and the other uricase, or three isolated enzymes (urease, uricase and creatininase) capable of degrading, respectively, urea, uric acid and creatinine. The combination capsules were effective both *in vitro* and *in vivo* in a rodent model with chemically induced renal failure.

15.6 Future trends

With advancements in the research field, the remotely feasible ideas of yesteryear such as artificial cells are increasingly becoming a reality today. This idea may potentially replace conventional treatment protocols, especially since an astronomical amount of resources is currently invested in dialysis, yet not without secondary complications. One important area of research is the application of artificial cells in renal failure. Artificial cells refer to synthetic microscopic structures that possess some functional properties of biological cells. As the goal of the artificial cell is to mimic some biological processes of the real cell, it may be developed as a partial substitute for human tissues or organs. The advantages of oral administration may result in the popularization of this mode of therapy, with the incorporation of biologically active agents such as bacteria in therapeutics or advanced food systems, although obstacles in the delivery of viable agents to therapeutically relevant and favourable sites require further studies. For the realization of such applications, it is pertinent to improve microcapsule membrane designs, methods for improved cell harvest, the mass production of artificial cell microcapsules, and cost effective storage and overall clinical efficacy.

15.7 Acknowledgements

The authors would like to acknowledge the support of research grants from the Canadian Institute of Health Research (CIHR) and the Natural Sciences and Engineering Research Council (NSERC) of Canada.

15.8 References

1. de, M.A. *et al.* Acute renal failure in the ICU: risk factors and outcome evaluated by the SOFA score. *Intensive Care Med.* **26**, 915–921 (2000).
2. Chertow, G.M., Levy, E.M., Hammermeister, K.E., Grover, F., & Daley, J. Independent association between acute renal failure and mortality following cardiac surgery. *Am. J. Med.* **104**, 343–348 (1998).
3. Iaina, A., Silverberg, D.S., & Wexler, D. Therapy insight: congestive heart failure, chronic kidney disease and anemia, the cardio-renal-anemia syndrome. *Nat. Clin. Pract. Cardiovasc. Med.* **2**, 95–100 (2005).
4. National Institute of Diabetes and Digestive and Kidney Diseases. *U.S. Renal Data System 2005 Annual Data Report: Atlas of End-Stage Renal Disease in the United States*. 2005. Bethesda, MD, National Institute of Health. United States Renal Disease System.
5. Schreiner, G.E. & Maher, J.F. *Uremia; biochemistry, pathogenesis and treatment* (Thomas, Springfield, 1961).
6. Uehlinger, D.E. *et al.* Comparison of continuous and intermittent renal replacement therapy for acute renal failure. *Nephrol. Dial. Transplant.* **20**, 1630–1637 (2005).
7. Khan, I.H. & Catto, G.R.D. Complications of long-term dialysis: cellular immunity infection, and neoplasia, in *Complications of Long-term Dialysis* (eds. E. Brown & P. Parfrey) 53–71 (Oxford University Press, New York, 1999).
8. *2005 Annual Report of the U.S. Organ Procurement and Transplantation Network and the Scientific Registry of Transplant Recipients: Transplant Data 1994–2003.* eds. Department of Health and Human Services, United Network for Organ Sharing, R. V., & University Renal Research and Education Association, Ann Arbor, Michigan, 2005.

9. Excerpts from the USRDS 2004 Annual Data Report. *Am. J. Kidney Dis.* **45**, supplement 1, S1–S280 (2005).
10. Gurland, H.J. *et al.* Combined report on regular dialysis and transplantation in Europe, 3, 1972. *Proc. Eur. Dial. Transplant. Assoc.* **10**, XVII–LVII (1973).
11. Foley, R.N. & Parfrey, P.S. Cardiovascular disease and mortality in ESRD. *J. Nephrol.* **11**, 239–245 (1998).
12. Kooman, J.P. & Leunissen, K.M. Cardiovascular aspects in renal disease. *Curr. Opin. Nephrol. Hypertens.* **2**, 791–797 (1993).
13. Varma, R., Garrick, R., McClung, J., & Frishman, W.H. Chronic renal dysfunction as an independent risk factor for the development of cardiovascular disease. *Cardiol. Rev.* **13**, 98–107 (2005).
14. Disney, A.P. Demography and survival of patients receiving treatment for chronic renal failure in Australia and New Zealand: report on dialysis and renal transplantation treatment from the Australia and New Zealand Dialysis and Transplant Registry. *Am. J. Kidney Dis.* **25**, 165–175 (1995).
15. Mallick, N.P., Jones, E., & Selwood, N. The European (European Dialysis and Transplantation Association–European Renal Association) Registry. *Am. J. Kidney Dis.* **25**, 176–187 (1995).
16. Foley, R.N., Parfrey, P.S., & Sarnak, M.J. Epidemiology of cardiovascular disease in chronic renal disease. *J. Am. Soc. Nephrol.* **9**, S16–S23 (1998).
17. Oh, D.J. & Lee, K.J. The relation between hypoalbuminemia and compliance and intima-media thickness of carotid artery in continuous ambulatory peritoneal dialysis patients. *J. Korean Med. Sci.* **20**, 70–74 (2005).
18. Spittle, M. *et al.* Oxidative stress and inflammation in hemodialysis patients, in *Improving Prognosis for Kidney Disorders* (ed. M.M. Avram) 45–51 (Kluwer Academic, Boston, 2002).
19. Ohsawa, M. *et al.* Cardiovascular risk factors in hemodialysis patients: results from baseline data of kaleidoscopic approaches to patients with end-stage renal disease study. *J. Epidemiol.* **15**, 96–105 (2005).
20. Murray, J.E., Lang, S., Miller, B.F., & Dammin, G.J. Prolonged functional survival of renal autotransplants in the dog. *Surg. Gynecol. Obstet.* **103**, 15–22 (1956).
21. Smiddy, F.G. The influence of the uraemic state on lymphoid tissue with particular reference to the homograft reaction. *Surg. Clin. North Am* **49**, 523–531 (1969).
22. Goldblum, S.E. & Reed, W.P. Host defenses and immunologic alterations associated with chronic hemodialysis. *Ann. Intern. Med.* **93**, 597–613 (1980).
23. Casciani, C.U. *et al.* Immunological aspects of chronic uremia. *Kidney Int. Suppl* **8**, S49–S54 (1978).
24. Vanholder, R. & Dhondt, A. Renal failure as an immunodeficiency state, in *The Infectious Complications of Renal Disease* (eds. P. Sweny, R. Rubin, & N. Tolkoff-Rubin) 3–19 (Oxford University Press, New York, 2003).
25. Pesanti, E.L. Immunologic defects and vaccination in patients with chronic renal failure. *Infect. Dis. Clin. North Am.* **15**, 813–832 (2001).
26. Contin, C., Couzi, L., Moreau, J.F., Dechanet-Merville, J., & Merville, P. [Immune dysfunction of uremic patients: potential role for the soluble form of CD40]. *Nephrologie* **25**, 119–126 (2004).
27. Descamps-Latscha, B. & Chatenoud, L. T cells and B cells in chronic renal failure. *Semin. Nephrol.* **16**, 183–191 (1996).
28. Girndt, M., Sester, U., Kaul, H., & Kohler, H. Production of proinflammatory and regulatory monokines in hemodialysis patients shown at a single-cell level. *J. Am. Soc. Nephrol.* **9**, 1689–1696 (1998).

29. Vamvakas, S., Bahner, U., & Heidland, A. Cancer in end-stage renal disease: potential factors involved – editorial. *Am. J. Nephrol.* **18**, 89–95 (1998).
30. Higgins, R.M. Infections in a renal unit. *Q. J. Med.* **70**, 41–51 (1989).
31. Girndt, M. Humoral immune responses in uremia and the role of IL-10. *Blood Purif.* **20**, 485–488 (2002).
32. Koller, H. *et al.* Apoptosis of human polymorphonuclear neutrophils accelerated by dialysis membranes via the activation of the complement system. *Nephrol. Dial. Transplant.* **19**, 3104–3111 (2004).
33. Subramanian, S., Venkataraman, R., & Kellum, J.A. Influence of dialysis membranes on outcomes in acute renal failure: a meta-analysis. *Kidney Int.* **62**, 1819–1823 (2002).
34. Moore, M.A., Kaplan, D.S., Picciolo, G.L., Wallis, R.R., & Kowolik, M.J. Effect of cellulose acetate materials on the oxidative burst of human neutrophils. *J. Biomed. Mater. Res.* **55**, 257–265 (2001).
35. Crosnier, J. *et al.* Randomised placebo-controlled trial of hepatitis B surface antigen vaccine in French haemodialysis units: II, Haemodialysis patients. *Lancet* **1**, 797–800 (1981).
36. Kohler, H., Arnold, W., Renschin, G., Dormeyer, H.H., & Meyer zum Buschenfelde, K.H. Active hepatitis B vaccination of dialysis patients and medical staff. *Kidney Int.* **25**, 124–128 (1984).
37. Jungers, P. *et al.* Immunogenicity of the recombinant GenHevac B Pasteur vaccine against hepatitis B in chronic uremic patients. *J. Infect. Dis.* **169**, 399–402 (1994).
38. Benhamou, E. *et al.* Hepatitis B vaccine: randomized trial of immunogenicity in hemodialysis patients. *Clin. Nephrol.* **21**, 143–147 (1984).
39. Contin, C. *et al.* Potential role of soluble CD40 in the humoral immune response impairment of uraemic patients. *Immunology* **110**, 131–140 (2003).
40. Kaslow, R.A. & Zellner, S.R. Infection in patients on maintenance haemodialysis. *Lancet* **2**, 117–119 (1972).
41. Lewis, S.L. & Van Epps, D.E. Neutrophil and monocyte alterations in chronic dialysis patients. *Am. J. Kidney Dis.* **9**, 381–395 (1987).
42. Gabriel, R. Morbidity and mortality of long-term haemodialysis: a review. *J. R. Soc. Med.* **77**, 595–601 (1984).
43. Lowrie, E.G. & Lew, N.L. Death risk in hemodialysis patients: the predictive value of commonly measured variables and an evaluation of death rate differences between facilities. *Am. J. Kidney Dis.* **15**, 458–482 (1990).
44. Klein, E., Pass, T., Harding, G.B., Wright, R., & Million, C. Microbial and endotoxin contamination in water and dialysate in the central United States. *Artif. Organs* **14**, 85–94 (1990).
45. Harding, G.B., Klein, E., Pass, T., Wright, R., & Million, C. Endotoxin and bacterial contamination of dialysis center water and dialysate; a cross sectional survey. *Int. J. Artif. Organs* **13**, 39–43 (1990).
46. Baurmeister, U. *et al.* Dialysate contamination and back filtration may limit the use of high-flux dialysis membranes. *ASAIO Trans.* **35**, 519–522 (1989).
47. Marion-Ferey, K. *et al.* Endotoxin level measurement in hemodialysis biofilm using 'the whole blood assay'. *Artif. Organs* **29**, 475–481 (2005).
48. Shinoda, T. 'Clean dialysate' requires not only lower levels of endotoxin but also sterility of dilution water. *Blood Purif.* **22** Suppl 2, 78–80 (2004).
49. Ward, R.A. Ultrapure dialysate. *Semin. Dial.* **17**, 489–497 (2004).
50. Arizono, K. *et al.* Use of ultrapure dialysate in reduction of chronic inflammation during hemodialysis. *Blood Purif.* **22** Suppl 2, 26–29 (2004).

51. Klaus, G. Prevention and treatment of peritoneal dialysis-associated peritonitis in pediatric patients. *Perit. Dial. Int.* **25** *Suppl 3*, S117–S119 (2005).
52. Chow, K.M. *et al.* A risk analysis of continuous ambulatory peritoneal dialysis-related peritonitis. *Perit. Dial. Int.* **25**, 374–379 (2005).
53. Salzer, W. Antimicrobial-resistant gram-positive bacteria in PD peritonitis and the newer antibiotics used to treat them. *Perit. Dial. Int.* **25**, 313–319 (2005).
54. Prasad, N. & Gupta, A. Fungal peritonitis in peritoneal dialysis patients. *Perit. Dial. Int.* **25**, 207–222 (2005).
55. D'Haese, P.C. *et al.* Aluminum, iron, lead, cadmium, copper, zinc, chromium, magnesium, strontium, and calcium content in bone of end-stage renal failure patients. *Clin. Chem.* **45**, 1548–1556 (1999).
56. Nakazawa, R. [The K/DOQI guidelines on diagnosis and treatment of aluminum bone disease in hemodialysis patients]. *Clin. Calcium* **14**, 738–743 (2004).
57. Afifi, A. Renal osteodystrophy in developing countries. *Artif. Organs* **26**, 767–769 (2002).
58. Schrooten, I. *et al.* Strontium causes osteomalacia in chronic renal failure rats. *Kidney Int.* **54**, 448–456 (1998).
59. Reis, K.A. *et al.* Intravenous iron therapy as a possible risk factor for atherosclerosis in end-stage renal disease. *Int. Heart J.* **46**, 255–264 (2005).
60. Gyorffy, B., Kocsis, I., & Vasarhelyi, B. Biallelic genotype distributions in papers published in *Gut* between 1998 and 2003: altered conclusions after recalculating the Hardy–Weinberg equilibrium. *Gut* **53**, 614–615 (2004).
61. Gyorffy, B., Kocsis, I., & Vasarhelyi, B. Missed calculations and new conclusions: re-calculation of genotype distribution data published in *Journal of Investigative Dermatology*, 1998–2003. *J. Invest. Dermatol.* **122**, 644–646 (2004).
62. Fishbane, S., Miyawaki, N., & Masani, N. Hepatic iron in hemodialysis patients. *Kidney Int.* **66**, 1714–1715 (2004).
63. Kalantar-Zadeh, K., Don, B.R., Rodriguez, R.A., & Humphreys, M.H. Serum ferritin is a marker of morbidity and mortality in hemodialysis patients. *Am. J. Kidney Dis.* **37**, 564–572 (2001).
64. Fishbane, S., Kalantar-Zadeh, K., & Nissenson, A.R. Serum ferritin in chronic kidney disease: reconsidering the upper limit for iron treatment. *Semin. Dial.* **17**, 336–341 (2004).
65. Tessitore, N. *et al.* The role of iron status markers in predicting response to intravenous iron in haemodialysis patients on maintenance erythropoietin. *Nephrol. Dial. Transplant.* **16**, 1416–1423 (2001).
66. Fishbane, S., Kowalski, E.A., Imbriano, L.J., & Maesaka, J.K. The evaluation of iron status in hemodialysis patients. *J. Am. Soc. Nephrol.* **7**, 2654–2657 (1996).
67. Furuyoshi, S. *et al.* New adsorption column (Lixelle) to eliminate beta2-microglobulin for direct hemoperfusion. *Ther. Apher.* **2**, 13–17 (1998).
68. Guccion, J.G., Redman, R.S., & Winne, C.E. Hemodialysis-associated amyloidosis presenting as lingual nodules. *Oral Surg. Oral Med. Oral Pathol.* **68**, 618–623 (1989).
69. Campistol, J.M. *et al.* Visceral involvement of dialysis amyloidosis. *Am. J. Nephrol.* **7**, 390–393 (1987).
70. Suzuki, M. *et al.* Pleural involvement of dialysis-related amyloidosis. *Intern. Med.* **44**, 628–631 (2005).
71. Miyata, T., Inagi, R., & Kurokawa, K. Diagnosis, pathogenesis, and treatment of dialysis-related amyloidosis. *Miner. Electrolyte Metab.* **25**, 114–117 (1999).
72. Benz, R.L., Siegfried, J.W., & Teehan, B.P. Carpal tunnel syndrome in dialysis

patients: comparison between continuous ambulatory peritoneal dialysis and hemodialysis populations. *Am. J. Kidney Dis.* **11**, 473–476 (1988).
73. Jadoul, M., Garbar, C., & Vanholder, R. High prevalence of histological beta2-microglobulin amyloidosis in CAPD patients. *Nephrol., Dial., Transplant.* **12**, A175 (1997).
74. Jadoul, M. Dialysis-related amyloidosis: importance of biocompatibility and age. *Nephrol. Dial. Transplant.* **13** Suppl 7, 61–64 (1998).
75. van Ypersele, D.S., Jadoul, M., Malghem, J., Maldague, B., & Jamart, J. Effect of dialysis membrane and patient's age on signs of dialysis-related amyloidosis. The Working Party on Dialysis Amyloidosis. *Kidney Int.* **39**, 1012–1019 (1991).
76. Niwa, T. Dialysis-related amyloidosis: pathogenesis focusing on AGE modification. *Semin. Dial.* **14**, 123–126 (2001).
77. Wada, T., Miyata, T., Sakai, H., & Kurokawa, K. Beta2-microglobulin and renal bone disease. *Perit. Dial. Int.* **19** Suppl 2, S413–S416 (1999).
78. Haberal, M. *et al.* The impact of cyclosporine on the development of immunosuppressive therapy. *Transplant. Proc.* **36**, 143S–147S (2004).
79. Duncan, M.D. & Wilkes, D.S. Transplant-related Immunosuppression: a review of immunosuppression and pulmonary infections. *Proc. Am. Thorac. Soc.* **2**, 449–455 (2005).
80. Patel, R. & Paya, C.V. Infections in solid-organ transplant recipients. *Clin. Microbiol. Rev.* **10**, 86–124 (1997).
81. Snydman, D.R. Epidemiology of infections after solid-organ transplantation. *Clin. Infect. Dis.* **33** Suppl 1, S5–S8 (2001).
82. Chang, T.M. Semipermeable aqueous microcapsules ('artificial cells'): with emphasis on experiments in an extracorporeal shunt system. *Trans. Am. Soc. Artif. Intern. Organs* **12**, 13–19 (1966).
83. Chang, T.M. Therapeutic applications of polymeric artificial cells. *Nat. Rev. Drug Discov.* **4**, 221–235 (2005).
84. Chang, T.M. The role of artificial cells in cell and organ transplantation in regenerative medicine. *Panminerva Med.* **47**, 1–9 (2005).
85. Chang, T.M.S & Prakash, S. Therapeutic uses of microencapsulated genetically engineered cells. *Mol. Med. Today*, **4**(5), 221–227 (1998).
86. Chang, T.M.S. Artificial cells for replacement of metabolic organ functions. *Artif. Cells Blood Substit. Immobiliz. Biotechnol.* **31**, 151–161 (2003).
87. Di Campli, C. *et al.* Use of hemodialysis with albumin (MARS) for liver antibiotic toxicity in a patient with systemic tubercolosis. *Hepatology* **36**, 680A (2002).
88. Di Campli, C. *et al.* Advances in extracorporeal detoxification by MARS dialysis in patients with liver failure. *Current Med. Chem.* **10**, 341–348 (2003).
89. Gerlach, J.C. *et al.* Use of primary human liver cells originating from discarded grafts in a bioreactor for liver support therapy and the prospects of culturing adult liver stem cells in bioreactors: a morphologic study. *Transplantation* **76**, 781–786 (2003).
90. Gottschalk, C.W. & Fellner, S.K. History of the science of dialysis. *Am. J. Nephrol.* **17**, 289–298 (1997).
91. Cimino, J.E. & Brescia, M.J. Simple venipuncture for hemodialysis. *Nord. Hyg. Tidskr.* **267**, 608–609 (1962).
92. Chang, T.M., MacIntosh, F.C., & Mason, S.G. Semipermeable aqueous microcapsules. I. Preparation and properties. *Can. J. Physiol Pharmacol.* **44**, 115–128 (1966).
93. Chang, T.M. & Malave, N. The development and first clinical use of

semipermeable microcapsules (artificial cells) as a compact artificial kidney. *Trans. Am. Soc. Artif. Intern. Organs* **16**, 141–148 (1970).

94. Mehall, J.R., Koenig, J.L., Lindan, O., & Sparks, R.E. Screening study of adsorbents for urea removal from artificial kidney dialyzing fluid. *J. Biomed. Mater. Res.* **3**, 529–543 (1969).
95. Gardner, D.L. & Emmerlin, D. Microencapsulated Stabilized Urease. *Abstracts Papers Amer. Chem. Soc.* (162nd American Chemical Society Meeting, Washington DC, 1971).
96. Uludag, H., De, V.P., & Tresco, P.A. Technology of mammalian cell encapsulation. *Adv. Drug Deliv. Rev.* **42**, 29–64 (2000).
97. Wong, H. & Chang, T.M. A novel two step procedure for immobilizing living cells in microcapsules for improving xenograft survival. *Biomater. Artif. Cells Immobilization Biotechnol.* **19**, 687–697 (1991).
98. Yang, H. & Wright, J.R., Jr. Co-encapsulation of Sertoli enriched testicular cell fractions further prolongs fish-to-mouse islet xenograft survival. *Transplantation* **67**, 815–820 (1999).
99. Rahman, T.M., Diakanov, I., Selden, C., & Hodgson, H. Co-transplantation of encapsulated HepG2 and rat Sertoli cells improves outcome in a thioacetamide induced rat model of acute hepatic failure. *Transpl. Int.* **18**, 1001–1009 (2005).
100. Liu, Z.C. & Chang, T.M. Coencapsulation of hepatocytes and bone marrow stem cells: *in vitro* conversion of ammonia and *in vivo* lowering of bilirubin in hyperbilirubemia Gunn rats. *Int. J. Artif. Organs* **26**, 491–497 (2003).
101. Takebe, K. *et al.* Xenogeneic (pig to rat) fetal liver fragment transplantation using macrocapsules for immunoisolation. *Cell Transplant.* **5**, S31–S33 (1996).
102. Lanza, R.P. *et al.* Xenotransplantation of cells using biodegradable microcapsules. *Transplantation* **67**, 1105–1111 (1999).
103. Gomez, N. *et al.* Evidence for survival and metabolic activity of encapsulated xenogeneic hepatocytes transplanted without immunosuppression in Gunn rats. *Transplantation* **63**, 1718–1723 (1997).
104. Asher, W.J., Vogler, T.C., Bovee, K.C., Holtzapple, P.G., & Hamilton, R.W. Liquid membrane capsules for chronic uremia. *Trans. Am. Soc. Artif. Intern. Organs* **23**, 673–680 (1977).
105. Asher, W.J. *et al.* Liquid membrane system directed toward chronic uremia. *Kidney Int. Suppl.* **3**, 409–412 (1975).
106. Meriwether, L.S. & Kramer, H.M. *In vitro* reactivity of oxystarch and oxycellulose. *Kidney Int. Suppl.* **7**, S259–S265 (1976).
107. Giordano, C., Esposito, R., & Pluvio, M. Oxycellulose and ammonia-treated oxystarch as insoluble polyaldehydes in uremia. *Kidney Int. Suppl.* **3**, 380–382 (1975).
108. Friedman, E.A., Fastook, J., Beyer, M.M., Rattazzi, T., & Josephson, A.S. Potassium and nitrogen binding in the human gut by ingested oxidized starch (OS). *Trans. Am. Soc. Artif. Intern. Organs* **20A**, 161–167 (1974).
109. Sparks, R.E., Mason, N.S., Meier, P.M., Litt, M.H., & Lindan, O. Removal of uremic waste metabolites from the intestinal tract by encapsulated carbon and oxidized starch. *Trans. Am. Soc. Artif. Intern. Organs* **17**, 229–238 (1971).
110. Sparks, R.E., Salemme, R.M., Meier, P.M., Litt, M.H., & Lindan, O. Removal of waste metabolites in uremia by microencapsulated reactants. *Trans. Am. Soc. Artif. Intern. Organs* **15**, 353–359 (1969).
111. Sparks, R.E. Review of gastrointestinal perfusion in the treatment of uremia. *Clin. Nephrol.* **11**, 81–85 (1979).

112. Chang, T.M. Semipermeable microcapsules. *Science* **146**, 524–525 (1964).
113. Chang, T.M. Artificial cells containing detoxicants. 1972 [classical article]. *Artif. Cells Blood Substit. Immobil. Biotechnol.* **30**, 457–468 (2002).
114. Setala, K. Treating uremia with soil bacterial enzymes: further developments. *Clin. Nephrol.* **11**, 156–166 (1979).
115. Malchesky, P.S. & Nose, Y. Biological reactors as renal substitutes. *Artif. Organs* **3**, 8–10 (1979).
116. Malchesky, P.S. & Nose, Y. Biological reactors for renal support. *Trans. Am. Soc. Artif. Intern. Organs* **23**, 726–729 (1977).
117. Malchesky, P.S. & Nose, Y. Biological reactors as artificial organs. Concept and preliminary *in vitro* study. *Cleve. Clin. Q.* **42**, 267–271 (1975).
118. Setala, K. Bacterial enzymes in uremia management. *Kidney Int. Suppl.* S194–S202 (1978).
119. Gu, K.F. & Chang, T.M. Conversion of ammonia or urea into essential amino acids, L-leucine, L-valine, and L-isoleucine using artificial cells containing an immobilized multienzyme system and dextran-NAD. L-lactic dehydrogenase for coenzyme recycling. *Appl. Biochem. Biotechnol.* **26**, 115–124 (1990).
120. Prakash, S. & Chang, T.M. Microencapsulated genetically engineered live *E. coli* DH5 cells administered orally to maintain normal plasma urea level in uremic rats. *Nat. Med.* **2**, 883–887 (1996).
121. Prakash, S. & Chang, T.M. Microencapsulated genetically engineered *E. coli* DH5 cells for plasma urea and ammonia removal based on: 1. Column bioreactor and 2. Oral administration in uremic rats. *Artif. Cells Blood Substit. Immobil. Biotechnol.* **24**, 201–218 (1996).
122. Prakash, S. & Chang, T.M. *In vitro* and *in vivo* uric acid lowering by artificial cells containing microencapsulated genetically engineered *E. coli* DH5 cells. *Int. J. Artif. Organs* **23**, 429–435 (2000).
123. Chow, K.M., Liu, Z.C., Prakash, S., & Chang, T.M. Free and microencapsulated *Lactobacillus* and effects of metabolic induction on urea removal. *Artif. Cells Blood Substit. Immobil. Biotechnol.* **31**, 425–434 (2003).
124. O'Loughlin, J.A., Bruder, J.M., & Lysaght, M.J. Oral administration of biochemically active microcapsules to treat uremia: new insights into an old approach. *J. Biomater. Sci. Polym. Ed.* **15**, 1447–1461 (2004).
125. O'Loughlin, J.A., Bruder, J.M., & Lysaght, M.J. *In vivo* and *in vitro* degradation of urea and uric acid by encapsulated genetically modified microorganisms. *Tissue Eng.* **10**, 1446–1455 (2004).
126. O'Loughlin, J.A., Bruder, J.M., & Lysaght, M.J. Degradation of low molecular weight uremic solutes by oral delivery of encapsulated enzymes. *ASAIO J.* **50**, 253–260 (2004).

16

Engineered cells for treatment of diabetes

S EFRAT, Tel Aviv University, Israel

16.1 Introduction

The term diabetes mellitus encompasses two distinct diseases. Type 1 (insulin-dependent) diabetes, which afflicts about 0.5% of the population, is an autoimmune disease leading to destruction of the pancreatic islet insulin-producing β cells. Type 2 diabetes, which is 10 times more common than type 1, results from insulin resistance in key tissues, such as muscle, fat and liver, coupled with a diminishing ability of β cells to compensate for the increased insulin demands. The current treatments for diabetes fail to prevent the long-term complications of the disease, arising from vascular damage in major organs which are caused by hyperglycemia. Type 1 diabetes is treated by insulin administration; however, the difficulty of adjusting the precise insulin dosage in response to changing physiological conditions results in episodes of hypo- and hyperglycemia. Type 2 diabetes is treated initially by diet, exercise and drugs that stimulate insulin secretion from β cells, reduce hepatic glucose output and increase insulin sensitivity in target cells. However, a large proportion of patients with type 2 diabetes eventually become dependent on exogenous insulin. The incidence of type 2 diabetes is on the rise in epidemic proportions owing to modern lifestyle, placing an increasing burden on global health systems. This chapter focuses on restoration of a regulated insulin supply in type 1 diabetes. However, many of the approaches described here may be applicable to type 2 diabetic patients requiring insulin.

16.2 Approaches for restoring regulated insulin secretion

The only way for avoiding the long-term complications of hyperglycemia, as well as the acute risks of hypoglycemia, is restoration of a tightly regulated insulin release. This could take the form of surrogate insulin-producing cells that replace the function of the damaged cells, or a mechanical device for insulin supply. Efforts aimed at development of an insulin pump connected in a closed loop to a glucose sensor have fallen short so far from producing a working 'artificial pancreas'. Thus, cell replacement through regeneration or transplantation currently represents the most promising approach for a cure of type 1 diabetes.

In a normal adult pancreas a slow rate of β-cell renewal is responsible for maintenance of an adequate functional β-cell mass. This rate is accelerated in conditions of increased demands for insulin, such as pregnancy[1] and obesity.[2] It is not known whether β-cell renewal in the adult pancreas relies on replication of differentiated β cells, or neogenesis from precursor cells, which may reside in the islets, in pancreatic ducts, in the exocrine cells, or in an extra-pancreatic tissue. A recent report provided support for the former, demonstrating that β-cell renewal in a mouse model occurs from cells which already express insulin.[3] Conceivably, β-cell replacement can be achieved by stimulating this homeostatic process of β-cell renewal. This could be accomplished in principle by local delivery of proteins or genes involved in β-cell replication, or by systemic treatment with such factors, provided that specific expansion of the β-cell mass can occur without deleterious effects to other tissues. A number of hormones and growth factors are known to stimulate rodent pancreatic duct cell differentiation into islets and β-cell replication.[4] One such factor is exendin-4, a stable analog of glucagon-like peptide 1 (GLP-1), which has been shown to stimulate both β-cell neogenesis and replication in a rat model of type 2 diabetes involving partial pancreatectomy.[5] In a recent work using Goto-Kakizaki rats, a non-obese model of type 2 diabetes, injections of GLP-1 or exendin-4 increased the β-cell mass, resulting in long-term improvements in glycemia.[6] GLP-1 is a particularly attractive candidate because of its additional stimulatory effects on glucose-induced insulin secretion from β cells.[7] Another example is a combined treatment with epidermal growth factor (EGF) and gastrin, which was shown to increase β-cell mass and reduce hyperglycemia in streptozotocin-diabetic rats[8] and alloxan-treated mice.[9] Despite these encouraging results, such factors are unlikely to be suitable for systemic treatment owing to pleotropic effects in other tissues.

Our present understanding of the mechanisms involved in normal β-cell renewal, let alone the ability to manipulate it *in vivo*, remains quite limited and requires much additional work before islet regeneration can be considered for clinical application. In addition, regenerated β cells are likely to become targets for recurring autoimmunity. Therefore, any islet regeneration strategy must include approaches for prevention of islet destruction. Such approaches are quite far from being well developed. Thus, at present the most realistic β-cell replacement strategy is through transplantation. The introduction of β cells from an exogenous source may allow *ex vivo* manipulation of their immunogenicity and resistance to immune responses prior to transplantation, in ways that are not possible in islets regenerated *in vivo* (see Sections 16.8 and 16.9).

16.3 Islet and β-cell transplantation

Pancreas transplantation, although quite successful, represents a rather invasive intervention, which is restricted to patients with advanced complications, requires constant immunosuppression, and is severely limited by donor availability.[10] Progress in human islet isolation and in immunosuppression protocols resulted in restoration of euglycemia in patients who received islets from two or three pancreas donors.[11] However, a 5-year follow-up showed that only 10% of the transplanted patients maintained insulin independence.[12] This was due primarily to the difficulty of preserving islet function, and to toxicity of immunosuppression. The need for multiple donors to provide a sufficient number of islets for a single recipient further underscores the need for generation of an abundant source of β cells for transplantation.

One obvious approach for obtaining more islet cells is expansion of adult donor islets in tissue culture. However, despite their ability to expand *in vivo*, islet expansion *in vitro* has proven quite difficult. Adult human islet cells grown on HTB-9 matrix in the presence of hepatocyte growth factor were shown to proliferate at the most for 10–15 population doublings, after which they underwent senescence.[13–15] The replication span could not be extended by expression of the catalytic subunit of human telomerase (TERT), which was introduced into the cells with a retrovirus.[16] An even greater difficulty encountered in these studies has been the partial or complete loss of function of the expanded cells.[16–18] Two recent reports have described considerable expansion of human islet cells and the use of serum-free medium[19] and EGF and nerve growth factor[20] for cell redifferentiation. However, insulin amounts in the redifferentiated cells were low, and insulin secretion was not glucose-regulated. Generation of differentiated islets from cultured pancreatic duct cells has been demonstrated using both human[21,22] and mouse[23] cells. However, the expansion capacity of these cells *in vitro* was shown to be quite restricted. Similarly, culture of pluripotent cells obtained from human islets[24] and from adult mouse pancreas[25] resulted in limited expansion. The modest success in generation of abundant and functional β cells from pancreatic sources has prompted investigators to attempt differentiation of insulin-producing cells from embryonic stem cells and from progenitor or mature cells from non-pancreatic tissues.

16.4 Development of insulin-producing cells from stem/progenitor cells

Progress in stem cell research in recent years has raised hopes for the generation of surrogate β cells from stem/progenitor cells, which can be expanded in tissue culture with a relative ease. The derivation of pluripotent human embryonic stem (ES) cells,[26] and the reported broad differentiation spectrum of stem/progenitor cells from a number of fetal and adult tissues,[27–30] have encouraged investigators to attempt the generation of insulin-producing cells from non-pancreatic tissues. ES cells are capable of spontaneous differentiation into insulin-producing cells, among many other cell types.[31,32] In addition, they can be induced to preferentially differentiate into insulin-containing cells by tissue culture conditions[33,34] and genetic manipulations.[35,36] However, these cells express low levels of insulin and lack key β-cell properties, such as regulated insulin secretion in response to physiological secretagogues. In addition, these results have been challenged by findings that the insulin content may result from uptake from the culture medium, rather than synthesis.[37–39] Additional considerations involved in the use of ES cells are the risk of uncontrolled proliferation of residual undifferentiated cells, as well as the ethical controversy surrounding their utilization. Nevertheless, the potential of ES cells remains promising, and future work may advance it by following closely the normal developmental events in islet formation.[40]

In addition to the tutipotent ES cells, many fetal and adult tissues were shown to contain stem/progenitor cells, which are responsible for normal tissue repair and renewal.[41–44] These cells fulfill the criteria of stem cells as proposed by Weissman:[45] self-renewal, multipotency and capacity for tissue reconstitution. The growth capability of these cells is considered to be limited, compared with that of ES cells, and to diminish with age. Although these properties restrict their utility as an abundant source of cells for transplantation, it may represent a safety advantage. In addition to a lower risk of uncontrolled proliferation, tissue stem cells offer the possibility of employing autologous cells, with the likely advantage of improved graft tolerance, compared with an allograft.

The replication limit of somatic cells has been attributed to telomere shortening in the absence of TERT activity.[46] Activation or overexpression of TERT has been shown to be effective in extending the replication capacity of a number of human cell types,[47–49] without compromising their ability to undergo contact inhibition in culture, changing their karyotype, or increasing their neoplastic potential *in vivo*.[50–52] Introduction of the TERT gene into tissue stem cells may extend their replication capacity in tissue culture up to that of ES cells. However, since TERT activation is a hallmark of many tumors, the risks of neoplasia need to be carefully evaluated in each modified cell source.

Generation of β-like cells from non-pancreatic tissue is a tall order. The differentiated cells should be able to do much more than just produce large amounts of insulin. They should be capable of correct processing of proinsulin, storage of mature insulin and regulated secretion of appropriate amounts of insulin in response to physiological signals. Using a source of cells lacking these properties would not represent a significant advantage over insulin administration. Induction of these capabilities will probably require activation of multiple β-cell genes.

Achieving such a profound phenotypic change depends on the ability of tissue stem cells to undergo nuclear reprogramming in response to appropriate stimuli. Until recently tissue stem cells were thought to be committed and therefore restricted, compared with ES cells, with respect to the number of differentiated cell types that they can generate. In recent years this concept has been challenged by reports demonstrating that cells from adult organs can give rise to cell types found in other tissues, both *in vivo* and in culture.[27–30] At present the transdifferentiation potential of tissue stem cells remains poorly defined. It is unclear whether the different reports of transdifferentiation document (1) the direct change of a differentiated cell type A into a differentiated cell type B; (2) de-differentiation of a differentiated cell type A into a common progenitor cell type, followed by differentiation into cell type B; (3) *de novo* differentiation of pluripotent cells which persist in adult tissues; or (4) fusion of such cells with already differentiated cells. Future work using cell lineage tracing approaches may allow distinguishing between these possibilities.

Regardless of their physiological plasticity, recent work has demonstrated that cells from both fetal and adult tissues can undergo forced nuclear reprogramming both *in vivo* and *in vitro* following introduction of dominant transcription factor genes, or exposure to specific growth conditions and factors in the tissue culture medium. The cascade of transcription factors which are responsible for mouse endocrine pancreas development, as well as for maintaining the phenotype of differentiated islets, has been described in considerable detail.[53–55] Some of these transcription factors have been shown to be capable of switching on parts of the β-cell developmental program in non-pancreatic cells. In the case of one factor, pancreatic duodenal homeobox 1 (Pdx1) (which plays key roles in pancreas development and gene expression in mature β cells), ectopic expression was achieved not only by gene transfer, but also by using the capability of the protein itself to transfer into cells.[56] Thus far, insulin-producing cells have been derived from three major non-pancreatic tissues: liver, intestine and bone marrow.

16.5 Insulin-producing cells derived from liver cells

Mature hepatocytes are glucose-sensitive and share similarities in gene expression with mature β cells. In addition, hepatocytes are a primary target of insulin, which is delivered from the pancreas through the portal vein. However, hepatocytes are not equipped with a regulated secretory pathway.

Rodent liver cells have been shown to give rise to insulin-producing cells *in vivo*, and following genetic manipulations both *in vivo* and *in vitro*. Cells derived from mouse fetal liver differentiated *in vivo* into a number of hepatic, pancreatic, and intestinal cell types.[57] In addition, adult rat hepatic oval stem cells gave rise to pancreatic endocrine cells *in vitro*.[58] Mouse[59] and *Xenopus*[60] liver cells were shown to activate β-cell gene expression *in vivo* following expression of Pdx1. Similarly, expression of another β-cell transcription factor, NeuroD/Beta2, in mouse liver cells *in vivo* resulted in reversal of hyperglycemia.[61]

We utilized human fetal liver cells,[62] which express markers of hepatocytes, bile duct cells and oval cells, and are capable of differentiation into mature hepatocytes *in vivo*, in an attempt to direct them towards the β-cell phenotype. The replication capacity of these cells in tissue culture is less than 20 population doublings. This capacity was significantly extended by introduction of the *TERT* gene, without evidence for tumorigenicity *in vivo*.[63] We have recently shown[64] that these cells can be induced by ectopic expression of Pdx1 to produce and store mature insulin in significant amounts, release it in response to physiological glucose levels, and replace β-cell function in immunodeficient mice made diabetic by injection of streptozotocin (STZ), an agent with specific toxicity to β cells. The modified cells expressed multiple β-cell genes; however, they also activated genes expressed in other islet cells and exocrine pancreas, and continued to express some hepatic genes. Although multiple lines of evidence suggested that Pdx1 expression induced a regulated secretory pathway in the fetal liver cells, this remains to be directly demonstrated by ultrastructural and biochemical analyses. Manipulation of the culture conditions of the Pdx1-expressing human fetal liver cells was shown to further promote the differentiation of these cells towards the β-cell phenotype, as judged by gene expression and insulin content.[64,65] In these experiments the insulin content of the cells could be raised up to two-thirds of that of normal human islets.

16.6 Insulin-producing cells derived from intestine epithelial cells

Intestine epithelial cells, hepatocytes and pancreas cells share a common developmental origin in the primordial gut. The endocrine cells in the intestinal epithelium possess a regulated secretory pathway, which may allow storage of insulin and its release in response to physiological signals. Ectopic expression of Pdx1, in combination with treatment with betacellulin, a β-cell mitogen or co-expression of another β-cell transcription factor, Isl1, in a rat enterocyte cell line, IEC-6, resulted in activation of insulin expression.[66] However, the subcellular compartment in which insulin was stored in these cells was not determined. In another study using the same cells and inductive conditions, the presence of insulin secretory granules was shown by electron microscopy; however, insulin secretion was constitutive.[67] Finally, in a study utilizing primary mouse intestinal cells, insulin production was induced following treatment with GLP-1, which activated the expression of neurogenin-3 (Ngn3), a transcription factor that is responsible of endocrine pancreas development.[68] Transplantation of these cells into STZ-diabetic mice resulted in normalization of blood glucose levels 8 weeks later. This relatively long lag period implies that additional cell differentiation and/or proliferation were required *in vivo* to achieve sufficient insulin production. These studies suggest that intestinal cells represent a potential cell source for development of insulin-producing cells.

However, intestinal endocrine cells are quite rare, and the intestinal epithelium is not easily accessible for biopsy. Thus, realization of their potential will probably require *in vivo* gene targeting, or the use of allogeneic donor cells.

16.7 Insulin-producing cells derived from bone marrow

The adult bone marrow (BM) has been shown in recent years to constitute a promising source of tissue stem cells. BM cells contain at least two types of stem cells with pluripotent capacities, hematopoietic stem cells, and stromal or mesenchymal stem cells. BM transplantation resulted in differentiation of the transplanted cells into a variety of ectodermal, mesodermal and endodermal tissues, in both mice and humans.[69] Although a number of reports have challenged these results by demonstrating that they were probably caused by fusion of BM cells with differentiated cells, rather than a differentiation of BM cells themselves, other studies support the wide differentiation potential of BM stem cells. It is possible that adult BM cells normally serve as a source for continuous renewal of other tissue stem cells, including those in the pancreas. A recent report has suggested that insulin-positive cells, which appear in a number of tissues in STZ-treated mice, originate from BM.[70] BM cells represent an attractive source for autologous stem cells, since they can be biopsied with a relative ease.

An initial study suggested that grafted mouse BM cells could differentiate into endocrine pancreas cells.[71] However, subsequent work has challenged these findings, demonstrating that BM cells induce regeneration of endogenous islets in STZ-diabetic mice, rather than differentiate themselves into islet cells.[72] Both Choi *et al.*[73] and Lechner *et al.*,[74] using mouse BM cells labeled with green fluorescent protein, subsequently concluded that BM cells did not contribute to islet repopulation following STZ-induced damage to β cells. In addition to these *in vivo* models, Tang *et al.*[75] demonstrated that a clonal population of adult murine BM stem cells (probably mesenchymal) could be induced to differentiate *in vitro* into insulin-producing cells by culture in a high-glucose medium, followed by transfer to a low-glucose medium containing nicotineamide and exendin 4, two agents known to promote β-cell differentiation. Although their insulin content was quite low, these cells were capable of correcting hyperglycemia *in vivo*.[75] Similar results were reported by Oh *et al.*, using rat BM cells cultured under somewhat different conditions.[76] Taken together, these studies cautiously support the potential of BM cells to differentiate into insulin-producing cells. However, further work is clearly needed to directly demonstrate the derivation of such cells from BM cells, as well as for determining how close BM-derived insulin-producing cells are to functional, mature β cells.

16.8 Approaches for immunoprotection of transplanted insulin-producing cells

Transplanted surrogate β cells are likely to be exposed to recurring autoimmunity, as well as to allograft rejection. Generation of insulin-producing cells in tissue culture may offer a number of ways for avoiding the immune responses directed against β cells. The antigenic targets of the autoimmune response remain largely unknown. However, it is possible that non-β-cells induced to differentiate into insulin-producing cells will not express these antigens. In addition, it is possible that these cells will be more resistant, compared with normal β cells, to apoptosis induced by cytokines and free radicals. Beta cells are known to express relatively low levels of free-radical scavenging enzymes.[77] Moreover, propagation of these cells in tissue culture provides an opportunity for increasing their resistance to immune responses, by *ex vivo* gene transfer, or by cell encapsulation in semi-permeable membrane devices. Genes shown to provide protection from immune responses to β cells in culture and in mouse models of autoimmune diabetes include anti-apoptosis genes,[78–81] genes encoding antioxidant proteins,[82,83] and those encoding proteins interfering with cytokine receptor signal transduction pathways[84–87] or antigen presentation on the cell surface.[81,88–90] In addition, expression of inhibitory cytokines, which can be secreted from the cells and locally suppress the function of cells of the immune system, has also been shown to be effective in β-cell protection.[91–93]

16.9 Encapsulation of islets and insulin-producing cells

The concept of protecting transplanted islets by immuno-isolation in a semipermeable membrane has been repeatedly tested during the past two decades, using both macrocapsules[94] (either intravascular or extravascular) or microcapsules[95] (reviewed in ref. 96). However, while cell encapsulation proved successful with other cell types, the specific problems associated with islet cell encapsulation precluded significant progress in animal models of diabetes. The major obstacle has been poor cell survival caused by hypoxia, due to the high metabolic rate of β cells, and by small immune effector molecules, such as cytokines and free radicals, which can penetrate the capsule. In addition, the kinetics of secretagogue equilibration across the membrane was too slow for the rapid insulin secretory response required of functional β cells. However, recent studies with alginate/polyamino acids microcapsules suggested that these obstacles could be overcome,[97,98] raising new hopes for potential clinical applications.

16.10 Future trends

Much more work is clearly needed before cell replacement therapy for diabetes can be advanced to the clinic. The challenges center on the generation of an abundant source of regulated insulin-producing cells, as well as on developing ways for protecting the cells following transplantation. The difficulty of expanding mature β cells or their pancreatic precursors has prompted investigators to explore the potential of cells from other tissues, which can be more readily obtained and expanded, to develop into surrogate β cells. Work in recent years has confirmed the ability of embryonic stem cells and cells from non-pancreatic tissues to activate properties of β cells in response to a small number of genetic and/or epigenetic manipulations. However, in most studies utilizing fetal or adult tissues the precise identity of the cells that differentiated towards the β-cell phenotype remained unknown. Application of lineage tracing approaches is likely to provide clear answers to questions concerning cell origin and the differentiation pathway. Detailed phenotypic characterization of surrogate β cells, including gene expression profiling by microarrays and functional assays *in vitro* and *in vivo*, will allow focusing on the cells which resemble most normal β cells.

At present it remains unclear whether a cell population consisting primarily of β-like cells generated by such manipulations may be able to replace the function of intact islets. The evidence from small animal models tends to support this possibility; however, data from large animal studies and clinical trials is needed to convincingly establish it. At the current state of the art it is unrealistic to expect cell engineering strategies to generate normal islets, which are mini-organs consisting of multiple cell types and characterized by a typical architecture.

The use of autologous *vs.* allogeneic cells is another open question. The relative vigor of recurring autoimmunity against insulin-producing cells derived from autologous tissues, compared with that against allogeneic surrogate β cells, is unknown. The decision on the optimal approach will largely depend on the manipulations required to induce differentiation of non-β cells into surrogate β cells. Manipulations that can be performed *in vivo* will increase the attractiveness of autologous cells. In contrast, the need for genetic manipulations, which cannot be safely performed *in vivo*, and may disrupt the normal genetic composition of the cells even when performed *ex vivo*, will increase the appeal of a universal donor allograft, which can be thoroughly characterized in tissue culture and banked to serve a large number of recipients. Finally, if the manipulations involve complex gene transfer and quality control procedures, their application to cells from each patient may be impractically difficult and expensive.

Ex vivo cell manipulation provides opportunities for engineering protective devices to preempt graft rejection and recurring autoimmunity against the transplanted cells. Cell encapsulation, on its own or in combination with transfer of protective genes, may provide sufficient protection. However, the immuno-isolation capacity of encapsulation membranes is inversely proportional to their accessibility to nutrients, which is critical for long-term survival and function of encapsulated insulin-producing cells. Thus, the longevity of functional transplants will be determined, in addition to the lifespan of the insulin-producing cells themselves, by the balancing act of assuring sufficient access to nutrients while preventing contact with immune effector molecules. Ideally, the graft should last long enough to be replaced no more than once or twice a year. If the recent rapid progress in cell engineering is matched in the coming years by progress in cell protection techniques, one can expect clinical trials in diabetic patients involving an engineered cell replacement strategy in the not too distant future.

16.11 Sources of further information and advice

Comprehensive information and references on the biology of β cells, as well as on etiology, pathology and therapeutics of both types of diabetes can be found in *Diabetes Mellitus: A Fundamental and Clinical Text*, Third Edition (D LeRoith, S Taylor, JM Olefsky, eds.), Lippincott, Williams & Wilkins, Philadelphia, 2004. Updated information on research trends, events, and clinical trials can be found in the internet sites of the American Diabetes Association (www.diabetes.org), the European Association for the Study of Diabetes (www.easd.org), the Juvenile Diabetes Research Foundation (www.jdrf.org), and the National Institute of Diabetes, Digestive, and Kidney Diseases (www.niddk.nih.gov). The site of the Beta-Cell Biology Consortium (www.betacell.org) provides information on research tools and a database of gene expression profiles.

16.12 Acknowledgements

Work in my laboratory was funded by the Israel Science Foundation, the National Institutes of Health, the Juvenile Diabetes Research Foundation International, D-Cure, and the Beta Cell Therapy Consortium of the European Union.

16.13 References

1. Sorenson RL, Brelje TC. Adaptation of islets of Langerhans to pregnancy: beta-cell growth, enhanced insulin secretion and the role of lactogenic hormones. *Horm Metab Res* 1997, 29: 301–307.
2. Butler AE, Janson J, Bonner-Weir S, Ritzel R, Rizza RA, Butler PC. Beta-cell deficit and increased beta-cell apoptosis in humans with type 2 diabetes. *Diabetes* 2003, 52: 102–110.
3. Dor Y, Brown J, Martinez OI, Melton DA. Adult pancreatic beta-cells are formed by self-duplication rather than stem-cell differentiation. *Nature* 2004, 429: 41–46.
4. Nielsen JH, Galsgaard ED, Moldrup A, Friedrichsen BN, Billestrup N, Hansen JA, Lee YC, Carlsson C. Regulation of beta-cell mass by hormones and growth factors. *Diabetes* 2001, 50 (Suppl. 1): S25–S29.
5. Xu G, Stoffers DA, Habener JF, Bonner-Weir S. Exendin-4 stimulates both beta-cell replication and neogenesis, resulting in increased beta-cell mass and improved glucose tolerance in diabetic rats. *Diabetes* 1999, 48: 2270–2276.

6. Tourrel C, Bailbe D, Lacorne M, Meile MJ, Kergoat M, Portha B. Persistent improvement of type 2 diabetes in the Goto-Kakizaki rat model by expansion of the beta-cell mass during the prediabetic period with glucagon-like peptide-1 or exendin-4. *Diabetes* 2002, 51: 1443–1452.
7. Kieffer TJ, Habener JF. The glucagon-like peptides. *Endocr Rev* 1999, 20: 876–913.
8. Brand SJ, Tagerud S, Lambert P, Magil SG, Tartarkiewicz K, Doiron K, Yan Y. Pharmacological treatment of chronic diabetes by stimulating pancreatic β-cell regeneration with systemic co-administration of EGF and gastrin. *Pharmacol Toxicol* 2002, 91: 414–420.
9. Rooman I, Bouwens L. Combined gastrin and epidermal growth factor treatment induces islet regeneration and restores normoglycemia in C57BL6/J mice treated with alloxan. *Diabetologia* 2004, 47: 259–265.
10. Sutherland DE, Gruessner R, Kandswamy R, Humar A, Hering B, Gruessner A. Beta-cell replacement therapy (pancreas and islet transplantation) for treatment of diabetes mellitus: an integrated approach. *Transplant Proc* 2004, 36: 1697–1699.
11. Shapiro AM, Lakey JR, Ryan EA *et al.* Islet transplantation in seven patients with type 1 diabetes mellitus using a glucocorticoid-free immunosuppressive regimen. *New Engl J Med* 2000, 343: 230–238.
12. Ryan EA, Paty BW, Senior PA, Bigam D, Alfadhli E, Kneteman NM, Lakey JR, Shapiro AM. Five-year follow-up after clinical islet transplantation. *Diabetes* 2005, 54: 2060–2069.
13. Hayek A, Beattie GM, Cirulli V, Lopez AD, Ricordi C, Rubin JS. Growth factor/matrix-induced proliferation of human adult beta-cells. *Diabetes* 1995, 44: 1458–1460.
14. Beattie GM, Cirulli V, Lopez AD, Hayek A. *Ex vivo* expansion of human pancreatic endocrine cells. *J Clin Endocrinol Metab* 1997, 82: 1852–1856.
15. Beattie GM, Itkin-Ansari P, Cirulli V, Leibowitz G, Lopez AD, Bossie S, Mally MI, Levine F, Hayek A. Sustained proliferation of PDX-1+ cells derived from human islets. Accelerated telomere shortening and senescence in human pancreatic islet cells stimulated to divide *in vitro*. *Diabetes* 1999, 48: 1013–1019.
16. Halvorsen TL, Beattie GM, Lopez AD, Hayek A, Levine F. Accelerated telomere shortening and senescence in human pancreatic islet cells stimulated to divide *in vitro*. *J Endocrinol* 2000, 166: 103–109.
17. Beattie GM, Rubin JS, Mally MI, Otonkoski T, Hayek A. Regulation of proliferation and differentiation of human fetal pancreatic islet cells by extracellular matrix, hepatocyte growth factor, and cell–cell contact. *Diabetes* 1996, 45: 1223–1228.
18. Beattie GM, Montgomery AM, Lopez AD, Hao E, Perez B, Just ML, Lakey JR, Hart ME, Hayek A. A novel approach to increase human islet cell mass while preserving beta-cell function. *Diabetes* 2002, 51: 3435–3439.
19. Gershengorn MC, Hardikar AA, Wei C, Geras-Raaka E, Marcus-Samuels B, Raaka BM. Epithelial-to-mesenchymal transition generates proliferative human islet precursor cells. *Science* 2004, 306: 2261–2264.
20. Lechner A, Nolan AL, Blacken RA, Habener JF. Redifferentiation of insulin-secreting cells after *in vitro* expansion of adult human pancreatic islet tissue. *Biochem Biophys Res Commun* 2005, 327: 581–588.
21. Bonner-Weir S, Taneja M, Weir GC, Tatarkiewicz K, Song KH, Sharma A, O'Neil JJ. *In vitro* cultivation of human islets from expanded ductal tissue. *Proc Natl Acad Sci USA* 2000, 97: 7999–8004.

22. Gao R, Ustinov J, Pulkkinen MA, Lundin K, Korsgren O, Otonkoski T. Characterization of endocrine progenitor cells and critical factors for their differentiation in human adult pancreatic cell culture. *Diabetes* 2003, 52: 2007–2015.
23. Ramiya VK, Maraist M, Arfors KE, Schatz DA, Peck AB, Cornelius JG. Reversal of insulin-dependent diabetes using islets generated *in vitro* from pancreatic stem cells. *Nat Med* 2000, 6: 278–282.
24. Abraham EJ, Leech CA, Lin JC, Zulewski H, Habener JF. Insulinotropic hormone glucagon-like peptide-1 differentiation of human pancreatic islet-derived progenitor cells into insulin-producing cells. *Endocrinology* 2002, 143: 3152–3161.
25. Seaberg RM, Smukler SR, Kieffer TJ, Enikolopov G, Asghar Z, Wheeler MB, Korbutt G, Van Der Kooy D. Clonal identification of multipotent precursors from adult mouse pancreas that generate neural and pancreatic lineages. *Nat Biotechnol* 2004, 22: 1115–1124.
26. Thomson JA, Itskovitz-Eldor J, Shapiro SS *et al.* Embryonic stem cell lines derived from human blastocysts. *Science* 1998, 282: 1145–1147.
27. Wagers AJ, Weissman IL. Plasticity of adult stem cells. *Cell* 2004, 116: 639–648.
28. Tosh D, Slack JM. How cells change their phenotype. *Nat Rev Mol Cell Biol* 2002, 3: 187–194.
29. Wagers AJ, Weissman IL. Plasticity of adult stem cells. *Cell* 2004, 116: 639–648.
30. Fuchs E, Tumbar T, Guasch G. Socializing with the neighbors: stem cells and their niche. *Cell* 2004, 116: 769–778.
31. Soria B, Roche E, Berna G, Leon-Quinto T, Reig JA, Martin F. Insulin-secreting cells derived from embryonic stem cells normalize glycemia in streptozotocin-induced diabetic mice. *Diabetes* 2000, 49: 157–162.
32. Assady S, Maor G, Amit M, Itskovitz-Eldor J, Skorecki KL, Tzukerman M. Insulin production by human embryonic stem cells. *Diabetes* 2001, 50: 1691–1697.
33. Lumelsky N. Blondel O, Laeng P, Velasco I, Ravin R, McKay R. Differentiation of embryonic stem cells to insulin-secreting structures similar to pancreatic islets. *Science* 2001, 292: 1389–1394.
34. Segev H, Fishman B, Ziskind A, Shulman M, Itskovitz-Eldor J. Differentiation of human embryonic stem cells into insulin-producing clusters. *Stem Cells* 2004, 22: 265–274.
35. Blyszczuk P, Czyz J, Kania G, *et al.* Expression of Pax4 in embryonic stem cells promotes differentiation of nestin-positive progenitor and insulin-producing cells. *Proc Natl Acad Sci USA* 2003, 100: 998–1003.
36. Miyazaki S, Yamato E, Miyazaki J. Regulated expression of pdx-1 promotes *in vitro* differentiation of insulin-producing cells from embryonic stem cells. *Diabetes* 2004, 53: 1030–1037.
37. Rajagopal J, Anderson WJ, Kume S, Martinez OI, Melton DA. Insulin staining of ES cell progeny from insulin uptake. *Science* 2003, 299: 363.
38. Hansson M, Tonning A, Frandsen U, Petri A, Rajagopal J, Englund MC, Heller RS, Hakansson J, Fleckner J, Skold HN, Melton D, Semb H, Serup P. Artifactual insulin release from differentiated embryonic stem cells. *Diabetes* 2004, 53: 2603–2609.
39. Sipione S, Eshpeter A, Lyon JG, Korbutt GS, Bleackley RC. Insulin expressing cells from differentiated embryonic stem cells are not beta cells. *Diabetologia* 2004, 47: 499–508.
40. Dominguez-Bendala J, Klein D, Ribeiro M, Ricordi C, Inverardi L, Pastori R, Edlund H. TAT-mediated neurogenin 3 protein transduction stimulates pancreatic endocrine differentiation in vitro. *Diabetes* 2005, 54: 720–726.

41. Osawa M, Hanada K, Hamada H, Nakauchi H. Long-term lymphohematopoietic reconstitution by a single CD34-low/negative hematopoietic stem cell. *Science* 1996, 273: 242–245.
42. Gage FH. Mammalian neural stem cells. *Science* 2000: 1433–1438.
43. Prockop DJ. Marrow stromal cells as stem cells for nonhematopoietic tissues. *Science,* 1997, 276: 71–74.
44. Watt FM. Stem cell fate and patterning in mammalian epidermis. *Curr Opin Genet Dev* 2001, 11: 410–417.
45. Weissman IL. Translating stem and progenitor cell biology to the clinic: barriers and opportunities. *Science* 2000, 287: 1442–1446.
46. Harley CB, Futcher AB, Greider CW. Telomeres shorten during ageing of human fibroblasts. *Nature* 1990, 345: 458–460.
47. Bodnar AG, Ouellette M, Frolkis M *et al.* Extension of life-span by introduction of telomerase into normal human cells. *Science* 1998, 279: 349–352.
48. Vaziri H, Benchimol S. Reconstitution of telomerase activity in normal human cells leads to elongation of telomeres and extended replicative life span. *Curr Biol* 1998, 8: 279–282.
49. Yang J, Chang E, Cherry AM *et al.* Human endothelial cell life extension by telomerase expression. *J Biol Chem* 1999, 274: 26141–26148.
50. Jiang XR, Jimenez G, Chang E *et al.* Telomerase expression in human somatic cells does not induce changes associated with a transformed phenotype. *Nat Genet* 1999, 21: 111–114.
51. Morales CP, Holt SE, Ouellette M *et al.* Absence of cancer-associated changes in human fibroblasts immortalized with telomerase. *Nat Genet* 1999, 21: 115–118.
52. Harley CB. Telomerase is not an oncogene. *Oncogene* 2002, 2: 494–502.
53. Huang HP, Tsai MJ. Transcription factors involved in pancreatic islet development. *J Biomed Sci* 2000, 7: 27–34.
54. Kim SK, MacDonald RJ. Signaling and transcriptional control of pancreatic organogenesis. *Curr Opin Genet Dev* 2002, 12: 540–547.
55. Servitja JM, Ferrer J. Transcriptional networks controlling pancreatic development and beta cell function. *Diabetologia* 2004, 47: 597–613.
56. Noguchi H, Kaneto H, Weir GC, Bonner-Weir S. PDX-1 protein containing its own antennapedia-like protein transduction domain can transduce pancreatic duct and islet cells. *Diabetes* 2003, 52: 1732–1737.
57. Suzuki A, Zheng Yw YW, Kaneko S *et al.* Clonal identification and characterization of self-renewing pluripotent stem cells in the developing liver. *J Cell Biol* 2002, 156: 173–184.
58. Yang L, Li S, Hatch H, Ahrens K *et al. In vitro* trans-differentiation of adult hepatic stem cells into pancreatic endocrine hormone-producing cells. *Proc Natl Acad Sci USA* 2002, 99: 8078–8083.
59. Ferber S, Halkin A, Cohen H *et al.* Pancreatic and Duodenal homeobox gene 1 induces expression of insulin genes in liver and ameliorates streptozotocin-induced hyperglycemia. *Nat Med* 2000, 6: 568–572.
60. Horb ME, Shen CN, Tosh D, Slack JM. Experimental conversion of liver to pancreas. *Curr Biol* 2003, 13: 105–115.
61. Kojima H, Fujimiya M, Matsumura K *et al.* NeuroD-betacellulin gene therapy induces islet neogenesis in the liver and reverses diabetes in mice. *Nat Med* 2003, 9: 596–603.
62. Malhi H, Irani AN, Gagandeep S, Gupta S. Isolation of human progenitor liver epithelial cells with extensive replication capacity and differentiation into mature

hepatocytes. *J Cell Sci* 2002, 115: 2679–2688.

63. Wege H, Le HT, Chui MS *et al.* Telomerase reconstitution immortalizes human fetal hepatocytes without disrupting their differentiation potential. *Gastroenterology* 2003, 124: 432–444.
64. Zalzman M, Gupta S, Giri R *et al.* Reversal of hyperglycemia in mice by using human expandable insulin-producing cells differentiated from fetal liver progenitor cells. *Proc Natl Acad Sci USA* 2003, 100: 7253–7258.
65. Zalzman M, Anker-Kitai L, Efrat S. Differentiation of human liver-derived insulin-producing cells towards the beta-cell phenotype. *Diabetes* 2005 54: 2568–2575.
66. Kojima H, Nakamura T, Fujita Y *et al.* Combined expression of pancreatic duodenal homeobox 1 and islet factor 1 induces immature enterocytes to produce insulin. *Diabetes* 2002, 51: 1398–1408.
67. Yoshida S, Kajimoto Y, Yasuda T *et al.* PDX-1 induces differentiation of intestinal epithelioid IEC-6 into insulin-producing cells. *Diabetes* 2002, 51: 2505–2513.
68. Suzuki A, Nakauchi H, Taniguchi H. Glucagon-like peptide 1 (1-37) converts intestinal epithelial cells into insulin-producing cells. *Proc Natl Acad Sci USA* 2003, 100: 5034–5039.
69. Jiang Y, Jahagirdar BN, Reinhardt RL *et al.* Pluripotency of mesenchymal stem cells derived from adult marrow. *Nature* 2002, 418: 41–49.
70. Kojima H, Fujimiya M, Matsumura K *et al.* Extrapancreatic insulin-producing cells in multiple organs in diabetes. *Proc Natl Acad Sci USA* 2004, 101: 2458–2463.
71. Ianus A, Holz GG, Theise ND, Hussain MA. *In vivo* derivation of glucose-competent pancreatic endocrine cells from bone marrow without evidence of cell fusion. *J Clin Invest* 2003, 111: 843–850.
72. Hess D, Li L, Martin M *et al.* Bone marrow-derived stem cells initiate pancreatic regeneration. *Nat Biotechnol* 2003, 21: 763–770.
73. Choi JB, Uchino H, Azuma K *et al.* Little evidence of transdifferentiation of bone marrow-derived cells into pancreatic beta cells. *Diabetologia* 2003, 46: 1366–1374.
74. Lechner A, Yang YG, Blacken RA, Wang L, Nolan AL, Habener JF. No evidence for significant transdifferentiation of bone marrow into pancreatic beta-cells *in vivo*. *Diabetes* 2004, 53: 616–623.
75. Tang DQ, Cao LZ, Burkhardt BR *et al. In vivo* and *in vitro* characterization of insulin-producing cells obtained from murine bone marrow. *Diabetes* 2004, 53: 1721–1732.
76. Oh SH, Muzzonigro TM, Bae SH, LaPlante JM, Hatch HM, Petersen BE. Adult bone marrow-derived cells trans-differentiating into insulin-producing cells for the treatment of type I diabetes. *Lab Invest* 2004, 84: 607–617.
77. Lenzen S, Drinkgern J, Tiedge M. Low antioxidant enzyme gene expression in pancreatic islets compared with various other mouse tissues. *Free Radic Biol Med* 1996, 20: 463–466.
78. Rabinovitch A, Suarez-Pinzon W, Strynadka K, Ju Q, Edelstein D, Brownlee M, Korbutt GS, Rajotte RV. Transfection of human pancreatic islets with an anti-apoptotic gene (bcl-2) protects beta-cells from cytokine-induced destruction. *Diabetes* 1999, 48: 1223–1229.
79. Dupraz P, Rinsch C, Pralong WF, Rolland E, Zufferey R, Trono D, Thorens B. Lentivirus-mediated Bcl-2 expression in beta TC-tet cells improves resistance to hypoxia and cytokine-induced apoptosis while preserving *in vitro* and *in vivo* control of insulin secretion. *Gene Ther* 1999, 6: 1160–1169.
80. Grey ST, Arvelo MB, Hasenkamp W, Bach FH, Ferran C. A20 inhibits cytokine-induced apoptosis and nuclear factor kB-dependent gene activation in islets. *J Exp*

Med 1999, 190: 1135–1146.

81. Efrat S, Serezze D, Svetlanov M, Post CM, Johnson EA, Herold K, Horwitz MS. Adenovirus early region 3 (E3) immunomodulatory genes decrease the incidence of autoimmune diabetes in nonobese diabetic (NOD) mice. *Diabetes* 2001, 50: 980–984.
82. Benhamou PY, Moriscot C, Richard MJ, Beatrix O, Badet L, Pattou F, Kerr-Conte J, Chroboczek J, Lemarchand P, Halimi S. Adenovirus-mediated catalase gene transfer reduces oxidant stress in human, porcine, and rat pancreatic islets. *Diabetologia* 1998, 41: 1093–1100.
83. Hotta M, Tashiro F, Ikegami H, Niwa H, Ogihara T, Yodoi J, Miyazaki J. Pancreatic β cell-specific expression of thioredoxin, an antioxidative and antiapoptotic protein, prevents autoimmune and streptozotocin-induced diabetes. *J Exp Med* 1998, 188: 1445–1451.
84. Dupraz P, Cottet S, Hamburger F, Dolci W, Felley-Bosco E, Thorens B. Dominant negative MyD88 proteins inhibit interleukin-1beta/interferon-gamma-mediated induction of nuclear factor kappa B-dependent nitrite production and apoptosis in beta cells. *J Biol Chem* 2000, 275: 37672–37678.
85. Cottet S, Dupraz P, Hamburger F, Dolci W, Jaquet M, Thorens B. SOCS-1 protein prevents Janus Kinase/STAT-dependent inhibition of beta cell insulin gene transcription and secretion in response to interferon-gamma. *J Biol Chem* 2001, 276: 25862–25870.
86. Chong MM, Chen Y, Darwiche R, Dudek NL, Irawaty W, Santamaria P, Allison J, Kay TW, Thomas HE. Suppressor of cytokine signaling-1 overexpression protects pancreatic beta cells from CD8+ T cell-mediated autoimmune destruction. *J Immunol* 2004, 172: 5714–5721.
87. Flodstrom-Tullberg M, Yadav D, Hagerkvist R, Tsai D, Secrest P, Stotland A, Sarvetnick N. Target cell expression of suppressor of cytokine signaling-1 prevents diabetes in the NOD mouse. *Diabetes* 2003, 52: 2696–2700.
88. Efrat S, Fejer G, Brownlee M, Horwitz MS. Prolonged survival of murine pancreatic islet allografts mediated by adenovirus early region 3 immunoregulatory transgenes. *Proc Natl Acad Sci USA* 1995, 92: 6947–6951.
89. von Herrath MG, Efrat S, Oldstone MBA, Horwitz MS. Expression of adenoviral E3 transgenes in β cells prevents autoimmune diabetes. *Proc Natl Acad Sci USA* 1997, 94: 9808–9813.
90. Pierce MA, Chapman HD, Post CM, Svetlanov A, Efrat S, Horwitz MS, Serreze DV. Adenovirus early region 3 (E3) anti-apoptotic 10.4K, 14.5K, and 14.7K genes decrease the incidence of autoimmune diabetes in NOD mice. *Diabetes* 2003, 52: 1119–1127.
91. Gallichan WS, Kafri T, Krahl T, Verma IM, Sarvetnick N. Lentivirus-mediated transduction of islet grafts with interleukin 4 results in sustained gene expression and protection from insulitis. *Hum Gene Ther* 1998, 9: 2717–2726.
92. Moritani M, Yoshimoto K, Wong SF, Tanaka C, Yamaoka T, Sano T, Komagata Y, Miyazaki J, Kikutani H, Itakura M. Abrogation of autoimmune diabetes in nonobese diabetic mice and protection against effector lymphocytes by transgenic TGF-β1. *J Clin Invest* 1998, 102: 499–506.
93. Goudy K, Song S, Wasserfall C, Zhang YC, Kapturczak M, Muir A, Powers M, Scott-Jorgensen M, Campbell-Thompson M, Crawford JM, Ellis TM, Flotte TR, Atkinson MA. Adeno-associated virus vector-mediated IL-10 gene delivery prevents type 1 diabetes in NOD mice. *Proc Natl Acad Sci USA* 2001, 98: 13913–13918.

94. Lacy PE, Hegre OD, Gerasimidi-Vazeou A, Gentile FT, Dionne KE. Maintenance of normoglycemia in diabetic mice by subcutaneous xenografts of encapsulated islets. *Science* 1991, 254: 1782–1784.
95. Sun Y, Ma X, Zhou D, Vacek I, Sun AM. Normalization of diabetes in spontaneously diabetic cynomologus monkeys by xenografts of microencapsulated porcine islets without immunosuppression. *J Clin Invest* 1996, 98: 1417–1422.
96. de Vos P, Hamel AF, Tatarkiewicz K. Considerations for successful transplantation of encapsulated pancreatic islets. *Diabetologia* 2002, 45: 159–173.
97. Duvivier-Kali VF, Omer A, Parent RJ, O'Neil JJ, Weir GC. Complete protection of islets against allorejection and autoimmunity by a simple barium-alginate membrane. *Diabetes* 2001, 50: 1698–1705.
98. Luca G, Basta G, Calafiore R, Rossi C, Giovagnoli S, Esposito E, Nastruzzi C. Multifunctional microcapsules for pancreatic islet cell entrapment: design, preparation and *in vitro* characterization. *Biomaterials* 2003, 24: 3101–3114.

17

Drug delivery system for active brain targeting

X-G MEI and M-Y YANG, Beijing Institute of Pharmacology and Toxicology, People's Republic of China

17.1 Introduction

17.1.1 Brain disease

Many brain problems such as neurodegenerative diseases, Alzheimer's disease, stroke (ischemia or hemorrhage), trauma, infections of brain and brain cancer are becoming more prevalent especially as populations become older. Among this, tumors of the central nervous system (CNS) represent one of the most devastating forms of human illness. In the United States alone, approximately 16 800 people are diagnosed with primary brain tumors each year, and 13 100 Americans die from these lesions. Doctors who treat patients with such disease struggle with one seemingly intractable challenge: how to deliver drugs quickly to the brain. Despite major advances in neuroscience, the blood–brain barrier (BBB) ensures that many potential therapeutic agents cannot reach the CNS.

17.1.2 The blood–brain barrier and blood–cerebrospinal fluid barrier

As is well known, there are various barriers protecting the animal and human life systems from changes in environment and invasion of toxic foreign molecules, such as the gastrointestinal (GI) barrier, blood–retina barrier and the BBB. The concept of the BBB was firstly raised by the German bacteriologist Paul Ehrlich in the late 19th century, when he found that colored dyes injected into the circulation of animals stained all tissues except for the brain. This concept was confirmed lately by using electron microscopy, which showed that the BBB involves a single layer of tile-like endothelial cells, with tight intercellular junctions. The junctions line the inner surface of brain capillaries and exist between the blood and the CNS. The functions of the BBB are to provide neurons with their precisely controlled nutritional requirements, to maintain a proper balance of ions and other chemical constituents, and to isolate the CNS from certain toxic chemicals in the blood. Also, there are closely associated glial cells. The other physiological and pharmacological barrier is blood–cerebrospinal fluid barrier

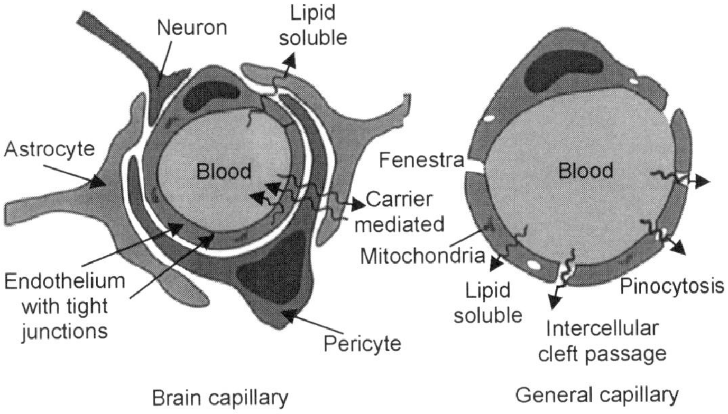

17.1 Schematic comparison between general and brain capillaries; see text for details.

(BCSFB), which is located at the choroid plexuses. Since the surface area of the human BBB is estimated to be 5000 times greater than that of the BCSFB, the BBB is considered to be the main region controlling the uptake of drugs into the brain parenchyma and the target for delivering drugs to the brain. Furthermore, the relatively high blood flow of the brain is also an obstacle.

For the above reasons, the brain is probably one of the least accessible organs. Small electrically neutral, lipid-soluble molecules can readily penetrate the BBB; however, many chemotherapeutic agents that are large, ionically charged or hydrophilic do not fall into this category, and thus are difficult to transport into the CNS (Fig. 17.1). It is, therefore, the major obstacle to drugs that combat diseases affecting the CNS because of these unique characteristics of the brain.

17.2 Strategies for delivering drugs to the brain

To the BBB, drugs are foreign molecules that are not allowed to enter the cerebrospinal fluid and gain access to brain cells. While the exclusion of foreign molecules is highly desirable under normal circumstances, cerebral drug delivery is faced with many obstacles owing to the unique anatomic and physiological characteristics of the brain. These present serious problems for medical treatment of cancer and other diseases of the brain. Thus, the development of efficient drugs and drug carriers capable of entering the brain has become a major challenge; various strategies have been devised to improve brain drug delivery and accumulation as follows:

- Preparation of lipid soluble analogs or prodrugs.
- Direct injection into brain or CSF.
- BBB modification.

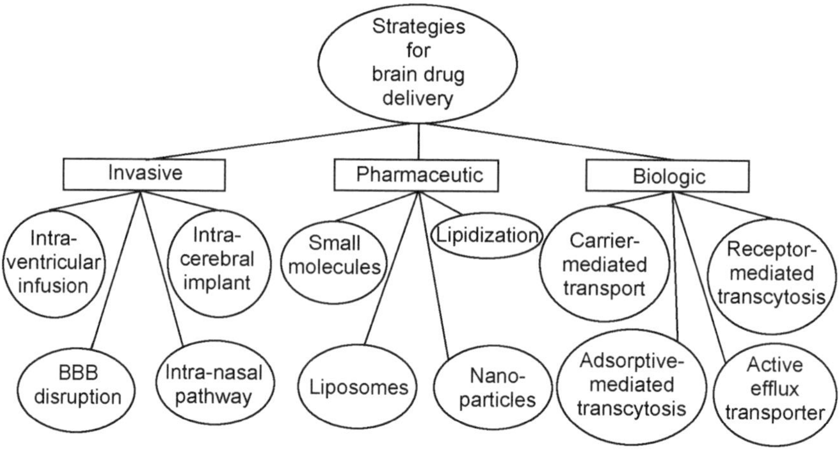

17.2 Strategies for brain drug delivery.

- Design of drugs that are taken up into brain by carrier-, receptor- or absorptive-mediated transport systems.
- Inhibition of active removal or breakdown systems.

Strategies for CNS drug delivery may be classified as being invasive or neurosurgically based, pharmaceutical and biochemical methods (Fig. 17.2).

17.2.1 Invasive delivery strategies

To enhance the local concentration of chemotherapeutic agents, a drug may be introduced to the brain by direct injection or infusion, or by implantation of a drug-loaded polymer matrix. This idea is to assure a sustained release of the drug to avoid recurrences within several centimeters of the initial location. Microspheres seem particularly suitable for direct implantation in a tumor because they spontaneously remain *in situ* in tumorized areas. Moreover, a multi-point administration can easily be performed during surgery or by stereotaxy. This type of administration yields the highest degree of targeting; it is however limited by invasiveness and diffusion restrictions. Intracerebroventricular and intrathecal administrations deliver the drug directly into the CSF compartment, either in the lateral ventricle or in the subarachnoid space; these routes of administration are less invasive than a direct intracerebral injection and allow access to a much wider area of the CNS through CSF circulation pathways. However, diffusional and cellular barriers for penetration into surrounding tissues and significant clearance of CSF into the venous and lymphatic circulation are also limiting factors.

Another invasive method relies on reversible BBB disruption. Intracarotid injection of an inert hypertonic solution such as mannitol has been employed to initiate endothelial cell shrinkage and opening of BBB tight junctions for a period of a few hours, and so allow the delivery of antineoplastic agents to the human brain. In contrast to osmotic disruption methods, a biochemical opening approach uses bradykinin receptor stimulation as a means to transiently increase BBB permeability. Although from a quantitative point of view these approaches look attractive, the risk of this barrier function loss may allow uncontrolled access of solutes into the brain.

An intravenous (injectable) drug delivery technology for CNS-active biopharmaceutical drugs will enhance treatment of many more brain disorders. Currently, there are, however, no such brain drug delivery technologies on the market for clinical use. One reason for this might be that these technologies still involve potential safety hazards, such as the obstruction of brain entry of essential compounds (like insulin or iron), or potentially dangerous interactions with endogenous substrates.

Moreover, two routes have been proposed for the direct passage of drugs from the nose to the brain: an intraneuronal and an extraneuronal pathway. This suggests that even large molecules such as peptides (such as nerve growth factor) could be transported from the nasal cavity to the CNS by nasal spray, and will help to treat brain problems and other CNS disorders. The main advantage of intranasal administration is the minimal invasiveness. It also offers several other advantages. For example, in imaging technologies, it can allow the rapid diagnosis and assessment of acute brain injury in stroke and trauma.

Although these modes of delivery are traditional, the risks of infection and neuropathological changes due to disruption of the BBB emphasize the need to develop new noninvasive delivery strategies.

17.2.2 Pharmaceutical strategies

Pharmacologic-based strategies include small molecules, lipidization strategies, liposomes and nanoparticles. Small molecules are delivered to the brain in proportion to the lipid solubility of the compound, providing there is no significant plasma protein binding of the drug. However, what is not generally appreciated is that there is a molecular mass threshold of lipid soluble small molecule transport through the BBB and this threshold has been estimated at 400–600 Da. Indeed, essentially all small molecule drugs that are currently in neuropathological practice fulfill the dual criteria of lipid-solubility and a molecular mass less than a threshold of 500 Da. Lipidization also increases its penetration in other tissues in the body and decreases systemic exposure. Another possibility to increase the passive diffusion of drugs through the BBB is the encapsulation of the compound into small liposomes or nanoparticles. Liposomes, even small unilamellar vesicles of 50 nm in diameter, are too large to undergo transport through the BBB. Although the mechanism is not fully understood, nanoparticle delivery to the brain may be a promisingly strategy for the treatment of brain tumors.

17.2.3 Biological strategies

Biological-based brain drug delivery strategies source from an understanding of the anatomy and physiology of the normal BBB endogenous transport process. This process may be classified as carrier-mediated transport, receptor-mediated transcytosis, adsorptive-mediated transcytosis and active efflux transporters. These brain-targeted chemical delivery systems represent a general and systematic method that can provide localized and sustained release for a variety of therapeutic agents including neuropeptides. These vectors can be peptides, modified proteins or monoclonal antibodies (Fig. 17.3). By using a sequential metabolism approach, they exploit the specific trafficking properties of the BBB and provide site-specific or site-enhanced delivery.

Carrier-mediated transport

There are some polar small molecule drugs that have a molecular structure mimicking nutrient that normally undergoes carrier-mediated transport through the BBB. Carrier-mediated systems include the large neutral amino acid transporter (LAT)-1: this transporter mediates the brain uptake of various drugs such as 1-DOPA, melphalan and baclofen. Carrier-mediated transport may be considered by drug development focusing on structural requirements that enable BBB transport via a carrier-mediated system.

Receptor-mediated transcytosis

The uptake of drugs by the brain can be improved by conjugation to an endogenous compound, which uses receptor-mediated transcytosis. Examples of endogenous receptor-mediated transcytosis include the BBB insulin receptor or the BBB transferring receptor. For example, transferrin receptors are abundant on the vascular endothelium of brain capillaries, and these receptors are internalized by the endothelial cells in a process designed to deliver iron to the brain. This allows the BBB to be bypassed using transferrin or transferrin-receptor antibodies as carriers of proteins such as antibodies and neuropeptides. A highly studied receptor-mediated transport vector is the OX26, a mouse monoclonal antibody to transferrin receptor; conjugation of this transport vector being facilitated with avidin/biodin technology.

Adsorptive-mediated transcytosis

A promising approach consisted of incorporating daunomycin in OX26 immuno-liposomes, thus greatly increasing the transport capacity of OX26. Cationized proteins such as cationized albumin and immunoglobin G move across the BBB by adsorption-mediated transcytosis that becomes saturated at higher concentrations than receptor-mediated transcytosis. Doxorubicin coupled

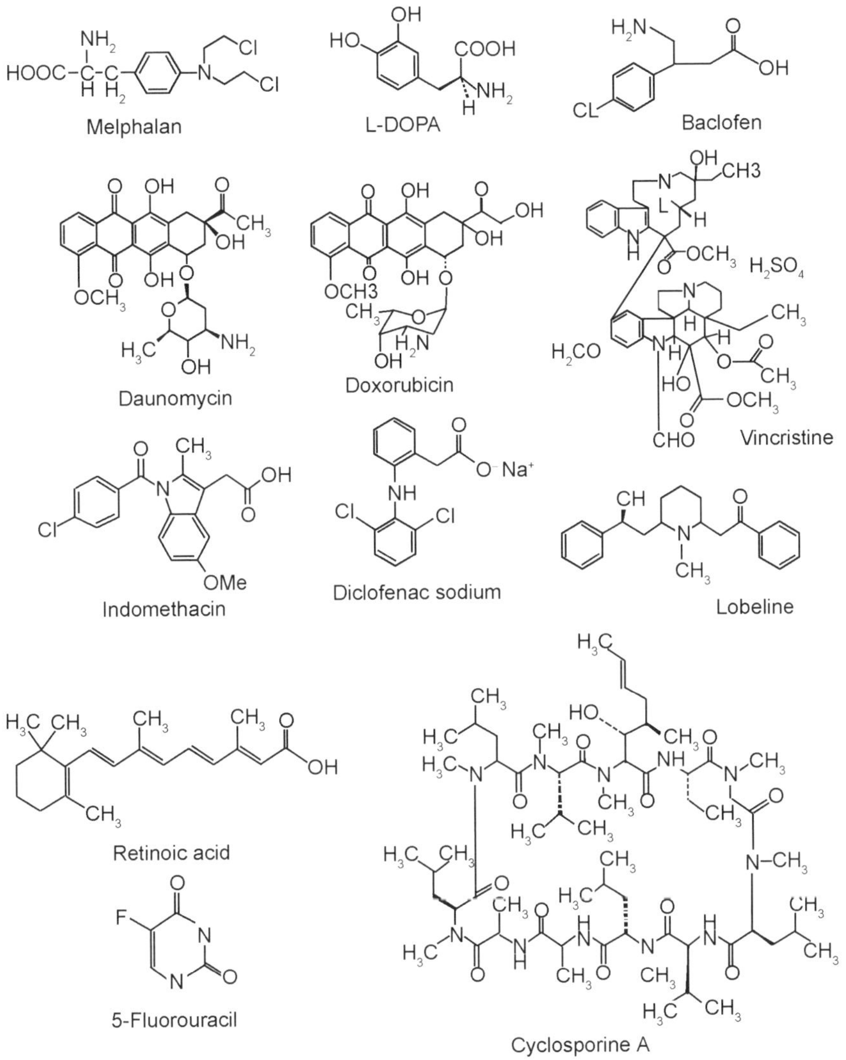

17.3 Structure of several drugs delivered to the brain.

to a small peptide vector – Syn B – appears to cross the BBB by adsorption-mediated endocytosis.

Active efflux transporters

The advantage of using brain influx transport systems is the specificity; however only limited amounts of drugs can be delivered by this way since the brain penetration is limited by the number and the carrying capacity of the transporters.

In the search for strategies that will increase cerebral delivery of drugs, the development of specific inhibitors to reduce drug efflux at the BBB and eventually at the BCSFB must be considered. Considering their broad substrate profiles and their demonstrated expression at both BBB and BCSFB, it is clear that these transporters represent an attractive target for brain delivery. Modulation of these efflux transporters by design of inhibitors and/or design of compounds having minimal affinity for these transporters may improve the treatment of CNS disorders. Important issues in the search for efflux inhibitors remain their ability to interact exclusively at the levels of BBB and BCSFB as well as the existence of multiple efflux transport systems at these barriers.

Each technique has its own application with specific advantages and limitations; the strategy used by pharmaceutical research should be a combination of different models in order to obtain the best predictability to the clinical situation. Following the identification of brain penetration characteristics and the selection of the most promising compounds, the success of a CNS drug development program may be increased by enhancing drug delivery to the CNS.

17.2.4 Surface modification

Surface modification of micro/nano-particles with hydrophilic non-immunogenic polymers has been successfully applied to obtain drug carriers with a long circulation time. Long circulation drug-entrapped liposomes have become the first clinically approved drug delivery system. Initial clinical trials showed encouraging results in terms of reduced toxicity and drug targeting to the tumor site. Some of the problems encountered with liposomal systems may be overcome with the use of more stable biodegradable and polymeric micro/nano-particles with a hydrophilic surface.

Homing devices on the carrier surface (Fig. 17.4) can also guide the carrier to the specific site where it performs its function. By applying proteins, including antibodies, antibody fragments and lipoproteins, lectins, hormones, mono-, oligo- and polysaccharides, as targeting moieties, microcapsules, microparticles, liposomes and micelles could be successfully used for targeted drug delivery. These targeted drug delivery systems may also prove particularly valuable to enable the use of a particular drug that would otherwise be ineffective or even toxic if delivered systemically, such as neural growth factors which need to cross the BBB or vaccines which need to be taken up by antigen-presenting cells. At the current pace of gene cloning and recombinant-protein production within the biopharmaceutical industry, many more site-specific drug delivery products will be clinically investigated and implemented in the near future.

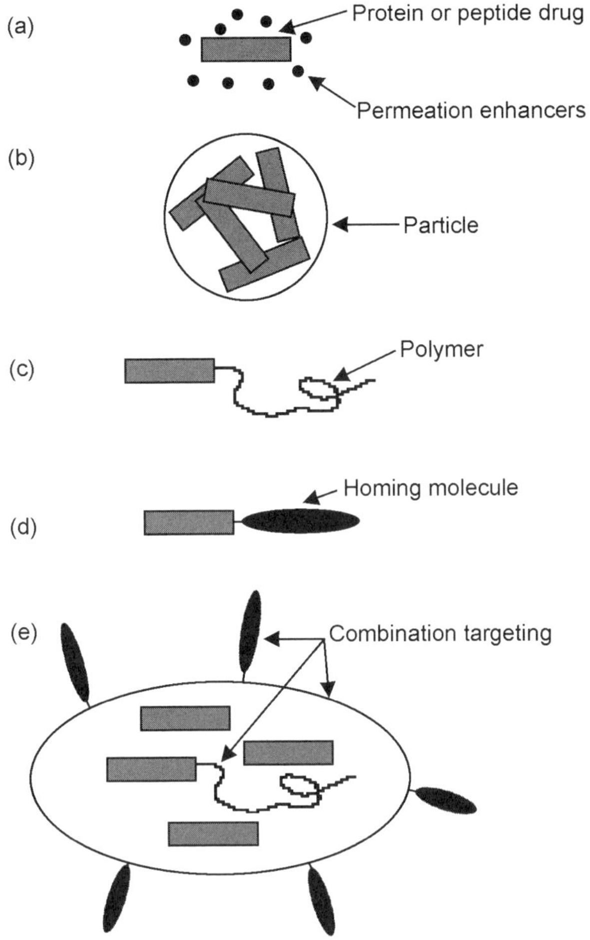

17.4 Schematic process of active targeting drug delivery system for brain drug delivery.

17.3 Different drug delivery systems for active brain targeting

After a brief description of the design principles, this section reviews a number of active-targeting brain delivery examples. Concerning the biomimetic approach to drug delivery, one may say that an ideal drug delivery system requires a non-protein adsorbing surface, a predictable drug release profile and an interaction with only the disease site. Artificial cell approaches may be a solution for this type of drug delivery system. Reconstruction of biomembranes and the transport of proteins over a polymeric capsule may find a place in the design of artificial cells.

17.3.1 Active targeting nanoparticles

Nanoparticles are defined as being submicrometer (10–1000 nm) colloidal systems generally made of polymers (biodegradable or not). They were first developed by Birrenbach and Speiser (1976). The drug is dissolved, entrapped, encapsulated or attached to a nanoparticle matrix. Depending upon the process used for the preparation of nanoparticles, nanospheres or nanocapsules can be obtained. Nanocapsules are vesicular systems in which the drug is confined to a cavity (an oil or aqueous core) surrounded by a unique polymeric membrane. Nanospheres are matrix systems in which the drug is physically and uniformly dispersed throughout the particles.

The BBB represents an insurmountable obstacle for the delivery of a large number of drugs including antibiotics, antineoplastic agents, and a variety of CNS-active drugs, especially neuropeptides to the CNS. One of the possibilities of overcoming this barrier is drug delivery to the brain using nanoparticles. Nanoparticles can cross the BBB because of their extremely small size. Appropriately coated nanoparticles may be able to pass the BBB and bring the therapeutic molecules into the brain. Furthermore, a controlled and targeted release of the drug from the nanoparticles can provide a desired concentration just to the tumor cells and thus greatly decrease the side effects.

In the past decade, effects of various drug-loaded, surfactant-coated nanoparticles of different polymers on the BBB have been investigated. *In vivo* experiments have shown that nanoparticle technology appears to have significant promise in delivering therapeutic molecules across the BBB.

In order to achieve a significant transport across the BBB, coating the nanoparticles with polysorbate 80 (Tween 80) or other polysorbates with 20 polyoxyethylene units was required. Other surfactants were less successful. The most promising results were obtained with doxorubicin for the treatment of brain tumors. Intravenous injection of polysorbate 80-coated nanoparticles loaded with doxorubicin (5 mg/kg) achieved very high brain levels of 6 Ag/g brain tissue while all the controls, including uncoated nanoparticles and doxorubicin solutions mixed with polysorbate, did not reach the analytical detection limit of 0.1 Ag/g. In rats with intracranially transplanted glioblastomas 101/8 (Fig. 17.5) that typically kill the rats within 10–20 days, intravenous injection of these doxorubicin-loaded polysorbate 80-coated nanoparticles led to 20–50% cure in different experimental runs. The dose schedule in these experiments was rather conservative, i.e. 3×1.5 mg/kg. Cure was proven by histology following sacrifice of these animals after 6 months. In the seven control groups only one other rat survived (doxorubicin in saline plus polysorbate 80) (Fig. 17.5).

The mechanism of the drug transport across the BBB with the nanoparticles appears to be endocytotic uptake by the brain capillary endothelial cells followed either by release of the drugs in these cells and diffusion into the brain or by transcytosis. After injection of the nanoparticles, apolipoprotein E (apo E)

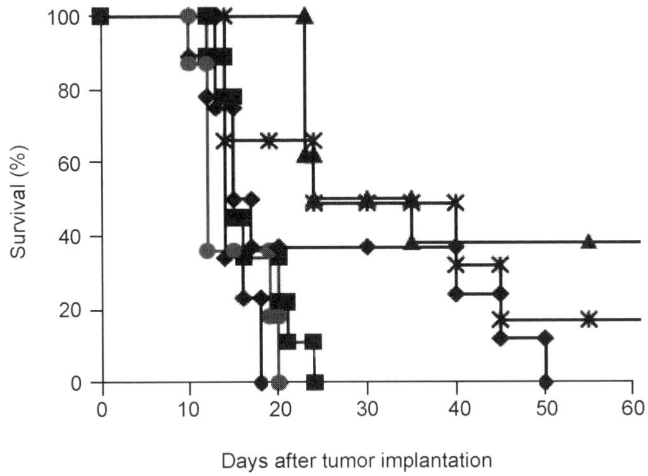

17.5 Survival of rats with an intracranially transplanted glioblastoma 101/8 after intravenous injection of doxorubicin (1.5 mg/kg on days 2, 5 and 8) after tumour transplantation using the following preparations: ◆ (left line) untreated controls, ● empty nanoparticles coated with polysorbate 80, ■ doxorubicin in saline, ◆ (right line) doxorubicin-loaded nanoparticles, * doxorubicin in polysorbate 80, ▲ doxorubicin-loaded nanoparticles coated with polysorbate 80.

or apo B adsorb on the particle surface and then seem to promote the interaction with the low-density lipoprotein (LDL) receptor followed by endocytotic uptake. The nanoparticles thus would mimic the uptake of naturally occurring lipoprotein particles. This hypothesis was supported by the achievement of an anti-nociceptive effect using dalargin-loaded poly(butyl cyanoacrylate) nanoparticles with adsorbed apo E or loperamide-loaded albumin nanoparticles with covalently bound apo E.

Recently, a novel nanoparticle (NP) comprised of emulsifying wax and Brij 78 was shown to have significant brain uptake using the *in situ* rat brain perfusion technique. To further these studies and to specifically target brain, Lockman *et al.* (2003) have incorporated thiamine as a surface ligand on the nanoparticles. The solid nanoparticles were prepared from oil-in-water microemulsion precursors. Comparison of NP brain uptake demonstrated that the thiamine-coated nanoparticles associated with the BBB thiamine transporter and had an increased K_{in} between 45 and 120 s. It was concluded that the thiamine ligand facilitated binding and/or association with BBB thiamine transporters, which may be a viable mechanism for nanoparticle mediated brain drug delivery.

Since cationic bovine serum albumin (CBSA) indicates a good accumulation profile in the brain. Lu *et al.* (2005) studied a novel drug carrier for brain delivery, CBSA conjugated with polyethyleneglycol–polylactide (PEG–PLA) nanoparticle (CBSA-NP) (Fig. 17.6). Transmission electron micrograph (TEM)

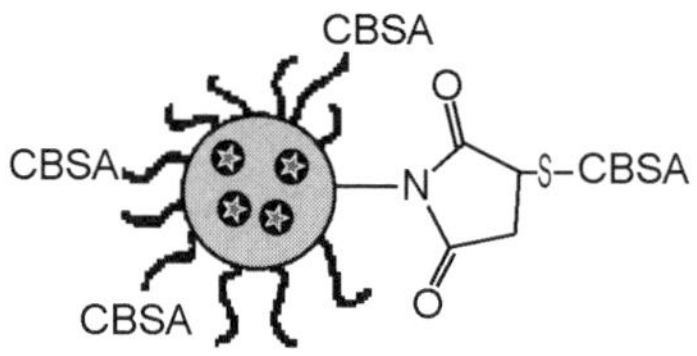

17.6 The schematic diagram of CBSA-NP.

showed the CBSA-NP had a round and regular shape with a mean diameter around 100 nm. To evaluate the effects of brain delivery, BSA conjugated with pegylated nanoparticles (BSA-NP) was used as the control group and 6-coumarin was incorporated into the nanoparticles as the fluorescent probe. The qualitative and quantitative results of CBSA-NP uptake experiment compared with those of BSA-NP showed that rat brain capillary endothelial cells (BCECs) took in much more CBSA-NP than BSA-NP at 37.8 °C, incubating at different concentrations and time. The significant results *in vitro* and *in vivo* showed that CBSA-NP was a promising brain drug delivery carrier with low toxicity. Furthermore, this study for the first time presented a new brain drug delivery system with CBSA as a brain specific target covalently conjugated with the maleimide function group at the distal of PEG surrounding the nanoparticles.

17.3.2 Active targeting liposomes

Recently, immunoliposomes, which are the brain-specific targets, were covalently conjugated to PEG-modified liposomes (pegylated liposomes) as drug carrier via the tips of its functional PEG strands, proved to be successful in brain drug delivery. Figure 17.7 shows its transcytosis procedure. Its advantages over the drug-target direct combination technique are the larger drug loading capacity, disguise of limiting characteristics of drugs with physical nature of the liposome and reduction of drug degradation *in vivo*. The surface modification of PEG enable the liposomes to escape the arrest of mononuclear phagocytic system (MPS) so as to prolong its half-life in plasma and increase the area under the concentration–time curve (AUC). Huwyler *et al.* (1996) developed mouse monoclonal antibody against the rat transferrin receptor, OX26, coupled with pegylated liposome to deliver the drug into the CNS through receptor-mediated transport (RMT) process. This immunoliposome succeeded in the delivery of small molecules such as daunomycin and plasmid DNA.

In research, Jain *et al.* (2003) first prepared negatively charged magnetic liposomes using soya lecithin (Soya PC), cholesterol and phosphatidyl serine (PS). Then small peptide domain Arg-Gly-Asp (RGD) was covalently coupled to the negatively charged liposomes via carbodiimide-mediated coupling. Results suggest that selective uptake of RGD-anchored magnetic liposomes by these cells imparts them magnetic property. A high level of model drug

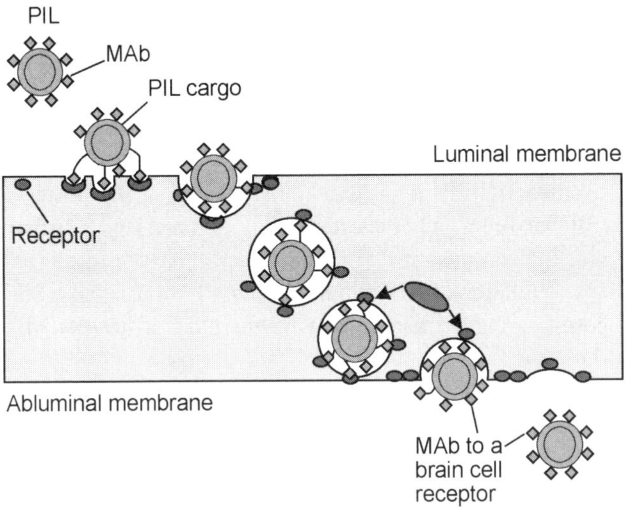

17.7 The putative multiple-receptor-binding model of polyethylene glycol (PEG)-coated immunoliposome (PIL) transcytosis. Multiple PEG-linked MAbs bind to their receptor targets on the luminal capillary membrane before the PIL undergoes endocytosis. Electron micrographs of lymphocyte transcytosis show that the vesicular membrane may be in contact with the liposome border. Exocytosis at the abluminal membrane is thought to be energy dependent. A second MAb may be attached to some of the PEG strands to target the PIL to a second receptor located on specific cells within the brain.

diclofenac sodium was quantified in target organ brain. In the case of negatively charged, uncoated magnetic liposomes, the brain levels of the drug were 5.95 times larger compared with the free drug and 7.58 times larger in comparison to non-magnetic formulation, while for RGD-coated magnetic liposomes this ratio was 9.1 times larger compared with the free drug solution, 6.62 times larger compared with non-magnetic RGD-coated liposomes and 1.5 times larger when compared with uncoated magnetic liposomes. Liver uptake was significantly bypassed (37.2% and 48.3% for uncoated and RGD-coated magnetic liposomes, respectively). This study suggests the potential of negatively charged and RGD-coated magnetic liposomes for monocyte/neutrophil-mediated active delivery of drugs to relatively inaccessible inflammatory sites, i.e. the brain. The study opens a new perspective of active delivery of drugs for a possible treatment of cerebrovascular diseases.

Glucose analogs are theoretically good candidates for drug transport through the BBB. Indeed, the large and uninterrupted energetic demand of the brain is provided almost exclusively by β-D-glucose. Furthermore, the glucose consumption of the brain amounts to about 30% of the total body glucose consumption. This high level of cerebral glucose uptake suggests that the facilitative β-D-glucose transporter GLUT1 might be a useful carrier for efficient and selective glucose-targeted drug delivery to the brain. Chemotherapeutic agents encapsulation in polysaccharide anchored liposomes could be delivered selectively to the site. Dufes *et al.* (2004) evaluated glucose-bearing niosomes as a brain targeted delivery system for the vasoactive intestinal peptide (VIP) by intravenous administration of VIP in solution or encapsulated in glucose-bearing niosomes or in control niosomes. Brain distribution of intact VIP after injection of glucose-bearing niosomes indicated that radioactivity was preferentially located in the posterior and the anterior parts of the brain, whereas it was homogeneously distributed in the whole brain after the administration of control vesicles. The results showed this novel vesicular formulation of VIP delivers intact VIP to particular brain regions in mice. Glucose-bearing vesicles might be therefore a novel tool to deliver drugs across the BBB.

Yagi and coworkers (1994) have developed and engineered liposomal constructer for brain targeting at human glioma. They employed sulfatides and MoAb as site directing devices to endow targetability to the liposomes. Targeted chemotherapy of brain tumor using polysaccharide-anchored liposomes loaded with the antitumor drug cisplation has been attempted by Ochi *et al.* (1990).

17.3.3 Hydrogel

The capsule wall contains 93% (w/w) water and can be classified as a hydrogel. Many hydrogels have gained general acceptance as being biocompatible materials. Furthermore, the microencapsulation technique can be utilized for cell immobilization and drug delivery systems.

A new family of nanoscale materials on the basis of dispersed networks of crosslinked ionic and non-ionic hydrophilic polymers is being developed. One example is the nanosized cationic network of crosslinked polyethylene oxide (PEO) and polyethyleneimine (PEI), PEO-cl-PEI nanogel. Interaction of anionic amphiphilic molecules or oligonucleotides with PEO-cl-PEI could result in formation of nanocomposite materials. Inside the material, the hydrophobic regions from polyion complexes are joined by the hydrophilic PEO chains. Formation of polyion complexes leads to the collapse of the dispersed gel particles. These kinds of system allow for immobilization of negatively charged biologically active compounds such as retinoic acid, indomethacin and oligonucleotides (bound to polycation chains) or hydrophobic molecules (incorporated into non-polar regions of polyion-surfactant complexes).

Furthermore, the nanogel particles carrying biological active compounds have been modified with polypeptide ligands to enhance receptor-mediated delivery. Efficient cellular uptake and intracellular release of oligonucleotides immobilized in PEO-cl-PEI nanogel have been demonstrated. Antisense activity of an oligonucleotide in a cell model was elevated as a result of formulation of oligonucleotide with the nanogel. This delivery system has a potential of enhancing oral and brain bioavailability of oligonucleotides as demonstrated using polarized epithelial and brain microvessel endothelial cell monolayers.

A multi-component drug delivery system that closely mimics the secretory granule, the lumen of which is composed of a crosslinked poly-anionic condensed polymer network encapsulated within a lipid membrane, was studied by Kiser *et al.* (2000). This lipid-coated hydrogel microparticle (microgel) was triggered to release doxorubicin content by using electroporation. When the lipid-coated microgels were electroporated in a saline solution, they swelled and disrupted their bilayer coating over a period of several seconds and exchanged doxorubicin with the external plasma saline over a period of several minutes.

17.3.4 Intranasal

Delivery of a drug to a required site in the body is an important application of a drug delivery system. It thus becomes possible to increase drug efficacy and decrease systemic adverse reactions. Direct administration of a drug to the affected site is the simplest and surest method. However, when a drug is injected in aqueous solution, its efficacy is quickly lost through dispersion from the site of application, in many cases making it impossible to obtain sufficient effects. Therefore, it is important to retain the drug at the affected site for the period required for onset of drug efficacy. In particular, at sites where frequent administration is difficult, it is essential to use a sustained release preparation that can maintain required local drug concentration.

The olfactory region of the nasal passages has unique anatomic and physiological attributes that provide both extracellular and intracellular pathways into the CNS that bypass the BBB. Olfactory sensory neurons are the only first-order neurons whose cell bodies are located in a distal epithelium. A direct extracellular pathway between the nasal passages and the brain was first conclusively demonstrated for horseradish peroxidase (HRP), a 40 kDa protein tracer. Electron microscopy showed that intranasally administered HRP migrates through open intercellular clefts of the olfactory epithelium to the olfactory bulbs of mice, rats and squirrel monkeys within minutes after application. Conversely, the intracellular pathway from the nasal passages to the brain has been demonstrated most convincingly for the lectin conjugate wheatgerm agglutinin-HRP (WGA-HRP). Additional support for the existence of a direct pathway connecting a submucosal compartment in the nasal passages to brain interstitial fluid or CSF has come from the demonstration that a substantial fraction of large molecular weight molecules are cleared from the CNS directly into the deep cervical lymph nodes, which receive afferent lymphatics from the nasal passages. Recent studies have shown nerve growth factor, fibroblast growth factor-2, insulin, vasoactive intestinal peptide and growth factor analogs are able to gain access to or have effects in brain tissue or CSF following intranasal administration. However, the precise mechanisms underlying nose to brain transport remain incompletely understood and have led to differing interpretations of experimental evidence for CNS delivery by the nasal route. Nevertheless, intranasal administration is associated with several advantages (non-invasiveness, ease of application, rapid termination of effects in the event of adverse reaction and avoidance of hepatic first-pass elimination) that encourage its study as a viable strategy for delivering proteins such as neurotrophic factors into the CNS.

Pharmaceutical applications of chitosan in the form of beads, microspheres and microcapsules were developed in the early 1990s. Large chitosan microspheres and beads have typically been used for the prolonged release of drugs and proteins such as BSA, DNA and brain-derived neurotrophic factor. Small particle size chitosan microspheres which containing anticancer agents such as 5-fluorouracil (5-FU) has been described for site-specific delivery. 5-Fluorouracil-loaded chitosan microspheres were prepared by Zheng *et al.* (2004) for intranasal administration. They used the liquid paraffin as the oil phase, and span 80 as the emuifier; 5-fluorouracil-loaded chitosan microspheres were achieved by the emulsion chemical crosslink technique. Microspheres have good shape and narrow size distribution. The drug release profile *in vitro* could be described by the Higuichi equation. The results show that chitosan is a good material for nasal preparation and has prospective development in the pharmaceutical field.

17.4 Future trends

The delivery of therapeutic molecules into the BBB has proven to be a major obstacle in treating brain disorders. In this chapter, many non-invasive CNS delivery techniques increasing brain uptake have been described. Coupling nanoparticle/liposome with the specific brain transport target results in absorptive mediated transcytosis or receptor mediated transcytosis. Applications of some carriers are shown in Table 17.1. Up to now, the strategies to realize the active brain targeting by drug delivery system have been focused on the therapeutic applications, such as thiamine-coated doxorubicin nanoparticles, ferritin IA modified cisplatin liposomes and 5-FU chitosan microspheres for brain cancer therapy. Moreover, amphotericin B liposomes modified by RMP-7 could be applied to meningitis therapy. There are also some reports about antisense gene therapy and other therapeutic agents such as anodyne active delivering to brain. In addition to therapeutic applications, there are other uses of these systems. For example, immunoliposomes might be used as diagnostic tools to localize tumor tissue or amyloid plaques in Alzheimer's disease. Such applications rely on brain delivery of quantitative amounts of contrast agents such as magnetoferritin or gadolinium. However, these techniques are complicated because of their chemical modifications and limited by their lower drug carrying capacity.

While improved brain delivery has been demonstrated, the mechanism of transport such as NP-BBB circumvention is still theoretical. Further studies are needed to establish the possible application of such systems in targeting of drugs

Table 17.1 Applications of some active brain targeting carriers

Applications	Surface ligand	Delivered drug	Subject	Results
Nanoparticles				
Tumor especially glioblastoma tumor	Polysorbate 80	DOX	Rats with glioblastoma 101/8	High brain drug concentration
Human breast cancer	Thiamine and radiolabeled	–	Balb/c mice	Associated with the BBB thiamine transporter
–	Cationic albumin and pegylated	–	Mice	A higher accumulation of CBSA-NP in the lateral ventricle, third ventricle and periventricular region than that of BSA-NP
Antinociceptive effect application	Polysorbate 80	Hexapeptide and dipeptide	Mice	Higher antinociceptive activity than solution
Antidepressant	Polysorbate 80	Amitriptyline	–	Improvement in brain AUC following intravenous injection
Microspheres				
–	Polysorbate 80 and magnetic	Dextran and cationic aminodextran	Male fisher 344 rats bearing rg-2 tumors	The tissue distribution results show significantly higher concentration in brain tumor
Active targeting liposomes				
Gene therapy	OX26	–	–	Optimum brain delivery
–	β-galactosidase	Therapeutic DNA	Rats	The galactosidase was highly expressed throughout the brain
Malignant brain tumours	Transferrin	–	Human	About half the patients showed tumor responses
–	PEG-coated	–	Human brain tumor model	Transport across the several biological membranes

Table 17.1 Continued

Applications	Surface ligand	Delivered drug	Subject	Results
Antitumor	–	EGF-receptor antisense gene	Mice with intracranial U87 human gliomas	The presence of brain tumors in surviving mice was confirmed at autopsy, and massive tumor growth was confirmed in the control mice
–	RGD peptide	Diclofenac sodium	Albino rats	High levels of diclofenac sodium in target organ brain
Niosomes				
Alzheimer's disease	Glucose	VIP	Male swiss mice	Brain distribution of intact VIP after injection of glucose-bearing niosomes
Nanosized cationic hydrogels				
–	Polypeptide	Retinoic acid, indomethacin	–	A potential of enhancing oral and brain bioavailability of oligonucleotides
Intranasal administration				
Neurodegenerative diseases or acute CNS injury	–	Insulin-like growth factor-I (IGF-I)	Rat	IGF-I results in rapid delivery to multiple areas of the CNS along olfactory and trigeminal pathways
Stroke, Alzheimer's disease, or traumatic brain or spinal cord injury	Spray	Melanocortin vasopressin	Human	Peptide concentrations were increased in the cerebrospinal fluid within minutes after nasal administration, with the exception of vasopressin
Boron drug-bound carriers				
Human Glioma Neutron Capture Therapy (BNCT)	–	LDL receptor-related protein	–	The implanted polymer device controls and optimizes the release of drug

to poorly accessible tumor sites. Moreover, studies are also required for effective quantitation and optimization of charge and ligand (such as RGD) on to the carrier surface. The strategy is seemingly useful in two ways: first, it negotiates targeting of drug and, second, it leads to the recruitment of leukocytes at the site of infection to combat infection by playing a major role in natural body defense mechanism. Many drugs may become effective for the treatment of brain diseases if strategies for brain delivery are to be developed efficiently.

17.5 Sources of further information and advice

Recently, in various strategies of brain drug delivery, nanoparticle transport across the BBB has been discussed even more. One of the advantages of nanoparticles over microparticles is their larger surface area for the same weight or volume. This facilitates the drug release from the nanoparticles resulting from either surface diffusion or surface erosion.

Key features of the drug-loaded nanoparticles in relation to movement across the BBB include the size and size distribution, surface and bulk morphology, surface chemistry, surface charge, drug encapsulation efficiency, physical and chemical status of the drug within the polymeric matrix, etc. Efficacy and transport mechanisms for nanoparticle delivery of chemotherapeutic agents and limitations due to physiological factors such as phagocytic activity of the reticuloendothelial system and protein opsonization were also discussed. In the past decade, effects of various drug-loaded, surfactant-coated nanoparticles of different polymers on the BBB have been investigated. *In vivo* experiments have shown that nanoparticle technology appears to have significant promise in delivering therapeutic molecules across the BBB.

17.6 References

Birrenbach, G., Speiser, P.P. Polymerized micelles and their use as adjuvants in immunology. *J. Pharm. Sci.*, 1976, 65: 1763–1766.

Bogunia-Kubik, K., Sugisaka, M. From molecular biology to nanotechnology and nanomedicine. *Biosystems*, 2002, 65: 123–138.

Chang, T.M. Pharmaceutical and therapeutic applications of artificial cells including microencapsulation. *Europ. J. Pharm. Biopharm.*, 1998, 45: 3–8.

Cornford, E.M., Cornford, M.E. New systems for delivery of drugs to the brain in neurological disease. *Lancet Neurol.*, 2002, 1(5): 306–315.

Dufes, C., Gaillard, F., Uchegbu, I.F. *et al.* Glucose-targeted niosomes deliver vasoactive intestinal peptide (VIP) to the brain. *Int. J. Pharm.*, 2004, 285(1–2): 77–85.

Gessner, A., Olbrich, C., Schroder, W., Kayser, O., Muller, R.H. The role of plasma proteins in brain targeting: species dependent protein adsorption patterns on brain-specific lipid drug conjugate (LDC) nanoparticles. *Int. J. Pharm.*, 2001, 214: 87–91.

Huwyler, J., Wu, D., Pardridge, W.M. Brain drug delivery of small molecules using immunoliposomes. *Proc. Natl. Acad. Sci. USA*, 1996, 93: 14164–14169.

Illum, L. Transport of drugs from the nasal cavity to the central nervous system. *Eur. J. Pharm. Sci.*, 2000, 11: 1–18.

Jain, S., Mishra, V., Singh, P. *et al.* RGD-anchored magnetic liposomes for monocytes/neutrophils-mediated brain targeting. *Int. J. Pharm.*, 2003, 261(1–2): 43–55.

King, T.W., Patrick, C.W. Jr. Development and *in vitro* characterization of vascular endothelial growth factor (VEGF)-loaded poly(DL-lacticco-glycolic acid)/poly(ethylene glycol) microspheres using a solid encapsulation/single emulsion/solvent extraction technique. *J. Biomed. Mater. Res.*, 2000, 51: 383–390.

Kiser, P.F., Wilson, G., Needham, D. Lipid-coated microgels for the triggered release of doxorubicin. *J. Control. Release*, 2000, 68: 9–22.

Kreuter, J., Nanoparticulate systems for brain delivery of drugs. *Adv. Drug Deliv. Rev.*, 2001, 47: 65–81.

Kreuter, J. Application of nanoparticles for the delivery of drugs to the brain. *International Congress Series*, 2005, 1277: 85–94.

Kreuter, J., Ramge, P., Petroy, V., *et al.* Direct evidence that polysorbate-80-coated poly (butylcyanoacrylate) nanoparticles deliver drugs to the CNS via specific mechanisms requiring prior binding of drug to thc nanoparticlcs. *Pharm Res.*, 2003, 20(3): 409–416.

Li, R.H. Materials for immunoisolated cell transplantation. *Adv. Drug Deliv. Rev.*, 1998, 33(1–2): 87–109.

Lockman, P.R., Oyewumi, M.O., Koziara, J.M. *et al.* Brain uptake of thiamine-coated nanoparticles. *J. Control. Release*, 2003, 93: 271–282.

Lu, W., Zhang, Y., Tan, Y-Z., Hu, K.L., *et al.* Cationic albumin-conjugated pegylated nanoparticles as novel drug carrier for brain delivery. *J. Control. Release*, 2005, 17: 428–448.

Mittal, S., Cohen, A., Maysinger, D. *In vitro* effects of brain derived neurotrophic factor released from microspheres. *Neuroreport*, 1994, 5: 2577–2582.

Noble, C.O., Kirpotin, D.B., Hayes, M.E., Mamot, C., Hong, K., Park, J.W., Benz, C.C., Marks, J.D., Drummond, D.C. Development of ligand targeted liposomes for cancer therapy. *Expert. Opin. Ther. Targets*, 2004, 8: 335–353.

Ochi, A., Shibata, S., Mori, K., Sato, T., Sunamoto, J. Targeting chemotherapy of brain tumor using liposome encapsulated cisplatin. Part 2: Pullulan anchored liposomes to target brain tumor. *Drug Delivery Syst.*, 1990, 5: 261–271.

Pettit, D.K., Gombotz, W.R. The development of site-specific drug-delivery systems for protein and peptide biopharmaceuticals. *Trends Biotechnol.*, 1998, 16: 343–349.

Rapoport, S.I. Modulation of the blood-brain barrier permeability, *J. Drug Target.*, 1996, 3: 417–425.

Roco, M.C. Nanotechnology: convergence with modern biology and medicine. *Curr. Opin. Biotechnol.*, 2003, 14: 337–346.

Schroeder, U., Sommerfeld, P., Ulrich, S., Sabel, B.A. Nanoparticle technology for delivery of drugs across the blood–brain barrier. *J. Pharm. Sci.*, 1998, 87: 1305–1307.

Shi, N., Pardridge, W.M. Noninvasive gene targeting to the brain. *Proc. Natl. Acad. Sci. USA*, 2000, 97: 7567–7572.

Shi, N., Zhang, Y., Zhu, C., Boado, R.J., Pardridge, W.M. Brain-specific expression of an exogenous gene after i.v. administration. *Proc. Natl. Acad. Sci. USA*, 2001, 98: 12754–12759.

Soppimath, K.S., Aminabhavi, T.M., Kulkarni, A.R., Rudzinski, W.E. Biodegradable polymeric nanoparticles as drug delivery devices. *J. Control. Release*, 2001, 70: 1–20.

Szoka, F. Jr., Papahadjopoulos, D. Procedure for preparation of liposomes with large internal aqueous space and high capture by reverse-phase evaporation. *Proc. Natl. Acad. Sci. USA*, 1978, 75: 4194–4198.

Vinogradov, S.V., Bronich, T.K., Kabanov, A.V. Nanosized cationic hydrogels for drug delivery: preparation, properties and interactions with cells. *Adv. Drug Deliv. Rev.*, 2002, 54(1): 135–147.

Yagi, K., Hayashi, Y., Ishida, N. *et al.* Interferon-beta endogenously produced by intratumoral injection of cationic liposome-encapsulated gene: cytocidal effect on glioma transplanted into nude mouse brain. *Biochem. Mol. Biol. Int.*, 1994, 32(1): 161–171.

Zhang, Y., Zhu, C., Pardridge, W.M. Antisense gene therapy of brain cancer with an artificial virus gene delivery system. *Molec. Ther.*, 2002, 6: 67–72.

Zhang, Y., Calon, F., Zhu, C., Boado, R.J., Pardridge, W.M. Intravenous nonviral gene therapy causes normalization of striatal tyrosine hydroxylase and reversal of motor impairment in experimental parkinsonism. *Hum. Gene Ther.*, 2003, 14: 1–12.

Zhang, Y., Schlachetzki, F., Zhang, Y.F., Boado, R.J., Pardridge, W.M. Normalization of striatal tyrosine hydroxylase and reversal of motor impairment in experimental parkinsonism with intravenous nonviral gene therapy and a brain-specific promoter. *Hum. Gene Ther.*, 2004, 15: 339–350.

Zhang, X.B., Yuan, S., Lei, P.C., Hou, X.P. Therapeutic efficiency of amphotericin B liposmome modified by RMP-7 to transport drug across blood brain barrier. *Acta Pharmaceutica Sinica*, 2004, 39(4): 292–295.

Zheng A.P., Yushao, S.Y., Zhao, Y. *et al.* Preparation and properties of 5-fluorouracil loaded chitosan microspheres for the intra nasal administration. *J. Peking Univ. (Healthsciences)*, 2004, 36(3): 300–304.

18

Artificial cells in enzyme therapy with emphasis on tyrosinase for melanoma in mice

B YU and T M S CHANG, McGill University, Canada

18.1 Introduction

The first report of artificial cells containing enzymes, intact cells, haemoglobin and other biologically active materials was published by Chang in 1964.[1] This has since then been extensively developed for blood substitutes, enzyme therapy, cell therapy, drug delivery and many other areas.[2–6] More recently, it has been extensively developed for blood substitutes and stem cell therapy.[5,6] This chapter describes the use of artificial cells in enzyme therapy with emphasis on artificial cells containing tyrosinase for melanoma as an example. In the case of enzyme therapy Chang's group started with artificial cells containing urease for the removal of systemic urea.[1,2,7,8] They showed that this is effective in removing systemic urea using implantation,[1,2] extraporeal hemoperfusion[7] and oral administration.[2,8] The oral administration approach was later developed by another group for clinical trial in patients.[9] This basic research was followed by further research by Chang's group showing that implanted artificial cells containing catalase can supplement the deficient enzyme in acatalesemic mice, animals with an inborn error in metabolism with deficiency in catalase.[10,11] This replaces the deficient enzymes and protects the animals from the damaging effects of oxidants. The artificial cells protect the enclosed enzyme from immunological reactions.[11]

Chang also showed that artificial cells containing asparaginase implanted into mice with lymphosarcoma delayed the onset and growth of lymphosarcoma.[12–14] This result led to a number of other studies using microencapsulated asparaginase.[15,16] For those conditions that need long-term enzyme therapy as in inborn errors of metabolism, the single problem preventing the clinical application of enzyme artificial cells is the need to repeatedly inject these enzyme artificial cells. To solve this problem oral administration of microencapsulated xanthine oxidase was used as an experimental therapy in a patient with Lesch–Nyhan disease.[17,18] In addition, it was found that microencapsulated phenylalanine ammonia lyase given orally can lower the elevated phenylalanine levels in phenylketonuria (PKU) rats.[19–22] This is because of a more recent finding from

Chang's group of an extensive recycling of amino acids between the body and the intestine.[23] This resulted in their proposal of the entero-recirculation theory[23] that allows for the oral administration of enzyme artificial cells to interrupt the entero-recirculation of a specific amino acid. Therefore, it solves the problem of the requirement for repeated injections. This is now being developed for clinical trials in PKU.

The rest of this review describes in detail our studies on the use of tyrosinase artificial cells of microscopic dimensions for oral administration and nano-dimension artificial cells for intravenous injection for the removal of systemic tyrosine for murine melanoma, a skin cancer.

18.2 Tyrosinase artificial cells and melanoma: background

Chang's group has carried out many studies on the use of tyrosinase artificial cells for the removal of systemic tyrosine as part of a liver support system.[24–26] The reason for the more recent studies on the use of tyrosinase artificial cells is as follows. At present, there is no effective treatment for melanoma, a fatal skin cancer, which is now the fifth most common type of cancer in North America. One unique characteristic of melanoma cells is that they need a higher concentration of tyrosine for growth than that for normal cells.[27] Dietary tyrosine restriction lowers systemic tyrosine and suppresses the growth of melanoma in mice, but this is not tolerated by humans resulting in nausea, vomiting and severe weight loss.[28,29] Other antitumor agents specific for malignant melanoma have been extensively studied by many research groups, such as 4-*S*-cysteaminylphenol, nitrogen mustard, melanoma-associated antigens, etc.[30–32] However, these agents have either severe side effects, are short-lasting *in vivo* or have limitations in drug-selective liberation. On the other hand, it is well documented that artificial cells containing enzymes can selectively remove amino acids when this specific amino acid is used as a substrate in the enzyme reactions.[14,23,24] In this chapter, we report the successful use of oral administration of polymeric microcapsules containing tyrosinase and intravenous injection of nanodimension polyhemoglobin–tyrosinase on lowering systemic tyrosine level in a melanoma animal model.

18.3 *In vitro* and *in vivo* enzyme kinetics of artificial cells encapsulated tyrosinase

In our earlier studies, we tested the enzyme's kinetics and its stability at different pH and temperature.[33] Our results showed that microencapsulated tyrosinase can catalyze tyrosine reaction as effectively as the free enzyme. In addition, the encapsulated enzyme is much more stable at different pH and body temperature than the free enzyme form because of the protection of the ultra-thin membrane.

In subsequent studies, we investigated the effects of encapsulated tyrosinase administered orally on lowering systemic tyrosine level in rats.[34] In this study, we used normal rats as the animal model, and orally fed them with different dosages of artificial cells containing tyrosinase for up to 22 days. We found that one dose a day was not effective in lowering plasma tyrosine level. In two doses daily, we observed that the systemic tyrosine level started to decrease after 1 week of oral administration. To optimize and increase the rate of removal of systemic tyrosine, we tried dosing three times a day. Our results showed that three doses daily can markedly lower systemic tyrosine level from day 4 to a level that would inhibit the growth of melanoma. Moreover, we monitored the body weight gain of all animals during the whole experiment period. No abnormal effect or behavior was observed in any animal group.

18.4 *In vitro* and *in vivo* enzyme kinetics of nanodimension artificial cells consisting of polyhemoglobin–tyrosinase

However, this oral administration of encapsulated tyrosinase in enzyme therapy takes a few days to reach a low tyrosine level in the body system. We tried to find a way to lower systemic tyrosine level as soon as possible once the enzyme had been delivered. In recent studies, it was found that intravenous injection of polyhemoglobin–tyrosinase (PolyHb–tyrosinase) could decrease tyrosine level within an hour. Before we have gone to any further studies in animal models, we carried out a series of enzyme studies *in vitro* and *in vivo* of PolyHb–tyrosinase.[35] In the crosslinking time study, we found that a longer crosslinking time results in more PolyHb–tyrosinase of high molecular weight (>100 kDa). The crosslinking time from 3.5 to 48 h had no adverse effect on the tyrosinase activity. At 37 C, PolyHb–tyrosinase retained a higher activity than free enzyme. This is because high concentrations of hemoglobin stabilize enzymes at 37 °C. Furthermore, PolyHb–tyrosinase has similar oxygen affinity characteristics (P_{50} = 21 mmHg) to non-crosslinked hemoglobin (P_{50} = 23 mmHg). This characteristic of PolyHb is very important for radiotherapy – being a solution, PolyHb–tyrosinase can easily reach the narrower capillaries of the melanoma to supply the additional oxygen needed in radiotherapy.

We next carried out a preliminary animal study on the effect of an intravenous injection of PolyHb–tyrosinase to lower the systemic tyrosine level. Our results showed that higher volumes of PolyHb–tyrosinase injection yielded the lowest level of systemic tyrosine within 30 min. However, the animals dramatically lost weight owing to severe starvation for tyrosine.[36] Therefore, we conclude that the optimal preparation is prepared from a crosslinking time of 24 h using 1 ml of PolyHb–tyrosinase for intravenous injection. This way it decreases the systemic tyrosine level without any adverse effect on maintaining body weight. The injection of PolyHb–tyrosinase could provide a potential promising approach to lower systemic tyrosine in malignant melanoma.

18.5 Effect of PolyHb–tyrosinase on melanoma mice model

In the following studies, we investigated the effect of PolyHb–tyrosinase on decreasing systemic tyrosine level in melanoma cell culture and melanoma mice model.[37,38] In the cell culture study (Table 18.1), we found that free tyrosinase and PolyHb–tyrosinase solution have the same ability to inhibit the growth of melanoma cells. In contrast, in the control groups of saline and PolyHb, no adverse effect on the growth of melanoma cells was observed. The melanoma cells in the control groups kept increasing quickly.

In the melanoma mice model, we gave an intravenous injection of PolyHb–tyrosinase daily after 9 days of tumor implantation. From our results we can conclude that 4 days after a daily intravenous injection of PolyHb–tyrosinase, the tumor volume was significantly lower than that in the control group. Nineteen days after the inoculation of the B16F10 melanoma cells, the tumor volume in the test group was only 45.28 ± 10.09% of that in the control group. Thus, our results suggest that PolyHb-tyrosinase retards the growth of B16F10 melanoma in mice (Table 18.2). We also followed the body weight of all mice

Table 18.1 B16F10 melanoma cell numbers (1×10^5) in cell culture after the addition of saline, PolyHb, free tyrosinase or PolyHb–tyrosinase to the medium.

Time (day)	Saline	PolyHb	Free tyrosinase	PolyHb–tyrosinase
0	0.7 ± 0.1	0.7 ± 0.04	0.8 ± 0.05	0.8 ± 0.1
4	29.5 ± 1.7	28.6 ± 3.9	0.4 ± 0.06	0.3 ± 0.03

Source: from Yu and Chang, *Artificial Cells, Blood Substitutes, and Biotechnology*, 32(2): 293–302, 2004 (with copyright permission from Marcel Dekker Publisher).

Table 18.2 Tumor growth (mm^3) of B16F10 melanoma in mice. Sham control group: no intravenous injection; Saline group: 0.1 ml intravenous saline daily; PolyHb–tyrosinase group: 0.1 ml intravenous PolyHb–tyrosinase daily. All values are represented as mean ± SEM.

Time (day)	Sham control (mm^3)	Saline (mm^3)	PolyHb–tyrosinase (mm^3)
9	123 ± 33	125 ± 24	124 ± 35
19	3263 ± 259	3190 ± 367	1444 ± 322

during the experiment period. There was no significant difference in weight gain/loss in all mice groups.

18.6 Effects of combined methods on lowering systemic tyrosine level in rats

As mentioned in the previous study, it takes a few days for oral administration of encapsulated tyrosinase to lower systemic tyrosine level. Although intravenous injection of PolyHb–tyrosinase can decrease the systemic tyrosine level to about 10% within an hour, the level increases towards normal after 24 h. Therefore, we designed a novel method of combining artificial cells containing tyrosinase and PolyHb–tyrosinase to lower systemic tyrosine level.[39] By optimizing this combined method for its efficiency, we carried out studies in animal experiments (Figs 18.1 and 18.2). Our results showed that two intravenous injections of PolyHb–tyrosinase followed by three times a day of oral administration of

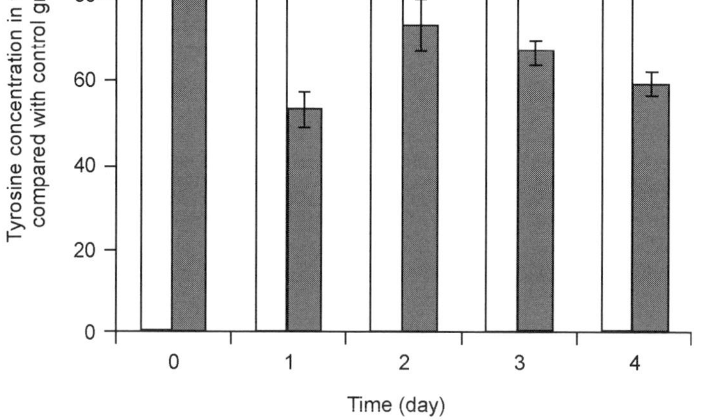

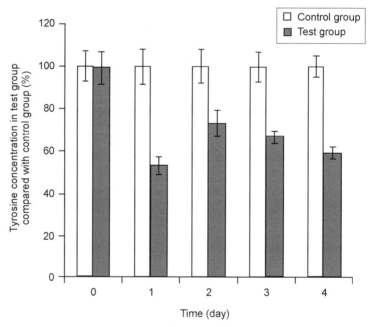

18.1 Tyrosine concentration in rat's plasma (%). For control group: oral administration of artificial cells containing no tyrosinase three times a day with one injection of 1 ml of PolyHb solution per 250 g body weight on day 1. For test group: oral administration of artificial cells containing tyrosinase three times a day with one injection of 1 ml of PolyHb–tyrosinase solution per 250 g body weight on day 1. (*Source*: from Yu and Chang, *Artificial Cells, Blood Substitutes, and Biotechnology*, 32(1): 129–148, 2004 with copyright permission from Marcel Dekker Publisher.)

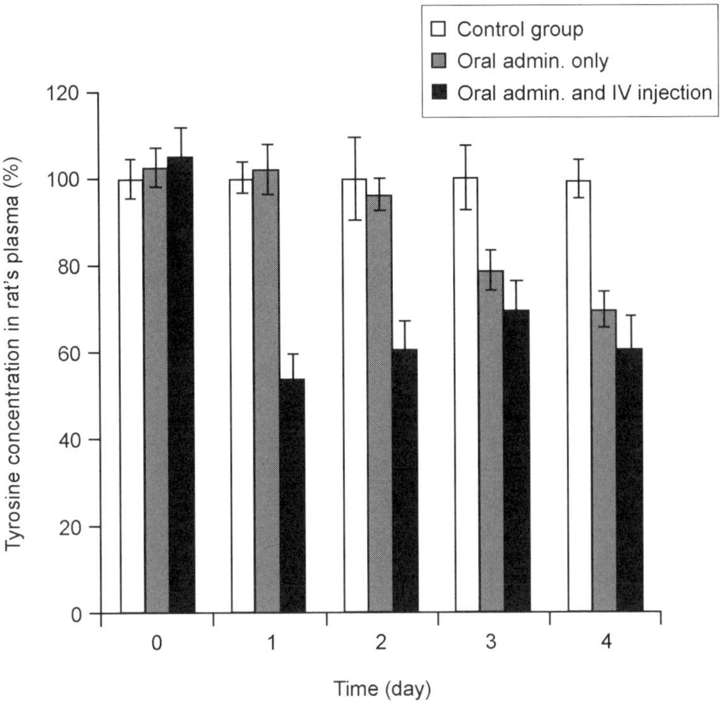

18.2 Tyrosine concentration in rat's plasma (%). For control group: oral administration of artificial cells containing no tyrosinase three times a day with one injection of 1 ml of PolyHb solution on day 1 and another half volume injection on day 2. For group of oral administration only: oral administration of artificial cells encapsulated tyrosinase three times a day. For test group of oral administration and intravenous (IV) injection: oral administration of artificial cells encapsulated tyrosinase three times a day with one injection of 1 ml PolyHb–tyrosinase solution on day 1 and another half volume injection on day 2. (*Source*: from Yu and Chang, *Artificial Cells, Blood Substitutes, and Biotechnology*, 32(1): 129–148, 2004 with copyright permission from Marcel Dekker Publisher.)

encapsulated tyrosinase could immediately lower the body tyrosine level and maintained this low level as long as the oral administration was continued.

18.7 Conclusions

It is well demonstrated that the oral administration of polymeric artificial cells containing enzymes is safe and effective, and offers several advantages over other methods.[5] The polymeric membrane artificial cells pass through the intestine and are excreted once they have carried out their functions; thus it is easier to ensure their safety in patients.

Our recent studies have shown that microencapsulated tyrosinase can effectively lower the systemic tyrosine level in rat's intestine.[34] In animal studies, we found that doses two or three times a day can reduce the systemic blood tyrosine level and keep it at that low level as long as the oral administration is continued. By introducing the concept of chemically crosslinking tyrosinase with hemoglobin (PolyHb–tyrosinase) using glutaraldehyde, we carried out extensive studies in structural and functional properties of PolyHb–tyrosinase *in vitro* and *in vivo*.[35] In a preliminary animal study of the effect of intravenous injection of PolyHb–tyrosinase, we conclude that this preparation can dramatically lower systemic tyrosine level within an hour and no adverse effect is observed on animal body weight. Our studies on the effects of this preparation on B16F10 melanoma cell culture and melanoma mice model[38] show that it not only inhibits the growth of melanoma cell culture, but also retards the growth of melanoma in B16F10 melanoma mice model by daily intravenous injection. In order to avoid the need for repeated intravenous injection we show that intravenous injection of PolyHb–tyrosinase combined with oral administration of encapsulated tyrosinase[39] can immediately lower the systemic tyrosine level with oral administration of encapsulated tyrosinase three times a day to maintain this low level. This combined method does not cause weight loss in animals, which is a main problem for dietary regimen in restriction of tyrosine. In conclusion, our research may provide a potential method to inhibit/retard the growth of malignant melanoma. However, further studies need to be carried out on the effects of encapsulated tyrosinase and PolyHb–tyrosinase on lowering tyrosine level in melanoma animal models.

18.8 Acknowledgements

The many years of operating grants from the Medical Research Council of Canada and the Canadian Institutes for Health Research to TMS Chang are gratefully acknowledged as is the MESS Virage Centre of Excellence in Biotechnology.

18.9 References

1. Chang, T.M.S. Semipermeable microcapsules. *Science* 146, 524–525, 1964.
2. Chang T.M.S. *Artificial Cells*. Charles C. Thomas, Springfield, Illinois, 1972.
3. Chang T.M.S. Therapeutic applications of polymeric artificial cells. *Nature Reviews* 4: 221–235, 2005.
4. Chang T.M.S. Artificial cells bioencapsulation in macro, micro, nano and molecular dimensions. *Artificial Cells, Blood Substitutes & Biotechnology* 32: 1–23, 2004.
5. Chang T.M.S. *Artificial Cells: Biotechnology, Nanotechnology, Blood Substitutes, Regenerative Medicine, Bioencapsulation, Cell/Stem Cell Therapy*. World Scientific Publishing, Singapore, 2007.
6. Liu Z.C., Chang T.M.S. Transdifferentiation of bioencapsulated bone marrow stem cells into hepatocyte-like cells in the 90% hepatectomized rat model. *Journal of Liver Transplantation* 12: 566–572, 2006.

7. Chang T.M.S. Semipermeable aqueous microcapsules (artificial cells): with emphasis on experiments in an extracorporeal shunt system. *Transactions – American Society for Artificial Internal Organs* 12: 13–19, 1966.
8. Chang T.M.S., Loa S.K. Urea removal by urease and ammonia absorbents in the intestine. *The Physiologist* 13(3), 70–72, 1970.
9. Gardner D.L., Falb R.D., Kim B.C., Emmerling D.C. Possible uremic detoxification via oral-ingested microcapsules. *Transactions – American Society for Artificial Internal Organs* 17: 239–245, 1971.
10. Chang T.M.S., Poznansky M.J. Semipermeable microcapsules containing catalase for enzyme replacement in acatalasaemic mice. *Nature* 218: 243–245, 1968.
11. Poznansky M.J., Chang T.M.S. Comparison of the enzyme kinetics and immunological properties of catalase immobilized by microencapsulating and catalase in free solution for enzyme replacement. *Biochimica et Biophysica Acta* 334: 103–115, 1974.
12. Chang T.M.S. The *in vivo* effects of semipermeable microcapsules containing L-asparaginase on 6C3HED lymphosarcoma. *Nature* 229, 117–118, 1971.
13. Chang T.M.S. L-Asparaginase immobilized with semipermeable microcapsules: *in vitro* and *in vivo* stability. *Enzyme* 14(2): 95–104, 1973.
14. Siu Chong E.D., Chang T.M.S. *In vivo* effects of intraperitoneally injected L-asparaginase solution and L-asparaginase immobilized within semipermeable nylon microcapsules with emphasis on blood L-asparaginase, body L-asparaginase, and plasma L-asparagine levels. *Enzyme* 18: 218–239, 1974.
15. Mori T., Tosa T., Chibata I. Enzymatic properties of microcapsules containing asparaginase. *Biochimica et Biophysica Acta* 653–661, 1973.
16. Mori T., Sato T., Matuo Y., Tosa T., Chibata I. Preparation and characteristics of microcapsules containing asparaginase. *Biotechnology and Bioengineering* 14: 663–673, 1972.
17. Chang T.M.S. Preparation and characterization of xanthine oxidase immobilized by microencapsulation in artificial cells for the removal of hypoxanthine. *Biomaterials, Artificial Cells, and Artificial Organs* 17: 611–616, 1989.
18. Palmour R.M., Goodyer P., Reade T., Chang T.M.S. Microencapsulated xanthine oxidase as experimental therapy in Lesch–Nyhan disease. *Lancet* 2(8664): 687–688, 1989.
19. Bourget L., Chang T.M.S. Artificial cell microencapsulated phenylalanine ammonia-lyase. *Applied Biochemistry and Biotechnology* 10: 57–59, 1984.
20. Bourget L., Chang T.M.S. Phenylalanine ammonia-lyase immobilized in semipermeable microcapsules for enzyme replacement in phenylketonuria. *Federation of European Biochemical Societies (FEBS Letters)* 180(1): 5–8, 1985.
21. Bourget L., Chang T.M.S. Phenylalanine ammonia-lyase immobilized in microcapsules for the depletion of phenylalanine in plasma in phenylketonuric rat model. *Biochimica et Biophysica Acta* 883: 432–438, 1986.
22. Bourget L., Chang T.M.S. Effects of oral administration of artificial cells immobilized phenylalanine ammonia-lyase on intestinal amino acids of phenylketonuric rats. *Biomaterials, Artificial Cells, and Artificial Organs* 17: 161–182, 1989.
23. Chang T.M.S., Bourget L., Lister C. A new theory of enterorecirculation of amino acids and its use for depleting unwanted amino acids using oral enzyme-artificial cells, as in removing phenylalanine in phenylketonuria. *Artificial Cells, Blood Substitutes, and Immobilization Biotechnology* 25: 1–23, 1995.
24. Shi Z.Q., Chang T.M.S. The effects of hemoperfusion using coated charcoal or

tyrosinase artificial cells on middle molecules and tyrosine in brain and serum of hepatic coma rats. *Transactions – American Society for Artificial Internal Organs* 28: 205–209, 1982.

25. Shu C.D., Chang T.M.S. Tyrosinase immobilized within artificial cells for detoxification in liver failure: I. Preparation and *in vitro* studies. *International Journal of Artificial Organs* 3(5): 287–291, 1980.
26. Shu C.D., Chang T.M.S. Tyrosinase immobilized within artificial cells for detoxification in liver failure: II. *In vivo* studies in fulminant hepatic failure rats. *International Journal of Artificial Organs* 4(2): 82–85, 1981.
27. Elmer G.W., Meadows G.G., Linden C., Digiovanni J., Holcenberg J.S. Influence of tyrosine phenolyase on growth of B-16 melanoma. *Pigment Cell* 2: 339–346, 1976.
28. Uhlenkott C.E., Huijzer J.C., Cardeiro D.J., Elstad C.A., Meadows G.G. Attachment, invasion, chemotaxis, and proteinase expression of B16-BL6 melanoma cells exhibiting a low metastatic phenotype after exposure to dietary restriction of tyrosine and phenylalanine. *Clinical and Experimental Metastasis* 14: 125–137, 1996.
29. Pelayo B.A., Fu Y.M., Meadows G.G. Decreased tissue plasminogen activator and increased activator protein-1 and specific promoter 1 are associated with inhibition of invasion in human A375 melanoma deprived of tyrosine and phenylalanine. *International Journal of Oncology* 18: 877–883, 2001.
30. Inoue S., Hasegawa K., Wakamatsu K., Ito S. Comparison of antimelanoma effects of 4-*S*-cysteaminylphenol and its homologues. *Melanoma Research* 8: 105–112, 1998.
31. Jordan A.M., Khan T.H., Osborn H., Photiou A., Riley P.A. Melanocyte-directed enzyme prodrug therapy (MDEPT): development of a targeted treatment for malignant melanoma. *Bioorganic & Medicinal Chemistry* 7: 1775–1780, 1999.
32. Castelli C., Rivoltini L., Andreola G., Carrabba M., Renkvist N., Parmiani G. T-cell recognition of melanoma-associated antigens. *Journal of Cellular Physiology* 182: 323–331, 2000.
33. Yu B., Chang T.M.S. *In vitro* enzyme kinetics of microencapsulated tyrosinase. *Artificial Cells, Blood Substitutes, and Immobilization Biotechnology* 30(5&6): 533–546, 2002.
34. Yu B., Chang T.M.S. Effects of long term oral administration of microencapsulated tyrosinase on maintaining decreased systemic tyrosine levels in rats. *Journal of Pharmaceutical Sciences* 93: 831–837, 2004.
35. Yu B., Chang T.M.S. *In vitro* and *in vivo* enzyme studies of polyhemoglobin-tyrosinase. *Biotechnology and Bioengineering* 86(7): 835–841, 2004.
36. Meadows G.G., Oeser D.E. Response of B16 melanoma-bearing mice to varying dietary levels of phenylalanine and tyrosine. *Nutrition Reports International* 28: 1073–1082, 1983.
37. Yu B., Chang T.M.S. Polyhemoglobin–tyrosinase, an oxygen carrier with murine B16F10 melanoma suppression properties: a preliminary report. *Artificial Cells, Blood Substitutes, and Biotechnology* 32(2): 293–302, 2004.
38. Yu B., Chang T.M.S. *In vitro* and *in vivo* effects of polyhemoglobin–tyrosinase on murine B16F10 melanoma. *Melanoma Research* 14(3): 197–202, 2004.
39. Yu B., Chang T.M.S. Effects of combined oral administration and intravenous injection on maintaining decreased systemic tyrosine levels in rats. *Artificial Cells, Blood Substitutes, and Biotechnology* 32(1): 129–148, 2004.

19

Stem cell and regenerative medicine: commercial and pharmaceutical implications

L EDUARDO CRUZ and S P AZEVEDO, Cryopraxis Criobiologia Limited, Brazil

19.1 Introduction

Life expectancy reached a record high of 76.9 years at birth in the United States in 2005 (Bureau of Census) and the overall life expectancy for both sexes in Canada was 79.7 years. About a third of the deaths in this country in 2002 were caused by diseases of the circulatory system. Nearly another third were caused by cancer. This proportion has been the same for several years, with deaths due to circulatory diseases decreasing and deaths due to cancer increasing. Over half of the cancer deaths were due to four types of malignant neoplasms – lung, colorectal, female breast and male prostate.[1]

Diabetes mellitus is an example of a degenerative disease that shows a rate increase with population age. The 2002 mortality statistics indicate a dramatic increase in deaths due to diabetes mellitus in the world. Deaths due to diabetes were up nearly 11% over 2001, and nearly 75% over 1992.[1]

The overall population of the world is getting older. People live longer and look forward to living better. Degenerative disease becomes a grand concern and a tremendous challenge to governments and healthcare institutions. In the United States myocardial degenerative disease and stroke are the leading cause of death and represent outstanding medicare costs. New drugs have made heart attack death rates decrease considerably in the last 30 years, but still millions of surviving patients will proceed to heart failure status (congestive heart failure, for example).[2] Tissue engineering and cell therapy will play a major role in this challenging but promising new era of citizens of a planet where being over 85 will mean having a joyful life.

Stem cell, called 'progenitor' – a direct ancestor of all cell lines, and regenerative medicine – therapeutics aimed to form or grow as a replacement for organ tissue lost, are paradoxically innovative. They might be the most relevant window of opportunity in medicare and health-related businesses at the beginning of the 21st century. The notion of off-the-shelf cell lineage flasks produced, released and available as medicines or replacement parts of human organs and tissues of synthetic, natural or semi-synthetic materials – or more recently, tissue engineered constructs – have captured the imagination of physicians, scientists, engineers, investors, pharmaceutical and healthcare private groups, governmental agencies and public health institutions.

Stem cells typically originate from embryo, fetus, umbilical cord blood, baby tooth, bone marrow, adipose tissue. Stem cells from umbilical cord, baby tooth, bone marrow and adipose tissue are called 'adult stem cell'.[3] There is still tremendous ethical debate and controversy of sacrificing tiny human embryos in order to collect the pluripotent stem cells for other therapeutic applications. Recent data shows enormous progress in coaxing the same life-saving potential out of stem cells that have been harmlessly taken from adult donors.[4] The collection of umbilical cord blood, for instance, risks no harm to the patient or the baby. Already, researchers used stem cells to successfully grow, in experimental preclinical models, new spinal tissue in dogs. Several phase 1 clinical trials have been reported using cell therapy in human patients with spinal cord injuries.[5] Still there is an ethical debate of whether or not to use embryonic stem cell lines. Aside, there are many reasons to focus on adult stem cells for tissue engineering. Unlike embryonic stem cells, adult stem cells can be taken from the patient him- or herself, or be harvested from human umbilical cord blood (hUCB).[6] When for autologous usage there is no fear that the body will reject the new tissues, and some protocols have shown promising immunogenic characteristics of hUCB. Indeed tissue grown from these adult stem cells may prove safer in the long run because it is more predictable and less likely to turn cancerous – a proven real risk when embryonic stem cell is cultivated.[7]

It seems that to begin with, the adult stem cells will lead the race over embryonic ones, as they are less controversial, technically safe and feasible, even though somewhat a less powerful alternative (?). The question mark is a necessary linguistic resource here, as some interesting studies recently suggested the advantages of hUCB or even of those progenitor cells collected from baby teeth or other sources.

In 2004, Sanberg and coworkers showed that an infusion of hUCB is 'dose-dependent' when rescuing behavioral deficits in a rat model of stroke and that it also reduced infarct volume.[8] That work might provide vital information leading to product or protocol for clinical application.

Biomaterials, bone substitutes, tissue engineering (TE) and cell therapy of autologous, allogeneic and/or artificial cell origins, although some dated back to the beginning of the 1980s, are in their debut as an industry. Just a few products or validated medical procedures have reached the market. As a whole it was, for many years, considered to be a sub-sector of the health industry combining elements of medical device, pharmaceuticals, biotechnology and tissue manipulation. Basically, start-ups were the only business category dedicated to this field.

Stem cells are being studied for cardiac, liver, lung, pancreas and neuron regeneration. Tissue engineering and regenerative medicine is an emerging intersecting ground of research, repair or replacement of failing tissues.

What it means is the application of the advanced biological knowledge of cell plasticity, tissues and organs behavior under stress or aging; and the potential of engineered materials of synthetic or natural origin to interact with the body and remold, repair and regenerate lesion areas. And that whether a product or a medical maneuver, it has to prove it is a safe, reproducible, validated and standard operational procedure – something that could be named Tissue and Cell Engineering for Therapeutics (TCEt).

This industry covers cell tissue engineered products, biomaterials science and technology, allograft-derived tissue substitutes, cell collection, transportation, manipulation, testing, culturing, storing and thawing. It goes as far as genetics. In all senses it comprises the challenges of bridging the safety gap that has always being there when a chemical entity is delivered to a biological environment. Yes, TCEt has to be proved to be as good as or safer than what has been the state-of-art in therapeutics: a chemical compound.

Most of the companies involved in TCEt are small, highly sophisticated in terms of qualified staff, and privately owned and funded. It is known that an innovative product will demand several millions of dollars and take from 8 to 10 years from laboratory to the bedside. Somehow in the early years of the 21st century larger pharmaceutical groups turned their attention to this promising land.

19.2 Impact of innovative technologies in healthcare

Innovative technology and, basically, medications have always played an important position in patient care. In the last decade, the US Food and Drug Administration (FDA) alone approved more than 300 new pharmaceutical compounds.[9] As a result of new medications and technologies, patients live longer and better. The overall population of the elderly has increased over the past years. Diseases such as diabetes, stroke and neurological disorders, cancer and heart disease will be an even more important factor in medicare. Several new medicines have been considered for cancer, heart, diabetes and even rare disorders such as cystic fibrosis and sickle cell anemia, although there have not been so many for stroke or neurological lesions.[10]

It takes a lot of time and investment for an innovation to become a generally accepted practice by physicians. Pharmaceutical companies have long acknowledged the simple fact that medical doctors, mainly key opinion leaders, will not exploit a new therapeutic tool unless it is proven safe and scientifically consistent. In addition, healthcare interventions that improve clinical outcomes are achieved at some cost. Tissue engineering and cell therapy will have to prove good enough to be reimbursed if it is to be evaluated as an evidence-based care practice that improves management of acute and chronic disease.

As an example of, on the one hand, the importance of validated new medical routines and also, on the other hand, the difficulties in maintaining them as a current policy, beta-blocker therapy – a very simple and well-known medicine – resulted in increased survival of 0.3 years per patient, reduced societal costs by $3959 per patient over 5 years and medicare costs declined by $6064 due primarily to lower hospitalization rates.[11] Using this perspective, medicare would gain the most if more heart failure patients were treated with beta-blockers. Effective and economically feasible methods of improving prescribing patterns in the community have not been established and have proved dependent upon educational interventions, monitoring and feedback activities. Last but not least, the limiting factor seems to be the increase in complexity of outpatient management. There is the question of what to do as to the clinical and economic consequences of chronic heart failure when 50% of the patients die within 5 years[12] and that more and more patients survive a myocardial infarction due to better pharmacologic treatments. But some studies suggest that treatments may be cost-effective once adopted, but less economically attractive when costs of achieving higher rates of adoption and adherence are considered. Cardiac diseases correspond to more than 30% of death rates all over the world.[13] Given these insights on the challenges of utilizing a very well-known and safe drug such as beta-blockers, one can imagine the hurdles faced when launching an absolutely new concept.

Pharmaceutical science and technology have evolved tremendously in the past 50 years. That, added to better education and sanitary and nutritional conditions, have resulted in an increase in those individuals of over 85 years of age. Neurological disorders, an example of a health issue that might be of even harder impact in social costs as the population gets older, already affect about 1.5 billion people worldwide and result in outstanding health expenses. Brain-related illness generates more healthcare-related costs and lost income than any other therapeutic area: an estimated $1.0 trillion worldwide and $350 billion annually only in the United States. Twenty per cent of Americans over 65 have Alzheimer's disease as well as 50% of those above 85 years of age. Alzheimer's alone accounts for a health cost in the United States of over $100 billion (Harvard Medical School data).

As of the year 2005 biopharmaceutical companies were developing about 146 medicines for heart disease and stroke. Of those, about 17 are meant for stroke, a health issue that affect about 700 000 Americans each year; 16 for congestive heart failure, which kills more than 50 000 Americans a year; 10 for high pressure, a leading risk factor for both heart disease and stroke; 8 for heart attacks; and 13 for arrhythmias.

Cutting-edge technologies might lead to new scientific approaches to dealing with the heart's metabolism to require less oxygen (reducing episodes of pain and allowing angina patients to be more active); the promotion of vessel growth (angiogenesis showed to be absolutely necessary in cell therapy approaches); or a vaccine that may be able to promote good cholesterol (2005 Survey, Medicines in Development for Heart Disease and Stroke, PhRMA). All the above will lead to more patients surviving the heart attack, or living longer and indeed increase the odds that degenerative disorders, of neurological or cardiac origin affect them.

Liver disease is dramatically on the increase – something doctors mostly blame on burgeoning lifestyles of excess. Deaths from alcoholic liver disease have doubled in the last 10 years, with figures for the condition in young people increasing eight-fold due to binge drinking. Add to that a growing obesity problem and a predicted trebling of the disorder from viral hepatitis in the next 20 years and it is clear that this condition is becoming a major challenge. London's Hammersmith Hospital pioneered attempting to reverse cirrhosis of the liver by harnessing and enhancing the body's own repair mechanism using adult stem cells extracted from patient's bone marrow to generate new tissue in damaged areas. Also in Japan and Brazil scientists are testing stem cells to regenerate liver fibrosis.[14,15]

19.3 Tissue engineering and cell therapy: selected applications

An approach aimed at regenerative medicine is needed owing to demographic and social challenges. There is a progressive increase of the mean age of populations, in particular of those in urban areas. This phenomenon affects medicare greatly; that is, feasible costs and innovative management of social and security agents. According to US Census Bureau (Fig. 19.1) it is estimated that the number of senior citizens (65–85 years old) will more than double by the

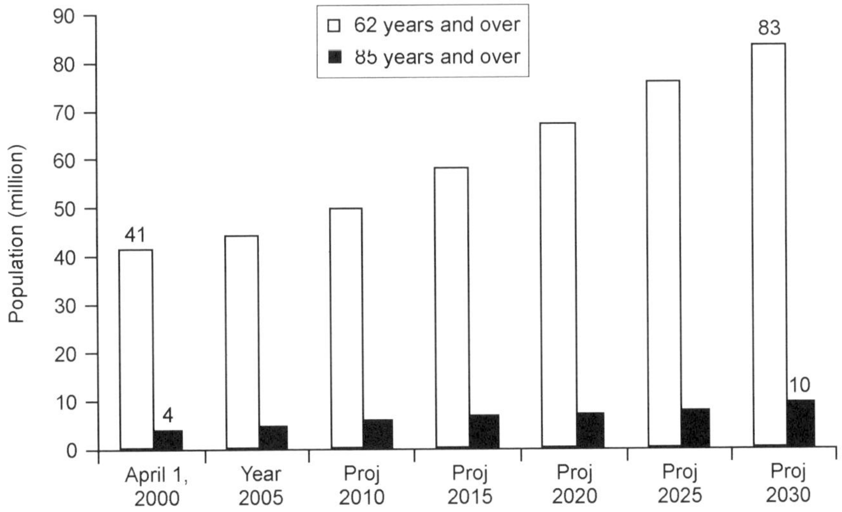

19.1 Interim projections of the population by selected age groups for the United States (US Bureau Sensus, Internet date release 21 April 2005).

year 2030, and the number of citizens in advanced senescence (>85 years old) will be 2.5-fold higher.[14] Those over 85 require more medical assistance resources and twice as many hospitalization days than the others.[15]

In their turn, young people living in areas of high population density are more exposed to traumatic lesions. Although less significant in numbers, these lesions highly impact health costs, generating debilitated young people and/or individuals permanently incapacitated for normal, productive lives. There are an estimated 10 000–12 000 spinal cord injuries every year in the United States; a quarter of a million Americans are currently living with spinal cord injuries. The cost of managing the care of spinal cord injury patients approaches $4 billion each year. About 38.5% of all spinal cord injuries happen during car accidents. Almost a quarter, 24.5%, are the result of injuries relating to violent encounters, often involving guns and knives. The rest are due to sporting accidents, falls, and work-related accidents. Some 55% of spinal cord injury victims are between 16 and 30 years old. More than 80% of spinal cord injury patients are men.[16] And what is there, in terms of new drugs, surgical or clinical protocols, to this highly impacting health issue? In Brazil, a developing country of more than 170 million people, for example, traumatic lesions are the first cause of either hospitalization or death in the second and third decades of life.[17]

19.3.1 Cardiovascular disease (CVD)

The World Health Organization (WHO) calls it an epidemic of heart attack and stroke. WHO numbers are that 17 million people die of CVD, particularly heart attacks and strokes, every year (Fig. 19.2). A substantial number of these deaths can be attributed to tobacco smoking, which increases the risk of dying from coronary heart disease and cerebrovascular disease two or threefold. Physical inactivity and unhealthy diet are other main risk factors that increase individual risks to cardiovascular diseases.[18] One of the strategies to respond to the challenges to population health and well-being due to the global epidemic of heart attack and stroke is to provide actionable information for development and implementation of appropriate policies. But still, as the overall population grows older, coronary heart disease and stroke might be defeated by new drugs, as already mentioned, and a suitable program of actions, but chagasic, hypertrophic and dilated cardiovascular disease might represent the next area for action. There, cell therapy, to a specific lesion area, or as a multipurpose, aimed at a person's regenerative cell reservoir mobilizing strategy, might be the solution.

It is estimated that there are over 70 000 heart attacks in Canada, for example, each year. In 2001 (the latest year for which statistics are available from Statistics Canada) 19 000 Canadians died from heart attacks. Over 80% of heart attack patients admitted to hospital survived.[19]

The economic impact of heart failure in the United States is outstanding. Nearly 5 million Americans are living with heart failure, and 550 000 new cases

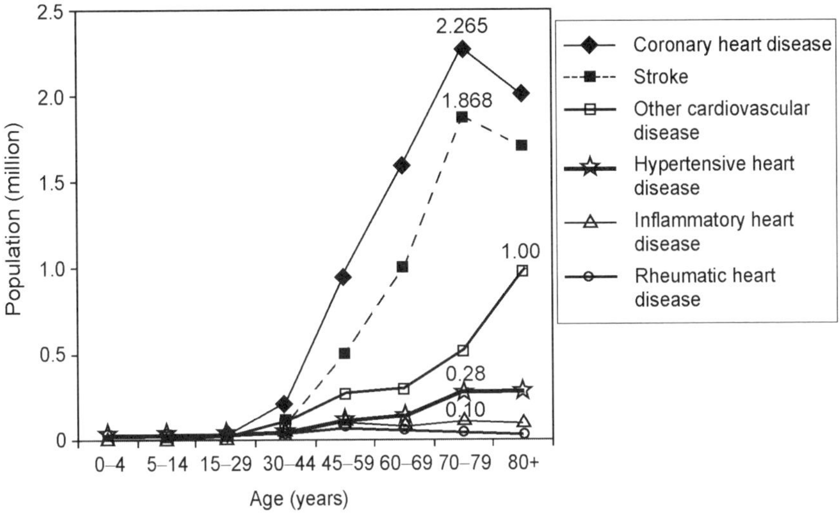

19.2 Deaths from cardiovascular diseases, WHO, number of deaths globally per year from different types of CVD by age (highest numbers shown).

are diagnosed each year. Heart failure is usually a chronic disease. That means it is a long-term condition that tends to gradually become worse. By the time someone is diagnosed, the chances are that the heart has been losing pumping capacity little by little for quite a while. At first the heart tries to make up for this by enlarging, developing more muscles and/or pumping faster. All these are meant to pump more blood and increase the heart's output.

The body also tries to compensate in other ways. The blood vessels narrow to keep blood pressure up, trying to make up for the heart's loss of power. The body diverts blood away from less important tissues and organs to maintain flow to the most vital organs, the heart and brain. These temporary measures mask the problem of heart failure, but they do not solve it. This helps explain why some people may not become aware of their condition until years after their heart begins its decline. Eventually the heart and body cannot keep up, and the person experiences the fatigue, breathing problems or other symptoms that usually prompt a visit to the doctor.[20]

The American Heart Association states that

> stem cell research offers great promise. It could be used to develop dramatic new procedures and techniques to reverse degenerative heart disease. For example, it may help generate new, healthy heart tissue, valves and other vital tissues and structures. About 128 million people suffer from diseases that might be cured or treated through stem cell research. About 58 million of these people suffer from cardiovascular disease.[21]

In 2005 biopharmaceutical companies were developing 146 medicines for heart disease and stroke all of them already in clinical trials or just waiting for

FDA approval. Just in the past three decades new drugs helped cut deaths from heart disease and stroke in half. Some 98% of Americans are under 85-years-old and since 1900 this has been the top killer disease (with the exception of the influenza epidemic that killed 450 000 in 1918 and in 2002 when cancer surpassed it). Stroke is the third leading cause of mortality. Still, death rates from cardiac degenerative disease and stroke are falling.[22] According to Dr Eugene Braunwald of Harvard Medical School, this reduction rate is 'one of the great triumphs of medicine in the past 50 years'.

Although decreased mortality rates are a gifted outcome of innovative drug discovery, many people who survive heart attacks develop congestive heart failure, a chronic disease that affects nearly 5 million Americans. The cost of cardiovascular disease in the United States is about U$400 billion; the hospitalization cost alone of those who survive a heart attack increased 175% from 1979 to 2002.[22]

There are several trials on the potential usage of stem cell on congestive heart disease. One multicenter, randomized, double blind, placebo-controlled study, sponsored by the Brazilian Ministry of Health, began in 2005 and will evaluate 1200 patients in four different cardiac disease. According to Dr Antonio C.C. de Carvalho (national coordinator of the Brazilian Multicenter Trial), 'in three years we should have results leading to confirm or not the efficacy of cell therapy in treatment of four cardiac diseases: (a) chronic ischemic cardiopathy, (b) acute myocardial infarction, (c) Chagasic cardiomyopathy and (d) dilated cardiomyopathy'. Chagas, a parasitic disease, affects more than 6 million people in Brazil alone; one-third will present congestive heart disease.[23] According to WHO, the control strategy for the elimination of Chagas disease over the period 1996–2010 is based on interruption of transmission by the vector, and the systematic screening of blood donors. Notice that the screening for *Trypanosome cruzi* antibodies in blood donors is not a customary procedure in the United States.

Cardiomyocytes regenerate very little: the infarcted lesion area will heal to a fibrous scar, which leads to structural dysfunction. Cell therapy aims to recover myocardial function, as well as revascularization. But little is known, in terms of cell therapy, about the proposed protocols regarding some basic pharmacological principles, that is: (i) optimal delivery methods; (ii) specific cell linage; (iii) optimal quantity of cells; (iv) route of administration; (v) when to start cell infusion; and, even before all that, a cell surveillance protocol aiming on toxic and oncogenic risks.

Henning *et al.*[24] investigated the use of human umbilical cord blood mononuclear progenitor cells (hUCBC) for the treatment of acute myocardial infarction and demonstrated that hUCBC substantially reduce infarction size in rats without requirements for immunosuppression. As a consequence, left ventricular (LV) function measurements, determined by LV ejection fraction, wall thickening and blood pressure and its rate of rise (dP/dt), were significantly greater than the same measurements in rats with untreated infarctions.[24]

The application of tissue-engineered heart valve substitute and the usage of adult and more recently cord blood-derived stem cells are promising therapeutic tools in the management of the degeneration of cardiac function (structure, muscle and new vascular formation). Indeed, there might be a chance for a combination therapy approach. Thus, to the young or the elderly, regenerative medicine will certainly play an important role in the healthcare industry.

19.4 The role of government, profit and non-profit institutions in realizing potential

We have discussed how relevant cardiac degenerative disease is. It has also been pointed out what innovative technologies can do in terms of medicine and that cell therapy might play an outstanding part in the near future in terms of bettering the lives of those who will live longer and more healthily. In modern society there are some epidemic, almost primitive, medical situations that will provide millions of potential patients who will benefit from the cell therapy industry. According to the WHO, Chagas disease, named after the Brazilian physician Carlos Chagas who first described it in 1909, exists only on the American Continent. It is caused by a flagellate protozoan parasite, *T. cruzi*, transmitted to humans by triatomine insects known popularly in the different countries as 'vinchuca', 'barbeiro', 'chipo', etc. The geographical distribution of the human *T. cruzi* infection extends from Mexico to the south of Argentina. The disease affects 16–18 million people, and some 100 million, i.e. about 25% of the population of Latin America, are at risk of acquiring it. After several years of an asymptomatic period, 27% of those infected develop severe cardiac symptoms which may lead to sudden death, 6% develop digestive damage, mainly megaviscera, and 3% will present peripheral nervous involvement. Between 1960 and 1989, the prevalence of infected blood in blood banks in selected cities of South America ranged from 1.7% in São Paulo, Brazil, to 53.0% in Santa Cruz, Bolivia, a percentage far higher than that of hepatitis or HIV infection. The authors could not confirm whether screening for Chagasic antibody would be introduced in blood banks in the United States or Europe.[25]

Almost 30% of those contaminated by T. *cruzi* will evolve to serious cardiac disease. In 2003 the first attempt to use stem cells to treat a Chagasic heart was performed in Brazil by Prof. Ricardo R. dos Santos.[26,27] It was a total success. Indeed as a result of that first reported case, a larger trial involving 250 patients started in 2005.

In addition to the above example of a poverty-related parasitic degenerative disease that might account for millions of patients living with severe heart malfunction, one must point out the awesome impact of the whole of degenerative disease in total healthcare. That is of particular interest to government and also to private institutions. It is clear that innovative, efficient, safe drugs, devices or clinical protocols, and healthy sanitary policies, might significantly contribute to diminishing morbidity and lead to less patient hospitalization and/or social costs. In all senses, even if there is an immediate increase in the primary costs of a certain therapeutic measure, if treatment results in the patient returning to a relatively normal life, society as a whole will benefit.

By 2001 stem cell therapy for heart disease was among the most highly motivated areas of interest. New treatments for heart failure – implantable heart devices and cell-grown tissues – were considered to be major gains, and were among the top 10 research advances in heart disease and stroke of the American Heart Association.

At the University Hospital Zurich in Switzerland scientists used human bone marrow cells to engineer heart valves in the laboratory. The cells were seeded on heart valve scaffolds made from bioabsorbable materials and grown in a pulse duplicator bioreactor system that mimics the blood circulation of humans. Heart valves worked properly to let blood flow in only one direction as it was pumped through the heart's chambers. The structures of each valve were flap-like (leaflets or cusps). These engineered human valves opened and closed synchronously in the laboratory setting. Microscopic images showed cell growth and mechanical function similar to natural human heart valves.

By 1999, the Zurich group was the first to make a complete heart valve in the laboratory using cells from sheep blood vessel walls. The valves showed excellent functional performance in blood circulation and strongly resembled natural heart valves.[28]

19.4.1 Human umbilical cord blood

Human umbilical cord blood is also a valuable source of endothelial progenitor cells (EPCs), providing novel cells for tissue engineering. Progenitor cells from hUCB were also used to structure endothelial layers in cardiovascular tissue engineering. EPCs were sourced from cord blood after a caesarian section and were culture-grown. These cells, when seeded and grown on a polymer scaffold, originated tissue strips that could be molded into different form (valve, vessel, patch, etc.). The cultures were treated with vascular endothelial growth factor (VEGF) and fibroblast growth factor (bFGF) to stimulate cell growth. After 2 weeks the cells formed capillary-like tubes, indicating the start of blood vessel formation.[28]

There is something interesting and exciting here: the possibilities for this cell source include 'banking' the cells for future use. One possible example would be that cord blood cells could potentially be used to create a tissue-engineered structure needed to correct a cardiac birth defect diagnosed prenatally. The new tissue could be ready to use when the baby is born – or even before birth for potential prenatal/fetal surgical repair.

In other cell transplant experiments, adult human cardiac myocytes (heart muscle cells) regenerated after heart attack. This means the heart may be able to replace damaged tissue by producing new functional cells. A subpopulation of myocytes that is not 'terminally differentiated' re-entered the cell cycle and divided after the infarction.[29] In similar research, adult stem cells derived from bone marrow regenerated, forming new functional heart cells when injected around the site of the heart attack.[30] Henning and coworkers also showed that there is a potential use of hUCB to regenerate myocytes. That is, for the autologous or allogeneic usage, there is a whole adventurous route ahead of medical science still to be traveled in cell therapy.[24] All these processes must prove a reproducible, validated protocol before regulatory authorities and physicians will adhere to them and they become routine.

19.4.2 Cell transplants offer promise for stroke recovery

The cerebral cortex is the mantle of gray substance that wraps each half of the brain. It is responsible for upper mental functions such as thought, memory and voluntary movement. This is the area most often injured by strokes, which are the second or third most significant causes of death or disability among elderly people all over the world.

Sanberg and coworkers[31] recently presented a technology platform using a mononuclear fraction (MNF) of hUCB which contains stem cells that have the innate ability to (1) target and engraft into areas of neurological injury, (2) induce motor and behavioral recovery, and (3) differentiate into cells of neuronal phenotype. Rat and human hUCB stem cells developed into neurons and other mature brain tissue when transplanted into normal and stroke-damaged adult and newborn rats. This suggests that brains and spinal cords can be repaired following trauma from stroke or other diseases.[31]

The works of Sanberg and others changed the previous belief that neurons or nerves did not regenerate or that the body's ability to repair lesions of the spinal cord or brain vanished completely soon after birth. Indeed stem cells have the potential and the flexibility to grow and differentiate into many kinds of cells. This ability seems to depend upon the cell age, type and lineage and a sort of physiological window.

Tissue engineering is another window of opportunity. It is at the interface of the medical implant industry. As molecular and cell biology-related technologies and know-how continue to evolve, tissue engineering might have a great impact on the pharmaceutical business in the near future. Future products will be much more biologic in nature, taking advantage of and mobilizing the inherent regenerative abilities of physiological components. In cardiovascular applications, for example, vascular grafts with anti-thrombogenic endothelial lining and pediatric aortic valves that can grow and remodel over time are under development.

Figure 19.3 depicts the number of surgical procedures in which tissue engineering and cell therapy would be applicable. An ideal scaffold must meet several criteria. It must be biocompatible and biodegradable into non-toxic

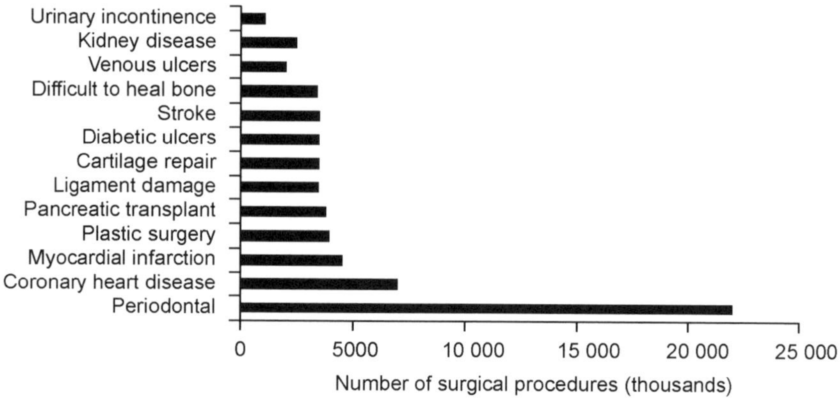

19.3 Potential for tissue engineering and cell therapy in selected applications.

products within the desired time frame. It should also be processed into a variety of shapes with different handling characteristics. And it should support and enhance cell proliferation with appropriate mechanical strength for the application it is to be.

Tissue should grow into the scaffold. At BIORIO Center researchers were able to develop a bone graft presenting cytocompatibility and osteoinduction, and exploited a formulation of porous substrata with intrinsic osteoinductive activity.[32] Human osteoblasts adhered and spread on the surface and through this bone graft. Cells seemed to find their way between adjacent particles.[32]

At Albert Einstein College of Medicine in New York scientists were able to harvest embryonic cortical cells (which come from the cerebral cortex – the outer layer of the brain) for transplants. In the laboratory these stem cells were injected into the brains of normal adult rats and adult rats damaged by stroke. The cells were marked and they could be seen in the damaged area, forming connections with neighboring cells. Blood vessels were also seen growing to nourish the transplanted cells. At 21–45 days after the transplants, most stem cells grew into mature neurons and other mature brain cells. This remarkable work of Dr Gaurav Gupta got the American Stroke Association Mordecai Y.T. Globus Young Investigator of the Year Award in 2001.[33]

In another milestone in stroke research at the animal level, intravenous administration of bone marrow cells reduced stroke-induced disability.[34] Another study showed that intravenous treatment with adult donor rat stromal cells (mature cells from bone marrow) allowed the rats to return to normal or near-normal function within 14 days of a stroke. An infusion of a stroke patient's own stromal cells may provide benefits and is easily given. If the treatment continues to show benefits in animals, it may provide new treatments in the future for stroke, brain trauma and spinal cord injury in humans. It may also be useful in treating Parkinson's disease, multiple sclerosis, Alzheimer's and other neurological diseases.

19.4.3 Latest controversies

There have been legislators backing and others banning human cloning – including techniques used to collect embryonic stem cells. Those in favor claim it would criminalize research that could benefit or even cure diseases such as Parkinson's, Alzheimer's, spinal cord injury and so on. It seems that those who authored the interdict, countered that the legislation was meant to set an ethical standard for research that would prohibit the creation of an embryo either for reproduction or to be destroyed for research. Others said it was confusing public opinion by claiming the ban was aimed at stem cell research, saying it would not impact works already prospering in those institutions that were already in advanced areas of stem cell and regenerative medicine research.

Politicians who oppose embryonic stem cell research question whether the use or approval of government funds, thus endorsing the creation of a human life and then the destruction of a human life, would be a hopeful potential or provide medical benefit. Others proffer that adequate ethical standards are already in place. The potential of the research to help, for example, children with juvenile diabetes is more important than concerns over 'cells in a dish.' It is claimed that respect for human life means you do not turn your back on cures that can save human lives.

At the time of writing (2005), reproductive cloning for human therapeutic means has been banned in the United States and most countries of the developed world at the claim that it would mean 'the creation of babies using existing human cells'. This means that it proscribes the so-called therapeutic cloning, in which embryos are created by injecting human eggs with a living person's cell to grow stem cells. The days-old embryo is destroyed in the process.

Still in the United States, some states forbid both procedures, while others have restrictions on reproductive cloning, according to the National Conference of State Legislatures. The FDA, which regulates human experiments, contends that its regulations inhibit human cloning without prior agency permission, which it seems having no intention of giving so far.[35]

Technically speaking, and still at embryonic stem cell research, if one is to think of a stem cell as an active principle that will be used in an off-the-shelf preparation, one has to reason out all the pros and cons of such a source. That is, will it be easy to use? Will it be safe? What is the hazard factors involved with such an entity? What about reproducibility? What are the toxic or neoplastic risks of such? Can it be withdrawn in cases of therapeutic harmful side effects?

19.5 Cells as products under FDA guidelines

It took about 10 years for an experimental drug to travel from discovery to patients. Of 5000 new proposed compounds of probable therapeutic usage, five are accepted for sale. Will it be the same when a cell derivative product is thought of in a regulatory point of view? And what about combination therapy of a cell protocol plus a biomaterial and/or a 'regular' medicine (chemical entity)? Will regulatory agencies evaluate, analyze and liberate for usage – that is 'register' – cell manipulation platform technologies in terms of quality assurance protocols, validation, safeness, reproducibility, risk assessment, in-process controls, characterization, current good tissue practices (cGTP), and so on? Should the same quality control and quality assurance procedures as in the pharmaceutical or biotechnology industries be applied to cell therapy? What about the off-the-shelf cell-containing flask?

It is probable that cell therapy and tissue engineering will involve the following associated steps:

- *Cell* – the organization of the standardized pool of stem cell (umbilical cord blood, bone marrow or embryo), that is: (i) collection, (ii) transportation, (iii) characterization, (iv) separation of specific cell lineage, (v) cryopreservation and/or expansion, (vi) storing and (vii) preparation of cell-working-bank.
- *Cell* – application of the resulting standardized cell pool in accordance to specificities of the regions where degeneration has occurred.
- *Biomaterials/scaffolds/grafts* – the introduction, into lesion regions, of structures similar to elements of the extra-cellular matrix associated or not with intercellular mediators – aimed to facilitate the recruitment, expansion and integration of *blast cell* colonies or even stem cells of autologous, allogeneic or even help congregate those of the endogenous environment,[36] promoting the repair of lesions or regeneration and renewal of degenerated tissues (biomimetics).

The stem cell is a key element to regenerative medicine, be it introduced off the shelf or from a previous collection (umbilical cord, bone marrow, etc.) or recruited among the patient's population. Another element is the technology involving biomaterials. In addition, a whole industry of diagnostic and even nutritional formulations will rise in parallel to this more individualized or focused medicine.

A stem cell could have the intrinsic ability to generate or regenerate all parts of the human body. Indeed the human body, as a whole, has the inherent ability to heal a variety of its organs and tissues as they are routinely damaged from both acute and long-term disease processes. It seems that the younger the stem cell (umbilical cord), the better it is – that is, its plasticity[3] or capability to regenerate is more accurate. Several scientists and private companies are developing cell collection and characterization protocols. Those that succeed will be the ones that prove safe, reproducible and validated. Throughout history, medicine has sought to facilitate or improve the intrinsic self-renewing ability of biological systems. Stem cell, regenerative medicine, cell and tissue engineering might be thought to be a window of opportunity in medical science and patient care.

That stem cells are supposed to be better the younger they are may result from a complex phenomenon involving the telomere structure – a region of DNA at the end of a chromosome that protects the start of the genetic coding sequence against shortening during successive replications. As the cell divides, a series of these structural components is lost until it is completely absent: the cell can then no longer divide and apoptosis follows. This, by some interesting meanings, could be also referred to as the biological clock.[37]

'This is all new biology, which could have an unlimited potential,' said Dr Paul Sanberg, director of the University of Southern Florida's center for aging and brain repair. 'Cord blood research is moving us into an era of regenerative medicine where we're going to be approaching chronic degenerative diseases with ways to repair them by generating new tissues.'

From benchtop to bedside, in a logical sequence, what is expected is that validated autologous cell therapy procedures alone or in combination with biomaterials become the first medical routine in regenerative medicine and TCEt. Then the road might diverge: one way to the off-the-shelf autologous and allogeneic cell, the other to allograft and tissue substitutes of synthetic or natural derivation. These different approaches will meet and connect to what shall path the future of TCEt, the combination therapy. Before being included as routine protocols in hospitals, TCEt will have to navigate the critical path of a new product or medical device under regulatory agencies worldwide (FDA, EC, ANVISA, and so on), as any other innovative technology in patient care.

As to cell therapy using autologous cells, there is no need to go through the long period required for approval of new drugs. Although regulatory agencies are issuing propositions to this matter, there are no formal regulations in this field. Still, although some declare that the only requirement for autologous cell treatments after a minimal *ex vivo* manipulation is to follow 'good laboratory procedures', that cannot be the rule. Standardized protocols must be searched and regulatory agencies will certainly realize that whatever cell manipulation there is, if one is meant to use it routinely, there has to be a safe, validated protocol. That is, in summary, each cell collection has to be seen and treated as if it were a pharmaceutical production lot. With all the quality control and quality assurance data, proceedings, etc. And when cells are more extensively manipulated *in vitro*, or when exogenous materials are used, standardized protocols, risk assessment, quality controls should be emphasized.

Unlike biotechnology, molecular biology and genomics, which already have a historical background of several decades, the organization and growth of tissue engineering and cell therapy industry is yet to be established. Synthetic materials would be the most interesting basis of tissue replacement products, but there are no synthetic materials that are totally accepted by and integrated into the body. Whatever synthetic structure that is implanted into the body will be encapsulated by the body's immune system and gradually lose effectiveness.[38]

19.5.1 Stem cell from bone marrow and cord blood regenerates heart parts

At the University of South Florida a protocol utilizing cord blood as a reasonable way to cope with the issue of a validated cell source was made public in 2004.[24] That same protocol might be consolidated as an example of a scale-up manufacturing process and the preservation of living-cell pharmaceutical product for off-the-shelf availability. This protocol shows the advantage of being potentially used in autologous cell therapy – at least for the first stage of this new era.

In cardiac surgery, replacement parts such as heart valves, blood vessels and vascular patches, are frequently needed. Issues of blood clots, tissue overgrowth, limited durability, contamination and inability to grow may all complicate the usage of replacement parts. Immune response might complicate the potential usage of biomaterials or allogeneic implants as it often rejects donor tissue. Immunosuppressive medication is a regular protocol in conventional transplantation. Tissue engineering using a patient's own blood or cells proffers the safe alternative. In cases of pediatric surgery, where the graft should ideally grow with the patient, tissue engineering becomes a most careful issue. When utilizing a bone substitute or a graft, the best will always be that one that promotes cell adherence and cell proliferation within the biomaterial. The best product will be that which works as a cell matrix, allowing osteoblasts, for example, to proliferate and differentiate into bone. A bone substitute was recently launched and it seems to meet those criteria.[32] In dilated and ischemic heart degenerative disease, total replacement (transplantation) is usually needed and seldom achieved.

19.6 Conclusion and future trends

Tissue engineering and cell therapy represent unprecedented opportunities in health-related research, development, clinical practice and business. It will encompass an array of interdisciplinary areas such as diagnostic, nutrition, training and a new industry of cell processing, storage, distribution and thawing.

In 1974 Knudtzon[39] reported the *in vitro* growth of granulocyte colonies from circulating cells in human cord blood. In 1989 Broxmeyer and colleagues demonstrated that umbilical cord blood (hUCB) was a rich source of hematopoietic stem/progenitor cells (HSPC) and that hUCB could be used in clinical settings for hematopoietic cell transplantation.[40] Since then, great interest has been generated on the biological characterization of these cells. Since 1990, several groups have focused on the study of hUCB HSPC, addressing different aspects, such as the frequency of these cells in hUCB, the identification of different HSPC subsets based on their immunophenotype, their ability to respond to hematopoietic cytokines, the factors that control their proliferation and expansion potentials, and their capacity to reconstitute hematopoiesis in animal models.[41–43] Most of these studies have shown that significant functional differences exist between HSPC from hUCB and adult bone marrow (i.e. the former possess higher proliferation and expansion potential than the latter). It is also noteworthy that genetic manipulation of hUCB HSPC has been achieved by several groups and that genetically modified hUCB cells have already been used in the clinic.[44]

In a regulatory point of view, these cells and whatever others that might be challenged to be a source of a validated therapeutic protocol will have to present enough data and basic safeguards before they become routine clinical applications. The use of controlled, standardized practices and procedures is fundamental. Some basic steps should account for:

- donor and/or specimen screening (infectious disease, molecular genetic testing, cell count and viability);
- validation of procedures that assures the prevention of the introduction, transmission or spread of communicable diseases and/or contamination during manufacturing;
- *in vitro* characterization;
- immunophenotyping;
- the commitment of certain cell lineages to the expected therapeutic effect;
- standard operating procedures that assure reproducibility in technical manipulation aiming at cell culture expansion;
- standard operating procedures that assure reproducibility and control in technical manipulation aiming for cell differentiation;
- derivation of certain cell lines assuring bioactivity and compliance to quality control protocols (for example, mesenchymal and endothelial cells from UCB $CD34^+$ cells);
- functional evaluation considered as part of in-process bioactivity and/or quality control protocol;
- *in vivo* characterization (e.g. post-implantation aiming at surveillance);
- pre- and post-implantation evaluation of possible genetic changes or chromosomal fragility related to manipulation, freezing and/or thawing.

Regenerative medicine presents the opportunity to make some of the current clinical treatments sound obsolete in the near future. As expected there are also unrealistic expectations – at least to our actual comprehension of the ways and meanings of the biological body. There is a general optimism. And why not? When a 60-year-old Chagasic patient, with an over-stressed, tired, malfunctioning heart, whose life expectancy was pretty bad, receives his own stem cell collected from his bone marrow, a couple of weeks later all basic clinical signs are better, and a couple of months later his and his family's life is also significantly better, how can one – even a scientist – not look at it as a 'miracle'? And what about reverse cirrhosis of the liver by harnessing and enhancing the body's own repair mechanism using adult stem cells?

Will there still be over-hyping? Yes. It is up to scientists, physicians, biologists, the media, and government officials and even private companies to help clarify the subject of tissue engineering and cell therapy so that risks and cutbacks are, at least, expected.

According to Ahsan and Nerem, in 2004 there were about 100 companies and 3000 people working in the industry of tissue engineering and cell therapy. The numbers may be much bigger than that, but the implant industry alone accounts for 300 000 jobs and estimated world sales of $200 billion a year. These numbers give us some clues on the potential market for regenerative disease.[45]

After the numerous experimental, preclinical trials that will be performed, a business model for cell therapy will be the one that will start with a reproducible protocol of cell manipulation (collection, transportation, processing, transplant, storage and thawing), to a need-to-be basis (patient by patient) (proof of concept), then to the manufacturing of large batches of a product. One might ask why, for cases of elective surgery, one needs off-the-shelf availability. Why should one adopt the approach of extracting cells from the patient, expanding them, and seeding the substitute that is then implanted? The answer is that, even for a case where the time of surgery is elective, unless there is an off-the-shelf availability this approach will not be used at the large variety of hospitals required to impact the wider patient population that is out there. Off-the-shelf availability seems to be crucial.

What we must understand is that basically we will be dealing with cell response, that is, the instigation of receptors, the transduction of whatever is perceived and the response itself, the 'induction' of the therapeutic action this 'product' is supposed to perform. The basics of biologics are translated to practical medical procedures.

It is not enough to be in front in the area of research: the science needs to be translated into enabling technology and this in turn into products. These products or strategies then will need to go through clinical trials, receive regulatory approval, key opinion leaders' faithfulness and finally be seeded in the hospital environment.

There is also the need for increased government funding for research and support for early stage product development. Perhaps, and because of the dynamics of cell therapy and tissue engineering, there is a need to a redesign of regulatory processes in order to bring new technologies to patients quicker. Last but not least, there needs to be a better understanding of the importance of reimbursement in the process of launching innovative technologies in healthcare.

It is quite probable that, due to today's amount of information and technical support, this new industry will take advantage of the so-called simultaneous engineering strategy, where cell biology, production, characterization and control protocols, the development of matrixes, pharmacological, toxic and immunological studies, regulatory affairs, commercial and business issues, promotional and marketing activities will all run in parallel in a team of multidisciplinary professionals. The successful ones will make history in this window of opportunity in human health.

19.7 References

1. http://canadaonline.about.com/od/statistics/a/lifedeathstats.htm
2. Huat SA, 'ACC/AHA 2005 guideline update for the diagnosis and management of chronic heart failure in the adult'. *Circulation*, 2005, 112: e154–e235.
3. Verfaillie CM, 'Stem cell plasticity'. *Graft*, 2000; 3: 296–298.
4. Orkin SH and Morrison SJ, 'Biomedicine: stem-cell competition'. *Nature*, 2002, 418: 25–27.
5. Chen N, Hudson JE, Walczak P *et al.*, 'Human umbilical cord blood progenitors: the potential of these hematopoietic cells to become neural'. *Stem Cells*, 2005, 23(1): 1560-1570.
6. Zhang L, Yang R, Han ZC, 'Transplantation of umbilical cord blood-derived endothelial progenitor cells: a promising method of therapeutic revascularisation'. *Eur. J. Haematol.*, 2006, 76: 1–8.
7. Newman MB, Misiuta I, Willing AE, Zigova T, Karl RC, Borlongan CV, Sanberg PR, 'Tumorigenicity issues of embryonic carcinoma-derived stem cells: relevance to surgical trials using NT2 and hNT neural cells'. *Stem Cells Review*, 2005, 14(1): 29–43.
8. Martina V, Jordan C, Newcomb J, Butler T, Pennypacker KR, Zigova T, Sanberg CD, Sanberg PR, and Willing AE, 'Infusion of human umbilical cord blood cells in a rat model of stroke dose-dependently rescues behavioral deficits and reduces infarct volume'. *Stroke*, 2004, 35: 2390–2395.
9. US Food and Drug Administration Department of Health and Human Services. Last updated: October 21, 2005 (www.fda.gov).
10. Lichtenberg F, 'Benefits and costs of newer drugs: an update', *National Bureau of Economic Research*, 2002, NBER Working Paper No. 8996.
11. Cowper PA, 'Economic effects of beta-blocker therapy in patients with heart failure', *Am J Medicine*, 2004, 116: 104–111.
12. Stewart S, Pearson S, Horowitz JD, 'Effects of a home-based intervention among patients with congestive heart failure discharged from acute hospital care'. *Arch Intern Med* 1998, 158(10): 1067–1072.
13. Mason J, Freemantle N, Nazareth I, *et al.*, 'When is it cost-effective to change the behavior of health professionals?'. *JAMA*, 2001, 286: 2988–2992.
14. http://news.bbc.co.uk/2/hi/health/4573453.stm
15. http://agenciact.mct.gov.br/index.php?action=/content/view&cod_objeto=27372
16. The National Spinal Cord Injury Statistical Center, Birmingham, University of Alabama, 2005.
17. MedMarkets Diligence, April 2005, Volume 4, Issue. Cadernos do Brasil, Secretaria Executiva do Ministério da Saúde, SIM/SINASC, 2004.
18. World Health Organization; Cardiovascular Disease Programme CH-1211 Geneva 27, Switzerland mendiss@who.int http://www.who.int/cardiovascular_diseases/en/
19. Heart and Stroke Foundation of Canada Annual Report Card on the Health of Canadians, 2001.
20. 'Learning about heart failure', American Heart Association, 2005.
21. www.aha.org (American Heart Association).

22. Medicines in Development for Heart Disease and Stroke, *Survey*; PhRMA, America's Pharmaceutical Companies, 2005.
23. Carod-Artal FJ *et al.*, 'Chagasic cardiomyopathy is independently associated with ischemic stroke in Chagas disease'. *Stroke*, 2005, 36: 965.
24. Henning RJ, Abu-Ali H, Balis JU, Morgan MB, Willing AE, Sanberg PR, 'Human umbilical cord blood mononuclear cells for the treatment of acute myocardial infarction'. *Cell Transplant.*, 2004, 13(7–8): 729–739.
25. Wendel S, Brener Z, Camargo ME, Rassi A, 'Chagas disease – American trypanosomiasis: its impact on transfusion and clinical medicine'. ISBT, São Paulo, Brazil, 2002.
26. Soares MBP *et al.*, 'Transplanted bone marrow cells repair heart tissue and reduce myocarditis in chronic Chagasic mice'; *Am. J. Pathol.*, 2004, 164: 441–447.
27. Vilas-Boas F, Feitosa GS, Soares MBP, Pinho Filho JA, Almeida A, Mota A, Carvalho HG, Oliveira ADD, Ribeiro-dos-Santos R, 'Bone marrow cell transplantation to the myocardium of a patient with heart failure due to Chagas cardiomyopathy. A case report'. *Arquivos Brasileiros de Cardiologia*, 2004, 82(2): 185–187.
28. Simon P, Hoerstrup MD, 'Tissue engineering of functional trileaflet heart valves from human marrow stromal cells'. *Circulation*, 2002, 106: I-143.
29. Beltrami AP *et al.*, 'Evidence that cardiac myocytes divide after myocardial infarction'. *New England J. Med.*, 2001, 344: 1750–1757.
30. Orlic D, Kajstura J, Chimenti S, Jakoniuk I, Anderson SM, Li B, Pickel J, McKay R, Nadal-Ginard B, Bodine DM, Leri A, Anversa P, 'Bone marrow cells regenerate infarcted myocardium'. *Nature*; 2001, 410: 701–705.
31. Sanberg PR, Willing AE, Garbuzova-Davis S, Saporta S, Liu G, Sanberg CD, Bickford PC, Klasko SK, El-Badri NS, 'Umbilical cord blood-derived stem cells and brain repair'. *Ann NY Acad Sci.*, 2005, 1049: 67–83.
32. http://www.silvestrelabs.com.br/extragraftxg13
33. American Heart Association, http://www.strokecenter.org/prof/awards.htm
34. Lindvall O, Kokaia Z, 'Recovery and rehabilitation in stroke: stem cells', *Stroke*, 2004, 35: 2691–2694.
35. http://www.duluthsuperior.com/mld/duluthsuperior/13074930.htm, Posted on 3 Nov. 2005, Gov. Jim Doyle: http://www.wisgov.state.wi.us/
36. Ahsan T, Nerem RM, 'Bioengineered tissues: the science, the technology, and the industry'. *Orthod. Craniofacial Res.*, 2005, 8: 134–140.
37. Lei M, Podell ER, Baumann P, Cech TR, 'DNA self-recognition in the structure of Pot1 bound to telomeric single-stranded DNA'. *Nature*; 2003, 426: 198.
38. Shirasugi N, Adams AB, Durham MM, Lukacher AE, Xu H, Rees P *et al.*, 'Prevention of chronic rejection in murine cardiac allografts: a comparison of chimerism and nonchimerism-inducing costimulation blockade-based tolerance induction regimens'. *J. Immunol.*, 2002, 169: 2677–2684.
39. Knudtzon S, '*In vitro* growth of granulocyte colonies from circulating cells in human cord blood'. *Blood*, 1974, 43: 357–361.
40. Broxmeyer HE, Douglas GW, Hangoc G *et al.*, 'Human umbilical cord blood as a potential source of transplantable hematopoietic stem/progenitor cells'. *Proc. Natl. Acad. Sci. USA*, 1989, 86: 3828–3832.
41. Gallacher L *et al.*, 'Isolation and characterization of human $CD34^{-}$ Lin^{-} and $CD34^{+}Lin^{-}$ hematopoietic stem cells using cell surface markers AC133 and CD7'. *Blood*, 2000, 95: 2813–2820.
42. Meagher RC, 'Human umbilical cord blood cells: how useful are they for the clinician?'. *J. Hematotherapy Stem Cell Res.*, 2002, 11(3): 445–448.

43. McGuckin CP *et al.*, 'Production of stem cells with embryonic characteristics from human umbilical cord blood'. *Cell Prolif.*, 2005, 38: 245–255.
44. Borlongan CV, Fournier C, Stahl CE, Yu G, Xu L, Matsukawa N, Newman M, Yasuhara T, Hara K, Hess DC, Sanberg PR, 'Gene therapy, cell transplantation and stroke'. *Front Biosci.*, 2006, 1(11): 1090–1101.
45. Ahsan T, Nerem RM, 'Bioengineered tissues: the science, the technology, and the industry'. *Orthodontics & Craniofacial Research*, 2005, 8(3): 134.

20

Inflammatory bowel diseases: current treatment strategies and potential for drug delivery using artificial cell microcapsules

D AMRE, University of Montreal, Canada and R D AMRE and S PRAKASH, McGill University, Canada

20.1 Introduction

Inflammatory bowel diseases (IBD), comprising the two main phenotypes Crohn's disease (CD) and ulcerative colitis (UC), are chronic relapsing inflammatory diseases of the gastrointestinal tract (GIT). Predominantly observed in Western countries, the incidence of, in particular, CD has been increasing. Countries such as Canada have the highest burden of disease in the world.[1] The etiology and pathogenesis of IBD has been an enigma. There is currently no cure for the disease, which is thought to represent a complex disease with multiple interacting pathways involving genetic, environmental and immunological parameters.

The disease commonly affects young populations (10–34 years of age) and is characterized by a heterogeneous clinical presentation that includes pain in the abdomen, diarrhea, fever and weight loss. Significant complications include growth delay, development of fistulas and abscesses, osteopenia/osteoporosis, etc. Of significance, most subjects in particular with CD require surgery at some point during the clinical course of disease.[2] The relapsing nature of the disease, the associated complications and the increasing burden (given the lack of cure) makes it an important public health priority for a search for effective therapeutic modalities.

Although recent extensive forays in understanding the pathogenesis of disease have been made, given the lack of precise information on the etiology, most therapeutic modalities are based on ensuring a 'control' of the disease. Most current therapies are either non-specific or target a limited number of molecules within the unknown chain of events leading to the chronic inflammation. Further compounded by the variable course of the disease, long-term effectiveness is limited and associated with severe complications. In the present chapter we will outline prevalent therapies for IBD, highlight more recent advances in the area, discuss their limitations and propose novel methods for improving the efficacy of existing or newer drugs in the management of the disease.

20.2 Therapeutic goals

IBD is a complex phenotype. In addition to the classification as either CD or UC, individual patients also differ according to other characteristics such as the age of onset, location of disease and disease behavior. In order to identify phenotypic classifications to ensure appropriate management, the Vienna classification[3] and more recently the Montreal classification have been formulated.[4] Currently these classifications are more likely to aid assessment of genotype-phenotype relationships, stratification for clinical trials and evaluation of markers for disease course and prognosis. Stratification for the purpose of establishing individualized therapeutic strategies would, however, require further research information. Currently, the goals of IBD therapy are the induction and/or maintenance of remission that in turn depend upon the extent and severity of disease, response to current or prior medication, and the presence of complications.

20.3 Current therapies

Conventional treatment of IBD consists of aminosalicylates (sulfasalazine, mesalazine (5-aminosalicylic acid)), corticosteroids and immunosuppressive drugs (azathioprine, mercapturine, methotrexate, cyclosporine). Aminosalicylates are considered first-line therapy for mild-moderate UC and CD both for induction and remission. Their mode of action is by non-specifically down-regulating intestinal inflammation. Sulfasalazine has been used as the prototype treatment for years. However, due to concerns over its toxicity (myelo-suppression in particular) and recognition that 5-ASA was the active moiety formulations with sulfa-free compounds have been developed. However, recent evidence suggests that for induction of remission 5-ASA are only slightly superior to Sulfasalazine and for maintenance of remission, 5-ASA are likely to be inferior.[5] Moreover the rates of adverse effects seem to be similar. In general, aminosalicylates, seem to be effective in UC, but their efficacy is more controversial in the treatment of CD.[6,7]

Corticosteroids are indicated for moderate-to-severe disease, or for those patients who are unresponsive to first-line therapy. They are, however, ineffective for maintenance therapy and long-term usage is associated with severe toxic effects and relapse is common. In addition, a high proportion of patients either are steroid resistant or become steroid dependent and need to undergo surgery frequently.[8,9]

Several immunosuppressive (azathioprine, 6-mercaptopurine) agents also show efficacy in IBD. They are effective in maintaining remission for steroid-dependent patients and for induction of remission among steroid-resistant patients.[10] Their onset of action is, however, slow and they are associated with severe side effects such as bone marrow suppression and require close monitoring.

Cyclosporine is a rapidly acting immunosuppressant shown to quite effective for IBD not responding to conventional treatment; however, it carries with it potential risks for hypertension, nephrotoxicity, electrolyte imbalance, encephalopathy, myelosuppression, etc.[11,12] Similar toxicity is associated with methotrexate. Antibiotics have been utilized for therapy considering the role of bacteria in the etio-pathogenesis of IBD. They seem to be effective only for CD.[13]

20.4 Newer therapies

Previous therapies for IBD were mostly non-specific suppression of the immune and/or inflammatory response. Although effective to a large extent in ameliorating symptoms, their ineffectiveness for long-term maintenance of remission, and their association with toxic effects has driven research for alternative therapeutic agents. Over the past decade substantial strides have been made in understanding the key players involved in the inflammatory cascade and efforts have been made to understand and delineate their complex interrelationships. Based on these advances, newer treatment modalities targeting these key elements have been propagated. From among these, biologic and probiotic therapy have received considerable attention. In the following sections we summarize these modalities highlighting their potential benefits and limitations in IBD therapy.

20.5 Biologic therapy

The term 'biologic therapy' encompasses a host of agents that have diverse modes of actions and is probably coined to emphasize the fact that application of these therapies is based on pathophysiological mechanisms underlying IBD. These therapies are based on targeting specific molecules that are considered important either in initiating, propagating or down-regulating inflammatory processes. The key targets of biologic therapy are the cytokines and/or their receptors. Cytokines are the cornerstone of the inflammatory cascade. Produced by activated T lymphocytes they are involved in cell-signaling mechanisms leading to cross-talk between immune cells, the end result of which in normal individuals is 'physiological inflammation'. In IBD, however, there is considerable evidence that alteration in the cytokine milieu is hampered, leading to disturbed immune balance. Although it is not clear what factors initiate the immune disturbances, the nature, level and activities of both so-called pro-inflammatory (IL-2, IFN, TNF) and immuno-regulatory (IL-10, TGF-beta) determine the clinical activity of the disease as well as distinguish to certain extent between different IBD phenotypes. Thus biological agents that target specific cytokines either for neutralizing or enhancing their impact are likely to reinstate immune balance. Toward this end, numerous agents that target key cytokines such as TNF-alpha, IL-10, IL-11, IL-12, IFN and others are either tested for their efficacy or are under various phases of development. From among these, antibodies to TNF-alpha are the only agents that have been approved for clinical use. In addition, given the immuno-regulatory effects of IL-10, much research has been focused on related agents for therapy. In the following section we briefly highlight the evidence for biologic agents that target these two key cytokines.

20.6 Anti-TNF

Monoclonal antibodies against TNF are the first line of biologic agents advocated for the treatment of IBD. From among the various antibodies Infliximab is registered for the treatment of steroid-refractory patients and patients with perianal fistulizing disease. It is a chimeric IgG1 monoclonal antibody comprising 75% human and 25% murine sequences. The mechanism of action of infliximab is elusive. Efficacy is thought to be related to a combination of effects including neutralization of TNF in the mucosa, antibody-dependent cytotoxicity, lysis of TNF-producing cells by complement fixation or T-cell apoptosis. Many large clinical trials have shown that Infliximab is efficacious for induction of remission as well as maintenance of remission for up to 55 weeks. Evidence for its utility for long-term maintenance is currently limited.[14]

Although well tolerated in the majority of patients, potential for mild and severe side effects remains. Studies have indicated potential for the occurrence of serious infections[15] demyelination, non-Hodgkin lymphomas, cardiac failure and death. Among the infections, re-activation of latent tuberculosis is a severe complication and evaluation and treatment of latent tuberculosis is a prerequisite prior to initiation of therapy.[16] One of the main risks associated with Infliximab is the potential for immunogenicity. Development of antibodies to Infliximab is observed in a high proportion of treated subjects leading to infusion-reactions and loss of response.[17] Concomitant use of immunosuppressive agents may be an alternative.[16] About 30–50% of patients do not respond to Infliximab. The underlying causes seem to be related to antibody formation, biological markers such as C-reactive protein and genetic variation.

20.7 IL-10

IL-10 is an immunoregulatory cytokine produced by T-helper-2 (Th2) cells, monocytes, macrophages, B cells, dendritic cells, mast cells, etc. Its myriad biological actions are related to the regulation of the immune response by blocking the secretion of pro-inflammatory cytokines such as IL-1-beta, TNF-alpha, IL-6, IL-4, IL-5,[18,19] inhibiting the expression of inflammatory enzymes inducible nitric oxide synthase (iNOS) and cyclooxgenase (COX-2), and blocking allergen presentation by mononuclear and dendritic cells. Given its wide immunoregulatory effects, and based on observations that IL-10 knock-out mice spontaneously develop colitis, IL-10 is thus a prime candidate for use in the therapy of IBD.

Early clinical trials demonstrated that IL-10 therapy was safe and could be well tolerated.[20] Trials on the efficacy of IL-10 have, however, shown mixed results.[21–23] Subcutaneous administration at lower doses shows modest clinical improvement but the effect is lost with higher doses. Similarly, induction of clinical remission is no better than those treated with placebo. In addition, higher concentration of IL-10 is associated with adverse effects such as headache, fever and anemia, effects thought to be related to the immunostimulatory effects of the cytokine at high doses. Thus the effectiveness of the cytokine is likely to depend on local concentrations of the drug and balance between its immunoregulatory and immunostimulatory effects.

20.8 Probiotics and prebiotics

Although the exact pathogenesis of IBD is not known, various lines of evidence suggest that the gut flora is likely to play a major role. The inability of genetically susceptible rodents to develop IBD in germ-free condition suggests that interactions between the gut flora and, in particular, the commensal gut flora, and the immune system may predispose to a dysregulation of the symbiotic ecosystem, leading to the chronic inflammation characteristic of the disease. Thus modulation of the gut flora would be a rational strategy for controlling IBD. This has led to research on the ability of probiotics and prebiotics in IBD therapy. Probiotics are viable microorganisms, consumption of which lead to therapeutic benefits. Originally acquired from cultured foods (in particular milk products) these bacteria and yeast include species such as lactobacillus and bifidobacterium (lactic acid bacteria), non-pathogenic strains of *Escherichia coli*, *Saccharomyces boulardi*, *Clostridium butyricum*, etc. Most of the reports on the utility of probiotics comes from animal models. *Lactobacillus* and *Bifidobacterium* species have been shown to attenuate experimental colitis in IL-10 knockout mice. Findings from these models reveal that different probiotics have different protective effects, the host background is important and that probiotics may be effective in maintaining rather than inducing remission. These experimental studies increased enthusiasm for clinical trials in IBD. However, evidence from the limited number of trials suggest that probiotics may only be moderately better than 'placebo' and are likely to be effective only among patients who have pouchitis and to some extent in UC patients.[24] Probiotics are thought to mediate their protective effects via either the inhibition of pathogenic enteric bacteria, maintaining the integrity of the mucosal barrier or via modulation of the immune response.

Current interest on the utility of prebiotics is growing. Prebiotics are dietary components usually carbohydrates that facilitate the growth of beneficial bacteria. These include lactosucrose, inulin, psyllium, lactosucrose and germinated barley. Prebiotics are thought to be beneficial in IBD via a number of mechanisms. These include regulating balance between aggressive and beneficial commensal bacteria, stimulating growth of protective lactic acid bacilli, mediating butyrate production (thus improving mucosal barrier function). There is some support for the efficacy of prebiotics in IBD from experimental models; however, evidence in human IBD is limited to a few controlled studied in small number of patients.[24] Current evidence suggests some effectiveness in UC, whereas studies in CD are not presently available.

20.9 Gene therapy

Although the exact etiology and pathogenesis of IBD is unknown, there is substantial information to suggest that genetic and immune-related factors play a role. Many current therapeutic strategies target the cytokine network either specifically or non-specifically. As these therapies are associated with unwanted adverse effects alternative strategies such as 'targeted gene therapy' hold promise. These strategies are still in their embryonic state, but nonetheless are of considerable interest. Many of the potential difficulties related to targeted gene therapy in the GIT are related to the presence of tight junctions, mucus, pH conditions, the survival of vectors and the rapid turnover of target cells reducing duration of action. Some of these limitations could be overcome by targeting intestinal stem cells. However, currently, definitive markers for identifying intestinal stem cells are not available.

Clinical trials for effectiveness of gene therapy in IBD are not available. Some interesting evidence from case studies and experimental studies has emerged. For example, Hogaboam *et al.*[25] studied the effectiveness of recombinant human adenovirus 5 (Ad5) vector-expressing murine IL-4 in preventing trinitrobenzesulfonic acid (TNBS) colitis in rats. After intraperitoneal injection, therapy led to lower tissue injury, lower colonic levels of pro-inflammatory cytokines such as IFN-gamma, and other inflammatory markers. Other studies[15,26,27] have examined gene transfer of IL-10 using Ad5 encoding IL-10. Colon inflammation was shown to be prevented in TNBS colitis in rats, mice or IL-10 deficient mice. Established disease was abrogated in only IL-$10^{-/-}$ deficient mice. Recently, Kitani *et al.*,[28] observed gut expression of recombinant TGF-beta 1 mRNA after intranasal administration of naked plasmid DNA encoding an expression cassette for the precursor form of TGF-B1. This expression was able to prevent and abrogate murine colitis.

20.10 Limitations of current therapies

Existing therapies, both conventional and newer, show promise and have proven efficacious for the treatment of IBD. Nonetheless however, given the complex nature of the disease, the heterogeneity in clinical presentation and course, the myriad of pathogenetic pathways leading to disease and the intricate network of the mediating molecules, make defining and developing adequate treatment strategies a gargantuan task. In addition, many of the treatment regimens include drugs that show short-term benefits, but long-term usage is hampered by associated complications. Even those drugs that show substantial short-term benefits such as corticosteroids cannot be used in the many patients who do not respond to these drugs due to unknown mechanisms. The introduction of biologic therapy has promulgated a new era in IBD therapy. However, the benefits of the most widely utilized, Infliximab, is limited owing to the associated complications, in particular antibody formation. Similarly, IL-10 therapy shown to be effective in *in vivo* studies does not show similar benefits in clinical trials because of an inability to regulate the appropriate dose to the appropriate targets. Probiotic formations are particularly of interest as they do not seem to be associated with substantial side effects, if any. Nonetheless, these latter therapies could likely be labeled as non-specific, with limited information on their mode and site of action.

20.11 Topical delivery methods

Current therapeutic modalities for IBD are aimed at reducing the extent and symptoms of the intestinal inflammation rather than curing the disease. Although there has been some success at disease control, the majority of patients do not respond adequately to the drugs, leading to complications such as abscesses and fistulas requiring surgery. Even post-surgery, current therapy is unable to ameliorate disease in previously unaffected regions of the GIT. In addition to the inability of current therapy to substantially modify disease course, most of the specific drugs are associated with adverse effects. Methods to reduce systemic absorption and targeting of the drug to affected regions of the GIT are of primary importance if benefit from drug therapy is to be expected. In the following sections we briefly outline current formulations that seek target delivery and describe newer methods that could enhance the efficacy of IBD medication.

20.12 Current strategies for oral delivery of IBD medication

The aim of these strategies is to enhance absorption of the drug in the diseased areas of the GIT. Approaches include chemical modifications that allow absorption only in the colon (for UC in particular), use of delayed/controlled formulations that deliver drug to the lower intestine and promotion of first-pass metabolism. Some of these strategies have been used with some success especially for the delivery of drugs such as 5-ASA and more recently corticosteroids and IL-10. Various formulations and chemical modifications are utilized that achieve either pH-dependent release or time-dependent release or are dependent on bacterial degradation. With regards to 5-ASA polymers of Eudragit L100, L100-

55, Eudragit S100 either singly or in combination has been shown to be useful in achieving either pH-dependent or time-dependent release of the drug to the intestine.[29,30] Chitosan capsules containing 5-ASA can be shown to release the drug exclusively to the colon,[31] whereas others have shown that guar gum matrix tablets or pectin matrixes can also effectively deliver the drug.[32–34] Similarly, Eudragit E and polyvinyacetal diethylaminoacetate polymer AEA have been studied for dexamethasone delivery.[35] A multiparticulate system consisting of drug-loaded cellulose acetate butyrate coated by Eudragit S has been successfully utilized for the targeted delivery of the corticosteroid budenoside.[36] Chemical modification strategies include attaching carrier molecules to the pro-drug of the parent drug have been demonstrated to effectively deliver either 5-ASA or dexamethasone to the target tissues. These and other delivery systems have recently been reviewed extensively by Kesisoglou and Zimmermann.[37]

20.13 Delivery using artificial cell microcapsules

One of the most promising approaches for local delivery of drugs in IBD is via the use of polymeric particulate systems, such as microspheres. Artificial cell microencapsulation is a technique to encapsulate biologically active materials in a specialized ultra-thin semipermeable polymer membrane.[38,39] They are, therefore, very useful as they can protect the encapsulated materials from the external environment while at the same time permitting selected materials to pass into and out of the microcapsules. In this manner, the enclosed material can be retained and separated from the undesirable external environment. Microcapsules are known to protect live cells, enzymes, DNA, etc. from immune rejection and other extreme ambiance and have a number of biomedical and clinical applications.[40]

The potential utility of using microcapsules for drug delivery in IBD has recently been researched by various authors. For example, Lamprecht *et al.*[41] demonstrated that microcapsules' adherence is high in inflamed intestinal segments. On the other hand, Nakase *et al.*[42] have reported that administration of dexamethasone encapsulated in microcapsules was efficient in treating dextran sodium sulfate colitis in mice. The therapeutic efficacy was high and systemic absorption was low. Microcapsules made of other formulations have also been shown to be effective for the local delivery of IL-10. This strategy when utilized in IL-10 deficient colitic mice (after rectal administration) successfully inhibited inflammation.[43]

The authors have recently carried out proof-of-principle studies for examining delivery of thalidomide for the treatment of CD. Thalidomide is a potent anti-inflammatory agent. Open-label studies and experimental studies have shown the effectiveness of the drug in resistant CD. However, given the potential toxicity of the drug, long-term use is not justified. Delivering thalidomide using microcapsules we have shown that the drug can be efficiently delivered to the different parts of the GIT[44] (Figs 20.1–20.3). In addition, we have

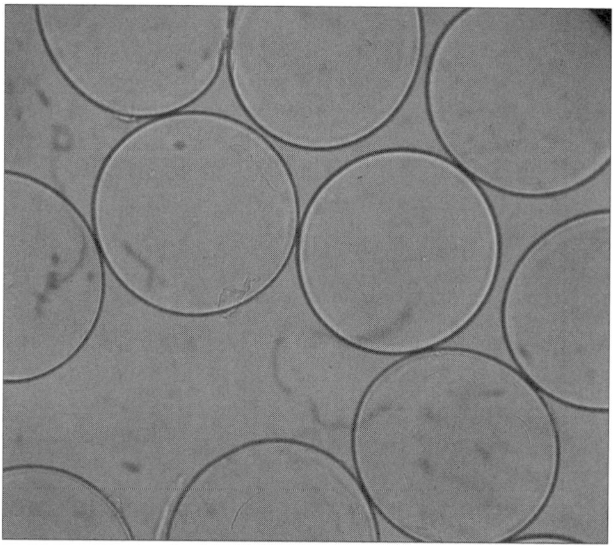

20.1 Photomicrographs of freshly prepared APA (alginate–poly-L-lysine–alginate) microcapsule thalidomide formulation (size 300 μm $\pm$ 50 μm, magnification 250×).

demonstrated that delivery of thalidomide in microspheres was efficient in inhibiting production of TNF-alpha (Figs 20.4 and 20.5) the potent pro-inflammatory cytokine implicated in IBD pathogenesis.[45]

These and other studies highlight the utility of microcapsule-based therapy for targeted delivery of drugs in IBD. The success of artificial cell microcapsule oral

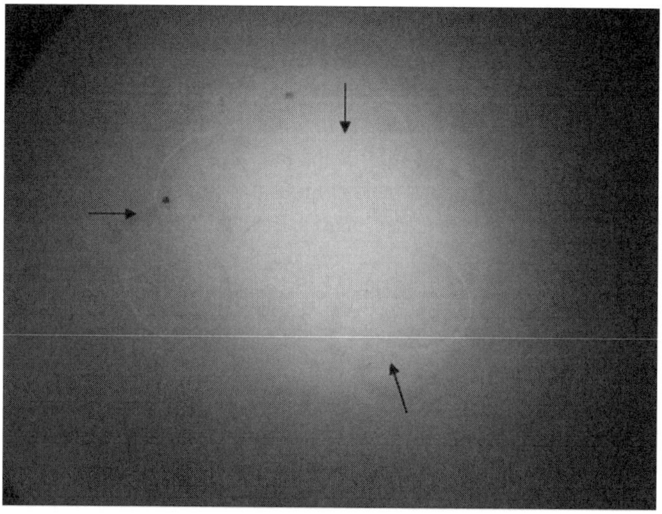

20.2 Photomicrograph of APA membranes after exposure to pH 7.5 conditions for 20 min (250× magnification).

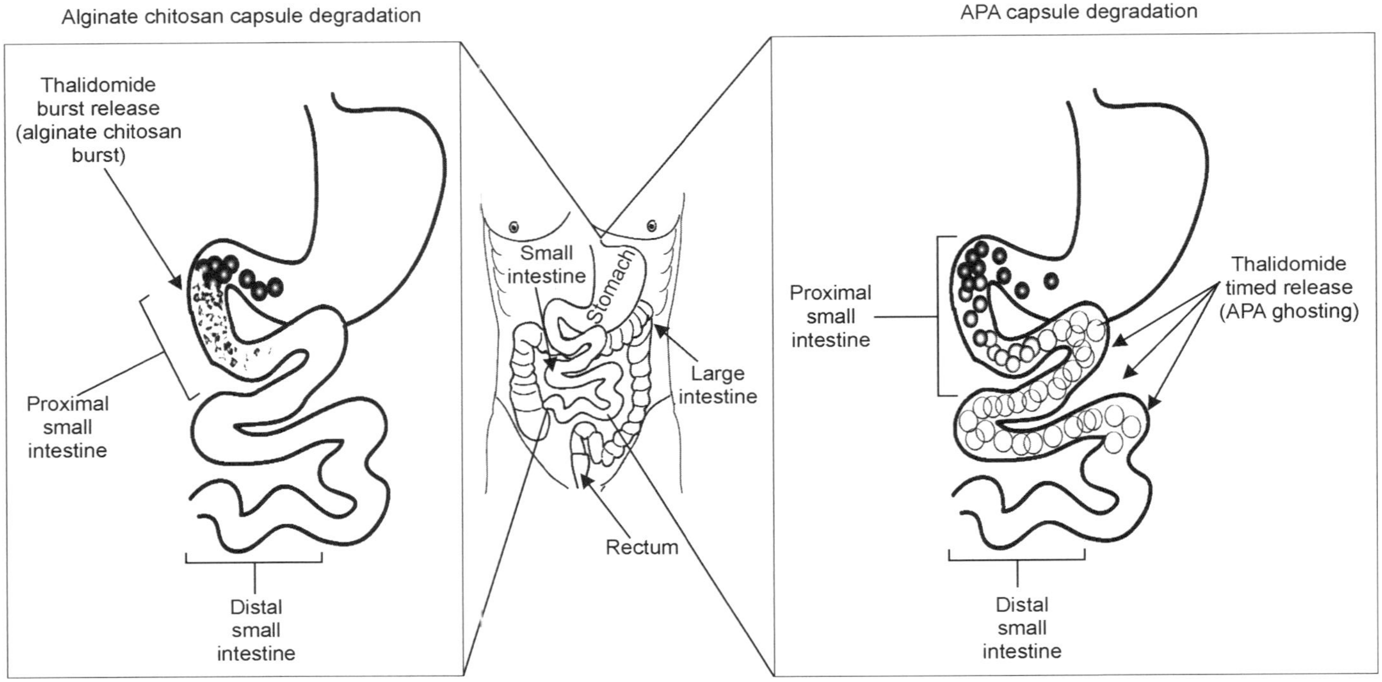

20.3 Thalidomide targeted delivery of AC and APA microcapsules.

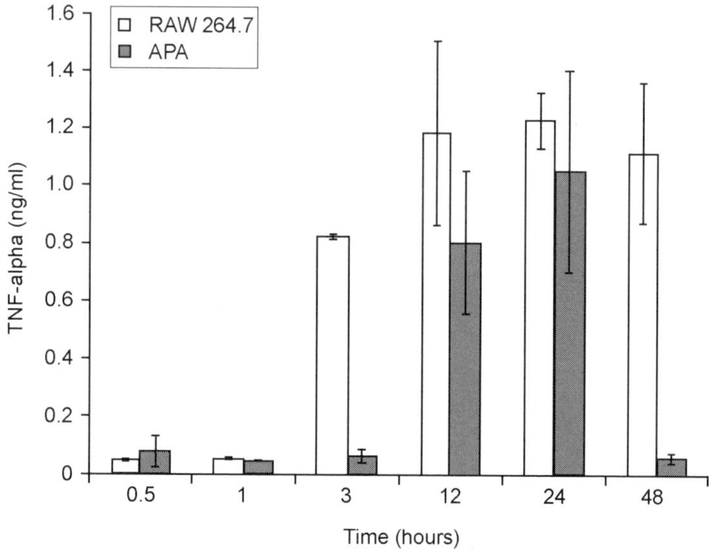

20.4 The concentration of TNF-α secretion from RAW 264.7 macrophage cells stimulated with 10 μg/ml of lipopolysaccharide in control and RAW 264.7 macrophage cells in the presence of APA-encapsulated thalidomide. Comparisons were made after incubation times of 0.5, 1, 3, 12, 24 and 48 h.

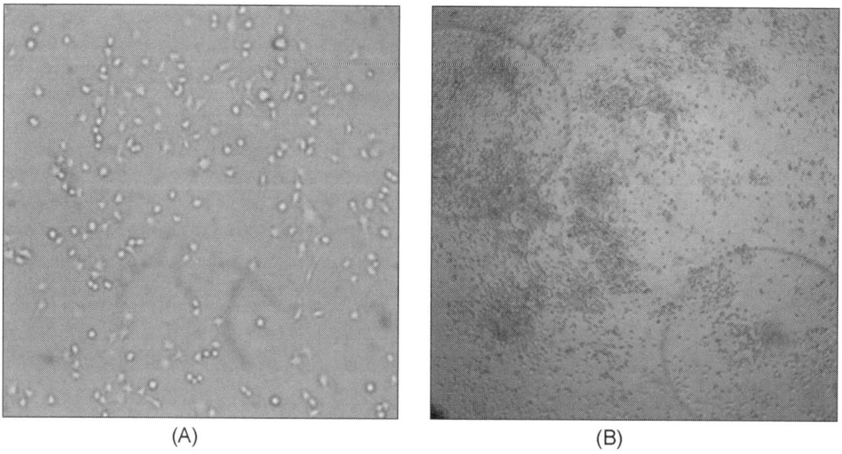

20.5 Photomicrographs of RAW 264.7 control macrophage cells and experimental RAW 264.7 macrophage cells after exposure to APA-encapsulated thalidomide formulations. (A) control macrophages (magnification of 200×); (B) after 48 h exposure to APA microcapsules thalidomide exposer (magnification 200×). APA microcapsule thalidomide formulations are also seen.

delivery for therapy, however, depends on the suitability of the microcapsule membrane for GI delivery. For example, a microcapsule can be disrupted by many different means during its intestinal passage; it may be fractured by the enzymatic action, chemical reaction, heat, pH, diffusion, mechanical pressure and other related physiological and biochemical stresses. The identification and characterization of the level of physiochemical changes in the microcapsules are, therefore, necessary. The microcapsule membrane must be provided with sufficient permeability for nutrients, and secretion and excretion products, to pass through, yet prevent the entry of hostile molecules or cells for the host, for example, products of the host's immune response, which could destroy the encapsulated material.[46–49] Therefore, future research is necessary to further examine the utility of this promising approach for IBD therapy.

20.14 Conclusions and future trends

Being a complex disease, arriving at a suitable therapeutic protocol for IBD is a huge task. Most current therapy target single molecules thought to be key in the pathogenesis of the disease. Although successful at ensuring short-term relief, most applications have to endure failure in the long run. Combination therapies and step-up-step-down therapies have all been advocated. Immunomodulators and biologicals show promise but their effectiveness in curing the disease remains a question. Probiotics show efficacy in animal models of inflammation, but evidence from human studies is limited. Artificial microcapsule encapsulation for delivery of drugs for IBD therapy is an exciting approach. Initial studies indicate tremendous potential for treatment especially for the delivery of drugs such as steroids and thalidomide that have extensive anti-inflammatory potential but whose utility is limited owing to inherent toxicities. Targeted delivery of these drugs to sites of inflammation would considerably enhance their efficacy in resulting in a cure for IBD. Further research on these delivery methods is therefore a priority if a cure for IBD is to be determined in the near future.

20.15 Acknowledgements

D. Amre would like to acknowledge the support of research grant from the Canadian Institute of Health Research (CIHR). S. Prakash acknowledges research grant support from the Canadian Institute of Health Research (CIHR).

20.16 References

1. Bernstein, CN; Blanchard, JF; Rawsthorne, P; Wajda, A. Epidemiology of Crohn's disease and ulcerative colitis in a central Canadian province: a population-based study. *Am J Epidemiol* 1999; 149(10): 916–924.

2. Farmer, RG; Whelan, G; Fazio, VW. Long-term follow-up of patients with Crohns-disease – relationship between the clinical-pattern and prognosis. *Gastroenterology* 1985; 88(6): 1818–1825.
3. Gasche, C; Scholmerich, J; Brynskov, J *et al.* A simple classification of Crohn's disease: Report of the Working Party for the World Congresses of Gastroenterology, Vienna 1998. *Inflamm Bowel Dis* 2000; 6(1): 8–15.
4. Satsangi, J; Silverberg, MS; Vermeire, S; Colombel, JF. The Montreal classification of inflammatory bowel disease: controversies, consensus, and implications. *Gut* 2006; 55(6): 749–753.
5. Camma, C; Giunta, M; Rosselli, M; Cottone, M. Mesalamine in the maintenance treatment of Crohn's disease: a meta-analysis adjusted for confounding variables. *Gastroenterology* 1997; 113(5): 1465–1473.
6. Akobeng, AK and Gardener, E. Oral 5-aminosalicylic acid for maintenance of medically-induced remission in Crohn's disease. *Cochrane Database Systematic Rev* 2005; 1: CD003715.
7. Lochs, H; Mayer, M; Fleig, WE *et al.* Prophylaxis of postoperative relapse in Crohn's disease with mesalamine: European Cooperative Crohn's disease Study VI. *Gastroenterology* 2000; 118(2): 264–273.
8. Faubion, WA; Loftus, EV; Harmsen, WS; Zinsmeister, AR; Sandborn, WJ. The natural history of corticosteroid therapy for inflammatory bowel disease: a population-based study. *Gastroenterology* 2001; 121(2): 255–260.
9. Munkholm, P; Langholz, E; Davidsen, M; Binder, V. Frequency of glucocorticoid resistance and dependency in Crohns-disease. *Gut* 1994; 35(3): 360–362.
10. Pearson, DC; May, GR; Fick, GH; Sutherland, LR. Azathioprine and 6-mercaptopurine in Crohn disease – a metaanalysis. *Ann Intern Med* 1995; 123(2): 132–142.
11. Friedman, S. General principles of medical therapy of inflammatory bowel disease. *Gastroenterol Clin North Am* 2004; 33(2): 191–208.
12. Hanauer, SB; Present, DH. The state of the art in the management of inflammatory bowel disease. *Rev Gastroenterol Disord* 2003; 3(2): 81–92.
13. Guslandi, M. Antibiotics for inflammatory bowel disease: do they work? *Eur J Gastroenterol Hepatol* 2005; 17(2): 145–147.
14. Ardizzone, S; Bianchi, PG. Biologic therapy for inflammatory bowel disease. *Drugs* 2005; 65(16): 2253–2286.
15. Ljung, T; Karlen, P; Schmidt, D *et al.* Infliximab in inflammatory bowel disease: clinical outcome in a population based cohort from Stockholm County. *Gut* 2004; 53(6): 849–853.
16. Rutgeerts, P; Van, AG; Vermeire, S. Review article: Infliximab therapy for inflammatory bowel disease – seven years on. *Aliment Pharmacol Ther* 2006; 23(4): 451–463.
17. Bratcher, JM; Korelitz, BI. Toxicity of Infliximab in the course of treatment of Crohns disease. *Exp Opin Drug Safety* 2006; 5(1): 9–16.
18. Chung, KF; Patel, HJ; Fadlon, EJ *et al.* Induction of eotaxin expression and release from human airway smooth muscle cells by IL-1beta and TNFalpha: effects of IL-10 and corticosteroids. *Br J Pharmacol* 1999; 127(5): 1145–1150.
19. Staples, KJ; Bergmann, M; Barnes, PJ; Newton, R. Stimulus-specific inhibition of IL-5 by cAMP-elevating agents and IL-10 reveals differential mechanisms of action. *Biochem Biophys Res Commun* 2000; 273(3): 811–815.
20. Colombel, JF; Rutgeerts, P; Malchow, H *et al.* Interleukin 10 (Tenovil) in the prevention of postoperative recurrence of Crohn's disease. *Gut* 2001; 49(1): 42–46.

21. Fedorak, RN; Gangl, A; Elson, CO *et al.* Recombinant human interleukin 10 in the treatment of patients with mild to moderately active Crohn's disease. The Interleukin 10 Inflammatory Bowel Disease Cooperative Study Group. *Gastroenterology* 2000; 119(6): 1473–1482.
22. Schreiber, S; Fedorak, RN; Nielsen, OH *et al.* Safety and efficacy of recombinant human interleukin 10 in chronic active Crohn's disease. Crohn's Disease IL-10 Cooperative Study Group. *Gastroenterology* 2000; 119(6): 1461–1472.
23. Tilg, H; van, MC; van den, EA *et al.* Treatment of Crohn's disease with recombinant human interleukin 10 induces the proinflammatory cytokine interferon gamma. *Gut* 2002; 50(2): 191–195.
24. Sartor, RB. Therapeutic manipulation of the enteric microflora in inflammatory bowel diseases: antibiotics, probiotics, and prebiotics. *Gastroenterology* 2004; 126(6): 1620–1633.
25. Hogaboam, CM; Vallance, BA; Kumar, A *et al.* Therapeutic effects of interleukin-4 gene transfer in experimental inflammatory bowel disease. *J Clin Invest* 1997; 100(11): 2766–2776.
26. Barbara, G; Xing, Z; Hogaboam, CM; Gauldie, J; Collins, SM. Interleukin 10 gene transfer prevents experimental colitis in rats. *Gut* 2000; 46(3): 344–349.
27. Rogy, MA; Beinhauer, BG; Reinisch, W; Huang, L; Pokieser, P. Transfer of interleukin-4 and interleukin-10 in patients with severe inflammatory bowel disease of the rectum. *Hum Gene Ther* 2000; 11(12): 1731–1741.
28. Kitani, A; Fuss, IJ; Nakamura, K *et al.* Treatment of experimental (trinitrobenzene sulfonic acid) colitis by intranasal administration of transforming growth factor (TGF)-beta1 plasmid: TGF-beta1-mediated suppression of T helper cell type 1 response occurs by interleukin (IL)-10 induction and IL-12 receptor beta2 chain downregulation. *J Exp Med* 2000; 192(1): 41–52.
29. Cheng, G; An, F; Zou, MJ *et al.* Time- and pH-dependent colon-specific drug delivery for orally administered diclofenac sodium and 5-aminosalicylic acid. *World J Gastroenterol* 2004; 10(12): 1769–1774.
30. Khan, MZ; Prebeg, Z; Kurjakovic, N. A pH-dependent colon targeted oral drug delivery system using methacrylic acid copolymers. I. Manipulation of drug release using Eudragit L100-55 and Eudragit S100 combinations. *J Control Rel* 1999; 58(2): 215–222.
31. Tozaki, H; Odoriba, T; Okada, N *et al.* Chitosan capsules for colon-specific drug delivery: enhanced localization of 5-aminosalicylic acid in the large intestine accelerates healing of TNBS-induced colitis in rats. *J Control Rel* 2002; 82(1): 51–61.
32. Tugcu-Demiroz, F; Acarturk, F; Takka, S; Konus-Boyunaga, O. *In-vitro* and *in-vivo* evaluation of mesalazine-guar gum matrix tablets for colonic drug delivery. *J Drug Target* 2004; 12(2): 105–112.
33. Krishnaiah, YS; Muzib, YI; Bhaskar, P; Satyanarayana, V; Latha, K. Pharmacokinetic evaluation of guar gum-based colon-targeted drug delivery systems of tinidazole in healthy human volunteers. *Drug Deliv* 2003; 10(4): 263–268.
34. Turkoglu, M; Ugurlu, T. *In vitro* evaluation of pectin-HPMC compression coated 5-aminosalicylic acid tablets for colonic delivery. *Eur J Pharm Biopharm* 2002; 53(1): 65–73.
35. Leopold, CS; Eikeler, D. Basic coating polymers for the colon-specific drug delivery in inflammatory bowel disease. *Drug Dev Ind Pharm* 2000; 26(12): 1239–1246.
36. Rodriguez, M; Antunez, JA; Taboada, C; Seijo, B; Torres, D. Colon-specific delivery of budesonide from microencapsulated cellulosic cores: evaluation of the

efficacy against colonic inflammation in rats. *J Pharm Pharmacol* 2001; 53(9): 1207–1215.

37. Kesisoglou, F; Zimmermann, EM. Novel drug delivery strategies for the treatment of inflammatory bowel disease. *Exp Opin Drug Deliv* 2005; 2(3): 451–463.
38. Chang, TM. Semipermeable microcapsules. *Science* 1964; 146: 524–525.
39. Chang, SJ; Lee, CH; Wang, YJ. Microcapsules prepared from alginate and a photosensitive poly(L-lysine). *J Biomater Sci Polym Ed* 1999; 10(5): 531–542.
40. Uludag, H; De, VP; Tresco, PA. Technology of mammalian cell encapsulation. *Adv Drug Deliv Rev* 2000; 42(1–2): 29–64.
41. Lamprecht, A; Schafer, U; Lehr, CM. Size-dependent bioadhesion of micro- and nanoparticulate carriers to the inflamed colonic mucosa. *Pharm Res* 2001; 18(6): 788–793.
42. Nakase, H; Okazaki, K; Tabata, Y; Chiba, T. Biodegradable microspheres targeting mucosal immune-regulating cells: new approach for treatment of inflammatory bowel disease. *J Gastroenterol* 2003; 38 Suppl 15: 59–62.
43. Nakase, H; Okazaki, K; Tabata, Y *et al.* New cytokine delivery system using gelatin microspheres containing interleukin-10 for experimental inflammatory bowel disease. *J Pharmacol Exp Ther* 2002; 301(1): 59–65.
44. Metz, T; Jones, ML; Chen, H *et al.* A new method for targeted drug delivery using polymeric microcapsules: implications for treatment of Crohn's disease. *Cell Biochem Biophys* 2005; 43(1): 77–85.
45. Metz, T; Haque, T; Chen, H; Amre, DK; Das, SK; Prakash, S. Preparation and *in vitro* analysis of microcapsule thalidomide formulation for targeted suppression of TNF-α. *Drug Del* 2006; 13: 1–7.
46. Rihova, B. Biocompatibility of biomaterials: hemocompatibility, immunocompatibility and biocompatibility of solid polymeric materials and soluble targetable polymeric carriers. *Adv Drug Deliv Rev* 1996; 21: 157–176.
47. Rihova, B. Immunocompatibility and biocompatibility of cell delivery systems. *Adv Drug Deliv Rev* 2000; 42: 65–80.
48. Ross, CJD; Ralph, M; Chang, PL. Delivery of recombinant gene products of the central nervous system with nonautologous cells in alginate microcapsules. *Hum Gene Therap* 1999; 10: 49–59.
49. Sun, AM; Goosen, MF; O'Shea, G. Microencapsulated cells as hormone delivery systems. *Crit Rev Ther Drug Carrier Syst* 1987; 4: 1–12.

21

Carrier-mediated and artificial-cell targeted cancer drug delivery

W C ZAMBONI, University of Pittsburgh, USA

21.1 Introduction

21.1.1 Cancer problem and potential: issues related to drug delivery in solid tumors

Major advances have been made in the use of cancer chemotherapy (Grever and Chabner, 2006). However most patients, especially patients diagnosed with solid tumors, fail to respond to initial treatment or relapse after an initial response (Grever and Chabner, 2006). Thus, there is a need to identify factors associated with lack of response and to develop new treatment strategies that address those factors. The development of effective chemotherapeutic agents for the treatment of solid tumors depends, in part, on the ability of those agents to achieve cytotoxic drug concentrations or exposure within the tumor (Jain, 1996; Zamboni *et al.*, 1999).

It is currently unclear why within a patient with solid tumors there can be a reduction in the size of some tumors while other tumors can progress during or after treatment, even though the genetic composition of the tumors is similar (Balch *et al.*, 2006). Such variable antitumor responses within a single patient may be associated with inherent differences in tumor vascularity, capillary permeability and/or tumor interstitial pressure that result in variable delivery of anticancer agents to different tumor sites (Jain, 1996; Zamboni *et al.*, 1999). However, studies evaluating the intra-tumoral concentration of anticancer agents and factors affecting tumor exposure in preclinical models and patients are rare (Blochl-Daum *et al.*, 1996; Muller *et al.*, 1997; Zamboni *et al.*, 1999). In addition, preclinical models evaluating tumor exposure of anticancer agents and factors affecting tumor exposure may not reflect the disposition of chemotherapeutic agents in patients with solid tumors due to differences in vascularity and lymphatic drainage (Jain, 1996; Zamboni *et al.*, 1999). Moreover, it is logistically difficult to perform the extensive studies required to evaluate the tumor disposition of anticancer agents and factors that determine the disposition in patients with solid tumors, especially in tumors that are not easily accessible.

Thus, there is impending need to develop and implement techniques and methodologies to evaluate the disposition and exposure of anticancer agents within the tumor matrix.

21.1.2 Carrier formulations of anticancer agents

Major advances in the use of carrier vehicles delivering pharmacologic agents and enzymes to sites of disease have occurred the past 10 years (Gradishar, 2006; Drummond *et al.*, 1999; Papahadjopoulos *et al.*, 1991; D'Emanuele and Attwood, 2005; Ouyang *et al.*, 2004). The primary types of carrier-mediated anticancer agents are liposomes and nanoparticles (Table 21.1). Nanoparticles are then subdivided into microspheres, dendrimers and conjugate formulations (Gradishar, 2006; Drummond *et al.*, 1999; Papahadjopoulos *et al.*, 1991; D'Emanuele and Attwood, 2005; Ouyang *et al.*, 2004). The theoretical advantages of liposomal and nanoparticle encapsulated and carrier-mediated drugs are increased solubility, prolonged duration of exposure, selective delivery of entrapped drug to the site of action, improved therapeutic index, and potentially overcoming resistance associated with the regular anticancer agent (Drummond *et al.*, 1999; Papahadjopoulos *et al.*, 1991). The process by which these agents preferentially accumulate in tumor and tissues is called enhanced permeation

Table 22.1 Summary of carrier-mediated chemotherapeutic agents

Liposomal agents		Nanoparticles	
Conventional	Pegylated (sterically stabilized)	Microspheres	Conjugates
LE-SN38	Doxil/Caelyx	Abraxane (ABI007)	Xyotax (PPX)
Lurtotecan/OSI-211	S-CKD602	Paclimer	DHA–paclitaxel
9NC		TOCOSOL-Paclitaxel	PEG–doxorubicin
Irinotecan		APA-Thalidomide	PEG–methotrexate
Paclitaxel		PX-NP	PEG–interferon
Doxorubicin			PEG–camptothecin
Daunorubicin			20-carbonate–camptothecin
Cytarabine			PL–ara-C
Topotecan			PL–gemcitabine
Vincristine (onco-TSC)			
FRL-doxorubicin : vincristine			
FRL-daunorubicin : cytarabine			
FRL-5-fluorouracil : irinotecan			
FRL-cisplatin : irinotecan			

FRL = fixed ratio liposomes; DHA = docosahexaenoic acid; PEG = polyethylene glycol; Ara-C = cytosine arabinoside; PL = phospholipids; APA = alginate-poly-lysine-alginate microcapsules; PX-NP = paclitaxel entrapped in cetyl alcohol/polysorbate nanoparticles and retention effect (Maeda *et al.*, 2000). Pegylated-Stealth™ liposomal doxorubicin (Doxil®, Caelyx®) and paclitaxel albumin-bound particles (Gradishar, 2006; Socinski, 2006) are the only two members of this relatively new class of agents that are FDA approved (Krown *et al.*, 2004; Markman *et al.*, 2004; Socinski, 2006). Doxil is approved for the treatment of refractory ovarian cancer and Kaposi's sarcoma (KS) (Krown *et al.*, 2004; Markman *et al.*, 2004; Rose, 2005). Non-pegylated liposomal formulations of doxorubicin, Myocet® and DaunoXome® are approved in Europe for the treatment of breast cancer and KS, respectively (Allen and Martin, 2004). Abraxane is approved for the treatment of refractory breast cancer. However, there are more than 50 other agents that are in preclinical and clinical development. Newer generations of liposomes containing two anticancer agents with a single liposome and antibody-targeted liposomes which may improve selective toxicity are in preclinical development (Abraham *et al.*, 2004; Laginha *et al.*, 2005; Park *et al.*, 2004). In addition, nanoparticle formulations, such as microspheres, dendrimers and conjugates provide a unique method to provide tumor-selective delivery of anticancer agents to tumors. As more existing anticancer agents go off patent these agents will be most likely be evaluated in some type of liposome or carrier-mediated formulation. In addition, anti-angiogenesis agents, antisense oligonucleotides and enzymes represent rational candidates for liposomal and nanoparticle formulations (Park *et al.*, 2004).

The pharmacokinetic disposition of these agents is dependent upon the carrier and not the parent-drug until the drug is released from the carrier (Laginha *et al.*, 2005). Thus, the pharmacology and pharmacokinetics of these agents are complex and detailed studies must be performed to evaluate the disposition of the encapsulated or conjugated form of the drug and the released active-drug (Zamboni *et al.*, 2004). The factors affecting the pharmacokinetic and pharmacodynamic variability of these agents remain unclear, but probably include the reticuloendothelial system (RES), which has also been called the mononuclear phagocyte system (MPS) (Laverman *et al.*, 2001; Litzinger *et al.*, 1994; Woodle and Lasic, 1992).

21.1.3 Methods for evaluation of carrier agents

The need to develop and readily gain information on the tumor disposition of agents may become more important with the increasing number of tumor targeting approaches, such as gene and antisense therapy, polyethylene glycol (PEG)-conjugated agents and liposomal delivery (Brunner and Muller, 2002; Zamboni *et al.*, 2004). In addition, methodology and study designs used to develop classic cytotoxic anticancer agents, such as platinum, taxane and camptothecin analogs, may not be appropriate for the new generations of anticancer therapy, such as angiogenesis inhibitors, antiproliferative agents and signal transduction inhibitors (Brunner and Muller, 2002; Gelmon *et al.*, 1999).

As these agents may not induce classic toxicities or any toxicity, it may be difficult to recommend a dose for future trials using the standard phase I dose escalation methods and end points (i.e. maximum tolerable dose and dose limiting toxicities). Alternatively, defining the dose for phase II studies could be based on the dose that achieves exposures associated with pharmacologic modulation, optimal biological exposure or cytotoxicity results from *in vitro* studies (Zamboni *et al.*, 1998a, 1999).

Historically, investigators have compared *in vitro* IC_{50} values with plasma concentrations in patients as a means to determine if sufficient exposure has been reached in clinical studies. However, the inherent tumor characteristics which influence tumor penetration and high intra- and inter-tumoral variability in tumor exposure makes this comparison highly unreliable (Boucher and Jain, 1992; Jain, 1996; Zamboni *et al.*, 1999), especially when the ratio of tumor exposure to plasma exposure may be approximately 0.2 to 0.5 (Ekstrom *et al.*, 1996, 1997b; Muller *et al.*, 1998; Zamboni *et al.*, 1999, 2002). Thus, comparing the *in vitro* exposures and plasma exposures in patients results in an overestimation of drug exposure in the tumor ECF, and thus the required exposure for effect may be insufficient. The use of methodologies that measure the exposure of anticancer agents within the tumor may improve the level of information needed to make informed decisions during the drug development process.

21.2 Carrier-mediated and artificial-cell targeted cancer drug delivery

21.2.1 Liposomal formulations of anticancer agents

Characteristics of liposomes

Liposomes are microscopic vesicles composed of a phospholipid bilayer that are capable of encapsulating the active drug. However, conventional liposomes are opsonized by plasma proteins, quickly recognized as foreign bodies, and rapidly removed by the MPS (Allen and Hansen, 1991; Allen and Stuart, 2005; Drummond *et al.*, 1999). Depending on the size and composition of the liposome, MPS uptake can occur within minutes after administration and remove the liposomes from the circulation. Studies evaluating the disposition and tumor penetration of liposomal and non-liposomal anticancer agents suggest liposomal agents have an extended systemic half-life and extravasate selectively into solid tumors through the capillaries of tumor neovasculature (Allen and Hansen, 1991; Drummond *et al.*, 1999). The exact mechanism of liposomal clearance is currently unclear. The mechanisms by which liposomes enter tissue and tumors, and release drug are not completely understood. In addition, the liposomes can be engineered to produce a complete spectrum of drug release rates that need to be evaluated in *in vivo* systems (Barenholz and Haran, 1993; Lasic *et al.*, 1992).

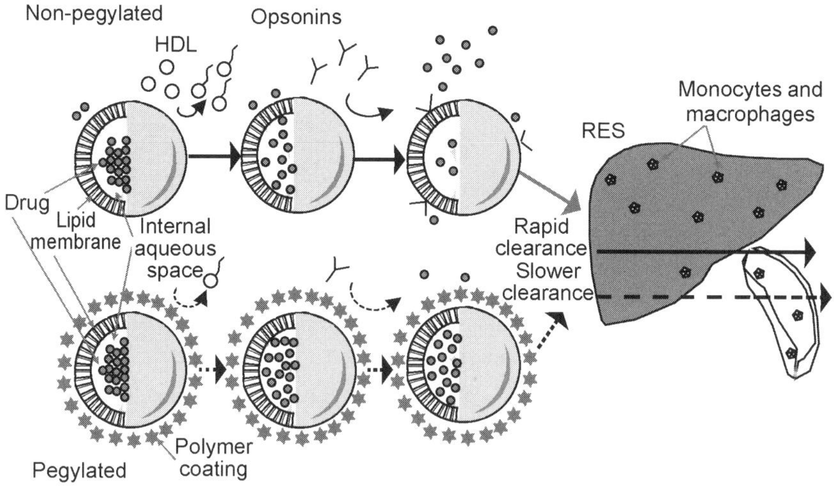

21.1 Clearance of pegylated (sterically stabilized) and non-pegylated (conventional) liposomes via the reticuloendothelial system (RES) in the liver and spleen. Non-pegylated liposomes undergo greater breakdown in blood and more rapid clearance via the RES compared with pegylated liposomes.

The development of Stealth™ liposomes, which contain lipids conjugated to PEG, was based on the theory that incorporation of PEG-lipids into liposomes would allow the liposome to evade the immune system and prolong the duration of exposure (Fig. 21.1) (Allen and Hansen, 1991; Allen and Stuart, 2005; Papahadjopoulos *et al.*, 1991; Woodle and Lasic, 1992). Stealth™ liposomes have a lipid bilayer membrane like conventional liposomes, but the surface contains surface-grafted linear segments of PEG extending 5 nm from the surface (Papahadjopoulos *et al.*, 1991; Woodle and Lasic, 1992). Stealth™ liposomes are relatively small, with an average diameter of approximately 100 nm. The size optimally balances drug-carrying capacity and circulation time, and allows extravasation through the endothelial gaps in the capillary bed of target tumors. Whether the drug is encapsulated in the core or in the bilayer of the liposome is dependent upon the characteristics of the drug and the encapsulation process. In general, water-soluble drugs are encapsulated within the central aqueous core, whereas lipid-soluble drugs are incorporated directly into the lipid membrane.

Systemic and tissue disposition of liposomes

Liposomes can alter both the tissue distribution and the rate of clearance of the drug by making the drug take on the pharmacokinetic characteristics of the carrier (Drummond *et al.*, 1999; Papahadjopoulos *et al.*, 1991; Woodle and Lasic, 1992). Pharmacokinetic parameters of the liposomes depend on the physiochemical characteristics of the liposomes, such as size, surface charge, membrane lipid packing, steric stabilization, dose and route of administration (Drummond *et al.*, 1999). The primary sites of accumulation of conventional liposomes are the tumor, liver and spleen compared with non-liposomal formulations (Allen and Hansen, 1991; Allen and Stuart, 2005; Drummond *et al.*, 1999; Laverman *et al.*, 2001; Litzinger *et al.*, 1994; Newman *et al.*, 1999; Working *et al.*, 1994). The development of Stealth^{TM} liposomes was based on the discovery that incorporation of PEG-lipids into liposomes yields preparations with superior tumor delivery compared with conventional liposomes composed of natural phospholipids (Allen and Hansen, 1991; Allen and Stuart, 2005; Drummond *et al.*, 1999). Incorporation of PEG-lipids, causes the liposome to remain in the blood circulation for extended periods of time (i.e. half-life > 40 hours) and distribute through an organism relatively evenly with most of the dose remaining in the central compartment (i.e. the blood) and only 10–15% of the dose being delivered to the liver (Allen and Hansen, 1991; Allen and Stuart, 2005; Newman *et al.*, 1999; Working *et al.*, 1994). This is a significant improvement over conventional liposomes where typically 80–90% of the liposome deposits in the liver.

The clearance of conventional liposomes has been proposed to occur by uptake of the liposomes by the MPS (Fig. 21.1) (Allen and Hansen, 1991; Drummond *et al.*, 1999). The MPS uptake of liposomes results in their rapid removal from the blood and accumulation in tissues involved in the MPS, such as the liver and spleen. Uptake by the MPS usually results in irreversible sequestering of the encapsulated drug in the MPS, where it can be degraded. In addition, the uptake of the liposomes by the MPS may result in acute impairment of the MPS and toxicity. Sterically stabilized liposomes, such as Stealth^{TM} liposomes, prolong the duration of exposure on the encapsulated liposome in the systemic circulation (Papahadjopoulos *et al.*, 1991; Woodle and Lasic, 1992). The presence of the PEG coating on the outside of the liposome does not prevent uptake by the RES, but simply reduces the rate of uptake (Fig. 21.1) (Allen and Hansen, 1991). The exact mechanism by which steric stabilization of liposomes decreases the rate of uptake by the RES is unclear (Drummond *et al.*, 1999; Mori *et al.*, 1991; Papahadjopoulos *et al.*, 1991; Woodle and Lasic, 1992).

Tumor delivery of liposomal agents

The development of effective chemotherapeutic agents for the treatment of solid tumors depends, in part, on the ability of those agents to achieve cytotoxic drug exposure within the tumor extracellular fluid (ECF) (Jain, 1996; Zamboni *et al.*, 1999). Solid tumors have several potential barriers to drug delivery that may limit drug penetration and provide inherent mechanisms of resistance (Jain, 1996). Moreover, factors affecting drug exposure in tissue, such as alteration in the distribution of blood vessels, blood flow, capillary permeability, interstitial pressure and lymphatic drainage, may be different in tumors and the surrounding normal tissue (Jain, 1996).

The accumulation of liposomes or large macromolecules in tumors is a result of the extended duration of exposure in the systemic circulation and the leaky microvasculature and impaired lymphatics supporting the tumor area (Drummond *et al.*, 1999; Jain, 1996; Newman *et al.*, 1999; Working *et al.*, 1994). Once in the tumor, the non-targeted Stealth™ liposomes are localized in the ECF surrounding the tumor cell, but do not enter the cell (Harrington, 2001a,b). Thus, for the liposomes to deliver the active form of the anticancer agent, such as doxorubicin, the drug must be released from the liposome into the ECF and then diffuse into the cell (Zamboni *et al.*, 2004). As a result, the ability of the liposome to carry the anticancer agent to the tumor and the ability to release it into the ECF are equally important factors in determining the anti-tumor effect of liposomal encapsulated anticancer agents. In general, the kinetics of this local release are unknown as it is difficult to differentiate between the liposomal-encapsulated and released forms of the drug in solid tissue, although with the development of microdialysis, as discussed below, this is becoming easier (Zamboni *et al.*, 2004).

Modification of toxicity with liposomal agents

Liposomal formulations can also modify the toxicity profile of a drug (e.g. Ambisome®; Veerareddy and Vobalaboina, 2004). This effect may be due to an alteration in tissue distribution associated with liposomal formulations (Allen and Hansen, 1991; Newman *et al.*, 1999; Working *et al.*, 1994; Zamboni *et al.*, 2004). Anthracyclines, such as doxorubicin, are active against many tumor types, but cardiotoxicity related to the cumulative dose may limit their use (Ewer *et al.*, 2004). Preclinical studies determined that liposomal anthracyclines reduced the incidence and severity of cumulative dose-related cardiomyopathy while preserving antitumor activity (Ewer *et al.*, 2004). There is also clinical evidence suggesting that Doxil is less cardiotoxic than conventional doxorubicin (Ewer *et al.*, 2004; Northfelt, 1994). Direct comparisons between Doxil or Caelyx and conventional doxorubicin showed comparable efficacy but significantly lower risk of cardiotoxicity with the Stealth™ liposomal formulations of doxorubicin (Ewer *et al.*, 2004). In addition, histologic examination of cardiac biopsies from patients who received cumulative doses of Doxil from 440 to 840 mg/m^2, and had no prior exposure to anthracyclines, revealed significantly less cardiac toxicity than in matched doxorubicin controls ($p < 0.001$) (Berry *et al.*, 1998). Administration of a drug in a liposomal may also result in new toxicities (Cattel *et al.*, 2004; Rose, 2005; Vail *et al.*, 2004). The most common adverse event associated with Doxil is hand–foot syndrome (HFS), also known as palmar–plantar erythrodysesthesia) and stomatitis, which have not been reported with conventional doxorubicin (Rose, 2005). The exact mechanisms associated with these toxicities are unknown, but are schedule and dose dependent. In general, Doxil is generally well tolerated and its side effect profile compares favorably with other chemotherapy used in the treatment of refractory ovarian cancer. Proper dosing and monitoring may further enhance tolerability while preserving efficacy; however these is still a need to identify factors associated HFS, which can be dose limiting in some patients (Rose, 2005).

Liposomal agents

Stealth™ (Doxil/Caelyx) and conventional (Myocet and DaunoXome) liposomal formulations of doxorubicin that are approved in the United States and Europe, respectively (Allen and Martin, 2004; Krown *et al.*, 2004; Markman *et al.*, 2004; Rose, 2005). Some of the other liposomal anticancer agents that are currently in development are SN-38 (LE-SN38) (Kraut *et al.*, 2005; Lei *et al.*, 2004; Pal *et al.*, 2005; Zhang *et al.*, 2004), lurtotecan (OSI-211) (Dark *et al.*, 2005; Gelmon *et al.*, 2004; Giles *et al.*, 2004; Kehrer *et al.*, 2002), 9NC (Knight *et al.*, 1999; Koshkina *et al.*, 1999; Verschraegen *et al.*, 2004), irintotecan (Drummond *et al.*, 2005; Messerer *et al.*, 2004), Stealth™ liposomal CKD-602 (S-CKD602) (Zamboni *et al.*, 2005a), paclitaxel (LEP-ETU) (Damajanov *et al.*, 2005), doxorubicin (Mendelson *et al.*, 2005) and vincristine (Sarris *et al.*, 2000). Liposomal encapsulation of camptothecins is an attractive formulation due to the solubility issues associated with most camptothecin analogues and the potential for prolonged exposure after administration of a single dose (Dark *et al.*, 2005; Kraut *et al.*, 2005; Zamboni *et al.*, 2005a). As compared with pegylated or coated liposomes, conventional liposomal formulations of camptothecin analogs, such as LE-SN38 and OSI-211, may result in the rapid release of the drug from the liposome in blood and thus act more as a IV formulation as compared to a tumor targeting agent (Dark *et al.*, 2005; Giles *et al.*, 2004; Kraut *et al.*, 2005; Lei *et al.*, 2004; Pal *et al.*, 2005; Zhang *et al.*, 2004). However, studies evaluating encapsulated and released drug in plasma and tumor have not been reported (Zamboni *et al.*, 2004).

A randomized phase II trial of OSI-211 in patients with relapsed ovarian cancer compared OSI-11 IV on days 1, 2 and 3 repeated every 3 weeks and OSI-11 IV on days 1 and 8 repeated every 3 weeks was performed (Dark *et al.*, 2005). OSI-211 daily for 3 days was declared the winner in terms of objective response. A phase I study of LE-SN38 was performed in which patients were prospectively assigned to cohorts based on UDP-glucuronosyltransferase 1A1 (UGT1A1) genotype (Kraut *et al.*, 2005). The MTD was not reached in the *28/*28 patients. The maximum tolerated dose (MTD) of LE-SN38 in the WT/WT cohort was 35 mg/m^2 IV over 90 min every weeks. The pharmacokinetic disposition of SN-38 was similar in the WT/WT and WT/*28 cohorts. Interestingly, there were no reports of acute or delayed diarrhea even though the exposures of SN-38 were several fold higher after administration of LE-SN38 compared to irinotecan.

Preliminary results of a phase I study S-CKD602 administered IV over 1 h every 21 days reported that the half-life was increased 4- to 8-fold and plasma exposure was increased approximately 50-fold higher after administration of S-CKD602 compared with non-liposomal CKD602 (Zamboni *et al.*, 2005a). The results of this study suggest that S-CKD602 exhibits the characteristics that are consistent with other Stealth™ liposomes and thus may have pharmacologic advantages over other liposomal formulations of camptothecin analogs (Zamboni *et al.*, 2005a). In addition, S-CKD602 has produced responses in patients with platinum-refractory ovarian cancer (Zamboni *et al.*, 2006). Aerosolized administration of liposomal 9NC was found to be feasible and safe in patients with advanced pulmonary malignancies and 9NC was detected in plasma shortly after the start of treatment (Knight *et al.*, 1999; Koshkina *et al.*, 1999; Verschraegen *et al.*, 2004). Liposomal irinotecan is currently in preclinical develop and in theory may provide targeted delivery of irinotecan to the tumor with subsequent conversion to SN-38 via tumor carboxyl esterase (Drummond *et al.*, 2005; Messerer *et al.*, 2004).

In a phase I study, the pharmacokinetic profile of paclitaxel was similar after administration of LEP-ETU and non-liposomal paclitaxel suggesting that paclitaxel is immediately released from the liposome after LEP-ETU administration (Damajanov *et al.*, 2005). In addition, it is unclear if LEP-ETU has pharmacologic or cytotoxic advantages over ABI-007 (Socinski, 2006; Damajanov *et al.*, 2005). Conventional liposomal formulations of doxorubicin do not appear to have a pharmacologic or cytotoxic advantages over Doxil (Ewer *et al.*, 2004; Mendelson *et al.*, 2005; Northfelt, 1994). Liposomal vincristine (Onco-TSC) has been evaluated in relapsed non-Hodgkins lymphoma (NHL) as a way to overcome the toxicity limitations and required dose reductions associated with the use of vincristine in this setting (Sarris *et al.*, 2000). Liposomal vincristine has a prolonged half-life and achieves higher exposures in tumors and lymph nodes compared with nerves. When administered at full doses liposomal vincristine appears to be less neurotoxic and more active compared with non-liposomal vincristine in preclinical models and in patients. The data suggest a potential role of liposomal vincristine in the combination therapy for NHL.

The future generations of liposomes will contain immunoliposomes, single liposomes that contain two anticancer agents, and liposomes that are thermosensitive (Abraham *et al.*, 2004; Laginha *et al.*, 2005; Mendelson *et al.*, 2005; Park *et al.*, 2004). Immunoliposomes combine antibody-mediated tumor recognition with liposomal delivery which are designed for target cell internalization and intracellular drug release (Park *et al.*, 2004). There are several liposomal formulations that contain fixed ratios of two anticancer agents, such as doxorubicin : vincristine, daunorubicin : cytarabine, cisplatin : irinotecan, and 5-fluorouracil : irinotecan that are currently in preclinical development (Abraham *et al.*, 2004; Johnstone *et al.*, 2005). Thermosensitive liposomes may provide a means of improving the tumor-specific delivery of anticancer agents by rapidly releasing drug from the liposome when hypothermia is applied to the tumor area (Mendelson *et al.*, 2005).

21.2.2 Nanoparticles

ABI-007 is the first protein-stabilized nanoparticle approved by the FDA (Gradishar, 2006; Socinski, 2006; Ibrahim *et al.*, 2002). ABI-007 is an albumin stabilized nanoparticle formulation of paclitaxel designed to overcome the solubility issues associated with paclitaxel that require the need for solvents such as cremophor, which have been associated with infusion-related reactions and requires the need for premedication and other incompatibilities with certain IV bags or tubing (Gradishar, 2006; Ibrahim *et al.*, 2002). The albumin-stabilized nanoparticle results in a more rapid distribution out of the vascular compartment and provides a tumor targeting mechanism. The albumin receptor-mediated transport on the endothelial cell wall with in blood vessels facilitates the passage of ABI-007 from the bloodstream into the underlying tumor tissue (Gradishar, 2006; Ibrahim *et al.*, 2002).

Similar to liposomal agents, the dosage of ABI-007 is determined by the paclitaxel content of the formulation (Gradishar, 2006; Ibrahim *et al.*, 2002). The approved regimen for ABI-007 is 260 mg/m^2 IV over 30 min every 3 weeks which is higher than the usual dose range for paclitaxel (i.e. 135–200 mg/m^2) (Gradishar, 2006; Socinski, 2006). In addition there was a lower incidence of myelosuppression after administration of ABI-007 than previously seen with similar doses of paclitaxel (Ibrahim *et al.*, 2002). The other toxicities associated with ABI-007 were similar to high-dose, short-infusion paclitaxel including sensory neuropathy and mucositis. Keratopathy, a relatively unique toxicity also associated with ABI-007 was reported by Ibrahim *et al.* (2002). Thus, as with liposomal formulations, administration of a drug in a nanoparticle formulation can alter the pharmacokinetic, tissue and tumor distribution, and toxicity pattern. Also similar to liposomal agents, the mechanism by which the albumin stabilized nanoparticle is catabolized and paclitaxel is released is unclear.

21.2.3 Microspheres

Microencapsulation is an emerging technology in the development of bio-artifical organs for drug, protein and delivery systems (Haque *et al.*, 2005). Bioencapsulation technology offers several advantages and has shown promising results for the treatment of diseases. For all of these applications, appropriate performance of the microcapsules is critically dependent on the properties of the capsular membrane {P2}. Several studies have encapsulated bacterial cells for potential therapeutic applications using oral administration, such as in kidney failure uremia, cancer therapy, diarrhea, cholesteremia and other diseases (Ouyang *et al.*, 2004). These microcapsule carriers may also be used for the oral administration of drugs and enzymes for cancer therapy. However, the success of microcapsule oral delivery depends on the suitability of the microcapsule membrane for oral delivery. The microcapsule can be disrupted by many different means during its intestinal passage and may be fractured by enzymatic action, chemical reactions, heat, pH, diffusion, mechanical pressure, and other related physiological and biochemical stresses. Microcapsule encapsulation of thalidomide allowed for the successful delivery of thalidomide in the gut and could prove beneficial in the treatment of Crohn's disease (Metz *et al.*, 2005). In addition, the use of microcapsules may be used for the delivery of anticancer agents in the treatment of gastric and colon cancer.

Patients with primary brain tumors or brain metastases have a very poor prognosis that is primarily attributed to the impermeability of the blood–brain barrier (BBB) to cytotoxic agents (Koziara *et al.*, 2004). Paclitaxel has shown activity against gliomas and other brain metastases; however its use in the treatment of brain tumor is limited due to low BBB penetration and side effects associated with IV administration. The lack of BBB penetration is believed to be associated with the p-glycoprotein (p-gp) efflux transporter. To overcome these issues, a formulation of paclitaxel entrapped in novel cetyl alcohol/polysorbate nanoparticles (PX-NP) was developed (Koziara *et al.*, 2004). PX-NP reduced efflux by p-gp, increased brain exposure and reduced toxicity compared with paclitaxel and thus may have potential in the treatment of brain tumors.

21.2.4 Dendrimers

A dendrimer is a nanoparticle of unimolecular micelles with a hydrophobic interior and hydrophilic exterior, which act as a drug carrier (Fig. 21.2) (D'Emanuele and Attwood, 2005). Dendrimers are a class of different fractal polymers prepared by a set of iterative reactions attached to a central core. Each pair of iterations defines a generation and while the diameter grows linearly, the number of surface functional groups grows geometrically. The interior host sites can shield the drug from the exterior biologic milieu and stabilize the drug. The exterior of the dendrimer can also be labeled with tumor specific-ligands, such as folate to provide tumor-selective delivery of anticancer agents to the tumor (Toffoli *et al.*, 1997).

Several types of interactions between dendrimers and drugs have been evaluated, which can be broadly subdivided into the entrapment of drugs within the dendritic architecture (involving electrostatic, hydrophobic and hydrogen bond interactions) and the interaction between a drug and surface of a dendrimer (electrostatic and covalent interactions) (D'Emanuele and Attwood, 2005). In addition, PEG has been added to dendrimers as a way to modify biocompatibility and biodistribution (Fig. 21.2) (Haba *et al.*, 2005; Namazi and Adeli, 2005; Pan *et al.*, 2005). The applications of these systems have been used

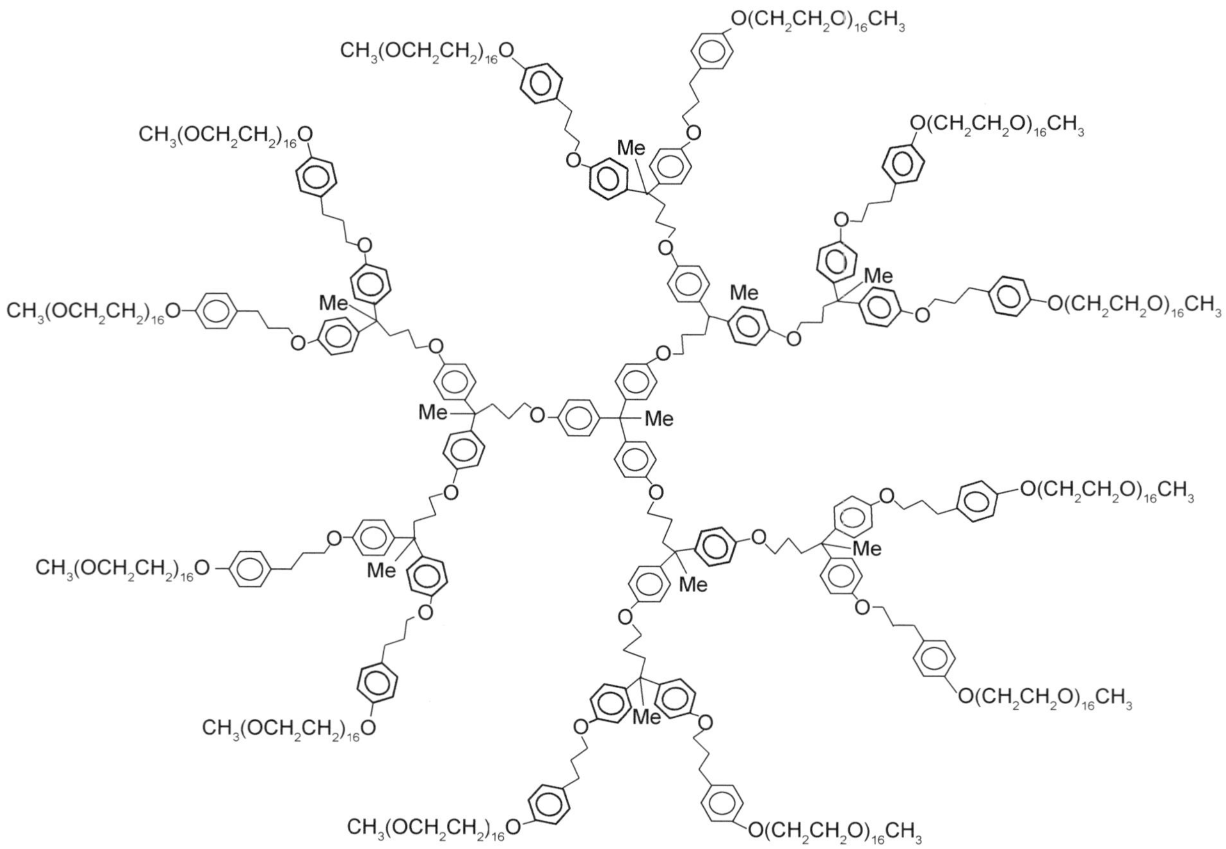

21.2 Structure of G2 water-soluable dendritic unimolecular micelles based on 4,4-bis(4′-hydroxyphenyl) pentanol building blocks and a surface shell of polyethylene glycol (PEG) chains.

to enhance drug solubility and bioavailability, prolong circulation time, act as release modifiers, and drug targeting. For example, an indomethacin-loaded dendrimer has provided sustained release of drug over 30 h. Dendrimer carriers of anticancer agents are currently in preclinical development.

21.2.5 Conjugates

Additional conjugate formulations of paclitaxel are in clinical and preclinical development. Paclitaxel poliglumex (PPX, Xyotax®), a macromolecular drug conjugate that links paclitaxel with a biodegradable polymer, poly-L-glutamic acid, has completed phase I studies (Takimoto *et al.*, 2005). PPX is a water-soluble formulations that also eliminates the need for cremophor in the formulation. Paclimer®, a microsphere formulation of paclitaxel, is currently in preclinical development (Dordunoo *et al.*, 2005). Paclimer microspheres contain paclitaxel in a polilactofate polymer microsphere and are designed to continuously deliver low-dose paclitaxel. Previous conjugates of paclitaxel have been stopped in clinical development and have been associated with potential pharmacologic and pharmacokinetic problems (Bradley *et al.*, 2001; Wolff *et al.*, 2003). Docosahexaenoic acid (DHA)–paclitaxel, a novel conjugate formed by covalently linking the natural fatty acid DHA to paclitaxel, was designed as a prodrug targeting intra-tumoral activation (Bradley *et al.*, 2001). At the MTD of DHA–paclitaxel (1100 mg/m^2), paclitaxel represented only 0.06% of the DHA–paclitaxel plasma exposure (Wolff *et al.*, 2003). However, the paclitaxel concentrations remained >0.01 μM for an average of 6 to 7 days and the paclitaxel AUC was correlated with neutropenia. The results of this study suggest that most of the drug remained in the inactive prodrug-conjugated form and that significant toxicity occurred only when released paclitaxel reached clinically relevant exposures. This depicts the need to perform detailed pharmacokinetic studies of conjugated and released drug in plasma and tumor.

During the past 10 years there has been a renaissance in the field of PEG-conjugated anticancer agents (Greenwald, 2001). This new development has been attributed to the use of higher molecular weight PEGs (>20 000) and especially with the use of PEG 40 000 which has an extended half-life in plasma and potential selective distribution to solid tumors (Greenwald, 2001). Various PEG-conjugates of anticancer agents, such as doxorubicin (Andersson *et al.*, 2005), methotrexate (Riebeseel *et al.*, 2002), interferon (Castells *et al.*, 2005; Derbala *et al.*, 2005) and camptothecin analogues (Paranjpe *et al.*, 2004; Rowinsky *et al.*, 2003), are currently in development (Andersson *et al.*, 2005; Castells *et al.*, 2005; Derbala *et al.*, 2005; Paranjpe *et al.*, 2004; Riebeseel *et al.*, 2002; Rowinsky *et al.*, 2003). PEG- and 20-carbonate-conjugates of camptothecin analogs, are especially interesting as the conjugated-prodrug forms highly water-soluble agents and significantly extend the duration of exposure after a single dose (de Groot *et al.*, 2002; Paranjpe *et al.*, 2004; Rowinsky *et al.*, 2003).

Hyaluronic acid conjugates of anticancer agents are also in development. Carrier-mediated conjugates of anticancer agents also have the same pharmacologic issues (the need to evaluate the pharmacokinetics of the prodrug conjugate and released drug) as liposomal and nanoparticle formulations and the overall clinical benefit of these agents remain unclear.

Nanoparticles have also been developed to improved the oral bioavailability of drugs (Alexander and Kucera, 2005). Phospholipid nucleoside conjugates and nucleosides with chemical additions to the amino moieties have been used since the 1970s in order to increase the biological activity of the parent compound. Synthetic phospholipid conjugates of cytosine arabinoside (ara-C) and gemcitabine have been developed (Alexander *et al.*, 2003, 2005). The novel ara-C conjugate has different systemic and cellular pharmacologic characteristics compared with the parent drug, such as decreased catabolism by cytidine deaminase, increased plasma half-life, penetrate the BBB, and release of nucleoside monophosphate, a reaction that bypasses the rate-limiting initial nucleoside phosphorylation (Alexander *et al.*, 2003). In contrast to gemcitabine, the gemcitabine conjugate did not enter the cell via the human equilibrative nucleoside transporter (hENT1) and was not a substrate for the multidrug resistance efflux pump, MDR-1 (Alexander *et al.*, 2005). These results suggest that gemcitabine conjugate may by superior to gemcitabine because of the conjugate's ability to bypass three resistance mechanisms and can be administered orally. These phospholipid nucleoside conjugates posses the potential to have superior anti-neoplastic cytotoxicity profiles with less toxicity than the parent compound.

21.3 Future trends

21.3.1 Background

Until recently, drug uptake into tissues and tumors have been described indirectly based on modeling from plasma pharmacokinetics or measured directly from tissue biopsies. As stated above, modeling of tumor exposure based on plasma exposures without incorporation of factors representing tumor heterogeneity is unreliable (Boucher and Jain, 1992; Jain, 1996; Zamboni *et al.*, 1999). The use of tissue or tumor biopsies is associated with several problems. Obtaining serial biopsies is most often logistically impossible, highly invasive and associated with patient discomfort (Front *et al.*, 1987; Presant *et al.*, 1994; Zamboni *et al.*, 1999). Thus, biopsies are usually available only for a single time point or measurement. Measurements of drug concentrations from biopsies are measured in tissue or tumor homogenates, where it may be difficult to control *ex vivo* catabolism and differentiate between various forms of the drug. Several new advanced techniques, such as magnetic resonance imaging (MRI), positron emission tomography (PET) and microdialysis, have been developed to quantify the concentrations of anticancer agents *in vivo* (Fischman *et al.*, 1997; Front *et al.*, 1987; Presant *et al.*, 1994). However, the use of MRI and PET is complicated by the lack of ability to differentiate between different forms and metabolites of a drug, availability, chemical synthesis of effective probes and cost (Fischman *et al.*, 1997; Front *et al.*, 1987), whereas the use of microdialysis to evaluate the disposition of anticancer agents in tumors and surrounding tissue is a methodology that has several advantages over other existing methods (Brunner and Muller, 2002; Jain, 1996; Johansen *et al.*, 1997; Muller *et al.*, 1995).

21.3.2 Introduction and advantages of microdialysis

Microdialysis is an *in vivo* sampling technique used to study the pharmacokinetics and drug metabolism in the blood and ECF of various tissues (Johansen *et al.*, 1997; Kehr, 1993; Muller *et al.*, 1995). The use of microdialysis methodology to evaluate the disposition of anticancer agents in tumors is relatively new (Blochl-Daum *et al.*, 1996; Muller *et al.*, 1997; Zamboni *et al.*, 1999). Microdialysis has been used to evaluate the tumor disposition of 5-fluorouracil and carboplatin in patients with primary breast cancer lesions and melanoma, respectively (Blochl-Daum *et al.*, 1996; Muller *et al.*, 1997). These studies depict the clinical utility of microdialysis in evaluating the tumor disposition of anticancer agents in patients with accessible tumors. Microdialysis is based on the diffusion of non-protein-bound drugs from interstitial fluid across the semipermeable membrane of the microdialysis probe (Johansen *et al.*, 1997; Kehr, 1993; Muller *et al.*, 1995). A schematic representation of a microdialysis probe in subcutaneous tissue or tumor is depicted in Fig. 21.3. Microdialysis provides a means to obtain samples from tumor ECF samples from which a concentration versus time profile can be determined within a single tumor (Blochl-Daum *et al.*, 1996; Johansen *et al.*, 1997; Muller *et al.*, 1997; Zamboni *et al.*, 1999).

Microdialysis provides several advantages over autoradiographic studies of tumor biopsies as a method to evaluate anticancer drug concentrations in tumor tissue. With microdialysis techniques it is possible to obtain serial sampling of anticancer drugs from the ECF of a single tumor with minimal tissue damage or alteration of fluid balance (Johansen *et al.*, 1997; Muller *et al.*, 1995; Zamboni *et al.*, 1999). The microdialysis probe can remain in peripheral or central nervous system (CNS) tissue for up to 72 h without complications, such as increased risk of infection, inflammation or alteration in probe recovery. Samples can be immediately obtained and analyzed from a single probe that allows for the real time evaluation of physiologic, pharmacologic and pharmacokinetic changes (Ekstrom *et al.*, 1997a; Ettinger *et al.*, 2001; Leggas *et al.*, 2004; Muller *et al.*, 1995). In addition, a single microdialysis probe can simultaneously sample several analytes of interest, thus allowing for the measurement of drug concentrations and pharmacologic end points that are required for pharmacodynamic

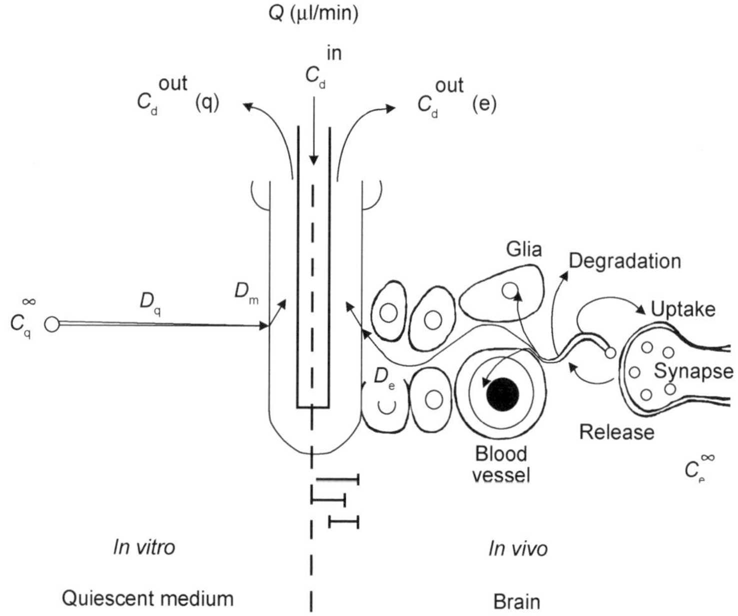

Q = flow rate of dialysate
C_d^{in} = concentration of drug in dialysate during calibration
C_d^{out} = concentration of drug recovered in dialysate during the calibration at sites Q and E
C_q = concentration of drug in tissue or tumor extracellular fluid
D_q = relative amount of drug recovered from tissue or tumor extracellular fluid
D_m = absolute amount of drug recovered in microdialysis probe
C_e = estimated concentration of drug in tissue or tumor extracellular fluid
D_e = estimated amount of drug in tissue or tumor extracellular fluid

21.3 A schematic representation of a microdialysis probe in subcutaneous tissue or tumor.

studies. Furthermore, the drug concentration can be measured specifically rather than quantitating radioactivity, which may be nonspecific. Owing to the pore cut-off size (20 kDa) of the semipermeable membrane, the use of microdialysis allows for the differentiation between liposomal encapsulated, conjugated-drugs, protein-bound drugs, and active-unbound drug in the tumor ECF (Brunner and Muller, 2002; Thompson *et al.*, 2001). Using microdialysis techniques, serial sampling of the non-protein-bound, active form of anticancer agents can be obtained from a single sight in a brain tumor, peripheral tumor or surrounding tissues. In addition, multiple microdialysis probes can be placed in a single tumor to evaluate intra-tumoral variability of the analyte of interest (Zamboni *et al.*, 1999, 2002). Thus, the data obtained with microdialysis techniques may more closely reflect the disposition of the active form of the drug within the tumor ECF (Conley *et al.*, 1999; Muller *et al.*, 1995; Zamboni *et al.*, 2002).

21.3.3 Microdialysis system and set-up

The principles of microdialysis sampling have been reviewed in detail previously (Bungay *et al.*, 1990; Conley *et al.*, 1999; Johansen *et al.*, 1997; Kehr, 1993). Briefly, a short length of hollow dialysis fiber is continuously perfused with a physiological solution. The presence of the analyte of interest in the ECF and its absence in the perfusate leads to a concentration gradient across the dialysis membrane. The analyte diffuses through the dialysis membrane, and is collected for analysis. This process is performed *in vivo* through the use of a microdialysis probe that is implanted into tissue, and continuously perfused with a physiologic solution at a low flow rate (0.5–10 μl/min). After the probe has been implanted into tumor tissue, substances are filtered by diffusion from the extracellular space through the semipermeable membrane into the perfusion medium, and carried via microtubing into the collection vials.

Commercially available microdialysis probes, microperfusion pumps and microfraction collectors are available. The type of microdialysis probe used depends on the sight or tissues of interest (e.g. subcutaneous tumor or tissue, brain or liver), size of the tumor and the analyte of interest (Zamboni *et al.*, 1999, 2002). A microdialysis probe (CMA 20, Stockholm, Sweden) with a molecular cut-off of 20 kDa, membrane length of 4 mm and outer diameter of 0.5 mm is the standard used for most pharmacokinetic studies of drugs in peripheral tissue and tumors (Blochl-Daum *et al.*, 1996; Muller *et al.*, 1997; Zamboni *et al.*, 1999, 2002). The molecular weight cut-off (i.e. 20 kDa) of the semipermeable membrane of probe prevents albumin-bound drug from crossing the membrane. However, small plasma protein, such as α-1-acid-glycoprotein, can pass through the semipermeable membrane. Thus, depending on the protein-binding characteristics of a drug, the recovery may not be limited to unbound drug. The microdialysis probe is perfused by a microperfusion pump and dialysate samples are collected by the microfraction collector. Ringers solution (USP) is the standard perfusion solution because it is similar to the makeup of ECF. Alternatively, 0.9% NaCl (USP) can be used for tissue and CNS studies.

21.3.4 Microdialysis methodology and study design

Only a fairly small fraction (i.e. 10–50%) of the analyte can cross the probe's semipermeable membrane and the percentage that crosses can vary between probes, drug type and flow rate (Bungay *et al.*, 1990; Conley *et al.*, 1999; Johansen *et al.*, 1997; Kehr, 1993). Thus, prior to *in vivo* studies it is standard to characterize the transfer rate, relative recovery and the optimal flow rate of the drug and probe used in the studies. The recovery of drug across the membrane is concentration independent (Bungay *et al.*, 1990; Conley *et al.*, 1999; Johansen *et al.*, 1997; Kehr, 1993; Zamboni *et al.*, 1999). The objective is to use the lowest flow rate that achieves sufficient recovery of the analyte that can be detected by the analytical system. High flow rates should be avoided owing to the propensity to alter the fluid balance in the tumors. The flow rate and collection interval are then modified to attain the needed sample volume required by the analytical system.

Microdialysis probes can be placed in any accessible tumor and tissue. However, areas of the tumor with pooled blood should be avoided to prevent false results. Probe placement can be confirmed by ultrasound or after tumor or tissue removal in animal studies. Dual probe studies can also be performed to evaluate the intra-tumoral disposition of the analyte or drug.

After probe placement, a short period (i.e. 45–60 min) is allowed for probe and tumor ECF equilibration prior to the start of calibration (Blochl-Daum *et al.*, 1996; Muller *et al.*, 1995; Zamboni *et al.*, 1999, 2002). Although use of microdialysis technique results in less tissue damage compared with other sampling methods (e.g. biopsy), insertion of the microdialysis probe into the tumor does induce some tissue damage and immune reactivity. Thus, samples collected immediately after probe insertion may not reflect basal tumor conditions because of acute tissue damage and changes in blood flow associated with probe insertion. Therefore, it is necessary to allow time for the probe and tumor ECF to equilibrate prior to the start of the probe calibration studies.

After probe placement, calibration and washout procedures are performed. Owing to variability in recovery for various probes at various sites, the calibration procedure is performed to determine the extent of recovery for each probe at each site. The washout period is performed to remove any drug introduced into the ECF during retrocalibration. The length of the washout period is determined by concentration of drug introduced in the ECF during calibration and the half-life of the drug in the ECF. After the washout period the drug is administered or the procedure is started, and the sample recovery procedure is performed.

21.3.5 *In vivo* calibration and recovery

In vitro recovery may be substantially different from the *in vivo* (i.e. tumor ECF) recovery (Johansen *et al.*, 1997; Kehr, 1993; Muller *et al.*, 1995). In addition, recovery can vary between probes, drug type, flow rate, and tissue or tumor site. An *in vivo* microdialysis study is a dynamic process in which substances are continuously removed from the tumor ECF by diffusion into the probe. Consequently, the concentration of drug in the perfusate does not reach equilibrium with the tumor ECF. However, under constant conditions (i.e. perfusate flow rate) a steady-state percentage recovery, which represents a constant fraction of the ECF concentration, will be reached. Thus, the *in vivo* recovery value is determined for each probe in each tumor or tissue, and is specific for that single procedure. This provides the advantage of accounting for processes that affect recovery in tissues and tumors. The *in vitro* recovery values can be calculated by retrodialysis calibration, reference or marker compound, and point of zero net flux methods (Bungay *et al.*, 1990; Johansen *et al.*, 1997; Kehr, 1993; Le Quellec *et al.*, 1995).

Retrodialysis calibration method can be used to estimate the steady-state percentage recovery (Bungay *et al.*, 1990; Ekstrom *et al.*, 1997b; Ettinger *et al.*, 2001; Le Quellec *et al.*, 1995; Muller *et al.*, 1998; Zamboni *et al.*, 1999, 2002). Retrodialysis quantification of *in vivo* recovery is based on the principle that the diffusion process across the microdialysis semipermeable membrane is equal in both directions. Therefore, the analyte of interest can be included in the perfusion medium and the disappearance from the perfusate into the tumor ECF is used as an estimation of *in vivo* recovery and the be used to estimate the tumor ECF concentration (Bungay *et al.*, 1990; Kehr, 1993; Leggas *et al.*, 2004; Zamboni *et al.*, 1999).

One limitation of the retrodialysis method is the time required to perform the calibration studies (i.e., four to five samples which lasts 1–1.5 h) and washout (three to four samples which lasts approximately 1 h). Alternatively, if the retrodialysis calibration studies could be performed at the same time as you collect samples, you could increase the ratio of sample number to study duration. This can be performed by using a reference or marker compound that has the same recovery characteristics as your analyte of interest. This processes occurs by placing the reference compound in the dialysis solution during sampling of the analyte of interest. The analyte of interest diffuses from the ECF into the probe at the same time, rate and extent as the reference compound diffuses out of the probe and into the ECF. The *in vivo* recovery of the reference compound is determined using the standard retrodialysis procedure and calculations. The *in vivo* recovery value for the reference compound is then used to calculate the estimated tumor concentration. The point of zero net flux is an alternative calibration method that determines relative recovery of a drug or analyte by varying the concentrations of the drug included in the perfusate solution (Bungay *et al.*, 1990; Fox *et al.*, 2002; Le Quellec *et al.*, 1995).

21.3.6 Preclinical studies of tumor and tissue distribution

Studies comparing plasma and tumor ECF exposure associated with response in preclinical models have used microdialysis methodology (Ekstrom *et al.*, 1997b; Zamboni *et al.*, 1999). Investigators reported a six-fold difference in dose and plasma exposure of topotecan associated with a complete response in mice bearing human neuroblastoma xenografts NB1691 (2.0 mg/kg, and 290 ng/ml h, respectively) as compared to NB1643 (0.36 mg/kg, and 52 ng/ml h, respectively) (Zamboni *et al.*, 1998b). However, factors related to the difference in topotecan response in the two neuroblastoma xenograft lines were not identified. Moreover, macro-tumor related factors affecting sensitivity and the relationship between tumor ECF exposure to topotecan and antitumor activity in the xenograft model had not been established. As a result, the tumor ECF disposition of topotecan using microdialysis methodology was evaluated and the relationship between topotecan tumor ECF exposure and antitumor response in mice bearing the relatively resistant (NB1691) and sensitive (NB1643) human neuroblastoma xenografts was evaluated (Zamboni *et al.*, 1999).

The concentration versus time profiles of topotecan in plasma and tumor ECF in NB1643 and NB1691 human tumor xenografts are presented in Fig. 21.4.

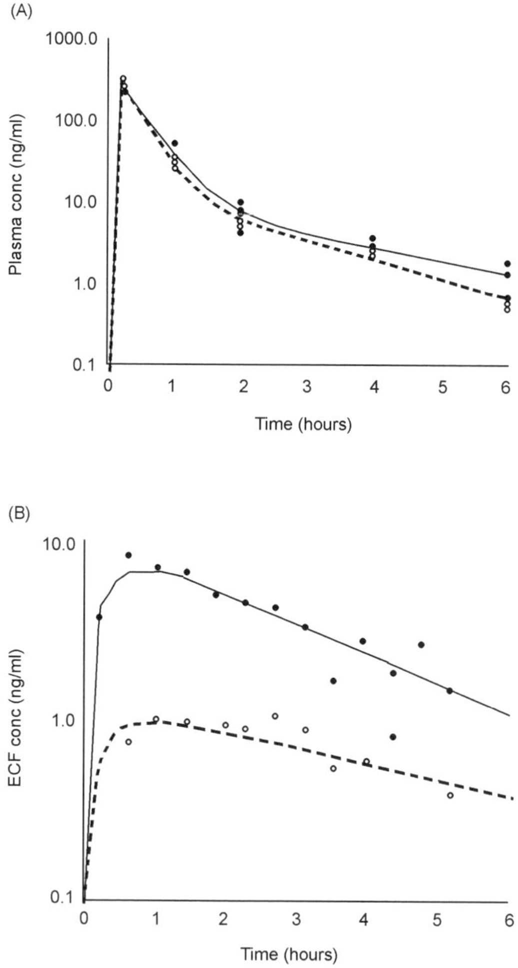

21.4 Topotecan lactone concentration versus time profiles in plasma and tumor ECF in resistant and sensitive neuroblastoma tumor xenografts. Representative topotecan plasma (A) and tumor extracellular fluid (ECF) (B) concentration–time plots in mice bearing NB-1691 (○, ----) and NB-1643 (●, ——). Individual data points and best fit line of the data are shown. The topotecan lactone tumor extracellular fluid AUC 0 to 5 hours for the representative NB-1643 and NB-1691 tumor xenografts were 22.3 and 9.1 ng/ml h, respectively.

There was a 3.5-fold difference in tumor ECF exposure and penetration in NB1643 (25.6 ± 19.6 ng/ml h and 0.15 ± 0.11, respectively) and NB1691 (7.3 ± 6.1 ng/ml h and 0.04 ± 0.04, respectively) ($p < 0.05$), which was consistent with the difference in sensitivity of these xenografts based on dose and plasma exposure. These results suggest that topotecan tumor penetration may be one factor associated with neuroblastoma antitumor response. Moreover, these data suggest inherent differences in tumor vascularity, capillary permeability and/or tumor interstitial pressure between the sensitive and resistant neuroblastoma tumor xenografts. This was the first study reporting a relationship between the exposure of an anticancer agent in tumor ECF and antitumor response.

The significance of ECF as an important exposure for pharmacologic effect of anticancer agents and the inter- and intratumor variability was evaluated in preclinical studies of cisplatin using microdialysis (Zamboni *et al.*, 2002). The relationship between unbound platinum in tumor ECF, total platinum in tumor homogenates, and the formation of platinum-DNA (Pt-DNA) adducts were evaluated after administration of cisplatin in mice bearing B16 murine melanoma tumors. Intra-tumor variability in platinum disposition was evaluated by placing two probes (A and B) in the same tumor. At the end of the 2 h sample period, tumor tissue was obtained at each probe site and analyzed for total-platinum and bifunctional intra-strand DNA-adducts between platinum and two adjacent guanines (Pt-GG), and platinum and adenine and guanine (Pt-AG).

The concentration of unbound-platinum in tumor ECF of B16 tumors was detectable from 12 min to 120 min after administration. In addition, the concentration versus time profile of unbound-platinum in tumor ECF did not follow the plasma concentration–time profile, suggesting that the clearance of a drug from the tumor may be the primary factor affecting drug accumulation within a tumor. The median (range, %CV) AUC_{ECF} and tumor penetration were 0.42 μg/ml h (0.05 to 1.57, 78%) and 0.16 (0.02 to 0.62, 77%), respectively.

The relationship between unbound-platinum AUC in tumor ECF from probe A and probe B is presented in Fig. 21.5. The median (range, %CV) AUC_{ECF} from probe A to probe B was 1.9 (1.3 to 5.5, 55%). The median (range, %CV) concentration of total platinum obtained at the end of the 2 h microdialysis procedure from probe A to probe B was 1.1 (1.0–2.0, 27%). Using an Emax model to describe the relationship between drug exposure and platinum-DNA adduct formation, there was a better correlation between unbound-platinum AUC_{ECF} ($R^2 = 0.69$ and 0.63, respectively) and Pt-GG and Pt-AG compared to total-platinum in tumor extracts ($R^2 = 0.29$ and 0.41, respectively). In addition, there was a poor correlation between unbound-platinum AUC_{ECF} and total-platinum in tumor extracts ($R^2 = 0.26$).

These results suggest there is relatively high inter- (approximately a 30-fold range) and low intra-tumor (approximately a 4-fold range) variability in unbound- and total-platinum in B16 murine melanoma tumors, and a poor relationship between unbound- and total-platinum. In addition, these results

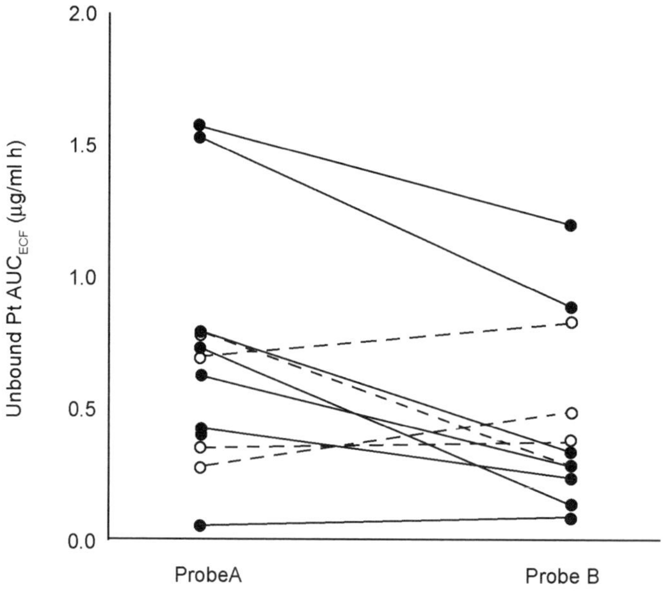

21.5 Low intra- and high inter-tumoral disposition of cisplatin in murine melanoma tumors. Inter- and intra-tumoral variability in unbound Pt AUCECF in mice bearing B16 tumors and in mice bearing H23 xenografts after administration of cisplatin at 10 mg/kg. In mice bearing B16 murine melanoma tumors, individual AUCs are represented by ●, and AUCs within the same tumor are connected by ——. In mice bearing H23 human NSCLC xenografts, individual AUCs are represented by ○, and AUCs within the same tumor are connected by ----.

suggest unbound-platinum in tumor ECF is a better correlate of platinum-DNA adduct formation compared to total-platinum measured in tumor extracts.

21.3.7 Evaluation of liposomal anticancer agents

The theoretical advantages of encapsulated liposomal drugs are prolonged duration of exposure and selective delivery of entrapped drug to the site of action (Allen and Stuart, 2005; Harrington *et al.*, 2001; Newman *et al.*, 1999; Woodle and Lasic, 1992). Major advances in the use of liposomes as vehicles delivering encapsulated pharmacologic agents and enzymes to sites of disease have occurred over the past 10 years. Moreover, liposomal-encapsulated drugs, such as liposomal-doxorubicin (Doxil®) are FDA-approved and have documented activity and decreased toxicity (Muggia *et al.*, 1997; Stewart *et al.*, 1998). Studies evaluating the disposition and tumor penetration of liposomal and non-liposomal anticancer agents suggest liposomal agents extravasate selectively into solid tumors through the capillaries of tumor neovasculature (Harrington *et al.*, 2000, 2001a,b). However, the mechanisms by which liposomes enter tissue and tumors, and release drug are not completely understood. In addition, the liposomes can be engineered to produce a complete spectrum of drug release rates which need to be evaluated in *in vivo* systems.

SPI-077 (ALZA Pharmaceuticals, Inc.) is cisplatin encapsulated in long-circulating Stealth® liposome. The disposition of liposomal-cisplatin is dependent on the liposomal vehicle (De Mario *et al.*, 1998; Harrington *et al.*, 2001a; Veal *et al.*, 2001). Once the cisplatin is released from the liposome, its disposition follows cisplatin pharmacology. SPI-077 has shown antitumor activity against a wide range of solid tumor xenografts, including murine colon tumors. In a study comparing SPI-077 and cisplatin tumor disposition in mice bearing murine colon tumors, the platinum exposure was several fold higher and prolonged after SPI-077 as compared with cisplatin administration (Newman *et al.*, 1999). However, because the platinum exposure was measured in tumor extracts, it is unclear whether the platinum measured was encapsulated, protein-bound platinum or unbound-platinum. In addition, it is unclear whether the platinum exposure was intracellular or extracellular. Thus, it is currently unclear whether SPI-077 releases cisplatin into the tumor extracellular fluid (ECF), or penetrates into the cell as the liposome and then releases the cisplatin intracellularly.

Thus, the tumor disposition of platinum after administration of liposomal formulations of cisplatin (SPI-077) and non-liposomal cisplatin was evaluated using microdialysis in mice bearing B16 murine melanoma tumors (Thompson *et al.*, 2001). Owing to the pore cut-off size (20 kDa) of the semipermeable membrane of the microdialysis probe and the size of the liposome (100 nm), the microdialysis probe was only be able to sample unbound-platinum and allow the differentiation between liposomal-encapsulated cisplatin and cisplatin released into the tumor ECF.

After administration of cisplatin, the concentration of unbound-platinum in tumor ECF was detectable from 12 to 120 min after administration. However, there was no detectable unbound-platinum in the tumor ECF after administration of SPI-077. The total-platinum in tumor extracts, unbound-platinum in tumor ECF as measured by AUC, and formation of Pt-GG DNA adducts after administration of cisplatin and SPI-077 are presented in Fig. 21.6 (Pluim *et al.*, 1999). The results of this study suggest SPI-077 distributes into tumors, but release significantly less platinum into tumor ECF which results in lower formation of Pt-DNA adducts compared with cisplatin. This was the first study using microdialysis methodology to evaluate the tumor disposition of liposomal encapsulated anticancer agents.

21.4 Conclusions

Liposomes and nanoparticles may be an effective carriers to deliver anticancer agents to tumors (Allen and Hansen, 1991; Allen and Stuart, 2005; Drummond *et al.*, 1999; Papahadjopoulos *et al.*, 1991; Zamboni *et al.*, 2004; D'Emanuele

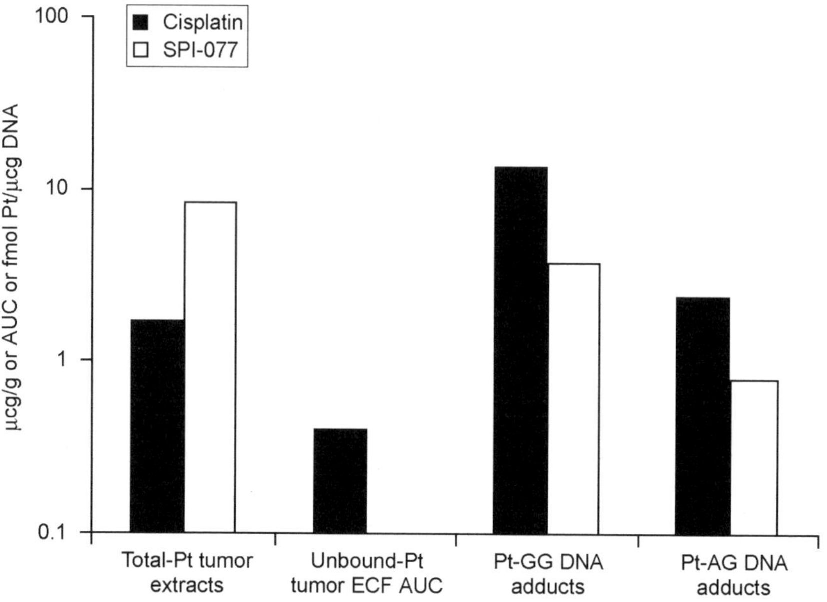

21.6 Unbound platinum (Pt) in tumor ECF, total-Pt in tumor homogenates, and Pt-GG and Pt-AG DNA adducts after administration of cisplatin and SPI-077 in murine melanoma tumors.

and Attwood, 2005; Ouyang *et al.*, 2004). However, for anticancer agents encapsulated or conjugated in liposomes or nanoparticles to be an effective treatment in patients with solid tumors, the active form of the anticancer agent must be released from the liposome into the tumor ECF or inside the cell (Zamboni *et al.*, 2004). As a result of this delivery process, new liposomal anticancer agents should be evaluated in preclinical models and early clinical trials to insure that adequate release of drug occurs at its site of action. Future immunoliposomes that contain an antibody conjugated to a liposome are being developed to provide targeted delivery to cancer cells expressing specific proteins (Abraham *et al.*, 2004; Papahadjopoulos *et al.*, 1999). For example, anti-HER2 immunoliposomes are being developed with the PEG liposomes linked to an anti-HER2 monoclonal antibody (Papahadjopoulos *et al.*, 1999). This technique may allow the entire liposome to be taken up by the cell and possibly avoid potential problems with the release of active drug into the tumor ECF (Abraham *et al.*, 2004; Papahadjopoulos *et al.*, 1999). It is unclear if a drug conjugated to PEG or other carriers, or drugs encapsulated in microspheres or protein-stabilized nanoparticles, must be released from the carrier in order to achieve cytotoxic effects. In addition, ligand-labeled nanoparticles, such as dendrimers, may provide even greater tumor-selectively delivery.

Future studies are needed to evaluate the mechanism of clearance of liposomes and nanoparticles and evaluate the factors associated with pharmacokinetic variability of these agents (Dark *et al.*, 2005; Giles *et al.*, 2004; Kraut *et al.*, 2005; Papahadjopoulos *et al.*, 1999; Working *et al.*, 1994; Zamboni *et al.*, 2005a). The elimination of these agents may be similar to antibodies and proteins and most likely associated with the RES (Allen and Hansen, 1991; Allen and Stuart, 2005; Drummond *et al.*, 1999). In addition, it is currently unclear what is the most appropriate preclinical model for toxicity, efficacy and pharmacokinetic studies (Zamboni *et al.*, 2005a,b).

21.5 Sources of further information and advice

In July 2004, the United States National Institutes of Health and the National Cancer Institute published a plan on the use of nanotechnology in cancer entitled 'A Strategic Initiative to Transform Clinical Oncology and Basic Research Through the Directed Application of Nanotechnology' (National Institutes of Health, 2004). Nanotechnology offers the unprecedented and paradigm-altering opportunity to evaluate and interact with normal and cancer cells in real time at the molecular and cellular level. This report outlined key opportunities for cancer nanotechnology, such as molecular imaging and early detection, *in vivo* imaging, reporters of efficacy, multifunctional therapeutics, prevention and control, and research enablers. In addition, this report discussed new strategies for cancer nanotechnology, such as center for cancer nanotechnology excellence, nanotechnology characterization laboratories, building research teams, creating cancer nanotechnology platforms through directed research programs, and the use of basic and applied initiatives for nanotechnology in cancer. The goal of this initiative is to use the power of nanotechnology to radically change the way we diagnose, treat, and prevent cancer and ultimately eliminate suffering and death from cancer.

21.6 References

Abraham, S. A., McKenzie, C., Masin, D., Ng, R., Harasym, T. O., Mayer, L. D., & Bally, M. B. (2004), '*In vitro* and *in vivo* characterization of doxorubicin and vincristine coencapsulated within liposomes through use of transition metal ion complexation and pH gradient loading', *Clin Cancer Res*, 10 (2), 728–738.

Alexander, R. L. & Kucera, G. L. (2005), 'Lipid nucleoside conjugates for the treatment of cancer', *Current Pharmaceut Design*, 11 (9), 1079–1089.

Alexander, R. L., Morris-Natschke, S. L., Ishaq, K. S., Fleming, R. A., & Kucera, G. L. (2003), 'Synthesis and cytotoxic activity of two novel 1-dodecylthio-2-decyloxypropyl-3-phosphatidic acid conjugates with gemcitabine and cytosine arabinoside', *J Med Chem*, 46 (19), 4205–4208.

Alexander, R. L., Greene, B. T., Torti, S. V., & Kucera, G. L. (2005), 'A novel phospholipid gemcitabine conjugate is able to bypass three drug-resistance mechanisms', *Cancer Chemother Pharmacol*, 56 (1), 15–21.

Allen, T. M. & Hansen, C. (1991), 'Pharmacokinetics of stealth versus conventional liposomes: effect of dose', *Biochim Biophys Acta*, 1068 (2), 133–141.

Allen, T. M. & Martin, F. J. (2004), 'Advantages of liposomal delivery systems for anthracyclines', *Semin Oncol*, 31 (6 Suppl 13), 5–15.

Allen, T. M. & Stuart, D. D. (2005), 'Liposomal pharmacokinetics. Classical, sterically-stabilized, cationic liposomes and immunoliposomes,' in *Liposomes: Rational Design*, A.S. Janoff, ed., New York, Marcel Dekker, Inc., pp. 63–87.

Andersson, L., Davies, J., Duncan, R., Ferruti, P., Ford, J., Kneller, S., Mendichi, R., Pasut, G., Schiavon, O., Summerford, C., Tirk, A., Veronese, F. M., Vincenzi, V., & Wu, G. (2005), 'Poly(ethylene glycol)–poly(ester-carbonate) block copolymers carrying PEG–peptidyl–doxorubicin pendant side chains: synthesis and evaluation as anticancer conjugates', *Biomacromolecules*, 6 (2), 914–926.

Balch, C. M., Reintgen, D. S., Kirkwood, J. M., *et al.* (2006), 'Cutaneous melanoma,' in *Cancer: Principles and Practice in Oncology*, 5th Edition, V. T. Devita, S. Hellman, & S. A. Rosenberg, eds., Philadelphia, Lippincott-Raven, p. 1947.

Barenholz, Y. & Haran, G. (1993), *Method of amphophilic drug loading into liposomes by pH gradient*, 5 192 549 (patent).

Berry, G., Billingham, M., Alderman, E., Richardson, P., Torti, F., Lum, B., Patek, A., & Martin, F. J. (1998), 'The use of cardiac biopsy to demonstrate reduced cardiotoxicity in AIDS Kaposi's sarcoma patients treated with pegylated liposomal doxorubicin', *Ann Oncol*, 9 (7), 711–716.

Blochl-Daum, B., Muller, M., Meisinger, V., Eichler, H. G., Fassolt, A., & Pehamberger, H. (1996), 'Measurement of extracellular fluid carboplatin kinetics in melanoma metastases with microdialysis', *Br J Cancer*, 73 (7), 920–924.

Boucher, Y. & Jain, R. K. (1992), 'Microvascular pressure is the principal driving force for interstitial hypertension in solid tumors: implications for vascular collapse', *Cancer Res*, 52 (18), 5110–5114.

Bradley, M. O., Swindell, C. S., Anthony, F. H., Witman, P. A., Devanesan, P., Webb, N. L., Baker, S. D., Wolff, A. C., & Donehower, R. C. (2001), 'Tumor targeting by conjugation of DHA to paclitaxel', *J Control Rel*, 74 (1–3), 233–236.

Brunner, M. & Muller, M. (2002), 'Microdialysis: an *in vivo* approach for measuring drug delivery in oncology', *Eur J Clin Pharmacol*, 58 (4), 227–234.

Bungay, P. M., Morrison, P. F., & Dedrick, R. L. (1990), 'Steady-state theory for quantitative microdialysis of solutes and water *in vivo* and *in vitro*', *Life Sci*, 46 (2), 105–119.

Castells, L., Vargas, V., Allende, H., Bilbao, I., Luis, L. J., Margarit, C., Esteban, R., & Guardia, J. (2005), 'Combined treatment with pegylated interferon (alpha-2b) and ribavirin in the acute phase of hepatitis C virus recurrence after liver transplantation', *J Hepatol.*, 43 (1), 53–59.

Cattel, L., Ceruti, M., & Dosio, F. (2004), 'From conventional to stealth liposomes: a new Frontier in cancer chemotherapy', *J Chemother*, 16 Suppl 4, 94–97.

Conley, B. A., Ramsland, T. S., Sentz, D. L., Wu, S., Rosen, D. M., Wollman, M., & Eiseman, J. L. (1999), 'Antitumor activity, distribution, and metabolism of 13-*cis*-retinoic acid as a single agent or in combination with tamoxifen in established human MCF-7 xenografts in mice', *Cancer Chemother Pharmacol*, 43 (3), 183–197.

Damajanov, N., Fishman, M. N., Steinberg, J. L., Fetterly, G. J., Haas, A., Lauay, C., Dul, J. L., Sherman, J. W., & Rubin, E. H. (2005), 'Final results of a Phase I study of liposome entrapped paclitaxel (LEP-ETU) in patients with advanced cancer', *Proc Am Soc Clin Oncol*, 23, 147s.

Dark, G. G., Calvert, A. H., Grimshaw, R., Poole, C., Swenerton, K., Kaye, S., Coleman, R., Jayson, G., Le, T., Ellard, S., Trudeau, M., Vasey, P., Hamilton, M., Cameron, T., Barrett, E., Walsh, W., McIntosh, L., & Eisenhauer, E. A. (2005), 'Randomized trial of two intravenous schedules of the topoisomerase I inhibitor liposomal lurtotecan in women with relapsed epithelial ovarian cancer: a trial of the national cancer institute of Canada clinical trials group', *J Clin Oncol*, 23 (9), 1859–1866.

de Groot, F. M., Busscher, G. F., Aben, R. W., & Scheeren, H. W. (2002), 'Novel 20-carbonate linked prodrugs of camptothecin and 9-aminocamptothecin designed for activation by tumour-associated plasmin', *Bioorg Med Chem Lett*, 12 (17), 2371–2376.

D'Emanuele, A. & Attwood, D. (2005), 'Dendrimer-drug interactions', *Adv Drug Deliv Rev*, 57 (15), 2147–2162.

De Mario, M. D., Vogelzang, N. J., Janisch, L., Tonda, M., Amantea, M. A., Pendyala, L., *et al.* (1998), 'A Phase I study of liposome-formulated cisplatin (SPI-077) given every 3 weeks in patients with advanced cancer', *Proc Am Soc Clin Oncol*, 17, 883.

Derbala, M., Amer, A., Bener, A., Lopez, A. C., Omar, M., & El Ghannam, M. (2005), 'Pegylated interferon-alpha 2b-ribavirin combination in Egyptian patients with genotype 4 chronic hepatitis', *J Viral Hepat*, 12 (4), 380–385.

Dordunoo, S. K., Vineck, W., Hoover, R., & Dang, W. (2005), 'Sustained release of paclitaxel from PACLIMER® microspheres', *Proc Am Assoc Cancer Res*, 46, 985.

Drummond, D. C., Meyer, O., Hong, K., Kirpotin, D. B., & Papahadjopoulos, D. (1999), 'Optimizing liposomes for delivery of chemotherapeutic agents to solid tumors', *Pharmacol Rev.*, 51 (4), 691–743.

Drummond, D. C., Noble, C. O., Guo, Z., Hayes, M. E., Hong, K., Park, J. W., & Kirpotin, D. B. (2005), 'Development of a highly stable liposomal irininotecan with low toxicity and potent antitumor efficacy', *Proc Am Assoc Cancer Res*, 46, 330.

Ekstrom, P. O., Andersen, A., Warren, D. J., Giercksky, K. E., & Slordal, L. (1996), 'Determination of extracellular methotrexate tissue levels by microdialysis in a rat model', *Cancer Chemother Pharmacol*, 37 (5), 394–400.

Ekstrom, P. O., Andersen, A., Saeter, G., Giercksky, K. E., & Slordal, L. (1997a), 'Continuous intratumoral microdialysis during high-dose methotrexate therapy in a patient with malignant fibrous histiocytoma of the femur: a case report', *Cancer Chemother Pharmacol*, 39 (3), 267–272.

Ekstrom, P. O., Giercksky, K. E., Andersen, A., Bruland, O. S., & Slordal, L. (1997b), 'Intratumoral differences in methotrexate levels within human osteosarcoma xenografts studied by microdialysis', *Life Sci*, 61 (19), L275–L280.

Ettinger, S. N., Poellmann, C. C., Wisniewski, N. A., Gaskin, A. A., Shoemaker, J. S., Poulson, J. M., Dewhirst, M. W., & Klitzman, B. (2001), 'Urea as a recovery marker for quantitative assessment of tumor interstitial solutes with microdialysis', *Cancer Res*, 61 (21), 7964–7970.

Ewer, M. S., Martin, F. J., Henderson, C., Shapiro, C. L., Benjamin, R. S., & Gabizon, A. A. (2004), 'Cardiac safety of liposomal anthracyclines', *Semin Oncol*, 31 (6 Suppl 13), 161–181.

Fischman, A. J., Alpert, N. M., Babich, J. W., & Rubin, R. H. (1997), 'The role of positron emission tomography in pharmacokinetic analysis', *Drug Metab Rev*, 29 (4), 923–956.

Fox, E., Bungay, P. M., Bacher, J., McCully, C. L., Dedrick, R. L., & Balis, F. M. (2002), 'Zidovudine concentration in brain extracellular fluid measured by microdialysis: steady-state and transient results in rhesus monkey', *J Pharmacol Exp Ther*, 301 (3), 1003–1011.

Front, D., Israel, O., Iosilevsky, G., Even-Sapir, E., Frenkel, A., Peleg, H., Steiner, M., Kuten, A., & Kolodny, G. M. (1987), 'Human lung tumors: SPECT quantitation of differences in Co-57 bleomycin uptake', *Radiology*, 165 (1), 129–133.

Gelmon, K. A., Eisenhauer, E. A., Harris, A. L., Ratain, M. J., & Workman, P. (1999), 'Anticancer agents targeting signaling molecules and cancer cell environment: challenges for drug development?', *J Natl Cancer Inst*, 91 (15), 1281–1287.

Gelmon, K., Hirte, H., Fisher, B., Walsh, W., Ptaszynski, M., Hamilton, M., Onetto, N., & Eisenhauer, E. (2004), 'A phase 1 study of OSI-211 given as an intravenous infusion days 1, 2, and 3 every three weeks in patients with solid cancers', *Invest New Drugs*, 22 (3), 263–275.

Giles, F. J., Tallman, M. S., Garcia-Manero, G., Cortes, J. E., Thomas, D. A., Wierda, W. G., Verstovsek, S., Hamilton, M., Barrett, E., Albitar, M., & Kantarjian, H. M. (2004), 'Phase I and pharmacokinetic study of a low-clearance, unilamellar liposomal formulation of lurtotecan, a topoisomerase 1 inhibitor, in patients with advanced leukemia', *Cancer*, 100 (7), 1449–1458.

Gradisher, W. J. (2006), 'Albumin-bound paclitaxel: a next generation taxane', *Expert Opin Pharmacother*, 7 (8), 1041–1053.

Greenwald, R. B. (2001), 'PEG drugs: an overview', *J Control Rel*, 74 (1–3), 159–171.

Grever, M. R. & Chabner, B. A. (2006), 'Cancer Drug Development,' in *Cancer: Principles and Practice in Oncology*, 5th Edition, V. T. Devita, S. Hellman, & S. A. Rosenberg, eds., Philadelphia, Lippincott-Raven, p. 385.

Haba, Y., Harada, A., Takagishi, T., & Kono, K. (2005), 'Synthesis of biocompatible dendrimers with a peripheral network formed by linking of polymerizable groups', *Polymer*, 46 (6), 1813–1820.

Haque, T., Chen, H., Ouyang, W., Martoni, C., Lawuyi, B., Urbanska, A. M., & Prakash, S. (2005), 'Superior cell delivery features of poly(ethylene glycol) incorporated alginate, chitosan, and poly-L-lysine microcapsules', *Mol Pharm*, 2 (1), 29–36.

Harrington, K. J., Rowlinson-Busza, G., Syrigos, K. N., Abra, R. M., Uster, P. S., Peters, A. M., & Stewart, J. S. (2000), 'Influence of tumour size on uptake of (111)In-DTPA-labelled pegylated liposomes in a human tumour xenograft model', *Br J Cancer*, 83 (5), 684–688.

Harrington, K. J., Lewanski, C. R., Northcote, A. D., Whittaker, J., Wellbank, H., Vile, R. G., Peters, A. M., & Stewart, J. S. (2001a), 'Phase I-II study of pegylated liposomal cisplatin (SPI-077) in patients with inoperable head and neck cancer', *Ann Oncol*, 12 (4), 493–496.

Harrington, K. J., Mohammadtaghi, S., Uster, P. S., Glass, D., Peters, A. M., Vile, R. G., & Stewart, J. S. (2001b), 'Effective targeting of solid tumors in patients with locally advanced cancers by radiolabeled pegylated liposomes', *Clin Cancer Res*, 7 (2), 243–254.

Ibrahim, N. K., Desai, N., Legha, S., Soon-Shiong, P., Theriault, R. L., Rivera, E., Esmaeli, B., Ring, S. E., Bedikian, A., Hortobagyi, G. N., & Ellerhorst, J. A. (2002), 'Phase I and pharmacokinetic study of ABI-007, a Cremophor-free, protein-stabilized, nanoparticle formulation of paclitaxel', *Clin Cancer Res*, 8 (5), 1038–1044.

Jain, R. K. (1996), 'Delivery of molecular medicine to solid tumors', *Science*, 271 (5252), 1079–1080.

Johansen, M. J., Newman, R. A., & Madden, T. (1997), 'The use of microdialysis in pharmacokinetics and pharmacodynamics', *Pharmacotherapy*, 17 (3), 464–481.

Johnstone, S., Harvie, P., Shew, C., Kadhim, S., Harasym, T., Tardi, P., Harasym, N., Bally, M., & Mayer, L. D. (2005), 'Synergistic antitumor activity observed for a fixed ratio liposome formulation of Cytarabine (Cyt) : Daunorubicin (Daun) against preclinical leukemia models', *Proc Am Assoc Cancer Res*, 46, 9.

Kehr, J. (1993), 'A survey on quantitative microdialysis: theoretical models and practical implications', *J Neurosci Methods*, 48 (3), 251–261.

Kehrer, D. F., Bos, A. M., Verweij, J., Groen, H. J., Loos, W. J., Sparreboom, A., de Jonge, M. J., Hamilton, M., Cameron, T., & de Vries, E. G. (2002), 'Phase I and pharmacologic study of liposomal lurtotecan, NX 211: urinary excretion predicts hematologic toxicity', *J Clin Oncol*, 20 (5), 1222–1231.

Knight, V., Koshkina, N. V., Waldrep, J. C., Giovanella, B. C., & Gilbert, B. E. (1999), 'Anticancer effect of 9-nitrocamptothecin liposome aerosol on human cancer xenografts in nude mice', *Cancer Chemother Pharmacol*, 44 (3), 177–186.

Koshkina, N. V., Gilbert, B. E., Waldrep, J. C., Seryshev, A., & Knight, V. (1999), 'Distribution of camptothecin after delivery as a liposome aerosol or following intramuscular injection in mice', *Cancer Chemother Pharmacol*, 44 (3), 187–192.

Koziara, J. M., Lockman, P. R., Allen, D. D., & Mumper, R. J. (2004), 'Paclitaxel nanoparticles for the potential treatment of brain tumors', *J Control Rel*, 99 (2), 259–269.

Kraut, E. H., Fishman, M. N., LoRusso, P. M., Gordon, M. S., Rubin, E. H., Haas, A., Fetterly, G. J., Cullinan, P., Dul, J. L., & Steinberg, J. L. (2005), 'Final results of a phase I study of liposome encapsulated SN-38 (LE-SN38): safety, pharmacogenomics, pharmacokinetics, and tumor response', *Proc Am Soc Clin Oncol*, 23, 139s.

Krown, S. E., Northfelt, D. W., Osoba, D., & Stewart, J. S. (2004), 'Use of liposomal anthracyclines in Kaposi's sarcoma', *Semin Oncol*, 31 (6 Suppl 13), 36–52.

Laginha, K., Mumbengegwi, D., & Allen, T. (2005), 'Liposomes targeted via two different antibodies: assay, B-cell binding and cytotoxicity', *Biochim Biophys Acta*, 1711 (1), 25–32.

Lasic, D. D., Frederik, P. M., Stuart, M. C., Barenholz, Y., & McIntosh, T. J. (1992), 'Gelation of liposome interior. A novel method for drug encapsulation', *FEBS Lett*, 312 (2-3), 255–258.

Laverman, P., Carstens, M. G., Boerman, O. C., Dams, E. T., Oyen, W. J., van Rooijen, N., Corstens, F. H., & Storm, G. (2001), 'Factors affecting the accelerated blood clearance of polyethylene glycol-liposomes upon repeated injection', *J Pharmacol Exp Ther*, 298 (2), 607–612.

Leggas, M., Zhang, Y., Welden, J., Self, Z., Waters, C. M., & Stewart, C. F. (2004), 'Microbore HPLC method with online microdialysis for measurement of topotecan lactone and carboxylate in murine CSF', *J Pharm Sci*, 93 (9), 2284–2295.

Lei, S., Chien, P. Y., Sheikh, S., Zhang, A., Ali, S., & Ahmad, I. (2004), 'Enhanced therapeutic efficacy of a novel liposome-based formulation of SN-38 against human tumor models in SCID mice', *Anticancer Drugs*, 15 (8), 773–778.

Le Quellec, A., Dupin, S., Genissel, P., Saivin, S., Marchand, B., & Houin, G. (1995), 'Microdialysis probes calibration: gradient and tissue dependent changes in no net flux and reverse dialysis methods', *J Pharmacol Toxicol Methods*, 33 (1), 11–16.

Litzinger, D. C., Buiting, A. M., van Rooijen, N., & Huang, L. (1994), 'Effect of liposome size on the circulation time and intraorgan distribution of amphipathic poly(ethylene glycol)-containing liposomes', *Biochim Biophys Acta*, 1190 (1), 99–107.

Maeda, H., Wu, J., Sawa, T., Matsumura, Y., & Hori, K. (2000), 'Tumor vascular permeability and the EPR effect in macromolecular therapeutics: a review', *J Control Rel*, 65 (1–2), 271–284.

Markman, M., Gordon, A. N., McGuire, W. P., & Muggia, F. M. (2004), 'Liposomal anthracycline treatment for ovarian cancer', *Semin Oncol*, 31 (6 Suppl 13), 91–105.

Mendelson, D. S., Brewer, M., Janiceck, M., Breitenbach, E., Barrett, E., Hamilton, M., Ptazynski, M., van Duym, C., & Gordon, M. (2005), 'Phase I study of OSI-211 (liposomal lurtotecan) in combination with liposomal doxorubicin (LD) every 3 weeks in patients (pts) with advanced solid tumors; final analysis suggests benefit in refractory ovarian cancer.', *Proc Am Soc Clin Oncol*, 23, 153s.

Messerer, C. L., Ramsay, E. C., Waterhouse, D., Ng, R., Simms, E. M., Harasym, N., Tardi, P., Mayer, L. D., & Bally, M. B. (2004), 'Liposomal irinotecan: formulation development and therapeutic assessment in murine xenograft models of colorectal cancer', *Clin Cancer Res*, 10 (19), 6638–6649.

Metz, T., Jones, M. L., Chen, H., Halim, T., Mirzaei, M., Haque, T., Amre, D., Das, S. K., & Prakash, S. (2005), 'A new method for targeted drug delivery using polymeric microcapsules: implications for treatment of Crohn's disease', *Cell Biochem Biophys*, 43 (1), 77–85.

Mori, A., Klibanov, A. L., Torchilin, V. P., & Huang, L. (1991), 'Influence of the steric barrier activity of amphipathic poly(ethyleneglycol) and ganglioside GM1 on the circulation time of liposomes and on the target binding of immunoliposomes *in vivo*', *FEBS Lett*, 284 (2), 263–266.

Muggia, F. M., Hainsworth, J. D., Jeffers, S., Miller, P., Groshen, S., Tan, M., Roman, L., Uziely, B., Muderspach, L., Garcia, A., Burnett, A., Greco, F. A., Morrow, C. P., Paradiso, L. J., & Liang, L. J. (1997), 'Phase II study of liposomal doxorubicin in refractory ovarian cancer: antitumor activity and toxicity modification by liposomal encapsulation', *J Clin Oncol*, 15 (3), 987–993.

Muller, M., Schmid, R., Georgopoulos, A., Buxbaum, A., Wasicek, C., & Eichler, H. G. (1995), 'Application of microdialysis to clinical pharmacokinetics in humans', *Clin Pharmacol Ther*, 57 (4), 371–380.

Muller, M., Mader, R. M., Steiner, B., Steger, G. G., Jansen, B., Gnant, M., Helbich, T., Jakesz, R., Eichler, H. G., & Blochl-Daum, B. (1997), '5-Fluorouracil kinetics in the interstitial tumor space: clinical response in breast cancer patients', *Cancer Res*, 57 (13), 2598–2601.

Muller, M., Brunner, M., Schmid, R., Mader, R. M., Bockenheimer, J., Steger, G. G., Steiner, B., Eichler, H. G., & Blochl-Daum, B. (1998), 'Interstitial methotrexate kinetics in primary breast cancer lesions', *Cancer Res*, 58 (14), 2982–2985.

Namazi, H. & Adeli, M. (2005), 'Dendrimers of citric acid and poly (ethylene glycol) as the new drug-delivery agents', *Biomaterials*, 26 (10), 1175–1183.

National Institutes of Health (2004), 'Cancer Nanotechnology Plan: A Strategic Initiative To Transform Clinical Oncology and Basic Research Through Directed Application of Nanotechnology', http://nano.cancer.gov/alliance_cancer_nanotechnology_plan.pdf

Newman, M. S., Colbern, G. T., Working, P. K., Engbers, C., & Amantea, M. A. (1999), 'Comparative pharmacokinetics, tissue distribution, and therapeutic effectiveness of cisplatin encapsulated in long-circulating, pegylated liposomes (SPI-077) in tumor-bearing mice', *Cancer Chemother Pharmacol*, 43 (1), 1–7.

Northfelt, D. W. (1994), 'STEALTH liposomal doxorubicin (SLD) delivers more DOX to AIDS-Kaposi's Sarcoma lesions than to normal skin.', *Proc Am Soc Oncol*, 13, 51.

Ouyang, W., Chen, H., Jones, M. L., Metz, T., Haque, T., Martoni, C., & Prakash, S. (2004), 'Artificial cell microcapsule for oral delivery of live bacterial cells for therapy: design, preparation, and *in-vitro* characterization', *J Pharm Pharm Sci*, 7 (3), 315–324.

Pal, A., Khan, S., Wang, Y. F., Kamath, N., Sarkar, A. K., Ahmad, A., Sheikh, S., Ali, S., Carbonaro, D., Zhang, A., & Ahmad, I. (2005), 'Preclinical safety, pharmacokinetics and antitumor efficacy profile of liposome-entrapped SN-38 formulation', *Anticancer Res*, 25 (1A), 331–341.

Pan, G. F., Lemmouchi, Y., Akala, E. O., & Bakare, O. (2005), 'Studies on PEGylated and drug-loaded PAMAM dendrimers', *J Bioactive Compatible Polym*, 20 (1), 113–128.

Papahadjopoulos, D., Allen, T. M., Gabizon, A., Mayhew, E., Matthay, K., Huang, S. K., Lee, K. D., Woodle, M. C., Lasic, D. D., & Redemann, C., (1991), 'Sterically stabilized liposomes: improvements in pharmacokinetics and antitumor therapeutic efficacy', *Proc Natl Acad Sci USA*, 88 (24), 11460–11464.

Papahadjopoulos, D., Kirpotin, D. B., Park, J. W., *et al.* (1999), 'Targeting of drugs to solid tumors using anti-HER2 immunoliposomes', *J Liposomes Res*, 8, 425–442.

Paranjpe, P. V., Chen, Y., Kholodovych, V., Welsh, W., Stein, S., & Sinko, P. J. (2004), 'Tumor-targeted bioconjugate based delivery of camptothecin: design, synthesis and *in vitro* evaluation', *J Control Rel*, 100 (2), 275–292.

Park, J. W., Benz, C. C., & Martin, F. J. (2004), 'Future directions of liposome- and immunoliposome-based cancer therapeutics', *Semin Oncol*, 31 (6 Suppl 13), 196–205.

Pluim, D., Maliepaard, M., van Waardenburg, R. C., Beijnen, J. H., & Schellens, J. H. (1999), '32P-postlabeling assay for the quantification of the major platinum–DNA adducts', *Anal Biochem*, 275 (1), 30–38.

Presant, C. A., Wolf, W., Waluch, V., Wiseman, C., Kennedy, P., Blayney, D., & Brechner, R. R. (1994), 'Association of intratumoral pharmacokinetics of fluorouracil with clinical response', *Lancet*, 343 (8907), 1184–1187.

Riebeseel, K., Biedermann, E., Loser, R., Breiter, N., Hanselmann, R., Mulhaupt, R., Unger, C., & Kratz, F. (2002), 'Polyethylene glycol conjugates of methotrexate varying in their molecular weight from MW 750 to MW 40000: synthesis, characterization, and structure-activity relationships *in vitro* and *in vivo*', *Bioconjug Chem*, 13 (4), 773–785.

Rose, P. G. (2005), 'Pegylated liposomal doxorubicin: optimizing the dosing schedule in ovarian cancer', *Oncologist*, 10 (3), 205–214.

Rowinsky, E. K., Rizzo, J., Ochoa, L., Takimoto, C. H., Forouzesh, B., Schwartz, G., Hammond, L. A., Patnaik, A., Kwiatek, J., Goetz, A., Denis, L., McGuire, J., & Tolcher, A. W. (2003), 'A phase I and pharmacokinetic study of pegylated camptothecin as a 1-hour infusion every 3 weeks in patients with advanced solid malignancies', *J Clin Oncol*, 21 (1), 148–157.

Sarris, A. H., Hagemeister, F., Romaguera, J., Rodriguez, M. A., McLaughlin, P., Tsimberidou, A. M., Medeiros, L. J., Samuels, B., Pate, O., Oholendt, M., Kantarjian, H., Burge, C., & Cabanillas, F. (2000), 'Liposomal vincristine in relapsed non-Hodgkin's lymphomas: early results of an ongoing phase II trial', *Ann Oncol*, 11 (1), 69–72.

Socinski, M. (2006), 'Update on nanoparticle albumin-bound paclitaxel', *Clin Adv Hematol Oncol*, 4 (10), 745–746.

Stewart, S., Jablonowski, H., Goebel, F. D., Arasteh, K., Spittle, M., Rios, A., Aboulafia, D., Galleshaw, J., & Dezube, B. J. (1998), 'Randomized comparative trial of pegylated liposomal doxorubicin versus bleomycin and vincristine in the treatment of AIDS-related Kaposi's sarcoma. International Pegylated Liposomal Doxorubicin Study Group', *J Clin Oncol*, 16 (2), 683–691.

Takimoto, C. H., Schwartz, G., Romero, O., Patnaik, A., Tolcher, A., Garrison, M., Oldham, F. B., Bernareggi, A., & Rowinsky, E. (2005), 'Phase I evaluation of paclitaxel poliglumex (PPX) administered weekly for patients with advanced cancer', *Proc Am Soc Clin Oncol*, 23, 145s.

Thompson, J. F., Siebert, G. A., Anissimov, Y. G., Smithers, B. M., Doubrovsky, A.,

Anderson, C. D., & Roberts, M. S. (2001), 'Microdialysis and response during regional chemotherapy by isolated limb infusion of melphalan for limb malignancies', *Br J Cancer*, 85 (2), 157–165.

Toffoli, G., Cernigoi, C., Russo, A., Gallo, A., Bagnoli, M., & Boiocchi, M. (1997), 'Overexpression of folate binding protein in ovarian cancers', *Int J Cancer*, 74 (2), 193–198.

Vail, D. M., Amantea, M. A., Colbern, G. T., Martin, F. J., Hilger, R. A., & Working, P. K. (2004), 'Pegylated liposomal doxorubicin: proof of principle using preclinical animal models and pharmacokinetic studies', *Semin Oncol*, 31 (6 Suppl 13), 16–35.

Veal, G. J., Griffin, M. J., Price, E., Parry, A., Dick, G. S., Little, M. A., Yule, S. M., Morland, B., Estlin, E. J., Hale, J. P., Pearson, A. D., Welbank, H., & Boddy, A. V. (2001), 'A phase I study in paediatric patients to evaluate the safety and pharmacokinetics of SPI-77, a liposome encapsulated formulation of cisplatin', *Br J Cancer*, 84 (8), 1029–1035.

Veerareddy, P. R. & Vobalaboina, V. (2004), 'Lipid-based formulations of amphotericin B', *Drugs Today (Barc.)*, 40 (2), 133–145.

Verschraegen, C. F., Gilbert, B. E., Loyer, E., Huaringa, A., Walsh, G., Newman, R. A., & Knight, V. (2004), 'Clinical evaluation of the delivery and safety of aerosolized liposomal 9-nitro-20(s)-camptothecin in patients with advanced pulmonary malignancies', *Clin Cancer Res*, 10 (7), 2319–2326.

Wolff, A. C., Donehower, R. C., Carducci, M. K., Carducci, M. A., Brahmer, J. R., Zabelina, Y., Bradley, M. O., Anthony, F. H., Swindell, C. S., Witman, P. A., Webb, N. L., & Baker, S. D. (2003), 'Phase I study of docosahexaenoic acid-paclitaxel: a taxane-fatty acid conjugate with a unique pharmacology and toxicity profile', *Clin Cancer Res*, 9 (10 Pt 1), 3589–3597.

Woodle, M. C. & Lasic, D. D. (1992), 'Sterically stabilized liposomes', *Biochim Biophys Acta*, 1113 (2), 171–199.

Working, P. K., Newman, M. S., Stuart, Y., *et al.* (1994), 'Pharmacokinetics, biodistribution and therapeutic efficacy of doxorubicin encapsulated in STEALTH liposomes', *Liposome Res*, 46, 667–687.

Zamboni, W. C., Gajjar, A. J., Mandrell, T. D., Einhaus, S. L., Danks, M. K., Rogers, W. P., Heideman, R. L., Houghton, P. J., & Stewart, C. F. (1998a), 'A four-hour topotecan infusion achieves cytotoxic exposure throughout the neuraxis in the nonhuman primate model: implications for treatment of children with metastatic medulloblastoma', *Clin Cancer Res*, 4 (10), 2537–2544.

Zamboni, W. C., Stewart, C. F., Thompson, J., Santana, V. M., Cheshire, P. J., Richmond, L. B., Luo, X., Poquette, C., Houghton, J. A., & Houghton, P. J. (1998b), 'Relationship between topotecan systemic exposure and tumor response in human neuroblastoma xenografts', *J Natl Cancer Inst*, 90 (7), 505–511.

Zamboni, W. C., Houghton, P. J., Hulstein, J. L., Kirstein, M., Walsh, J., Cheshire, P. J., Hanna, S. K., Danks, M. K., & Stewart, C. F. (1999), 'Relationship between tumor extracellular fluid exposure to topotecan and tumor response in human neuroblastoma xenograft and cell lines', *Cancer Chemother Pharmacol*, 43 (4), 269–276.

Zamboni, W. C., Gervais, A. C., Egorin, M. J., Schellens, J. H., Hamburger, D. R., Delauter, B. J., Grim, A., Zuhowski, E. G., Joseph, E., Pluim, D., Potter, D. M., & Eiseman, J. L. (2002), 'Inter- and intratumoral disposition of platinum in solid tumors after administration of cisplatin', *Clin Cancer Res*, 8 (9), 2992–2999.

Zamboni, W. C., Gervais, A. C., Egorin, M. J., Schellens, J. H., Zuhowski, E. G., Pluim, D., Joseph, E., Hamburger, D. R., Working, P. K., Colbern, G., Tonda, M. E., Potter,

D. M., & Eiseman, J. L. (2004), 'Systemic and tumor disposition of platinum after administration of cisplatin or STEALTH liposomal-cisplatin formulations (SPI-077 and SPI-077 B103) in a preclinical tumor model of melanoma', *Cancer Chemother Pharmacol*, 53 (4), 329–336.

Zamboni, W. C., Whitner, H., Potter, D. M., Ramanathan, R. K., Strychor, S., Tonda, M., Stewart, B., Modi, N., Engbers, C., & Dedrick, R. l. (2005a), 'Allometric scaling of STEALTH liposomal anticancer agents', *Proc Am Assoc Cancer Res*, 46, 326.

Zamboni, W. C., Ramalingam, S., Friedland, D. M., Belani, C. P., Stoller, R. G., Modi, N. B., Nath, R. P., Tonda, M. E., Strychor, S., & Ramanathan, R. K. (2005b), 'Phase I and pharmacokinetic (PK) study of STEALTH liposomal CKD-602 (S-CKD602) in patients with advanced solid tumors', *Proc Am Soc Clin Oncol*, 23, 152s.

Zamboni, W. C., Friedland, D. M., Ramalingam, S., Edwards, R. P., Stoller, R. G., Belani, C. P., Strychor, S., Ou, Y. C., Tonda, M. E., & Ramanathan, R. K. (2006), 'Final results of a phase I and pharmacokinetic study of STEALTH liposomal CKD-602 (S-CKD602) in patients with advanced solid tumors', *Proc Am Soc Clin Onc*, 24, 82s.

Zhang, J. A., Xuan, T., Parmar, M., Ma, L., Ugwu, S., Ali, S., & Ahmad, I. (2004), 'Development and characterization of a novel liposome-based formulation of SN-38', *Int J Pharm*, 270 (1–2), 93–107.

Index